W0227563

Motor Neuron Disease:

Biology and Management

Edited by
P.N. Leigh and M. Swash

With 86 Figures

Springer-Verlag
London Berlin Heidelberg New York
Paris Tokyo Hong Kong
Barcelona Budapest

P.N. Leigh, BSc, PhD, FRCP
Professor, University Department of Neurology, Institute of Psychiatry and Kings College School of Medicine and Dentistry, De Crespigny Park, London SE5 8AF, UK

Michael Swash, MD, FRCP, FRCPath
Consultant Neurologist, Department of Neurology, The Royal London Hospital, Whitechapel, London E1 1BB, UK

ISBN-13: 978-1-4471-1873-2 e-ISBN-13: 978-1-4471-1871-8
DOI: 10.1007/978-1-4471-1871-8

British Library Catologuing in Publication Data
A catalogue record for this book is available from the British Library

Library of Congress Cataloging-in-Publication Data
A catalog record for this book is available from the Library of Congress

Softcover reprint of the hardcover 1st edition 1995

First published 1995

Typeset by Concept Typesetting Ltd, Salisbury, England

28-3830-543210 Printed on acid-free paper

Foreword

Motor neuron disease is a diagnosis that strikes dread in the heart not only of those people unfortunate enough to develop it, but those making the diagnosis too. It is a conclusion often reached by exclusion of other possible diagnoses, and inevitably reached reluctantly since it amounts to a sentence of progressive paralysis and, ultimately, death for the patient. Doctors, whose training and professional culture is directed towards curing the sick, frequently feel they are helpless in the face of this condition.

Certainly, it is true that in spite of the advances in understanding the disease described in this book there is no treatment for the disease itself, nor any prospect of radical treatment or prevention in the immediate future. Nevertheless, much can be done to secure a higher quality of life for patients, to ensure that their dignity and personal autonomy is respected, and to help them to adjust physically and emotionally to the catastrophic series of changes that, sooner or later, will overtake them and their immediate circle of helpers.

Understanding the aetiology of motor neuron disease may help to find the key to other progressive neurological disorders. In this respect, as well as in bringing to an end the inexorable toll of human suffering, to say nothing of the social and economic cost of the disease, the urgency of finding means of prevention, treatment or cure cannot be doubted. The contribution of patients to research should not be overlooked. Many are powerfully motivated to fight the disease in any way open to them, and are eager to contribute to genetic and epidemiological studies, to take part in clinical trials, to report their experiences before and after diagnosis, to donate material by biopsy or other methods, and of course to contribute brain and spinal cord tissue after death. This motivation is a great asset to researchers and it requires little more than genuine recognition of the contribution and legacy of patients to the search for a cure, and regular feedback, to ensure it continues.

The distribution of motor neuron disease means that, worldwide, few medical practitioners have an opportunity at local level to see and learn from large numbers of cases. The development of regional centres of excellence, and recognition of the special contribution that can be made by the nursing and therapeutic professions, has done much to improve the quality of care for patients.

Much that can be done consists of apparently small and technically undemanding steps. Improvements in seating, in feeding and washing arrangements, in communication and security, require appropriate technology. But these modest steps need to be applied to the circumstances of the whole person, they need to anticipate and indeed "leapfrog" the loss of motor function. Advanced technology of course has a part to play, but cost and complexity often mean that it is hard to keep pace with the deterioration through the disease, and must be balanced against simplicity and availability.

The voluntary associations concerned with motor neuron disease in many countries of the world have an increasingly important role to play. Some are concerned with promoting research, some with providing direct care to patients, and some with both of these. However, they are all committed to ensuring that attention is paid to the disease and the people who endure it, and they are increasingly able to provide information and practical help to clinicians and researchers. But irrespective of the material help they are able to give, by their very existence they give to all those concerned with motor neuron disease the most important component of any treatment now or in future – hope.

January 1994

Peter Cardy
Director Motor Neurone Disease Association, UK
and Secretary-General, International Alliance of
MND/ALS Associations

Preface

Amyotrophic lateral sclerosis is a disease that has been known for a long time, but which attracted little scientific interest until very recently. As in most medical problems, interest in the disease was awakened by the application of new approaches to the study of the clinical and pathological manifestations. Charcot, Joffroy, and Cruveilhier, more than a hundred years ago, followed the same path, applying the most modern methodologies available to the study of the disease they had themselves so recently recognised. As a result the basic pathological changes were recognised soon after the clinical manifestations were characterised.

The more recent investigations have established diagnostic criteria for the core syndrome, and allowed recognition of variant forms of the disease. Electrophysiological studies have emphasised the early involvement of myotomes at a distance from the apparent focal presentation, and have suggested that there is relatively widespread disease even at the onset, raising the question as to how long a preclinical phase of the disease is present before clinical presentation and diagnosis. This concept is important in devising any therapy since it implies that the earlier treatment is made available the greater the probability of success in arresting the disease process. Unfortunately relatively little is yet known about the fundamental processes underlying cell neuronal death in the disease, and biological treatments, although of great current interest, remain essentially speculative. Pharmacological treatments themselves remain relatively neglected at present, although it is entirely possible that some of the currently exciting neurotrophic factors may be effective by causing an increase in muscle bulk, or in motor end plate sprouting, rather than by an improvement in neuronal survival or "health".

The currently emerging era of clinical trials, therefore, is fraught with difficulty, since the search for clinical effect precedes understanding of pathophysiology, a circumstance that is perhaps not so unusual as might be thought in medical science. One of Osler's more famous aphorisms states that "As is our pathology, so is our medicine". While this remains true, the pace of advance is such that presently there is no sense of order in understanding.

In this volume we have asked our co-authors to summarise current understanding of the major topics relevant to understanding the causation, pathophysiology, genetics, pharmacology, and neurophysiology of the disease, and to consider possible avenues for therapy in the light of this new knowledge. We hope that readers will find this approach lively and stimulating. We have included reviews of current supportive management, believing that there is a lack of provision of the very best aspects of care in many cases and, to illustrate the personal problems that result in the face of an inexorable, progressive wasting disease such as ALS, we have been privileged to be able to include a personal account of a carer as a concluding chapter. Every reader of this book should read this chapter with humility.

We dedicate this book to the care of patients with ALS, in the expectation that understanding of the aetiology of the disease, and the development of effective treatment, or prevention, will be forthcoming in the near future.

London, May 1994

P.N. Leigh
M. Swash

Contents

3 Familial Motor Neuron Disease

4 Pathology of Motor System Disorder

13 Somatic Motoneurons and Descending Motor Pathways. Limbic and Non-limbic Components

14 Neurophysiological Changes in Motor Neuron Disease

Contributors

Stanley H. Appel, MD
Professor and Chairman, Department of Neurology, Baylor College of Medicine, 6501 Fannin NB302, Houston, TX 77030, USA

Samuel M. Chou, MD, PhD
Section of Neuropathology, The Cleveland Clinic Foundation, 9500 Euclid Avenue, Cleveland, OH 44195, USA

J. S. de Belleroche, BA, MSc, PhD
Department of Biochemistry, Charing Cross and Westminster Medical School, Fulham Palace Road, London W6 8RF, UK

Jozsef I. Engelhardt, MD
Department of Neurology, Baylor College of Medicine, 6501 Fannin NB302, Houston, TX 77030, USA

Lorenz J. Finison, PhD
Department of Community Medicine, Tufts University, 750 Washington Street, Boston, MA 02111, USA

O. Garofalo, PhD
University Department of Neurology, Institute of Psychiatry,
De Crespigny Park, London SE5 8AF, UK

Roberto J. Guiloff, MD, FRCP
Westminster Hospital, Dean Ryle Street, London SW1P 2AP, UK

R. Langton Hewer, MD, FRCP
Professor of Neurology, Bristol University, Department of Neurology, Frenchay Hospital, Bristol BS16 1LE, UK

Dave Hollander, MD
Research Assistant, New England Medical Center, 750 Washington Street, Boston, MA 02111, USA

Gert Holstege, MD
Department of Anatomy and Embryology, Rijksuniversiteit Groningen, Oostersingel 69, 9713 EZ Groningen, The Netherlands

K. Kondo, MD
Professor and Chairman of Public Health, Hokkaido University, School of Medicine, Kitaku Sapporo 060, Japan

P. N. Leigh, BSc, PhD, FRCP
University Department of Neurology, Institute of Psychiatry and Kings College School of Medicine and Dentistry, De Crespigny Park, London SE5 8AF, UK

S. Malessa, MD
Institüt für Neuropathologie, Thalkirchner Straße 36, 8000 Munich 2, Germany

Joanne E. Martin MB, BS, MRCPath
Department of Neuropathology, The Institute of Pathology, The Royal London Hospital, Whitechapel, London E1 1BB, UK

Hiroshi Mitsumoto, MD
Director, Neuromuscular Disease Program, Department of Neurology, The Cleveland Clinic Foundation, 9500 Euclid Avenue, Cleveland, OH 44195, USA

Theodore L. Munsat, MD
Professor of Neurology and Pharmacology, Tufts University School of Medicine, 750 Washington Street, Boston, MA 02111, USA

Peter B. Nunn, PhD
Department of Biochemistry, Division of Biomolecular Sciences, King's College London, Strand, London WC2R 2LS, UK

F. Clifford Rose, MD, FRCP
Department of Biochemistry, Charing Cross and Westminster Medical School, Fulham Palace Road, London W6 8RF, UK

Martin S. Schwartz, MD
Consultant Neurophysiologist, Atkinson Morley's Hospital, 31 Copse Hill, Wimbledon, London SW20 0NE, UK

R.G. Smith
Baylor College of Medicine, 6501 Fannin NB302, Houston, TX 77030, USA

J. C. Steele MD
Micronesian Health Study, University of Guam, Mangilao 96923, Guam

E. Stefani
Baylor College of Medicine, 6501 Fannin NB302, Houston, TX 77030, USA

Michael Swash, MD, FRCP, FRCPath
Consultant Neurologist, Department of Neurology, The Royal London Hospital, Whitechapel, London E1 1BB, UK

D. B. Williams, MB, FRACP
Departments of Neurology and Health Sciences Research, Mayo Clinic, Rochester, MN, USA

1 Motor Neuron Disease: The Clinical Syndrome

M. Swash and M. S. Schwartz

The nomenclature of disorders of anterior horn cells is confused. This confusion reflects ignorance of the underlying causes of these syndromes; indeed, the current terminology is based on clinical or pathological descriptions of the syndromes themselves, rather than basic mechanisms. Rowland (1982, 1988) has pointed out the importance of distinguishing between motor neuron diseases and motor neuron disease in discussions of anterior horn cell disorders. The confusion in terminology can be resolved by appreciation of the historical development of the concept of anterior horn cell disease as a cause of muscle wasting.

Progressive muscular wasting was a clinical syndrome well known to physicians in the early nineteenth century. McHenry (1969) notes descriptions by Sir Charles Bell, Marshall Hall and Todd; at about this time the motor function of the ventral spinal roots and the sensory function of the posterior roots were defined independently by Bell and Magendie. The term *progressive muscular atrophy* was used by Aran (1850), who believed this syndrome was a muscular disorder. Duchenne (1849) also gave a description of this disorder. Thus, by the middle of the nineteenth century there were two conflicting views. Bell, supported by Cruveilhier (1853), who noted the thinness of the anterior spinal roots, regarded progressive muscular atrophy as a myelopathic disorder, whereas Aran and Duchenne favoured a muscular cause. Degeneration of anterior horn cells in the grey matter of the spinal cord was recognised independently by Luys (1860) in Paris, and by Lockhart Clarke in London. Charcot (1869) brought together these observations by studying the clinical and pathological features of the disease and described the involvement of the corticospinal tract. Charcot proposed the term *amyotrophic lateral sclerosis* (ALS) and recognised a clinical syndrome consisting of progressive muscular atrophy, often beginning in the hands and involving bulbar muscles, fibrillary contractions, especially during the period of active muscular atrophy, and permanent spasmodic contraction. Further, he noted the absence of sensory loss and that the disease was not complicated by paralysis of bladder or rectum, and that there was no tendency to the formation of bedsores. The myogenic origin of other cases of progressive muscular wasting, e.g. limb-girdle muscular dystrophy, was defined subsequently by von Leyden, Landouzy and Dejerine, and Erb (1891) (see McHenry 1969). *Progressive bulbar palsy* (primary labio-glosso-laryngeal paralysis) was described by Duchenne (1860). Charcot & Joffroy (1869) recognised its relationship to amyotrophic lateral sclerosis when loss of motor neurons was noted in the bulbar motor nuclei in pathological studies at the Hôpital Salpêtrière. Pure syndromes of myelopathic muscular atrophy without corticospinal involvement, and of *primary lateral sclerosis* (Spiller 1904) but without muscular atrophy are rare, as noted by Kinnier Wilson (1940) but these syndromes, nonetheless, have always been regarded as related to the core syndrome of amyotrophic lateral sclerosis. The term *motor neuron disease* was introduced by Brain in recognition of the relation between the syndromes of progressive muscular atrophy, amyotrophic lateral sclerosis and progressive bulbar palsy, as shown by the clinical variation of involvement of upper and lower motor neurons and by the topography of the anterior horn cell loss and thus of the muscular wasting (see Brain 1962). This term has become commonly used in the United Kingdom, although Charcot's designation amyotrophic lateral sclerosis is preferred in French-speaking countries, and in the United States.

Rowland (1982) recognised the utility of the term motor neuron disease (MND) in describing the whole clinical syndrome but stressed the importance of retaining the general usage of the term motor neuron diseases (plural) to describe all the diseases of the anterior horn cells and motor system, including the inherited spinal muscular atrophies which are clinically and pathologically distinct from motor neuron disease (MND) itself. Similarly, heredo-familial diseases causing upper and lower motor neuron involvement, e.g. familial spastic paraplegia, do not form part of the MND syndrome itself, but are separate entities. The clinical syndrome of MND is sufficiently distinct to allow recognition of familial cases of MND, but uncertainty still arises in considering certain atypical syndromes, such as monomelic motor neuron diseases, and some juvenile onset cases. These problems in classification are considered below.

Classification of Motor Neuron Diseases

Diseases affecting anterior horn cells and bulbar motor neurons may present at any age from infancy to the senium. Those diseases beginning in infancy, childhood or adolescence are usually limited to the anterior horn cells, but in adults other parts of the motor system, including the upper motor neuron, may be involved, as in MND itself. Certain viruses, particularly poliomyelitis, show a predilection to infect anterior horn cells but other viruses, e.g. Herpes zoster and Coxsackie viruses may also affect anterior horn cells. In most motor neuron diseases motor nuclei in the brainstem are also involved; the term *spinal muscular atrophy* (SMA), usually used to denote a familial disorder without corticospinal tract involvement, does not therefore exclude bulbar involvement.

The cardinal features of neurogenic disorders are muscular weakness and wasting. In addition, disorders of anterior horn cells and bulbar motor nuclei are often characterised by prominent fasciculation at rest. This is particularly evident in more rapidly progressive disorders, e.g. MND in adults and Werdnig–Hoffmann disease (SMA Type 1) in infants. When there is involvement only of the lower motor neuron the tendon reflexes are reduced or absent but in MND itself, in which lower and upper motor neuron degeneration coexists, the tendon reflexes are characteristically brisker than normal, even in wasted muscles. The major forms of spinal muscular atrophy and MND are clearly defined clinically and pathologically; and in spinal muscular atrophies also by their pattern of inheritance.

Spinal Muscular Atrophies

Spinal muscular atrophy consists of a syndrome of progressive muscular atrophy and weakness due to anterior horn cell degeneration without involvement of corticospinal or sensory tracts. Weakness is usually predominantly proximal, and ocular and sphincter muscles are clinically uninvolved. The spinal muscular atrophies are classified into different but related syndromes according to their age of onset, progression and outcome. They are all inherited, with only a rare exception, as autosomal recessive disorders (Table 1.1).

A number of other clinical types of spinal muscular atrophy have been described (Table 1.2). The classification of these disorders under this heading is less certain in many instances, being dependent principally on the distribution of neurogenic

Table 1.1. Spinal muscular atrophies (from Swash and Schwartz 1988)

SMA	Inheritance	Age of onset	Course and outcome
Type I Werdnig–Hoffmann disease	Autosomal-recessive	30% in utero 100% by 5 months	Severe, generalised and bulbar weakness; 95% dead by 18 months
Type II Intermediate SMA	Autosomal-recessive	3–24 months	Severe disability in infancy; little progression, later unable to walk; scoliosis
Type III Kugelberg–Welander disease	Autosomal-recessive	2–17 years	Variable; slowly progressive over many years; mild or severe course
Type IV adult-onset SMA	Autosomal-recessive,	40–60 years	Unable to walk after 20 years.
	Dominant	20–50 years	Slowly progressive proximal weakness.
	X-linked (Kennedy)	20–40 years	Bulbospinal weakness; gynaecomastia common

Table 1.2. Spinal muscular atrophies: other forms (modified from Swash and Oxbury (1991) Clinical neurology. Churchill Livingstone, Edinburgh)

Distal	Dominant or recessive	2–20 yr	Mild, normal life expectancy
Chronic asymmetrical	Recessive	16–45 yr	Mild
Monomelic	?	Variable	Mild
Scapuloperoneal	Recessive	40–50 yr	Moderate; may shorten life
Bulbar SMA with deafness	?	Childhood	Severe or mild
Bulbar SMA (Fazio–Londe)	Recessive	2–12 yr	Death in 10 yr from onset
Oculopharyngeal	Dominant	4th decade	Mild
Facio-scapulo-humeral	Dominant	2nd decade	May shorten life

weakness, the absence of corticospinal or sensory pathway involvement, and exclusion of other causes, especially axonal motor neuropathies. The latter may be especially difficult, especially in the syndromes of distal spinal muscular atrophy, and scapuloperoneal muscular atrophy (Dyck 1982). Clearly, features inconsistent with anterior horn cell loss, or loss of bulbar motor neurons are important in establishing diagnosis in these syndromes.

Motor Neuron Disease (Amyotrophic Lateral Sclerosis)

Motor neuron disease is a progressive disorder, characterised by muscular weakness and wasting, with fasciculation, and by spasticity, hyperreflexia and extensor plantar responses. Bulbar involvement is frequently prominent, leading to dysphagia and dysarthria, but external ocular muscles and pelvic sphincter muscles are almost invariably spared. Sensory symptoms, especially paraesthesiae, may occur but there are no objective sensory signs. The first symptom is usually weakness, although some patients note fasciculations as an early feature and sometimes muscle cramp may antedate other symptoms by several months. The onset of the disease is difficult to delineate since the clinical manifestations develop gradually; retrospective analysis may suggest that the disease began several years before presentation with muscular weakness or wasting (Swash and Ingram 1988), perhaps with a long preclinical phase during which the disease, although active, is not symptomatic. Clinical presentation with weakness or atrophy thus implies failure of compensatory reinnervation, and loss of substantial numbers of motor cells from the relevant part of the motor system. The onset and progression of this, clinically evident phase of the disease has been documented in the case of Lou Gehrig, the famous American baseball player, who died of the disease in 1941, from a study of his batting averages in the 1938 and 1939 seasons (Kasarskis & Winslow 1989).

Clinical Features

Common symptoms at presentation are undue fatiguability, cramps, dragging of a leg usually due to foot drop, or difficulty manipulating objects with the fingers of one hand. These early symptoms are usually asymmetrical and frequently involve only one limb, although even at this time examination may reveal features of more widespread neurogenic abnormality. Two-thirds of patients present in this way, with weakness of one limb (Gubbay et al. 1985; Jokelainen 1977). Jokelainen (1977) found that weakness of a leg was slightly more common than weakness of an arm. Monoparesis affected 23%, and hemiparesis 4% of the 318 patients reported by Gubbay et al. (1985); weakness of both legs was the presenting feature in 20% and bulbar features in 22%. Other presenting features include muscle atrophy (10%), muscle pains and cramps (9%), fasciculations (4%) and stiffness of gait (1%). Weight loss and paraesthesiae are rare presenting complaints. Exertional dyspnoea is a rare presenting feature. When there is a relative predominance of upper motor neuron features, spasticity, stiffness and even ankle clonus, may be the major clinical features at presentation. Bulbar involvement usually consists of a combination of upper and lower motor neuron involvement, causing hoarseness of the voice, subtle changes in articulation and a hypernasal speech timbre.

The tongue is usually symmetrically involved; there is symmetrical weakness, slowed movement, fasciculations and atrophy. In the early stages there is only slight weakness and speech articulation is relatively preserved. The development of weakness and spasticity results in reduced rate, range and force of articulatory tongue movements. Dysphagia develops as the weakness and spasticity progresses, leading to inability to progress a food bolus to the back of the mouth, choking, dribbling and inability to swallow. Sometimes spasticity of the cricopharyngeus muscle leads to obstruction of swallowing during contraction of the pharyngeal muscles, but surgical myotomy of the cricopharyngeus muscle is only useful in carefully selected patients, and has a high mortality due to respiratory complications (Lebo et al. 1976; Mady 1984). Cha and Patten (1989) used magnetic resonance imaging (MRI) to show that the tongue may be reduced in size by 60%, that it becomes rectangular rather than curved in shape, and that its internal structure becomes disorganised as it atrophies.

Fatiguability is a common early symptom (Swash and Ingram 1988) but is not specific since it is common in other denervating conditions and especially in multiple sclerosis. Muscle cramp is a frequent complaint and is often observed at clinical examination, when maximal muscle contraction may induce a sustained painful cramp in the muscle tested. Fasciculations are less commonly observed by the patient but are a cardinal feature of the disease, although they also occur in other myelopathies, e.g. cervical spondylotic myelopathy or in old poliomyelitis. Cramps may be severe and painful, leading to persistent aching painful discomfort in some patients (Newrick and Langton Hewer, 1985). Newrick and Langton Hewer (1985) noted that pain, often pricking, burning or dysaesthetic in quality, was a significant feature in 27 of 42 patients at some stage in the illness, and pointed out that Charcot described this in 1895. Cramps and even fasciculation may precede the development of weakness and wasting by many years (Fleet and Watson 1986). Fasciculations are common in normal people (Reed and Kurland 1963), but in MND they occur randomly in different parts of muscles, whereas benign fasciculations tend to occur repetitively in the same fascicle of a muscle.

Although sensory abnormalities are not normally considered part of the clinical syndrome of MND, paraesthesiae, coldness and even numbness are sometimes noted in the disease. However, objective sensory abnormalities are virtually never found in the disease and, indeed, would be considered grounds for excluding the diagnosis by most neurologists (Mulder 1982). Nonetheless, Shahani et al. (1971) reported abnormal resistance of peripheral sensory nerves to ischaemia in MND, and Jamal et al. (1985) noted that thermal thresholds were increased in 80% of 40 patients with MND, suggesting an abnormality in small sensory fibres in the peripheral nervous system. Mulder et al. (1983) found an abnormality in vibration thresholds in 18% of patients.

Even in the most advanced stages of the disease certain motor systems show little evidence of involvement. Thus urinary and anorectal continence is almost invariably unaffected, despite severe paralysis and wasting of skeletal muscles, including gluteal and other sacral myotomes. Absence of involvement of the pelvic sphincter and perineal muscles is evident clinically, by electromyelography (EMG) and by preservation of the motor neurons in the Onuf nucleus of the anterior horn of the spinal cord at S2 and S3 (Mannen et al. 1977). This relative resistance of sphincter muscles to denervation in MND contrasts with the selective vulnerability of these neurons in progressive autonomic failure (Sakata et al. 1978; Chalmers and Swash 1987). Extraocular movements are similarly uninvolved in MND but abnormalities of pursuit movements, consisting of saccadic interruption and of smooth tracking movements (Jacobs et al. 1981), and even partial ophthalmoplegia have been described as rare, atypical manifestations with loss of neurons in the motor nuclei subserving ocular muscles (Harvey et al. 1979). However, as a clinical rule, even the totally paralysed and bedfast patient remains continent and can move his eyes, and any impairment of these functions should raise questions as to the diagnosis.

Autonomic function is also generally normal, at least to clinical examination, although sympathetic hyperfunction with vagal hypofunction has been reported in cardiovascular tests (Chida et al. 1989); these findings accord with slight reduction in numbers of intermediolateral neurons noted by Kennedy and Duchen (1985). Patients with MND do not develop bedsores, even in the terminal stages of the disease, a feature consistent with sparing of sensation and autonomic regulation of skin blood flow. Ono et al. (1986) reported reduction in numbers and changes in arrangement and diameter of collagen fibrils in the dermis, and a delay in return of skin turgor after displacement, e.g. a pinch.

Dementia has long been recognised as an occasional associated feature in MND, but its frequency is difficult to define. David and Gillham (1986) found that patients with MND were slightly impaired in tests of cognitive function, particularly with picture recall, learning novel material and card sorting. These findings correlated with cerebral atrophy on computed tomography scanning. In both sporadic and familial MND the association with dementia has often been attributed to co-existent Alzheimer's disease. However, it usually develops late in the course of MND and is thus then probably directly associated. Hudson (1981) found that MND was associated with dementia in 15% of cases. Interpretation of this association is difficult because of the known relation of these two features in Western Pacific ALS (see Chapter 9), and in Creutzfeldt–Jakob disease. Brownell et al. (1970) found that there was vacuolar degeneration of cortical layers in 2 of 45 cases of MND, raising the possibility of clinical overlap in recognition of these syndromes.

Other Clinical Syndromes of MND

While amyotrophic lateral sclerosis accounts for 80% of cases of idiopathic MND, other clinical forms of the disease occur (Table 1.3). Progressive bulbar palsy is the presenting syndrome in 10%, progressive muscular atrophy in 8% and primary lateral sclerosis in 2% (Caroscio et al. 1984).

In *progressive bulbar palsy* the main presenting features are dysarthria, dysphonia, difficulty in chewing, salivation and dysphagia. The tongue is weak and atrophic and fasciculations are prominent. Facial weakness is sometimes present. There is usually clinical evidence of both upper and lower motor neuron involvement of the bulbar musculature, with spasticity of the tongue and a brisk jaw jerk, and similar abnormalities usually develop in the limbs during the course of the disease. At presentation, however, there may be no symptoms or signs of limb involvement. The prognosis is determined, as in amyotrophic lateral sclerosis, by

Table 1.3. Motor neuron diseases

Spinal muscular atrophies (see Tables 1.1 and 1.2)
Idiopathic motor neuron disease
Amyotrophic lateral sclerosis
Progressive bulbar palsy
Progressive muscular atrophy
Primary lateral sclerosis
Juvenile motor neuron disease
Monomelic motor neuron disease
Familial motor neuron disease
Excitotoxin-induced motor neuron disease
Guam-type MND (Western Pacific ALS), with or without Parkinson-Dementia Complex
Lathyrism
Metabolic and immunological
Hexosaminidase deficiency
Monoclonal gammopathy
Viral and transmissible
Poliomyelitis
Creutzfeldt–Jakob disease
AIDS
Post-encephalitic myelopathy
Herpes zoster myelitis (segmental)
? Other viruses, e.g. Coxsackie virus infection
System degenerations
Spino-cerebellar degenerations
Shy–Drager syndrome
Olivo-ponto-cerebellar degeneration
Joseph-Machado disease
Huntington's disease
Heavy metal poisoning
Lead and other heavy metals
Mercury
Manganese
Others
Stiff man syndrome
Post-traumatic
Syringomyelia
Post-irradiation syndromes
Electric shock and lightening injuries

the severity of bulbar involvement. *Progressive muscular atrophy* consists of lower motor neuron involvement of limb muscles, without definite evidence of upper motor neuron features. Nonetheless, the tendon reflexes are brisk even in the wasted muscles, a feature that seems to differentiate this rather poorly defined syndrome from progressive spinal muscular atrophy, in which the tendon reflexes are usually reduced. The progressive muscular atrophy variant of MND commences after age 20 years, and is not associated with a family history of a similar disorder (Mortara et al. 1984).

Primary lateral sclerosis consists of a slowly progressive spastic paraparesis, beginning in adult life, with no signs on examination other than those of cortico-spinal tract involvement, including sparing of the sphincters. In addition, familial spastic paraplegia is excluded by the absence of a family history of the disorder, and investigation reveals no other abnormality. EMG evaluation reveals no evidence of lower motor neuron disturbance. Multiple sclerosis, HTLV-1 associated tropical spastic paraparesis and AIDS myelopathy must be excluded by MRI and serological tests. In some patients there is an associated pseudobulbar palsy, and there are usually mild corticospinal abnormalities in the upper limbs. The disorder progresses slowly, for up to 20 years in some studies (Younger et al. 1988; Norris et al. 1993).

A predominantly distal form of MND, of juvenile onset, usually confined to one upper limb has been recognised (Hirayama et al. 1963). It is particularly associated with Japan (Hirayama et al. 1963; Sobue et al. 1978), Sri Lanka (Peiris et al. 1989) and India, and only a few cases have been recognised in Europeans (Meadows and Marsden 1969; Chaine et al. 1988). In India these cases may show involvement of both arms, or of a leg. There is a marked male predilection, especially for cases involving the upper limb (12 : 1; Virmani and Mohan 1985). This disorder begins before the age of 30 years and is often of juvenile onset, being below the age of 20 years in 40% of the Southern Indian cases described by Virmani and Mohan (1985). All the 102 cases of Peiris et al. (1989) began before age 29 years, and 90% were male. Ben Hamida et al. (1988) reviewed 101 patients from Tunisia with chronic proximal neurogenic muscular atrophy, 54% with onset before 5 years. Two-thirds of the latter patients showed autosomal recessive inheritance and these showed no predominance of males, although 70% of the sporadic cases were male. In contrast to this experience the Indian and Japanese patients showed distal involvement, affecting one limb, and without evidence of a genetic origin.

This disorder is usually confined to one upper limb for a period of about 2 years before progression occurs with involvement of proximal forearm muscles, and of the opposite arm. Tremor develops in the affected limbs and the tendon reflexes are reduced. In 15% of cases the disease arrests within 5 years of its onset. This syndrome is sporadic, without any recognised genetic basis. The period of progression distinguishes this disorder from poliomyelitis (Chopra et al. 1984). The Madras form of MND is a sporadic disorder of young adults, with generalised neurogenic atrophy, bulbar palsy and sensorineural deafness, but with a benign course (Jagganathan 1973; Sayeed et al. 1975). The nosological position of this disorder in relation to the similar autosomal recessive syndrome described by Vialetto (1936) and Van Laere (1966) is uncertain (Summers et al. 1987) since detailed family studies of sporadic cases are lacking.

Outcome of MND

Although the prognosis of MND is generally regarded as poor, it is important to recognise differences between the clinical forms of the disease and that long survival is possible in some patients. The duration of the disease from diagnosis is generally 1 to 5 years, with a median survival of about 2 years (Mulder and Howard 1976; Juergens et al. 1980). Mulder and Howard (1976) and Mortara et al. (1984) found that 20% of patients were still alive 5 years after onset of the disorder and 10% at 10 years. Jokelainen (1977) noted that some patients were living 12 years after the onset, and Mackay (1963) reported survival ranging from 6 months to 21 years. Patients with predominantly spinal forms of the disease, i.e. amyotrophic lateral sclerosis, had a mean survival of 3.3 years, and those with bulbar symptoms survived 2.2 years (Jokelainen 1977). Generally, the earlier age of onset the longer the survival, and atypical juvenile-onset, monomelic MND has a benign course.

Progressive muscular atrophy and primary lateral sclerosis also have a relatively benign outcome with survival for 10 years or more (Mortara et al. 1984; Younger et al. 1988). There is some evidence that survival in MND might be increasing (Caroscio et al. 1984). Mulder and Howard (1976) described a case in which the disease appeared to arrest and remit, but this exceptional clinical course is so exceptional that it must be suggested that the diagnosis may have been other than MND. However, plateaux in the progressive course of the disease have been recognised (Tyler 1979; Swash and Schwartz 1984). Jablecki et al. (1989) have suggested an algorithm for prediction of the approximate survival time of individual patients based on the age of onset, the duration of the weakness and an estimate of the clinical disability at the time of diagnosis.

Possible Associations of MDN with Other Diseases

A number of clinical associations of MND with other diseases, or with previous exposures, have been put forward sometimes in the context of causation and sometimes as apparently related disorders. In most instances it is clear that these apparent associations are spurious, the other disorder simply mimicking the clinical syndrome of MND/ALS, but in others there remains a possibility of a real association that may be relevant in considering causation and clinical management.

Poliomyelitis

Since the polio virus exhibits a specific neurotropic effect on motor neurons in the brain stem and spinal cord it has for long been thought that MND might in some way be related to previous poliomyelitis (Potts 1903; Salmon and Riley 1935; Zilkha 1962). Modern epidemiological investigations have not strongly supported this concept. For example, Poskanzer et al. (1969) found only 2.5% of patients with MND/ALS, and Juergens et al. (1980) found no evidence of previous poliomyelitis infection in 35 consecutive cases. More recently, Martyn et al. (1988) noted that the geographical and social clan distribution of motor neuron disease in England and Wales, based on death certification, resembled that of poliomyelitis 30–40 years earlier, and suggested that this might be evidence that MND might be a delayed effect of earlier subclinical poliovirus infection. However, no evidence of persistent or past infection with polio inpatients with MND has yet been ascertained (see Chapter 6).

The term "post-polio syndrome" (Brooks et al. 1978) describes new muscle weakness and atrophy developing either in muscles previously affected or previously spared during the acute phase of poliomyelitis infection. The rate of decline of muscle strength is very slow, and some patients show periods of stabilisation. The syndrome develops at least 15 years after the acute disease. Muscle aches and pains, and muscle fatigue are common additional features. This syndrome differs from MND in that rapid progression, bulbar weakness and signs of corticospinal tract involvement are absent. Fasciculations are common in patients with old poliomyelitis, especially after exercise, but do not denote progression of the disease. It has been suggested that the post-polio syndrome results from loss of individual motor end plates and nerve terminals rather than death of anterior horn cells, together with age-related impairment of the capacity for collateral sprouting, resulting in fractional loss of parts of motor units (Dalakas et al. 1986). There is no evidence of an increased incidence of MND in these patients, despite its frequency in affecting about a quarter of patients with poliomyelitis.

Hexosaminidase Deficiency (GM_2 Gangliosidosis)

In this autosomal recessive disorder the enzyme deficiency is associated with a heterogeneous clinical syndrome consisting of encephalopathy, dementia, ataxia and a disorder resembling motor neuron disease. Indeed, neurogenic muscular atrophy is a feature of the disease. Both upper and lower motor neurons may be involved, but in some cases only the lower motor neuron is affected. Complex repetitive discharges have been noted in EMG recordings. The disorder progresses as part of the multisystem degeneration associated with ganglioside storage (Mitsumoto et al. 1985). The enzyme defect is likely to present in younger patients (less than 50 years of age) with lower motor neuron syndromes, and investigation should be limited to this age group.

MND and Plasma Cell Dyscrasia

Paraproteinaemia, consisting of IgG or IgM protein, has been associated with MND in about 5% of patients presenting with the typical syndrome of MND/ALS. The paraproteinaemia in these patients is benign and is usually present only in low concentration (Shy et al. 1986). In some additional patients with MND the blood IgM level is raised but the protein is not associated with plasma cell dyscrasia (Shy et al. 1989). The clinical syndrome in these patients is non-specific, consisting of a predominantly lower motor neuron disturbance. Assessment of these patients is difficult since motor neuropathy, or sensorimotor neuropathy, is a well-recognised feature of plasma cell dyscrasia associated with monoclonal gammopathy, and the recognition of this syndrome in a patient with progressive proximal and distal muscle wasting may not be easy (Kelly et al. 1981; Rowland et al. 1982; Rowland 1988). The finding of slowed motor conduction does not necessarily exclude anterior horn cell involvement, but disturbance of sensory conduction is an important clue to the presence of the underlying neuropathy (Gherardi et al. 1988). No specific immunocytopathological relation between benign paraproteinaemia and MND has yet been demonstrated (Doherty et al. 1986), although Shy et al. (1989) have shown that 59% of patients with MND have IgG antibodies to ganglioside GM_1, a feature of only 25% of control subjects. A causal relation between this immune reaction and MND remains to be demonstrated. Treatment of the paraproteinaemia is of no proven benefit in patients with MND associated with paraproteinaemia.

Hyperparathyroidism and Thyrotoxicosis

In severe hyperparathyroidism a syndrome resembling motor neuron disease may develop, with weakness, fatiguability and hyperreflexia. Atrophy and fasciculation may also occur. This functional disorder is rapidly reversible, with improvement in hours or days following surgical hyperparathyroidism (Patten and Engel 1982). This syndrome may be difficult to differentiate from MND itself on clinical and electrophysiological grounds, but the findings of a raised blood calcium and alkaline phosphatase are important clues.

Thyrotoxicosis may rarely be associated with clinical features suggestive of fasciculation.

AIDS

A vacuolar myelopathy, often causing paraparesis, is common in patients with AIDS, and HIV has been isolated from the cerebrospinal fluid (CSF) of a patient with AIDS who developed MND (Hoffman et al. 1985). This observation is of uncertain significance, but a related retrovirus, HLTV-1 is the causative agent associated with progressive tropical spastic paraplegia (Johnson 1987), a disorder not clearly associated with lower motor neuron features.

Creutzfeldt–Jakob Disease

Fasciculation, weakness and muscular atrophy are features of some patients with this syndrome (Allen et al. 1971), well known for its transmissibility by brain inocula into primates, and for the characteristic spongiform degeneration found in cerebral cortex. The amyotrophic form of this disease is less clearly associated with spongiform change (Salazar 1982) but spongiform change in cortical layers occurs in some patients with otherwise typical MND (Brownell et al. 1970). Connolly et al. (1988) have reported transmission of the amyotrophic form of Creutzfeldt–Jakob disease to primates by brain inoculation after a long latent interval, of more than a year. Another presumed viral infection, encephalitis lethargica, was associated with amyotrophy in some survivors (Greenfield and Matthews, 1954).

Other Disorders

Neurogenic atrophy is associated with other degenerative diseases, but these conditions can be separated on clinical or other grounds from MND itself, and appear unrelated to the latter disease, despite the presence of motor neuron involvement (Table 1.3).

Diagnosis and Investigation

The diagnosis of MND is essentially clinical, since there is no specific diagnostic test. Definitive diagnosis can only be made after death by autopsy examination of the brain, spinal cord and muscles. The characteristic clinical features for diagnosis are determined by the clinical limits of the disease, including the well-recognised classical forms, and the rarer variants discussed above. The major features suggesting the diagnosis are the presence of upper and lower motor neuron signs in a distribution extending beyond a discrete spinal level, without sensory abnormality, and usually with spontaneous fasciculation. The combination of upper and lower motor neuron signs in the limbs with fasciculation and wasting of the tongue is especially characteristic (Li et al. 1986). Patients with clinical syndromes limited to bulbar muscles, or to upper neuron disturbance, pose special problems in diagnosis.

The implication of a diagnosis of MND is that the disorder will progress to death, and that all forms of treatment currently available will prove ineffective. It is therefore important to establish the diagnosis firmly by excluding other diagnoses, especially those that are amenable to medical or surgical treatment, by full and appropriate investigation. The diagnosis can then be fully discussed with the patient and family, and appropriate management decisions can be made, or planned for. Motor neuron disease affects people at an age at which other diseases are quite likely to occur as coincidental problems, including disorders that also result in muscle weakness and wasting such as peripheral nerve entrapment syndromes, peripheral neuropathies and cervical myelopathy due to spondylosis. These possibilities must be considered and treated as necessary.

Electrophysiological, radiological, biochemical, immunological and histopathological investigations may be necessary to firmly exclude alternative diagnoses. Electrophysiological investigation is particularly useful since it can be used to establish the distribution of neurogenic change, even in the early stages of the disease, and can provide information about the effectiveness of collateral sprouting and reinnervation that is useful in assessing prognosis (Swash and Schwartz 1984; Stalberg and Sanders 1984). Electrophysiological evidence of denervation and reinnervation in muscles in at least two limbs, not corresponding to a single nerve root or peripheral nerve distribution, is strongly supportive of a diagnosis of MND. Fasciculations can be recorded by EMG, and occur at a slow rate, usually not faster than 0.3 Hz in MND, and tend to fire randomly in single units rather than at a regular rate (Trojaborg and Buchthal 1965). Although motor conduction may be slightly slowed in recordings made in the innervation of atrophic muscles, sensory conduction is normal.

Most biochemical tests are normal in MND, but the CSF protein is often mildly raised; in fact this investigation does not contribute to the management of the disease. The blood creatine kinase (CK) level is raised to two or three times the normal level in about 50% of patients (Williams and Bruford 1970), and this abnormality is associated with the presence of regenerating fibres in the biopsy, a feature that is one component of the secondary myopathic change that develops in chronic partial denervation of muscle (Achari and Anderson 1974; Schwartz et al. 1976). Another muscle-derived enzyme, carbonic anhydrase III (CAIII) is a more sensitive index of this abnormality (Heath et al. 1983). The relation between abnormalities in immunoglobulins and MND is discussed above.

Radiological investigations are useful in excluding other diagnoses. Myelography has been much used in the past in order to exclude lumbosacral spondylosis with cord and root compression and the presence of tumours or developmental anomalies at the foramen magnum or in the high cervical region. This investigation is currently being superseded by MRI scanning. MRI or CT scanning of muscle can be useful in differentiating neurogenic muscular atrophy from myopathic disease, and also is capable of demonstrating the distribution of the abnormality and, for example, sparing of the perineal muscles in MND. The scope of modern imaging techniques, and their non-invasive methodology, is such that investigations of this type are indicated in order to be quite certain that treatable lesions have been excluded.

Differential Diagnosis

A number of diseases may present with clinical manifestations that, in some respects, resemble the clinical syndrome of classical MND/ALS. For example, wasting of upper limb and bulbar musculature occurs in *syringomyelia*, and in *cervical spondylosis* there may be a combination of upper and lower motor neuron lesions in upper limb muscles. However, in these conditions sensory disturbances are frequent and these exclude a diagnosis of MND. Furthermore, investigation discloses the abnormality in the spinal cord, or in the spinal canal. Similarly, in *motor neuropathies* there are abnormalities in motor nerve conduction, and the clinical features generally suggest the diagnosis, especially when tendon reflexes are absent and there are subtle sensory abnormalities. The diagnosis is particularly difficult when weakness and atrophy are restricted to one limb, as in presentations with a wasted hand, or with foot drop. In some such patients the clinical syndrome may be restricted in distribution and in progression for many years before changes suggestive of more generalised involvement develop (Swash and Ingram 1988). The diagnosis of *primary lateral sclerosis* is very much a matter of exclusion of other causes, including compressive, demyelinating and other degenerative conditions, especially familial spastic paraplegia (Younger et al. 1988). *Hyperthyroid myopathy* may present with weakness and with bulbar involvement, and fasciculations may occur in this disorder; however, there are no corticospinal signs and the clinical diagnosis of hyperthyroidism can usually be recognised. Hyperparathyroidism is also said to be present, on occasion, with fasciculation. *Myopathies*, although not associated with fasciculation, cause generalised atrophy and weakness with a proximal predilection, and without the marked asymmetry or hyperreflexia so often found in MND/ALS. *Spinal muscular atrophy* differs from MND in that there are no corticospinal abnormalities, the disorder is usually only slowly progressive, and there may be a family history of a similar disorder. *Multiple entrapment neuropathies* can pose a difficult diagnostic problem, but the presence of sensory disturbance, pain and Tinel's sign are important clinical features that serve to distinguish this syndrome from MND/ALS. Electrophysiological investigation should clearly suggest this diagnosis by the abnormalities in nerve conduction. *Multiple sclerosis* presents with other features, especially optic neuritis, diplopia and disturbances of ocular movement, and cerebellar signs, that usually allow ready diagnosis. In cases of difficulty investigations utilising evoked potential studies and MR imaging of the brain disclose evidence of demyelinating lesions in the brain and brain stem. *Multifocal vascular disease* may cause a pseudobulbar palsy with a spastic quadriparesis, without detectable sensory abnormalities, but the history usually suggests response progression associated with recurrent minor strokes and there are often signs of abnormality in ocular movement. CT or MR imaging of the brain is important in diagnosis, and electrophysiological tests fail to disclose the characteristic combination of denervation and reinnervation found in MND. *In the elderly*, a combination of peripheral nerve palsies, e.g. median or ulnar entrapment syndromes, with coexistent lumbosacral radiculopathy and canal stenosis, or with multifocal vascular disease, can present a syndrome closely resembling MND. Only careful history taking and investigation allows a diagnosis to be made. *Post-polio syndrome* (see above) can also be confused with MND.

References

Achari AN, Anderson MS (1974) Myopathic changes in amyotrophic lateral sclerosis. Neurology 24:477–481

Allen IV, Dermot E, Connolly JH, Hurwitz LJ (1971) A study of a patient with the amyotrophic form of Creutzfeldt–Jakob disease. Brain 94:715–724

Aran FA (1850) Recherches sur une maladie non encore décrite du système musculaire (Atrophie musculaire progressive). Arch Gen Med 24:5–35, 172–214

Ben Hamida M, Hentati F, Chebbi N et al. (1988) Amyotrophies spinales proximales et chroniques en Tunisie. Rev Neurol 144:737–747

Brain WR (1962) Diseases of the nervous system, 6th ed. Oxford University Press, Oxford, p 531

Brain WR, Croft PB, Wilkinson M (1969) The course and outcome of motor neuron disease. In: Norris FG, Kurland LT (eds) Motor neuron disease. Grune and Stratton, New York, pp 20–27

Brooks BR, Kurent J, Madden D, Sever J, Engel WK (1978) Cerebrospinal fluid anti-viral antibody titres in ALS and late post-poliomyelitis progressive muscular atrophy (LPPPMA). Neurology 28:388 (abstract)

Brownell B, Oppenheimer DR, Hughes JT (1970) The central nervous system in motor neuron disease. J Neurol Neurosurg Psychiatr 33:338–357

Caroscio JT, Calhoun WF, Yahr MD (1984) Prognostic factors in motor neuron disease – a prospective study of longevity. In: Rose FC (ed) Research progress in motor neuron disease. Pitman, London, pp 34–43

Cha CH, Patten BM (1989) Amyotrophic lateral sclerosis: abnormalities of the tongue on magnetic resonance imaging. Ann Neurol 25:468–472

Chaine P, Bouche P, Leger JM et al. (1988) Atrophie musculaire progressive localisée à la main. Rev Neurol 144: 759–763

Chalmers D, Swash M (1987) Selective vulnerability of urinary ONUF motoneurons in Shy-Drager syndrome. J Neurol 234:259–260

Charcot JM, Joffroy A (1869) Deux cas d'atrophie musculaire progressive avec lésions de la substance grise et des faisceaux antéro-latéraux de la moelle épinière. Arch Physiol Neurol Path 2:744

Chida K, Sakamaki S, Takasu T (1989) Alteration in autonomic function and cardiovascular regulation in amyotrophic lateral sclerosis. J Neurol 236:127–130

Chopra JS, Prabhakar S, Bannerjee AK (1984) The wasted leg syndrome: clinical, electrophysiological and histopathological studies. In: Rose FC (ed) Research progress in motor neuron disease. Pitman, London, pp 422–431

Connolly JH, Allen IV, Dermott E (1988) Transmissible agent in the amyotrophic form of Creutzfeldt–Jakob disease. J Neurol Neurosurg Psychiatr 51:1459–1460

Cruveilhier J (1852–1853) Sur la paralysie musculaire, progressive, atrophique. Bull Acad Med (Paris) 18:490, 546

Dalakas MC, Elder G, Hallett M et al. (1986) A long-term follow-up study of patients with post-poliomyelitis neuromuscular symptoms. N Engl J Med 314:959–963

David AS, Gillham RA (1986) Neuropsychological study of motor neuron disease. Psychosomatics 27:441–445

Doherty, P, Dickson JG, Flawgan TP et al. (1986) Effects of amyotrophic lateral sclerosis serum shown on cultured spinal neurons. Neurology 36:1330–1334

Duchenne de Boulogne GBA (1849) Recherches faites à l'orde des galvanisine sur l'état de la contractilité et de la sensibilité électromusculaires dans les paralysies des membres supérieurs. CR Acad Sci (Paris) 29:667

Duchenne de Boulogne GBA (1860) Paralysie musculaire progressive de la langue, du voile du palais et des lèvres: affection non encore décrite comme espèce morbide distincte. Arch Gén Méd 16:283, 431

Dyck PJ (1982) Are motor neuropathies and motor neuron diseases separable? In: Rowland LP (ed) Human motor neuron diseases. Raven Press, New York, pp 105–114

Erb WH (1891) Dystrophie muscularis progressiva: Klinische und pathologisch – anatomische studien. Dtsch Nervenheilk 1:13–94; 173–261

Fleet WS, Watson RT (1986) From benign fasciculations and cramps to motor neuron disease. Neurology 36:997–998

Gherardi R, Zuber M, Viard JP (1988) Mise en point: les neuropathies dysglobuliniques. Rev Neurol 144:391–408

Greenfield JG, Matthews WB (1954) Post-encephalitic Parkinsonism and amyotrophy. J Neurol Neurosurg Psychiatr 17:50–56

Gubbay SS, Kahana E, Zilber N, Cooper G, Pintov S, Leibowitz Y (1985) Amyotrophic lateral sclerosis. A study of its presentation and prognosis. J Neurol 232:295–300

Harvey DG, Torack RM, Rosenbaum HE (1979) Amyotrophic lateral sclerosis with ophthalmoplegia: a clinicopathologic study. Arch Neurol 36:615–617

Heath R, Schwartz MS, Brown IRF et al. (1983) Carbonic anhydrase III in neuromuscular disorders. J Neurol Sci 59:383–388

Hirayama K, Tsubaki T, Toyokura Y et al. (1963) Juvenile muscular atrophy of unilateral upper extremity. Neurology 13:317–380

Hoffman PM, Festoff BW, Giron LT Jr, Hallenbeck LC, Garruto RM, Ruscetti FW (1985) Isolation of LAV/HTLV III from a patient with amyotrophic lateral sclerosis. N Engl J Med 313:324–325

Hudson AJ (1981) Amyotrophic lateral sclerosis and its association with dementia in Parkinsonism and other neurological disorders: a review. Brain 104:217–247

Jablecki CF, Berg C, Leach J (1989) Survival prediction in amyotrophic lateral sclerosis. Neurology 12:833–841

Jacobs C, Bozran D, Heffner RR Jr et al. (1981) An eye movement disorder in amyotrophic lateral sclerosis. Neurology 31:1282–1287

Jagganathan K (1973) Juvenile motor neurone disease. In: Spillane JD (ed) Tropical neurology. Oxford University Press, London, pp 127–130

Jamal GA, Weir AI, Hansen S, Ballantyne JP (1985) Sensory involvement in motor neuron disease: further evidence from automated thermal threshold determination. J Neurol Neurosurg Psychiatr 48:906–910

Johnson RT (1987) Myelopathies and retroviral infections. Ann Neurol 21:113–116

Jokelainen M (1977) Amyotrophic lateral sclerosis in Finland. II. Clinical characteristics. Acta Neurol Scand 56:194–204

Juergens SM, Kurland LT, Okazaki H, Mulder DW (1980) ALS in Rochester, Minnesota; 1925–1977. Neurology (Minneap) 30:463–470

Kasarskis EJ, Winslow M (1989) When did Lou Gehrig's personal illness begin? Neurology 39:1243–1245

Kelly JJ Jr, Kyle RA, Miles JM, O'Brien PC, Dyck PJ (1981) The spectrum of peripheral neuropathy in myeloma. Neurology 31:24–31

Kennedy PGE, Duchen LW (1985) A quantitative study of intermediolateral column cells in motor neurone disease and the Shy-Drager syndrome. J Neurol Neurosurg Psychiatr 48:1103–1106

Lebo CD, Sang KU, Norris FH Jr (1976) Cricopharyngeal myotomy in amyotrophic lateral sclerosis. Laryngoscope 86:862–868

Li T-M, Alberman E, Swash M (1986) A suggested approach to the differential diagnosis of motor neuron disease from other neurological conditions. Lancet ii:731–732

Luys JB (1860) Atrophie musculaire progressive. Lésions histologiques de la substance grise de la moelle épinière. Gaz Med (Paris) 15:505

Mackay RP (1963) Course and prognosis in amyotrophic lateral sclerosis. Arch Neurol 8:117–127

Mady S (1984) Surgery for dysphagia in motor neurone disease. In: Rose FC (ed) Research progress in motor neurone disease. Pitman Medical, London, pp 443–447

Mannen T, Iwata M, Toyokura Y, Nagashima K (1977) Preservation of a certain motoneuron group of the sacral cord in amyotrophic lateral sclerosis: its clinical significance. J Neurol Neurosurg Psychiatr 40:464–469

Martyn CN, Barker DJP, Osmond C (1988) Motoneuron disease and post-poliomyelitis in England and Wales. Lancet i:1319–1322

McHenry LC (1969) Garrison's history of neurology. CC Thomas, Springfield, Illinois.

Meadows JC, Marsden CD (1969) A distal form of chronic spinal muscular atrophy. Neurology 19:53–59

Mitsumoto H, Sliman RJ, Schafer IA et al. (1985) Motor neuron disease and adult hexosaminidase A deficiency in two families; evidence for multisystem degeneration. Ann Neurol 17:378–385

Mortara P, Chio A, Rossa MG et al. (1984) Motor neuron disease in the province of Turin, Italy 1966–1980. J Neurol Sci 66:165–173 lder DW (1982) Clinical limits of amyotrophic lateral sclerosis. In: Rowland LP (ed) Human motor neuron diseases. Raven Press, New York, pp 15–29

Mulder DW, Howard FM (1976) Patient resistance and prognosis in amyotrophic lateral sclerosis. Mayo Clin Proc 51:537–541

Mulder DW, Bushek W, Spring E et al. (1983) Motor neuron disease/ALS: Evaluation of detection thresholds of cutaneous sensation. Neurology 33:1625–1627

Newrick PG, Langton Hewer R (1985) Pain in motor neuron disease. J Neurol Neurosurg Psychiatr 48:838–840

Norris F, Shepherd R, Denys EUK, Mukai E, Elias L, Holden D, Norris H (1993) Onset, natural history and outcome in idiopathic adult motor neuron disease. J Neurol Sci 118:48–55

Ono S, Toyokura Y, Mannen T et al. (1986) Amyotrophic lateral sclerosis: histologic, histochemical and ultrastructural abnormalities of skin. Neurology 36:948–956

Patten BM, Engel WK (1982) Phosphate and parathyroid disorders associated with the syndrome of amyotrophic lateral sclerosis. In: Rowland LP (ed) Human motor neuron diseases. Raven Press, New York, pp 181–200

Peiris JB, Seneviratne KN, Wickremasinghe HR et al. (1989) Non-familial juvenile distal spinal muscular atrophy of upper extremity. J Neurol Neurosurg Psychiatr 52:314–319

Poskanzer DC, Cantor HM, Kaplan GS (1969) The frequency of preceding poliomyelitis in ALS. In: Norris FH, Kurland LT (eds) Motor neuron diseases. Grune and Stratton, New York, pp 286–290

Potts, CS (1903) A case of progressive muscular atrophy occurring in a man who had had acute poliomyelitis nineteen years previously. Univ Penn Med Bull 16:31–37

Reed DM, Kurland LT (1963) Muscle fasciculations in a healthy population. Arch Neurol 9:363–367

Rowland LP (1982) Diverse forms of motor neuron disease. In: Rowland LP (ed) Human motor neuron diseases. Raven Press, New York, pp 1–13

Rowland LP (1988) Research progress in motor neuron diseases. Rev Neurol 144:623–629

Rowland LP, Defendini R, Sherman W et al. (1982) Macroglobulinemia with peripheral neuropathy simulating motor neuron disease. Ann Neurol 11:532–536

Sakata MS, Nakanishi T, Toyokura Y (1978) Anal muscle electromyograms differ in amyotrophic lateral sclerosis and Shy–Drager syndrome.

Salazar AM (1982) Discussion. In: Rowland LP (ed) Human motor neuron diseases. Raven Press, New York, pp 179–180

Salmon LA, Riley HA (1935) The relation between chronic anterior poliomyelitis or progressive spinal muscular atrophy and an antecedent attack of acute anterior poliomyelitis. Bull Neurol Inst NY 4:35–63

Sayeed ZA, Velmurugendran CU, Arjunds G et al. (1975) Anterior horn cell disease seen in South India. J Neurol Sci 26:484–498

Schwartz MS, Sargeant M, Swash M (1976) Longitudinal fibre splitting in neurogenic muscular disorders: its relation to the pathogenesis of myopathic change. Brain 99:617–636

Shahani B, Davies-Jones GAB, Russell WR (1971) Motor neuron disease: further evidence for an abnormality of disease metabolism. J Neurol Neurosurg Psychiatr 34:185–191

Shy ME, Rowland LP, Smith TS et al. (1986) Motor neuron disease and plasma cell dyscrasia. Neurology 36:1429–1436

Shy ME, Evans VA, Lublin FD et al. (1989) Antibodies to GM1 and GD1b in patients with motor neuron disease without plasma cell dyscrasia. Ann Neurol 25:511–513

Sobue I, Saito N, Iida M et al. (1978) Juvenile type of distal and segmental muscular atrophy of upper extremities. Ann Neurol 3:429–432

Spiller WG (1904) Primary degeneration of the pyramidal tracts: a study of eight cases with necrophy. Univ Pa Med Bull 17:390–395

Stalberg E, Sanders DB (1984) The motor unit in ALS studied with different neurophysiological techniques. In: Rose FC (ed) Research progress in motor neuron disease. Pitman, London, pp 105–122

Summers BA, Swash M, Schwartz MS et al. (1987) Juvenile onset bulbospinal muscular atrophy with deafness: Vialetto–van Laere syndrome or Madras-type motor neurone disease? J Neurol 234:440–442

Swash M, Ingram DA (1988) Preclinical and subclinical events in motor neuron disease. J Neurol Neurosurg Psychiatr 51:165–168

Swash M, Schwartz MS (1984) Staging motor neurone disease: single fibre EMG studies of asymmetry, progression and compensatory reinnervation. In: Rose FC (ed) Research progress in motor neurone disease. Pitman, London, pp 123–140

Swash M, Schwartz MS (1988) Neuromuscular diseases: a practical approach to diagnosis and management. Springer, Berlin Heidelberg New York, p 456

Trojaborg W, Buchthal F (1965) Malignant and benign fasciculations. Acta Neurol Scand (Suppl 13) 41:251–254

Tyler JR (1979) Double-blind study of modified neurotoxin in motor neuron disease. Neurology 29:77–81

Van Laere JE (1966) Paralysie bulbo-pontine chronique progressive familiale avec surdité. Rev Neurol 115:289–295

Vialetto E (1936) Contributo alla forma ereditaria della paralisi bulbare progressiva. Riv Sper Frevi 40:1–24

Virmani V, Mohan PK (1985) Non-familial spinal segmental muscular atrophy in juvenile and young subjects. Acta Neurol Scand 72:336–340

Williams ER, Bruford A (1970) Creatine phosphokinase in motor neuron disease. Clin Chim Acta 27:53–56

Wilson SAK (1940) Neurology. Edward Arnold, London, pp 1006–1034

Younger DS, Chou S, Hay SAP et al. (1988) Primary lateral sclerosis: a clinical diagnosis re-emerges. Arch Neurol 45:1304–1307

Zilkha KJ (1962) Discussion. Proc R Soc Med 55:1028–1029

2 Epidemiology of Motor Neuron Disease

K. Kondo

Epidemiology can give fundamental information about the frequency and other determinants of disease, including biological, geographical, and sociological risk factors. These data may give clues regarding both aetiology and prevention. In this chapter current epidemiological data on motor neuron disease are reviewed in relation to ideas concerning possible causation of the disease.

Some Important Clinical Patterns of Motor Neuron Disease/Amyotrophic Lateral Sclerosis (MND/ALS)

The clinical classification and nosology of MND/ALS is still a matter for debate. In most epidemiological studies, and in this chapter the term MND/ALS is used to define three major clinical entities. These are amyotrophic lateral sclerosis itself (ALS), progressive bulbar palsy (PBP) and spinal progressive muscular atrophy (SPMA). The term MND, however, is often used as equivalent to ALS, especially in the United Kingdom and related countries, covering ALS in its narrower sense together with PBP and SPMA. In a study of patients dying of MND/ALS in Japan there were 403 cases of ALS, 10 of PBP and 102 of SPMA (Kondo 1984a,b). MND/ALS is defined both clinically and pathologically in the absence of aetiological information. Its distinction from diverse forms of amyotrophies is sometimes confusing. Of these, the clinically recognisable primary degenerative diseases of motor neurons are summarised in Table 2.1. In this chapter the classical sporadic forms of MND/ALS, listed in group 1 in Table 2.1 are considered from the epidemiological standpoint, and evidence is presented that supports the distinction of motor neuron disease with posterior column involvement from so called familial MND/ALS. Many investigators still consider that the Guamanian form of motor neuron disease has a different causation from other sporadic forms of motor neuron disease, but cases in the Kii peninsula in Japan can scarcely be distinguished from other Japanese cases.

MND/ALS is known to occur only in the human, in whom the voluntary pyramidal motor system is highly evolved. It often presents with distal involvement of the upper extremity but may also present with early involvement of the distal parts of one or other leg. The right hand is more likely to be involved than the left. The extraocular and pelvic sphincter muscles, and their motoneuronal innervations are almost invariably spared, but these clinical points remain unexplained. It is tempting to suggest that motor neurons utilised in more skilled movements are more liable to degeneration in MND/ALS than other motor neurons.

Table 2.1. Proposed classification of primary degenerative diseases of motor neurons

1. (Classical) motor neuron disease
   Amyotrophic lateral sclerosis
   Progressive bulbar paralysis
   Spinal progressive muscular atrophy
2. Motor neuron disease with parkinsonism
   (a) Sporadic
   (b) Familial (AD?)
3. Motor neuron disease with dementia
   (a) Sporadic
   (b) Familial
      i. Adult onset (AD?)
      ii. Juvenile onset (AR)
4. Guam motor neuron disease
5. Motor neuron disease with posterior column involvement (AD)
6. Proximal spinal muscular atrophies
   Kugelberg–Welander disease (AD, AR, XR)
   Werdnig–Hoffmann disease (AR)
7. Others

AD, autosomal dominant; AR, autosomal recessive; XR, X-linked recessive.

So-called familial motor neuron disease includes Group 5 and a part of Group 1 which run in families.

Of the clinical features, the duration of the disease is especially important. It is commonly stated that MND/ALS is steadily progressive. A life table analysis, applied to 379 fatal cases, however, disclosed that life expectancies, given that a patient had already lived a number of months after the onset of the disease, decreased until about 20 months after the onset but then, somewhat surprisingly, increased (Fig. 2.1). MND/ALS, therefore, follows the classical teaching only in the initial stages but seems to have an increasingly benign prognosis once the initial dangerous period is safely passed (Kondo and Hemmi 1984). This finding raises the hope that if it were possible to improve the natural history in the early stages, by medical care, rehabilitation, attention to day-to-day living, etc., it might be possible to help patients to survive the initial months and therefore to enter the subsequent, apparently safer period of the disease.

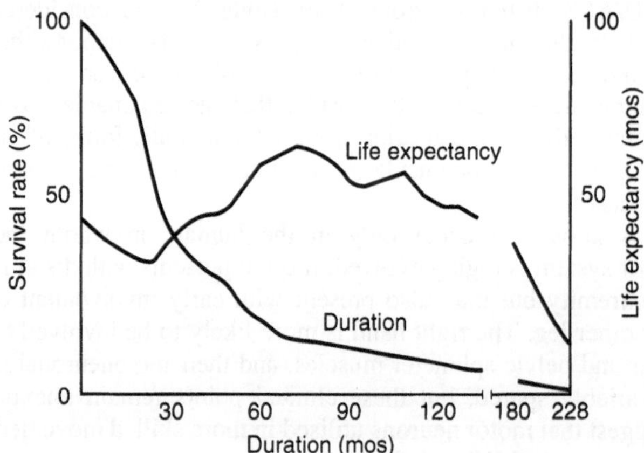

Fig. 2.1. Survival rates and average life expectancy of males with amyotrophic lateral sclerosis.

Of the determinants of duration (Y months), age (X years) is an important factor. In 515 cases, Kondo and Hemmi (1984) reported that:

$Y = 0.6732X + 77.8$ in ALS males
$Y = -0.4500X + 62.3$ in ALS females
$Y = -2.592X + 188.1$ in SPMA, sexes pooled.

Frequency of MND/ALS and its Determinants

Descriptive epidemiology measures the magnitude and distribution of a disease under study.

Prevalence Rate

The point prevalence rate, or the number of surviving patients ascertained at an agreed time in a given population, is used as a convenient measure in chronic incurable diseases such as MND/ALS. Table 2.2 summarises the available crude data for the disease classified according to the reports in the literature. In Japan there are 17 reports coming from parts of the country not known to be of particularly high risk for MND/ALS, with a range 0.8–19.7. The values reported are relatively consistent, even in areas with widely varying climatic geographical socioeconomic conditions, regardless of the three major racial types, i.e. Caucasoid, Mongoloid and Negroid. When the values from Guam and from the Kii peninsula in Japan, as well as the extreme values reported from the areas of the world not known to be at high risk, are eliminated a range 1.0–13.4 (mean 4.09) is found, based on 33 reports, pooling males and females (Table 2.2).

Incidence Rate

This is calculated in one of two ways in MND/ALS. It can be calculated from data collected from new cases that occurred in the year prior to prevalence day in a population study or can be derived from the formula : prevalence rate = duration of disease × incidence rate. In the literature there are 21 reported crude values of incidence rate ranging from 0.1 to 0.58 per 100 000 per year for areas other than those known to be high risk foci of the disease. Eliminating the upper and lower extremes from these values the average was 1.36 per 100,000 per year.

Age-specific incidence rates are more difficult to calculate. The best available data are probably those from Israel (Kahana et al. 1976). In this study MND was found to be highly age-dependent, its incidence rate steadily increasing with age. There is some controversy concerning the age-specific incidence rates in the older age groups. Most studies have found a slightly reduced age-specific incidence rate in the 8th and 9th decades.

Table 2.2. Crude prevalence rate, motor neuron disease or amyotrophic lateral sclerosis, areas except Japan and Guam

Areas	Authors	Period of case finding	Prevalence rate per 100 000
Europe			
Iceland	Gudmundsson (1968)	1965	6.4[a]
Finland	Jokelainen (1976)	1969–73	3.6[a]
Finland, Middle	Murros & Fogelholm (1983)	1976–81	6.4[a]
Sweden, North	Forsgren et al. (1988)	1980	4.8
Scotland	Holloway & Emery (1982)	1968	M 5.26 F 3.30
–Lothian	Holloway & Mitchell (1986)	1977	M 4.84 F 2.58
England, Carlisle	Brewis et al. (1966)	1955–61	7.0
Germany, Westphalia	Haberlandt (1959)	1941–55	2.5
Switzerland, North	Lorez (1969)	1951–67	6.6
Poland, Poznan	Cendrowski et al. (1970)	1955–63	2.2
Rumania	Kreindler et al. (1964)	1950–62	3.7
Italy – Florence	Bracco et al. (1979)	1967–76	2.1[a]
– Messina	Domenico et al. (1988)	1976–85	2.5[a]
– Modena	Scarpa et al. (1988)	1976–86	2.4[a]
– Parma	Juvarra et al. (1983)	1960–80	2.3[a]
– Sardinia	Rosati et al. (1977)	1965	1.6[a]
USA, Canada and Mexico			
Rochester, MN	Kurland (1958)	1925–64	6.7
Lehigh, PA	Zack et al. (1977)	1968–75	6.4
Hawaii	Matsumoto et al. (1972)	1952–69	2.4
Ontario, Southwest	Hudson et al. (1986)	1978–82	4.9[a]
Mexico DF	Olivares et al. (1972)	1952–69	0.8
Other areas			
Israel	Kahana et al. (1976)	1960–70	3.0[a]
Libya, Benghazi	Radharkrishnan et al. (1986)	1980–85	M 4.4 F 2.5

[a]ALS; unmarked values are for MND; M, male; F, female.
References which were quoted by Kondo (1978) are not given.

Mortality Rate

While prevalence as well as incidence rates require population studies for measurement, mortality rates can be calculated from existing national vital statistics, since MND/ALS is always fatal and the validity of death certificates is well established (Kondo and Tsubaki 1977). MND/ALS has been specifically coded in death statistics since 1949 in the revisions of the International Classification of Diseases.

World Patterns of Mortality Rate

The patterns of MND/ALS mortality in different countries were reviewed by Goldberg and Kurland (1962) and later by Kondo (1978). The adjusted mortality rates were greater in the latter part of this period (1966/1971) when compared for those in the earlier period (1953/1958). The difference was modest and could have been due, in part, to improved death certification. In Scotland, Holloway and Emery (1982) reported a rising trend. A similar report was made by Buckley et al. (1983) in England and Wales, an observation that has been confirmed recently by Martyn

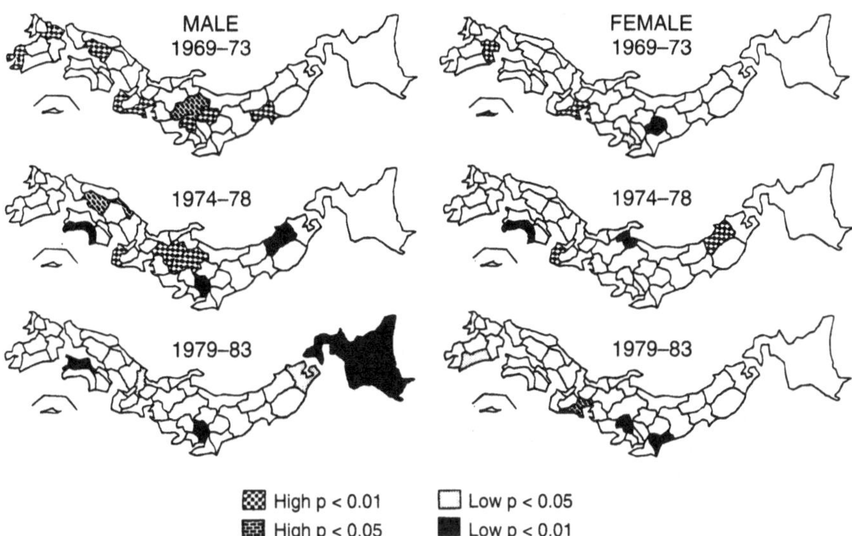

Fig. 2.2. Motor neuron disease standardised mortality ratio for Japan by sex and prefecture in the period 1969–1973, 1974–1978 and 1979–1983.

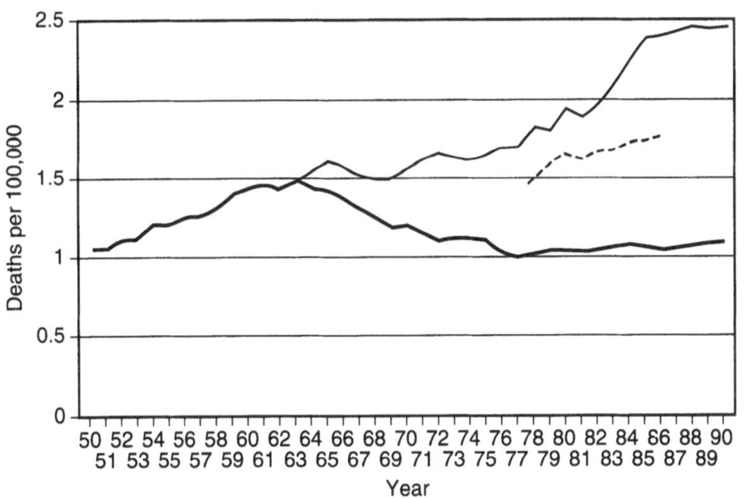

Fig. 2.3. Amyotrophic lateral sclerosis mortality rate 1950–1990, males at all ages age-adjusted to 1990 UK population. Bold line, Japan; medium line, England and Wales; dashed line, United States.

et al. (1988). In the United States Lilienfeld et al. (1989) observed that age-specific mortality rates rose in all areas from 1962 to 1984 in both sexes, whether white or non-white, suggesting an environmental aetiology for the disease. Increasing trends for mortality from MND/ALS had also been noted in Italy, Canada, Kentucky, Rochester, Minnesota and in Sweden (Stallones et al. 1989).

Japanese Experiences

The age and sex-adjusted mortality rate for MND/ALS, calculated annually in Japan, has disclosed a marked declining trend up to 1977. The increasing trend in mortality rate observed until 1960 might have been due to diagnostic practice at least partly, but the decline after this peak in such a short period seems unlikely to be due to purely genetic or purely environmental factors. The mean duration of the disease remained unchanged during this period. It is tempting to suggest that the determinants of such a change in mortality trends, and therefore in the underlying disease process in MND/ALS, is sensitive to socioeconomic change since during this period there were marked socioeconomic changes in Japan. The rise after 1977 is similar to the patterns seen in many other countries.

Changing patterns in mortality rate in MND/ALS have been further evaluated in Japan by analysing the data according to prefecture and by cohorts, using a standardised mortality ratio (SMR). The SMR represents a change in the mortality rate relative to the standard value. The values of SMR and their regional differences have declined in both sexes in Japan recently (Fig. 2.2). This decrease was more obvious in the younger cases, suggesting a cohort effect (Kondo and Minowa 1988).

Recently, mortality analyses in Japan were updated along with data from the United Kingdom and the United States (Neilson et al. 1993). As seen in Fig. 2.3, the declining trend stated above was reversed in 1977, as the trends in the two other countries showed. Analysis based on the Gompertz law indicated that the rise and fall in Japan only was due to environmental causes unique to that country, possibly due to the postwar difficulties.

High-Risk Foci

An increased occurrence of a disease might be caused by an increase in one or more of a number of factors; a causative agent, a change in host predisposition, or a change in modifying factors. Changes in the occurrence of MND/ALS in high-risk foci are particularly interesting in the assessment of the role of these factors in the disease.

As is well known there are three high-risk foci that have been well documented. These consist of a focus in the island of Guam, a focus in the few villages in the Kii peninsula of Japan, and a focus in West New Guinea (Gajdusek and Salazar 1982). These sites have been extensively studied. The villages involved are semi-closed ecosystems. The native residents are relatively primitive, impoverished, endogamous and engaged in local primary industries and in the production of their staple diet. MND/ALS in the first two foci consists of a disease that affects relatively young people at onset, is atypical clinically, and is associated with the presence of a parkinsonism-dementia complex occurring in different patients in the same areas. The cases in West New Guinea are not well documented and, to the author's knowledge, no autopsy has been reported from this region.

A number of foci of very high incidence in Guam and the Kii peninsula are now almost extinct, only a few decades after they were first recognised. Since the high incidence had been traced as far back as the nineteenth century, such a decline in frequency of the condition strongly suggests a role for changing socioeconomic variables, or at least for environmental factors closely associated with human activities, rather than purely natural–environmental and purely genetic variables. It may be that the traditional life of the indigenous residents contained risk factors which are still unknown, but which have been eliminated as a consequence of the recent profound socioeconomic changes in these isolated populations.

It is noteworthy that the clinical picture of parkinsonism-dementia has changed along with the decline of its incidence. While extrapyramidal symptoms ameliorated, the patients now increasingly resemble those with Alzheimer's disease.

Identification of the specific local factors leading to the development of MND/ALS in these high-risk foci has been extensively investigated in order to try to elucidate the mechanism of sporadic MND/ALS, and of the neurodegenerative conditions in general. However, such attempts have not proved very rewarding.

Ingestion of cycad nuts was suspected as the cause of Guamanian ALS. A recent resurgence of this hypothesis requires special comment. Cycad and its toxic element, β-amino-α-amino-methylaminopropionic acid (BMAA), has been proposed as a causative agent in both MND/ALS and parkinsonism-dementia complex in three high-risk foci (Spencer et al. 1987). However, Dr. K-M Chen, a neurologist in Guam, is aware of many patients who have never eaten cycad. Although BMAA produces acute neurological signs no chronic disorder was observed in juvenile Cynomolgus monkey fed with cycad combined with a low-calcium diet, and no synergism was evident between cycad and metal exposure in the environment (Garruto et al. 1979). This problem has been addressed by Steele and Williams elsewhere in this volume.

Tsubaki et al. (1963) observed a high incidence of ALS among 20 356 inhabitants in the South-Western part of Amami-Oshima, in South-Western Japan, but an investigation of the remainder of this part of South-Western Japan, and of three more adjacent islands in the Amami island chain disclosed no further cases of ALS but 32 cases of the Ryukyu-type spinal muscular atrophy in a total population of 148 904 inhabitants. The prevalence rate of ALS in this region of Japan was thus about 1.6 per 100 000 per year, a value which is similar to that pertaining to other regions of Japan. No increase in the frequency of ALS was observed by Tsubaki et al. (1963) or by Kondo et al. (1970) in those areas of Japan in which the cycad nut was widely ingested as part of the staple diet.

Risk Factors

Risk factor analysis in epidemiological studies can be important in detecting the cause, or at least detecting clues to the cause of the disease, in the identification of individuals at high risk, and perhaps in the prevention of the disease. Risk factors are identified by showing statistical associations of an event with the subsequent

Table 2.3. Case–control studies of motor neuron disease

Authors	Cases, controls and methods	Associated factors	Unassociated factors
Felmus et al. 1976	16M, 9F hospital cases[a], controls were age (± 7 yr) and sex-matched normals (also used diseased controls, but unquoted in this table), interview	Exposures to Pb and Hg, athletic participation, ingestion of large amounts of milk, fracture within 5 yr (32% cases, 12% controls)	Income, education, major trauma within 5 yr, surgery, Sabin vaccine, 5 child diseases, 23 adult diseases, dental filling, exposures to toxic elements
Campbell et al. 1970	74 hospital cases[a], age- and sex-matched controls, interview	Fracture within 5 yr (12/74 vs. 4/74)	Diseases of the axial skeleton, upper gastrointestinal disease, World War II POW, poliomyelitis, sarcoidosis
Hanish et al. 1976	109 cases[a] in Los Angeles ALS registry who returned questionnaires, nearest-neighbour controls matched for age and sex, questionnaires sent	Exposures to animal carcasses and hides	Urban–rural birthplace, service on Guam or Japan, outdoor activity, consumption of sheep, calf brain, hay fever, asthma, tonsillectomy
Kurtzke and Beebe 1980	504 male veterans[a], matched with like-aged veterans, record analysis	Pre-service mechanical injury and surgical operations, hospitalisation for trauma esp. fractures of the limbs in service, truck drivers, baseball participation	Race, birthplace, religion, intelligence score, sports, education, occupation/industry, residence, marital status, pre-service diseases, examination at induction, medical history during service
Kondo and Tsubaki 1981 (Study A)	458M, 254F cases from death certificates, 216 surviving husbands and 421 widows as controls, and informants for interview	Mechanical injury	Smoking, drinking, POW, stay in Guam, retention of fragmented shrapnel since World War II, atomic bombing, electric injury, surgical operation, occupation
Kondo and Tsubaki 1981 (Study B)	104M, 54F hospital cases[a], normal controls, individually matched for age and sex, interview	Mechanical injury	Smoking, drinking, residence, home space, drinking water, animals, POW, parental consanguinity, measles, polio, mumps, tuberculosis, rheumatism, total artificial denture, occupation, exposures to radiation, chemicals and gases
Pierce-Ruhland and Patten 1981	53M, 27F hospital cases, 52M, 26F friends as controls, aged ± 5 yr, from same area interview	Exposures to Pb and Hg, ingestion of milk (>3 glasses/day)	Allergy, operation, anaesthesia, injury (within 10 yrs; 37/80 vs. 28/78), dog ownership, PhD, MA, BA or BSc, education, athletic participation, smoking, drinking

Table 2.3. Continued.

Authors	Cases, controls and methods	Associated factors	Unassociated factors
Gawel et al. 1983	32M, 31F hospital cases[a] age- and sex-matched, 33M, 28F normals, questionnaires given	Injury to back (30/63 vs. 11/61), electric shock (15/63 vs. 5/61)	Injury to head, fractures, diseases
Kondo 1984b	430M, 203F cases from the autopsy records, 1964–78, control autopsies with myocardial infarction, pulmonary tuberculosis and craniocervical injuries, comparisons of complicating pathological findings	None	Complications %, by topography, morphology, aetiology, % neoplasms, % organ resections
Kondo and Fujiki 1984, 1989	188M, 105F cases from death certificates, 89 husbands and 136 wives as controls, Koseki (family registry) as materials	None	Parental age, birth order, sib size, interval between the birth of a case and his elder sib, season of birth
Roelofs–Iverson et al. 1984	58M, 47F who replied to interview among 145 hospital cases[a], three control groups were sibs, spouses and others (neighbours, friends, etc.)	Job exposures to As, Mn, Hg; other heavy metals in male	Family history, environmental, athletic, dietary histories, body height, weight, hair colour, eye colour, back/neck/other injuries, anaesthesia, blood transfusion, immunologic history
Deapen and Henderson 1986	An ALS society questionnaired patients[a] with 678 (85.6%) adequate answers, who also furnished "controls" with same age and sex (±5 yr), 518 (76.4%) answers were returned by the controls	Electric shock with unconsciousness, electricity-related jobs, physical trauma with unconsciousness	Surgical procedures, neurological diseases, exposures to Pb, Hg Cr, Ni, As, plastic manufacturing, pesticides, hides

[a]ALS; M, male; F, female.

occurrence of the disease. Much information has been gathered with this concept in mind in relation to MND/ALS (for review see Kondo 1987; Tandan and Bradley 1985). However, none of these observations, although provocative, have resulted in practical benefits in understanding, or preventing the disease.

Case–Control Studies

In this technique, the medical and socioeconomic history prior to the onset of the disease are compared between cases and controls. This methodology may be useful in providing a quick evaluation of a problem, but is often inaccurate, especially

Table 2.4. Characteristics, risk factors, aetiological hypotheses of Guam ALS and parkinsonism-dementia

Area	Small, isolated volcanic islands in the Marianas in the South-Western Pacific, monsoon climate
Ethnic background	Guamanian Chamorros are hosts of ALS/P-D. Chamorros in other Mariana islands show no increase in rates for ALS/PD (Yanagihara et al, 1983). Ethnically Guamian Chamorros are closer to South-East Asians than other Carolinians, but blood groups, haptoglobin and HLA data are insufficient to exclude the possibility that the two groups were of the same origin
Lifestyle characteristic	Patients lower socioeconomically, less educated, ate more homegrown foods and raw meats, have more contacts with animals (Reed and Brody 1975)
Epidemiology of ALS and PD	High prevalence in isolated villages in southern area, prevalent probably since 19th century, but less common recently. Age at onset older than sporadic ALS, and sex ratio nearly 1 : 1. Chamorros developed ALS 1–34 years even after they migrated from Guam. Filipinos six times more liable to ALS or PD, 1–29 years after their arrival at Guam yet 50% of the risk of the local Chamorros
Clinical characteristics	Combined features of ALS and PD. Compared with other countries, ALS younger onset, shows protracted course, even with bulbar onset, more pyramidal signs and dementia. Compared with Parkinson's disease, PD combined with dementia, often elevated TR, distal muscle atrophy, L-dopa effective for extrapyramidal symptoms.
Neuropathology	Characterised by neurofibrillary tangles and granulovacuolar inclusions. ALS and PD probably represent a spectrum of a single disease
Family patterns	ALS and PD familial but non-Mendelian, probably multifactorial (Reed et al. 1975). More recently, familial aggregation less evident
Virology	Titre and isolation studies negative for conventional viruses. Transmission experiments negative for slow virus (Gibbs and Gajdusek 1982)
Toxicology	See text for cycad data
Immunology	Blood groups and secretor factors unassociated with ALS/PD. Association of PD with HLA Bw 16. Elevated serum IgA and IgG in ALS and PD. ALS cases hyporesponsive to skin test, with low T cell mitogen responses. Patients with HLA Bw 35 showed a shorter clinical course
Possible cause	Low heavy metals in soils and water (Garruto et al. 1984), disturbances in calcium and vitamin D metabolism (Yanagihara et al. 1984), intraneuronal co-localisation of calcium, aluminium and silicon in affected CNS tissues (Perl et al. 1982), support a basic defect(s) in mineral metabolism and secondary hyperparathyroidism related to the intestinal absorption of toxic metals and their deposition possibly as hydroxyapatites and aluminosilicates in neurons. Such may disrupt the axonal transport resulting in excessive intraneuronal accumulation of the neurofilament protein and formation of neurofibrillary tangles. Mechanism of these changes in relation to early neuronal death are unknown

since it relies on memory. The major results are summarised in Table 2.3. The results are quite varied, although trauma appears frequently as a putative risk factor in sporadic MND/ALS from different countries.

Ecological Studies

In this approach the ecological characteristics of communities at risk are compared with those communities showing different frequencies of the disease under study.

This method is indirect, but it is useful for objective evaluation of the regional characteristics that might be associated with a disease, even if these do not appear as individual memories. This approach has been applied in two high-risk foci of MND/ALS in the Pacific (Table 2.4) but not in other areas, except for the study of Kondo and Minowa (1988) in Japan.

They studied death certificates from patients dying of MND/ALS for the 15-year period 1969–1983. Information was available according to age and sex for 2739 municipalities. A total of 600 statistics were collected for each of these municipalities. Standardised mortality ratios were calculated for each of the categories of a given item by pooling the deceased cases and the respective populations. Mountain residence was found to be a risk factor for MND/ALS. It is possible that traditional life in Japan, especially in mountain villages, carried with it certain risk factors for MND/ALS which have been eventually eliminated in more recent developments. The focus in the Kii peninsula may be an extreme example of such a situation.

Definite Risk Factors for MND/ALS

Of various suggestions put forward, three risk factors for MND/ALS are acceptable beyond doubt. These are as follows:

1. *Age*. This is the strongest known risk factor. In both sexes incidence rates rise sharply from around the age of 50 to a peak in the 8th decade of life. However, an elderly onset is correlated with a shorter duration of disease (Kondo and Hemmi 1984). A numerical decline of motor neurons in the elderly may underline such a pattern. Clinicians have been impressed that aged onset is increasing for MND/ALS of sporadic type, particularly in developing countries. Such a trend has also been observed in Guam, despite a gradual decline in the overall incidence rate.
2. *Sex*. This is a particular factor in MND/ALS because a male predominance in this disease appears unique among the neurodegenerative diseases as a whole. The sex ratio, the number of male cases divided by the number of female cases, was 1.42 in 384 fatal cases. It was 1.49 in ALS, 1.18 in PBP and 1.52 in SPMA. The sex ratio for cases manifesting in bulbar muscles was 1.21, but 1.88 in those manifested in one or both upper limbs and 1.75 in those manifested in upper and lower limbs simultaneously. For those beginning in the lower limbs it was 1.18 (Kondo 1975).

Males are more likely to sustain physical injury, a suspected risk factor for MND/ALS, but this factor alone does not explain male predominance because males also predominate among patients with MND/ALS not known to be injured (Kondo and Tsubaki 1981). On the other hand a decline has been noted in the sex ratio of Guamanian ALS, together with a decline in the incidence of this condition on the island of Guam, a finding suggesting that sex-related factors might be correlated with the socioeconomic or other changes underlying the recent change in incidence patterns of MND/ALS in Guam.

3. *Mechanical injuries*. One of Charcot's patients believed that his ALS was due to a fracture of the clavicle he had sustained 2 to 3 months before the onset of the disease. Post-traumatic cases were reviewed by Erb (1897) who accepted trauma as a possible causative agent. Bodechtel and Schrader (1953) considered that trauma might be causative in ALS if it was severe enough, if the neurological syndrome occurred within 3 to 4 months of the injury, and if the site of injury was correlated

with the site of origin of the first symptom of ALS. Peters (1954), in an extensive review of the neurodegenerative disorders after trauma, viewed from the standpoint of insurance medicine, emphasised the apparent significance of repeated injuries in the development of "post-traumatic" ALS. He reviewed more than 100 such post-traumatic cases. However, it is clear that such unilateral observations, lacking in control observations, are not adequate to establish trauma as a risk factor in the origin of this disease.

In two case–control studies, Kondo and Tsubaki (1981) found that mechanical injuries, not surgery, were associated with the subsequent development of ALS, both in men and women provided that the mechanical injury had occurred within 5 years prior to the onset of the motor syndrome. Injury to any particular part of the body was equally associated with MND/ALS, and no correlation between the site of the injury and the initial symptom attributed to the motor neuron disease syndrome. Subsequent reports by other authors are summarised in Table 2.3.

The odds ratio for a relationship between previous trauma and MND/ALS, based upon the above data, is about three, meaning that MND/ALS occurs about three times more frequently among a traumatised population. Thus, taking 1.0 as the annual incidence rate per 100 000 individuals, only three of 100 000 traumatised individuals would develop MND/ALS or only 15 at best during a 5-year period after such trauma. This analysis suggests that MND/ALS can never be considered a usual consequence of trauma. MND/ALS clearly occurs without trauma indicating that trauma can at best be only one of many factors precipitating the underlying disease process, not a cause of that process.

Familial Aggregation of MND/ALS and Separation of Genetic Entities

There is a family history of MND/ALS in 1.8% to 8.6% of cases, as shown in ten reports. Familial cases sometimes show atypical clinical-pathological features when compared with sporadic cases. Indeed, familial ALS may not represent a single disease entity and it is unclear whether such cases are otherwise entirely identical to sporadic cases. Twins with MND/ALS have only rarely been reported. In dizygotic twins Estrin (1977) noted that MND was concordant, but in monozygotic twins reported by Jokelainen et al. (1978) was discordant. Conjugal cases are also known (Chad et al. 1982; Paolino et al. 1983) but these rare incidences are no more significant than clusterings of cases in small areas.

The formal genetics of familial MND/ALS have been studied based on collected pedigree reports involving autopsies. There are a number of pedigrees of familial MND with classical pathology and others in which the classical pathology is combined with degeneration of the posterior columns, of the spinocerebellar tracts, and of Clarke's columns. The latter groups of cases tended to be younger at onset, to show a 1 : 1 sex ratio, to begin in the legs, and less frequently to have bulbar signs. These cases showed family patterns compatible with autosomal dominant inheritance.

In classical ALS, familial onset is very rare and the family patterns suggested a multifactorial hypothesis. Thus it seemed likely that so-called familial MND/ALS includes at least two groups of cases (Kondo 1989).

Aetiological Hypotheses

Any disease is a product of the interplay of agents, host factors and environmental factors. Apart from a few cases, classical MND/ALS is usually sporadic. The recent changes in its frequency in certain populations in a short period of time, particularly those formerly showing high-risk foci, suggested that the underlying process may be sensitive to environmental factors. However, specific identification of any such factors has not so far proved possible. Some aetiological hypotheses are summarised in Table 2.4. One of these concerns the relationship to disturbed parathyroid metabolism (Patten et al. 1974). However, this hypothesis, like many others, is essentially speculative, despite reports of abnormalities in aluminium and manganese metabolism in patients in some high-risk foci.

Since the time of Gowers, ageing has been assumed as a cause of the neurodegenerative conditions. Thus, cell death is a common factor in such diseases as a whole, including MND, Parkinson's disease, Alzheimer's disease, olivopontocerebellar atrophy. The molecular mechanisms of ageing are poorly understood at present. The changes observed in ageing neurons and in the age-dependent neurodegenerative conditions are not necessarily specific and show quantitative change in a number of variables, particularly relating to DNA and RNA metabolism (see Bradley and Krasin 1982).

Neurons are fixed post-mitotic cells that are no longer capable of undergoing mitosis, and thus have limited life spans. Cumulated somatic mutations may interfere with their longevity. With ageing, there is a reduction in their number, while surviving neurons may remain seemingly normal, or even become hypertrophied. The function of any neuronal system must therefore decline with age, at least in relation to a reduction in neuronal number, and also possibly in relation to declining neuronal function in the remaining neurons. A simple model illustrates how functional requirements per neuron drastically increase when neurons are subject to ageing (Kondo 1984a). The compensation apparently has its limits, but it is unknown what morphological or metabolic changes occur in the neurons involved, how these changes are manifest clinically, whether neurons rapidly die or whether neuronal death is a slow process. Furthermore, it must be asked whether such degenerative diseases are in fact caused in this way.

A multifactorial concept of causation in MND/ALS is useful when contributing factors are individually identified.

Acknowledgements. This work has been supported by funds from the Japan Ministry of Health and Welfare, particularly the research committee on neurodegenerative diseases. Cordial thanks are due to Miss Junko Mano for her technical assistance.

References

Bodechtel G, Schrader A (1953) Die Erkrankung des Rückenmarks. In: Mohr L, Staehelin R (eds) Handbuch der Inneren Medizin, vol. 2. Springer, Berlin Heidelberg New York, pp 504

Bracco L, Antuono P, Amaducci L (1979) Study of epidemiological and etiological factors of amyotrophic lateral sclerosis in the province of Florence, Italy. Acta Neurol Scand 60:112–124

Bradley WG, Krasin F (1982) DNA hypothesis of amyotrophic lateral scerosis. In: Rowland LP (ed) Human motor neuron diseases. Raven Press, New York, pp 493–500

Buckley J, Warlow C, Smith P, Hilton-Jones D, Irvine S, Tew JR (1983) Motor neuron disease in England and Wales, 1959–1979. J Neurol Neurosurg Psychiatr 46:197–205

Campbell AMG, Williams ER, Barltrop D (1970) Motor neuron disease and exposure to lead. J Neurol Neurosurg Psychiatr 37:877–885

Chad D, Mitsumoto H, Adelman LS (1982) Conjugal motor neuron disease. Neurology 32:306–307

Deapen DM, Henderson BE (1986) A case–control study of amyotrophic lateral sclerosis. Am J Epidemiol 123:790–799

Domenico PD, Malara CE, Marabello L et al. (1988) Amyotrophic lateral sclerosis: an epidemiological study in the province of Messina, Italy, 1976–1985. Neuroepidemiology 7:152–158

Erb W (1897) Zur Lehre von den Unfallserkrankungen des Rückenmarkes, über Poliomyelitis anterior chronica nach Trauma. Dtsch Z Nervenheilk 11:122–142

Estrin WJ (1977) Amyotrophic lateral sclerosis in dizygotic twins. Neurology 27:692–694

Felmus MT, Patten BM, Swanke L (1976) Antecedent events in amyotrophic lateral sclerosis. Neurology 26:167–172

Forsgren L, Almay BG, Holgren G, Wall S et al. (1988) Epidemiology of motor neuron disease in Northern Sweden. Acta Neurol Scand 68:20–29

Gajdusek DC, Salazar AM (1982) Amyotrophic lateral sclerosis and parkinsonian syndromes in high incidence among the Auyw and Jakai people of West New Guinea. Neurology 32:107–126

Garruto RM, Yanagihara R, Gajdusek DC (1979) Cycad and amyotrophic lateral sclerosis/parkinsonism dementia. Lancet II:1079

Garruto RM, Yanagihara R, Gajdusek DC, Arion DM et al. (1984) Concentration of heavy metal and essential minerals in garden soil and drinking water in the Western Pacific. In: Chen K-M, Yase Y (eds) Amyotrophic lateral sclerosis in Asia and Oceania. National Taiwan University, Taipei, pp 265–330

Gawel M, Zaiwalla A, Rose FC (1983) Antecedent events in motor neuron disease. J Neurol Neurosurg Psychiatr 46:1041–1043

Gibbs Jr CJ, Gajdusek DC (1982) An update on long-term in vivo and in vitro studies designed to identify a virus as the cause of amyotrophic lateral sclerosis, parkinsonism dementia, and Parkinson's disease. In: Rowland LP (ed) Human motor neuron disease, Raven Press, New York, pp 343–353

Goldberg ID, Kurland LT (1962) Mortality in 33 countries from diseases of the nervous system. World Neurol 3:444–465

Hanish R, Divorsky RL, Henderson BE (1976) A search for clues to the cause of amyotrophic lateral sclerosis. Arch Neurol 33:456–457

Holloway SM, Emery AEH (1982) The epidemiology of motor neuron disease in Scotland. Muscle Nerve 5:131–133

Holloway SM, Mitchell JD (1986) Motor neuron disease in the Lothian Region of Scotland, 1961–81. J Epidemiol Community Health 40:344–350

Hudson AJ, Davenport A, Hader WJ (1986) The incidence of amyotrophic lateral sclerosis in southwesten Ontario, Canada. Neurology 36:1524–1528

Jokelainen M (1976) The epidemiology of amyotrophic lateral sclerosis in Finland. J Neurol Sci 29:55–63

Jokelainen M, Palo J, Lokki J (1978) Monozygous twins discordant for amyotrophic lateral sclerosis. Eur Neurol 17:296–299

Juvarra G, Bettoni L, Bortone E, Garavelli A, Montanari E, Rocca M et al. (1983) Amyotrophic lateral sclerosis in the province of Parma, Italy: a clinical and epidemiological study in the period 1960–1980. Ital J Neurol Sci 4:473–478

Kahana E, Alter M, Feldman S (1976) Amyotrophic lateral sclerosis, a population study. J Neurol 212:205–213

Kondo K (1975) Clinical variability of motor neuron disease. Neurol Med (Jpn) 2:11–16

Kondo K (1978) Motor neuron disease; changing population patterns and clues for etiology. In: Schoenberg, BS (ed) Neurological epidemiology. Raven Press, New York, pp 509–542

Kondo K (1984a) Epidemiology of motor neurone disease; ageing and exhaustion hypotheses revisited. In: Rose FC (ed) Research progress in motor neurone disease. Pitman, London, pp 20–33

Kondo K (1984b) Motor neuron disease and Parkinson's disease are not associated with other disorders at autopsy. Neuroepidemiology 3:182–194

Kondo K (1987) Environmental factors in motor neurone disease. In: Gourie-Devi M (ed) Motor neurone disease. Oxford University Press, New Delhi, pp 54–60

Kondo K (1989) Genetic heterogeneity of familial motor neuron disease. Brain Nerve (Jpn) 41:255–228

Kondo K, Fujiki K (1984) Effects of parental age and birth order in motor neuron disease. Jpn J Hum Genet 29:45–50

Kondo K, Fujiki K (1989) Is risk to motor neuron disease influenced by the season of birth? Jpn J Hum Genet 34:243–246

Kondo K, Hemmi I (1984) Clinical statistics in 515 fatal cases of motor neuron disease. Neuroepidemiology 3:129–148

Kondo K, Minowa M (1988) Epidemiology of motor neuron disease in Japan: declining trends of the mortality rate. In: Tsubaki Y, Yase Y (eds) Amyotrophic lateral sclerosis. Elsevier, Amsterdam, pp 11–16

Kondo K, Tsubaki T (1977) Changing mortality patterns of motor neuron disease in Japan. J Neurol Sci 32:411–424

Kondo K, Tsubaki T (1981) Case–control studies of motor neuron disease; association with mechanical injuries. Arch Neurol 38:220–226

Kondo K, Tsubaki T, Sakamoto F (1970) The Ryukyuan muscular atrophy, an obscure heritable neuromuscular disease found in the islands of southern Japan. J Neurol Sci 11:359–382

Kurtzke JF, Beebe GW (1980) Epidemiology of amyotrophic lateral sclerosis. I. A case–control comparison based on ALS death. Neurology 30:453–462

Lilienfeld DE, Chan E, Ehland J, Godbold J, Lendrigan PJ, Marsh G et. al (1989) Rising mortality from motor neuron disease in the USA 1962–84. Lancet i:710–713

Martyn CN, Barker DJP, Osmond C (1988) Motor neuron disease and post-poliomyelitis in England and Wales. Lancet i:1319–1322

Murros K, Fogelholm R (1983) Amyotrophic lateral sclerosis in Middle-Finland: an epidemiological study. Acta Neurol Scand 67:41–47

Neilson S, Robinson I, Kondo K (1993) A new analysis of mortality from motor neurone disease in Japan 1950–1990; rise and fall in the postwar years. J Neurol Sci 117:46–53

Paolino E, Granieri E, Tola MR, Rosati G et al. (1983) Congual amyotrophic lateral sclerosis. Ann Neurol 14:699

Patten BM, Bilezikian JP, Mallette LE, Prince A, Engel WK, Aurbach GD et al. (1974) Neuromuscular disease in primary hyperparathyroidism. Ann Intern Med 80:182–193

Perl DP, Gajdusek DC, Garruto RM, Yanagihara RT, Gibbs CJ Jr et al. (1982) Intraneuronal aluminum accumulation in amyotrophic lateral sclerosis and parkinsonism-dementia on Guam. Science 217:1053–1055

Peters G (1954) Die häufigeren degenerativen Erkrankungen des Zentralnervensystems unter besonderer Berücksichtingung versorgungsärztlicher Gesichtspunkte. Fortschr Neurol Psychiatr 22:139–163

Pierce-Ruhland R, Patten BM (1981) Repeat study of antecedent events in motor neurone disease. Ann Clin Res 13:102–107

Radharkrishnan K, Ashok PP, Sridharan R, Mousa ME et al. (1986) Descriptive epidemiology of motor neuron disease in Benghazi, Libya. Neuroepidemiology 5:47–54

Reed DM, Brody JA (1975) Amyotrophic lateral sclerosis and parkinsonism-dementia on Guam, 1945–1972. I. Descriptive epidemiology. Am J Epidemiol 101:287–301

Reed DM, Torres JM, Brody JA (1975) Amyotrophic lateral sclerosis and parkinsonism-dementia on Guam, 1945–1972. II. Familial and genetic studies. Am J Epidemiol 101:302–310

Roelofs-Iverson RA, Mulder DW, Elveback LR, Kurland LT, Molgaard CA et. al (1984) ALS and heavy metal; a pilot case–control study. Neurology 34:393–395

Rosati G, Pinn L, Granieri E et al. (1977) Studies of epidemiological, clinical and etiological aspects of ALS disease in Sardinia, Southern Italy. Acta Neurol Scand 55:231–244

Scarpa M, Colombo A, Pancetti P, Sorgato P et. al (1988) Epidemiology of amyotrophic lateral sclerosis in the province of Modena, Italy. Influence of environmental exposure to lead. Acta Neurol Scand 77:456–460

Spencer PS, Nunn PB, Hugon J et al. (1987) Guam amyotrophic lateral sclerosis parkinsonism-dementia linked to plant excitant neurotoxin. Science 237:517–522

Stallones L, Kasarkis EJ, Stipanowich C, Snider G et al. (1989) Secular trends in mortality rates from motor neuron disease in Kentucky, 1964–1984. Neuroepidemiology 8:68–78

Tandan R, Bradley WG (1985) Amyotrophic lateral sclerosis; part 2, etiopathogenesis. Ann Neurol 18:419–431

Tsubaki T, Kondo K, Tukagoshi H et al. (1963) Study of neurological disorders in Amami-Oshima island. Clin Neurol (Jpn) 3:394–400

Yanagihara R, Garruto RM, Gajdusek DC et al. (1983) Epidemiological surveillance of amyotrophic lateral sclerosis and parkinsonism-dementia in the commonwealth of the Northern Mariana Islands. Ann Neurol 13:79–86

Yanagihara R, Grafton DA, Garruto RM, Gajdusek DC et al. (1984) Elemental content of scalp hair in Guamanian Chamorros with amyotrophic lateral sclerosis and parkinsonism-dementia. In: Chen K-M, Yase Y (eds) Amyotrophic lateral sclerosis in Asia and Oceania. National Taiwan University, Taipei, pp 331–336

Zack MM, Levitt LP, Schoenberg B (1977) Motor neuron disease in Lehigh County, Pennsylvania: an epidemiologic study. J Chron Dis 30:813–818

3 Familial Motor Neuron Disease

J. de Belleroche, P. N. Leigh and F. Clifford Rose

Introduction

Familial motor neuron disease (FMND) constitutes approximately 5%–10% of cases of motor neuron disease (MND). Whilst in most families the pattern of inheritance is consistent with an autosomal-dominant trait, with age-dependent penetrance, a few cases appear to show an autosomal-recessive mode of inheritance. Statistical analysis shows that the likelihood of chance aggregation is improbable because affected members span several generations, come from different environmental and geographical regions, and the condition does not develop in spouses.

Clinically, FMND can be distinguished from the sporadic form only by taking a family history. Although it is claimed that the onset of FMND is earlier, the same wide range in the age of onset is seen to occur in sporadic forms. Whilst the duration of illness is said to be shorter in FMND, it is usually between 1 and 5 years, as occurs in sporadic cases, with occasional cases living up to 20 years. Some familial cases show a distinctive pathology with degeneration of the posterior columns and spinocerebellar tracts, but this is not a pathognomonic feature of FMND. The similarity in the range of manifestations of both the familial and sporadic forms of MND indicates that both forms probably have a similar pathogenesis.

The Case for Studying Familial Motor Neuron Disease: Rapid Development of Techniques of Molecular Genetics and their Successful Application

The familial form of the disease offers the means of identifying a gene which is defective in FMND and vulnerable to acquire damage in sporadic cases. Enormous advances have occurred in recent years in molecular biology which have made it possible to identify defective genes in a number of fatal inherited disorders such as Duchenne muscular dystrophy and cystic fibrosis. In an even larger number of such conditions, the approximate location of a defective gene has been established by linkage of the disease locus to the positions of specific DNA markers. Once defective genes have been localised, prenatal and presymptomatic detection becomes possible, so that when the gene has been characterised and its function established, more specific targeting of treatment becomes feasible.

The Problems: Few Families, Clinical and Pathological Heterogeneity and Incomplete Penetrance

There is little doubt that FMND offers a challenge because of its rarity, the rapid progression of the disease, and the incomplete penetrance. Obligate gene carriers may live into their eighties without being affected. An additional confounding issue for linkage analysis is that the phenotypic variation may arise out of genetic heterogeneity, as occurs in hereditary sensory motor neuropathy, type 1. However, considerable variability in age of onset (24–70 years), duration of illness (3–11 years) and clinical presentation is found within individual families (Mulder et al. 1986; Siddique et al. 1989). Members of the same family can present with progressive bulbar palsy (PSB), amyotrophic lateral sclerosis (ALS) or progressive muscular atrophy (PMA) (Hawkes et al. 1984; Siddique et al. 1989; Veltema et al. 1990). Genetic heterogeneity may thus not necessarily explain the high degree of phenotypic variation, since this can be shown within individual families. Further, the appearance of ALS, PBP and PMA in the same family indicates that the pathogenesis of these conditions is likely to be very closely related. Hence segregation of these conditions into subtypes is not indicated in initial studies.

Variable expression of a single gene defect as clinically overlapping phenotypes is now thought to occur in some cases of spinal muscular atrophy. Although genetic heterogeneity was indicated for the two clinically distinct forms of childhood-onset chronic spinal muscular atrophy (intermediate SMA or SMA type II and Kugelberg–Welander or SMA type III) which show a variable age of onset (6 months to 17 years) and a degree of severity (death in adolescence to minor impairment of function), these conditions have now been shown to be genetically homogeneous, mapping to chromosome region 5q11.2–13.3 (Brzustowicz et al. 1990). These chronic forms are thus likely to occur as a result of allelic heterogeneity, as in Duchenne and Becker muscular dystrophies. There is an indication too that the more severe acute form of SMA (Werdnig–Hoffmann type), which is usually fatal within the first 4 years, may also be mapped to the same locus on chromosome 5q.

The early establishment of collaboration between research groups is clearly of vital importance for efficient future advancement in research on genetic factors in motor neuron disease. Currently about 300 families are known worldwide.

Profile of Familial Motor Neuron Disease

The typical picture of a familial MND pedigree is shown by the two examples in Fig. 3.1, where affected individuals are seen in at least two generations and there is an autosomal-dominant mode of inheritance. In a study of 35 families in the UK, affected individuals typically presented with both upper and lower motor neuron signs. In this series, the average age of onset was 51 years, and the male : female ratio was 0.9 : 1.0.

Estimates of the frequency of familial cases have varied between 0.8% in Finland (Jokelainen 1977) and 12% (Murros and Fogelholm 1983), but in most surveys, the prevalence of familial cases is around 5% (Kurland and Mulder 1955; Emery and Holloway 1982; Mulder et al. 1986; Li et al. 1988; Williams et al. 1988; Brown 1989; Chancellor et al. 1992; Chancellor and Warlow 1992). Most studies indicate that the age of onset in familial cases is generally earlier than in sporadic cases of MND, although some have found a similar age of onset for both forms of the disease (Williams et al. 1988). Several studies indicate that FMND is not homogeneous clinically, and that some families have a rapidly progressive form of the disease, whereas other families tend to show slow progression and long survival (Williams et al. 1988).

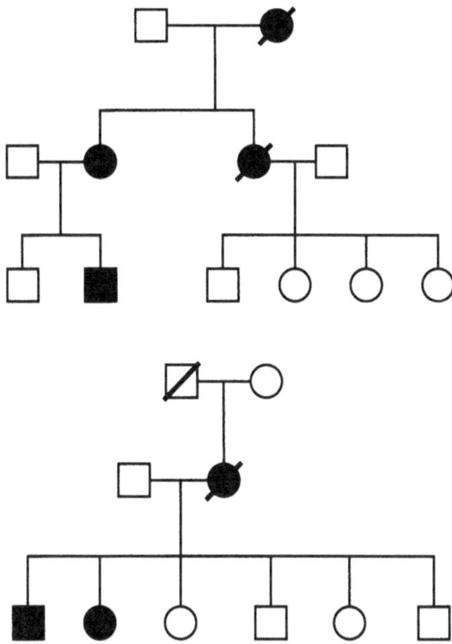

Fig. 3.1. Two typical familial MND pedigrees.

The clinical features of MND in familial and sporadic cases are virtually identical (Table 3.1). This suggests that survival is shorter in most FMND patients compared with the sporadic disease (Li et al. 1988) and between a half and two-thirds of patients will die within 2 years of onset.

Penetrance of typical autosomal-dominant FMND is high, particularly in those families with early onset. Emery and Holloway (1982) calculated that by the age of 55, 82% of familial cases compared with 43% of sporadic cases would have developed the disease. Siddique et al. (1991) estimated a penetrance of 90% by the age of 70. Variations in the age of onset within families make it inevitable that some obligate carriers survive into the fifth and sixth decades without manifesting the disease. Williams et al. (1988) found that the majority of the families they studied had low penetrance, with at least one obligate carrier dying after the age of 50 without evidence of disease. These families showed an average age of onset of 61 years, albeit with a wide range between 39 and 84 years. In contrast, families with high penetrance had earlier onset, with an average age of onset of 48 years, and a range of 24–71 years. These latter figures are very similar to those reported by Kurland and Mulder (1955) and Emery and Holloway (1982) for high-penetrance families. Low penetrance of FMND may lead to underestimation of the incidence of this condition.

Phenotypic Variation in FMND

As in sporadic MND, there is significant phenotypic variation in FMND. In the large kindred reported by Veltema et al. (1990), most individuals had the typical combination of upper and lower motor neuron signs, but some had only lower motor neuron signs (progressive muscular atrophy). This can cause problems with classification, because research diagnostic criteria tend to demand the combination of upper

Table 3.1. Clinical features of FMND compared with sporadic MND

	Emergy and Holloway 1982[a]		Mulder and Kurland 1986	Chio et al. 1987[a]	Li et al. 1988		Veltema et al. 1990[b]
	SMND	FMND	FMND	FMND	SMND	FNMD	FMND
Number of cases	967	231	103	27	553	27	13
Mean age of onset (±SD)							
Males	–	–	47.9 ± 12.9	–	–	–	32.8
Females	–	–	48.8 ± 10.2	–	–	–	28.5
Both	56.2 (11.8)	45.1 (10.5)**	48.3 (11.7)	50.3 (12.4)	56 (12.4)	52 (14.3) (NS)	31
Survival (years)	3.0	1.5 (median)	2.4 (median)	2.0 (median)	2.6 (3.1) (means ± SD)	1.1 (1.74) (NS)	1.7 (median)
Male:female ratio	1.57	1.2 (NS)	1.2	1.08	1.6	0.8 (NS)	1.3
Site of onset	(n=854)	(n=137)	(n=100)				
Legs	37.9%	48.2% } NS	48%	37.0%	37.3%	37.0% } NS	38.5%
Arms	41.7%	35.0% } NS	23%	48.1%	43.6%	48.1% } NS	61.5%
Bulbar	20.4%	16.8% } NS	29%	14.9%	19.0%	14.8% } NS	–
LMN signs	–	–	100%	–	–		100%
UMN signs	–	–	80%	–	–		69%

[a]Cases collected from the literature; [b]one family; [c]eight families; **p = <0.001; NS, not significant.

and lower motor neuron signs for a definite or probable diagnosis of MND (ALS) – for example, the European FALS group has adopted a synopsis of the World Federation of Neurology criteria (Swash and Leigh 1992). For the purposes of classifying families on clinical grounds, it is important that at least one affected member should have definite or probable MND (ALS) by these criteria. There are, however, reports of even more striking intrafamilial heterogeneity (Appelbaum et al. 1992). These authors reported four families with various combinations of what was termed late-onset spinal muscular atrophy (SMA 4) associated with progressive lateral sclerosis in a cousin; typical MND associated with SMA 3 in her granddaughter; typical MND (ALS) in a patient who had three children with SMA 2; and a family with suspected ALS associated with SMA 3 in the patient's niece. Shaw et al. (1992) described a patient with MND of PMA type progressing to death over 2 years, and associated with Werdnig–Hoffmann disease in four grandchildren. Similar associations have been reported by Camu and Billiard (1993) and by Harding et al. (1983). The significance of such associations is at present uncertain, but it seems unlikely that they can be attributed solely to chance.

Neuropathology

Most cases of FMND show similar pathology to that of sporadic cases, with marked loss of anterior horn cells and degeneration of the corticospinal tracts. It has been suggested, however, that cases of FMND more frequently show degeneration of the posterior columns, the spinocerebellar tracts, and Clarke's column (Engel et al. 1959; Hirano et al. 1967; Horton et al. 1976; Emery and Holloway 1982). It is now recognised, however, that involvement of the spinocerebellar tracts and the column of Clarke is common in sporadic MND (Swash et al. 1988; Williams et al. 1990). Pathological studies on patients who have been maintained alive with assisted ventilation indicate that pathological changes are often found in the posterior columns and spinocerebellar tracts in patients with sporadic MND. Indeed, in these cases there may be quite extensive involvement of subcortical structures, including the basal ganglia, thalamus, and various brainstem nuclei (Hayashi and Kato 1989; Mizutani et al. 1992). To our knowledge, no attempt has been made to carry out a systematic study of pathological involvement of structures other than the cortico-spinal tracts and lower motor neuron by comparing sporadic and familial MND cases matched for age of onset, duration of disease, and other relevant clinical features. Thus it remains uncertain whether more extensive or severe involvement of sensory systems really is a marker of FMND.

Dementia occurs in both sporadic and FMND, but it is said to be more common in the latter, as are extrapyramidal features (Hudson 1981). When dementia is associated with motor neuron disease, it is usually an "anterior" dementia (dementia of frontal lobe type), with a characteristic pathology that has been extensively documented in the Japanese and European literature (Kew and Leigh 1992). A minority of the cases reported have had a family history suggestive of an autosomal-dominant mode of inheritance (Robertson 1953; Finlayson et al. 1973; Gunarsson et al. 1991; Wightman et al. 1992). The pathology is virtually always that of mainly frontotemporal cortical atrophy associated with cell loss that is particularly striking in cortical layers II and III, and is associated with striking cortical and subcortical gliosis. There may also be neuronal loss in subcortical structures, particularly in the substantia nigra, although Lewy bodies are not found in such cases. Some reports suggest that MND and dementia may occur together or separately in individual families. For example, in a family described by Gunarsson et al. (1991), dementia occurred in three generations, and in the fourth generation, one individual was demented, but four individuals developed MND; in one of these the signs of MND were preceded by paranoid schizophrenia, and in three others there were changes in personality and cognitive function. Autopsy in one demented case showed the pathology referred to above. Most such cases have now been screened for mutations of the *prion* gene, and have been negative.

Immunocytochemistry with antibodies against ubiquitin has revealed characteristic ubiquitinated inclusions in lower motor neurons in both sporadic and familial cases (Leigh et al. 1991). These inclusions take the form of "skeins" or Lewy body-like inclusions. Both types of ubiquitinated inclusion have been detected in FMND (Leigh et al. 1991; Murayama et al. 1989; Kusaka et al. 1988). Neurofilamentous accumulations are also seen, both in sporadic MND and FMND, although they may be particularly abundant and striking in the latter (Hirano 1991).

Neurofilamentous accumulations in sporadic MND and FMND tend to stain intensely with antibodies directed against phosphorylated neurofilament epitopes. Although this may well be a secondary phenomenon, the observation has been made that overexpression of the 68 kDa neurofilament subunit (NF-L) in transgenic mice leads to accumulation of neurofilaments in spinal motor neurons, degeneration of motor axons, and evidence of muscle atrophy (Xu et al. 1993). Furthermore, Coté et al. (1993) also in transgenic experiments, have introduced the human 200 kDa neurofilament subunit (NF-H), and regulatory elements into mice, and have induced a delayed-onset form of motor neuron degeneration. There are thus good reasons to investigate further the role of neurofilaments and neurofilament genes in the pathogenesis both of sporadic MND and FMND.

Linkage Analysis in Familial Motor Neuron Disease

What is Involved in Locating a Defective Gene?

The human genome contains an estimated number of 100 000 or so genes, and about 2000 human genes have already been identified. Most of these genes have been identified quite recently, as a result of the phenomenal development in molecular biology over the last few years. The optimism about the speed at which the human genome can now be sequenced is indicated by the fact that the Human Genome Mapping Project aims to sequence the human genome in the next 15–20 years. Even if this is an optimistic estimate, it is certain that a lot of information about human genes will emerge in the near future. It is therefore highly appropriate to tackle linkage of as many human neurological disorders as possible, and the MND disease gene is therefore an important target for research.

The first stage in identifying a disease gene is to find the chromosomal location of that disease gene. This is carried out by linkage analysis in which the co-inheritance of genetic markers with the disease is studied. The closer the genetic marker is to the disease, the more likely it is to be co-inherited with the disease, and the marker is said to be linked to the disease. An extensive list of DNA markers has been used to screen MND families. The first probes that have been used are restriction fragment length polymorphism (RFLP) probes. These polymorphisms arise out of mutations in the genome which do not result in a significant alteration in function. When the mutation occurs at a site for restriction enzyme, then its presence can be detected following digestion of DNA with restriction enzymes (restriction endonucleases) and identification of the resulting fragments.

This approach has largely been replaced by the use of microsatellite polymorphisms for genome screening which has greatly improved the quality of DNA markers (discussed below).

Linkage to Chromosome 21 Markers

The first assignment of a chromosomal location to the MND/ALS gene has emerged recently with the publication in May 1991 (Siddique et al. 1991) of a joint paper produced from the collaboration of three laboratories in North America, those of Drs Siddique, Rouleau and Brown. Siddique has access to approximately 150 families, and the laboratories of Rouleau and Brown account for about 150 families as well, although there is thought to be some overlap between these families. In this study, 23 families were focused on for analysis of restriction fragment length polymorphisms on chromosome 21. In view of earlier indications of slight positive lod scores in this region, four polymorphisms were analysed. These were D21S52, D21S1/S11, amyloid precursor protein gene (APP) and D21S58. Overall, no single probe gave a significant result using two-point linkage analysis. The highest value obtained was 2.89 for D21S1/S11. However, these high positive values were sufficiently encouraging for analysis by two further approaches to be considered. First of all the data for all probes were subject to multipoint analysis. This type of analysis requires information about the order of probes on chromosome 21. The result obtained was a significant lod score value of 5.03 10 centimorgans telomeric to the DNA marker D21S58. The highest lod score for a single family was 2.58. The effect of the age curve was shown to affect markedly the multipoint lod score value, and to reduce it to 2.82 when the age of onset was shifted up by 20 years. This indicated that the linkage data were significantly influenced by older age of onset individuals.

This result prompted the authors to investigate whether the lack of significance in the earlier results was due to genetic heterogeneity. This was tested by the use of a programme known as HOMOG. Analysis of the multipoint lod scores showed a significant probability of heterogeneity with at least two genetic subgroups, the proportion of families with linkage to chromosome 21 being 0.55, while the confidence intervals (95%) were quite large, between 0.2 and 0.93.

Identification of Mutations in FMND

The link between FMND and markers on 21q has been amply confirmed with the identification of mutations of the gene encoding Cu,Zn superoxide dismutase (Cu,Zn SOD; SOD 1). Cu,Zn SOD was identified as a candidate gene because it serves to remove harmful superoxide ($O_2\cdot^-$) radicals, thus:

$$2O_2\cdot^- + 2H^+ \rightarrow H_2O_2 + O_2$$

Free radical damage has been implicated in the pathogenesis of many neurological disorders, including MND, although proof is lacking. When individuals from 150 families were screened for mutations in exons 2 and 4 of the Cu,Zn SOD gene, using single-strand conformational polymorphism (SSCP) analysis, 11 different missense mutations were identified in 13 families (Rosen et al. 1993) (Table 3.2).

Subsequently, Deng et al. (1993) reported these and other mutations in exon 1 (8 families) and in exon 5 (2 families) so that only in exon 3 have no mutations been found to date.

Mutations of Human Cu,Zn SOD in FMND

The mutations are likely to be causative, for the following reasons:

1. They have not been identified in several hundred normal controls, in a similar number of sporadic MND cases, or in patients with Parkinson's disease.
2. They are present in all families showing clear linkage to chromosome 21.
3. The mutations alter red cell Cu,Zn SOD enzyme activity, as predicted by structure–function analysis; the enzyme activity is reduced to less than half normal (Deng et al. 1993).

The Significance of Mutations of Cu,Zn SOD

It is now clear that only between 10% and 20% of all families with autosomal-dominant FMND can be attributed to mutations of the Cu,Zn SOD gene. This helps to confirm the longstanding impression that FMND is clinically and genetically heterogeneous. However, phenotypic variation may be just as wide in families with the various Cu,Zn gene mutations as between these families and those with other, as yet unknown mutations.

The Cu,Zn SOD enzyme is a homodimer, and Deng et al. (1993) have illustrated how the mutations in FMND could be predicted to destabilise folding of the subunit, or contact between the dimers. The most frequent mutation in exon 1 (Table 3.2) would do both. It is interesting that exon 3, in which no mutations have yet been found, encodes the zinc-binding loop of the active site. Presumably mutations in that region could lead to a completely inactive enzyme, incompatible with survival. Presumably the impaired activity of the Cu,Zn SOD enzyme in patients with the mutation leads to increased generation of superoxide radical, although this may not be so damaging to neurons as hydroxyl radicals, which could be generated by the Fenton and Haber vice reactions catalysed by transition metals (Halliwell et al. 1992; Coyle and Puttfarcken 1993). There is no convincing evidence that intracellular iron is increased in MND, although increased levels of selenium and manganese have been found in MND spinal cord (Mitchell and Jackson 1992). The enzyme glutathione peroxidase contains selenium, and another superoxide dismutase contains manganese (Mn SOD). Levels of these enzymes may be elevated in response to increased free radical production.

It has been suggested (Beckman et al. 1993) that the mutant Cu,Zn SOD enzyme may result in increased interaction between the active site of the molecule, nitric oxide in its reduced form, and superoxide radicals, with production of peroxynitrite. Peroxynitrite can decompose to OH·, the hydroxy radical, but it might also form nitronium intermediates which could react with tyrosine residues which are important in cellular transduction as components of tyrosine kinases. Furthermore, there is increasing evidence linking excitotoxicity to the generation of free radicals (McNamara and Fridovich 1993; Coyle and Puttfarcken 1993). Free radicals can be generated following activation of NMDA receptors, but also AMPA receptors, and the latter seem particularly pertinent to motor neuron death (Rothstein et al. 1993; Willis et al. 1993).

There are many possible pathways by which impaired removal of free radicals could lead to neuronal death, including lipid peroxidation leading to membrane damage, damage to DNA repair mechanisms, and damage to key neuronal proteins. Evidence for the latter has been found in sporadic MND; protein carbonyl content was elevated 85% in patients compared to controls (Bowling et al. 1993).

The Next Steps: Identification of Other MND Genes

Improvements in DNA Markers

RFLP markers have a limited use, being only diallelic, and are often completely uninformative for a high proportion of families. However, recently, significant improvements have occurred in the development of more useful probes with the identification of tandem-repetitive DNA sequences, e.g. $[TG]_n$ and variable number tandem repeats loci (VNTR) which arise out of uneven recombination at meiosis

Table 3.2. FMND mutations in human Cu,Zn SOD gene

Amino acid change	Point mutation	Structural location	Structural role
		Exon 1	
Ala^{4}→Val	GCC→GTC	β strand	β barrel and dimer packing
		Exon 2	
Gly^{37}→Arg	GGA→AGA	Greek key connection	Conserved left-handed Gly
Leu^{38}→Val	CTG→GTC	Greek key connection	β barrel plug
Gly^{41}→Ser	GGC→AGC	β strand	Conserved left-handed Gly
Gly^{41}→Asp	GGC→GAC	β strand	Conserved left-handed Gly
His^{43}→Arg	CAT→CGT	β strand	Hydrogen bonding
		Exon 4	
Gly^{85}→Arg	GGC→CGC	β strand	Conserved left-handed Gly
Gly^{93}→Ala	GGT→GCT	Loop 5	Conserved left-handed Gly
Gly^{93}→Cys	GGT→TGT	Loop 5	Conserved left-handed Gly
Glu^{100}→Gly	GAA→GGA	β strand	Stability for Greek key loop
Leu^{106}→Val	CTC→GTC	Greek key loop	Conserved β barrel plug
Ile^{113}→Thr	ATT→ATC	Greek key loop	Dimer interface packs self-symmetrically
		Exon 5	
Leu^{144}→Phe	TTG→TTC	β strand loop connection	Packs with leu^{38}
Val^{148}→Gly	GTA→GCA	β strand	Dimer interface packs self-symmetrically

Modified from Deng et al. (1993) with permission.

and give rise to variable numbers of the repeat sequence and hence multiple alleles, typically 6–8. Several hundred VNTR sequences have already been identified, and more than half show a heterozygosity of greater than 70% in the population. These are highly polymorphic markers, with substantial variability in the number of repeat units, and hence provide extremely informative markers which have been applied to linkage analysis (Nakamura et al. 1987), and to individual identification in forensic medicine (Jeffreys et al. 1985). The most abundant repetitive sequences are of the form $[CA]_n$ where $n = 6–30$, and it has been estimated that there are in the order of 50 000–10 000 $[CA]_n$ blocks in the human genome (Weber & May 1989), and hence these represent a vast pool of genetic markers.

The development of dinucleotide repeat (microsatellite) polymorphisms has revolutionised screening procedures, especially when combined with the polymerase chain reaction to amplify genomic DNA. For this method only small amounts of genomic DNA are necessary (10–200 ng), and the amplified product can be detected following polyacrylamide gel electrophoresis.

The majority of MND families do not have SOD-1 mutations and hence do not demonstrate linkage to chromosome 21 markers (King et al. 1993). The search for other MND genes is therefore being actively pursued using this approach.

Complementing the development of microsatellite probes has been the localisation and cloning of a number of candidate genes relevant to the function of the nervous system, e.g. GABA, glycine, nicotinic, muscarinic and adrenergic receptors and disease genes affecting anterior horn cells, e.g. SMA. We have recently excluded this locus and mouse models of MND from linkage to FMND (King et al. 1994). A single recombination between a DNA marker in a candidate gene and the disease is sufficient to eliminate the candidate as a disease-causing gene.

Choice of Families: Informative Families

Ascertainment of diagnosis is vital for any linkage analysis. When the affected individual has died, or when earlier family members are traced back, death certificates must be sought. Some definitions such as "bulbar and spinal paralysis" may be more acceptable than others, such as "wasting disease".

Statistical Analysis

Significant linkage is established by maximum likelihood analysis of the inheritance of marker alleles with the disease. Information from a large number of family members is required, their disease status and likelihood of accurate typing of normals obtained from age of onset curves. A 30-year-old "normal" individual cannot be as accurately typed as normal compared to a "normal" 80-year-old. Allele frequency also influences analysis. Analyses of this kind are very time-consuming, and are usually carried out using computer programs such as LINKAGE. An estimate of the significance of linkage is obtained when the log of the odds in favour of linkage (lod score) is 3 or more at a particular recombination fraction. A lod score of 3 represents the odds of 1000:1 in favour of linkage. It may be necessary to test out several hundred DNA markers before such a linkage is obtained. Multipoint lod scores can be calculated using multiple markers in close vicinity to one another on a chromosome.It often proves necessary to screen several hundred evenly distributed polymorphisms (2–5 cm apart) before linkage is obtained.

Candidate Genes

The finding of Cu,Zn SOD mutations in FMND immediately focuses attention on genes encoding enzymes important in free radical scavenging mechanisms. Mn SOD and extracellular Cu,Zn SOD have been excluded as candidates, and other enzymes such as glutathione peroxidase and catalase are under investigation. It is nevertheless important to continue the use of classical linkage techniques described above.

A second stage at which candidate genes can be useful is after linkage has been established. An example is seen in the case of SMA, where linkage to a region of chromosome 5 was established (Brzustowicz et al. 1990), which is known to coincide with the gene for hexosaminidase B, whose alpha and beta subunits have been shown to be deficient in chronic SMA. Another gene in close proximity to the SMA locus is that encoding the microtubule-associated protein 1B (MAP-1B) which has been mapped to the chromosomal location, 5q13. Genetic linkage analysis of SMA families using a dinucleotide repeat polymorphism 3' to the MAP-1B gene shows tight linkage to SMA mutations (Lien et al. 1991). A highly significant maximum lod score of +20.2 at a recombination fraction 0.02 was obtained between SMA and MAP-1B. A MAP-1B mutation would be a good candidate for SMA, since MAP-1B is involved in neuronal morphogenesis and localised in anterior horn cells. However, two possible recombinants between these loci have been detected, so it is unlikely that the MAP-1B gene is the site of mutations in SMA, and the mutations of the Hex B gene have now been excluded in SMA.

Other Motor Neuron Disorders: the Spinal Muscular Atrophies

The spinal muscular atrophies (SMA types 1, 2 and 3) have been linked to chromosome 5q13 (Brzustowicz et al. 1990; Melki et al. 1990a, b; Gilliam et al. 1990). The clinical variants of SMA are thus likely to be allelic disorders. At one stage, candidate genes located on chromosome 5q included hexosaminidase B, and the microtubule-associated protein 1B (MAP-1B) (Lien et al. 1991), but both these genes have now been excluded, although hexosaminidase deficiency can be associated with an SMA-like phenotype (Troost, 1991). Although the gene for autosomal-recessive SMA has not yet been identified, a mutation has been identified in another rare form of spinal muscular atrophy known variously as Kennedy's syndrome (Kennedy et al. 1968), X-linked recessive bulbospinal neuronopathy (Harding et al. 1982), or X-linked spinal and bulbar muscular atrophy (Arbizu et al. 1983). Typically, patients present in the fourth or fifth decades (although they may present in their teens, or in their sixties or seventies) with slowly progressive, mainly proximal weakness of the limbs associated with facial weakness and fasciculation, atrophy of the tongue, and often dysarthria and dysphagia, although the latter may only appear later in the course of the disease. Gynaecomastia is present in most but not all cases, and there may be testicular atrophy. Muscle cramps induced by exercise may precede the development of weakness, and many patients also notice postural tremor before the onset of weakness. The condition is sometimes associated with diabetes mellitus, and some patients are subfertile. Although patients do not usually have sensory symptoms, sensory action potentials are often abnormal, even without the co-existence of diabetes mellitus. Otherwise the EMG changes are those of chronic partial denervation, although muscle biopsy may show "secondary" myopathic changes, and there may be raised plasma creatine kinase levels. The prognosis is usually good, and the disease is compatible with a normal life span. Pathological findings include loss of brainstem and spinal motor neurons, with preservation of the oculomotor nuclei and the neurons of Onuf's nucleus in the sacral spinal cord (Sobue et al. 1989). Sural nerve biopsies show evidence of axonopathy which is mainly distal. Teased-fibre studies (Sobue et al. 1989) have shown evidence of segmental demyelination and remyelination in the sural nerves in most cases. Dorsal root ganglion cells are well preserved, but the distal axonopathy is presumably attributable to dysfunction of the primary sensory neurons.

The Kennedy's disease mutation has now been identified (La Spada et al. 1991). The abnormality consists of an expansion of a polyglutamine tract encoded by a series of CAG repeats in exon 1 of the gene. In normal individuals there are between 15 and 32 CAG repeats, with most individuals having between 20 and 29 repeats (Garofalo et al. 1993), but in Kennedy's syndrome there are usually 40 or more CAG repeats.

This expanded CAG repeat in exon 1 is not in the DNA or hormone-binding region of the gene, and probably exerts its deleterious effect on lower motor neurons by altering transcriptional regulation (Mhatre et al. 1993). It is unlikely that impaired androgen-binding activity is of major pathogenic significance in Kennedy's disease – indeed, androgen-binding activity in genital skin fibroblasts may be normal (Warner et al. 1992). The altered transregulatory performance of the abnormal androgen receptor presumably affects some normal function of the androgen receptor that is relatively specific to lower motor neurons.

Kennedy's disease joins the fragile X syndrome, myotonic dystrophy, Huntington's disease and spinocerebellar atrophy type 1 as a neurological disorder caused by expansion of trinucleotide repeats – so-called dynamic mutations (Richards and Sutherland 1992; Orr et al. 1993).

Several groups have noted positive correlations between the number of CAG repeats in the androgen receptor and clinical features of Kennedy's disease. Thus there are positive correlations between the age of onset and with age-adjusted disability scores (Igarashi et al. 1992; Doyu et al. 1992).

The identification of the androgen receptor mutation has important clinical implications. Firstly, it can be used to confirm or refute the diagnosis in suspected cases. Since the disease is compatible with a normal life span, and patients not infrequently are initially diagnosed as typical motor neuron disease, accurate diagnosis is of clinical benefit. At present, as in other forms of motor neuron disease, there is no treatment to alter the rate of progression, but if therapy does become available, then it may be important to detect presymptomatic men. It is of course possible to identify female carriers. Bearing in mind that this is usually a relatively mild disorder, it could be argued that there is little indication for pre-natal screening of pregnant mothers in order to terminate affected foetuses, but as Troy et al. (1990) have pointed out, there are severely affected individuals, and we cannot as yet predict exactly how severe the disease will be in an individual carrying the mutant gene. The key to such decisions must be detailed counselling before any test is carried out, so that a mother should be able to make a fully informed choice.

Androgen Receptors and Typical Motor Neuron Disease (ALS)

Most studies of sporadic MND show an excess of males, raising the possibility that androgens are important in some way in the pathogenesis of typical MND. This remains a possibility, but we found no evidence for an association between alleles coding for larger CAG repeats and MND, either in males or females (Garofalo et al. 1993; Figure 3.3a,b).

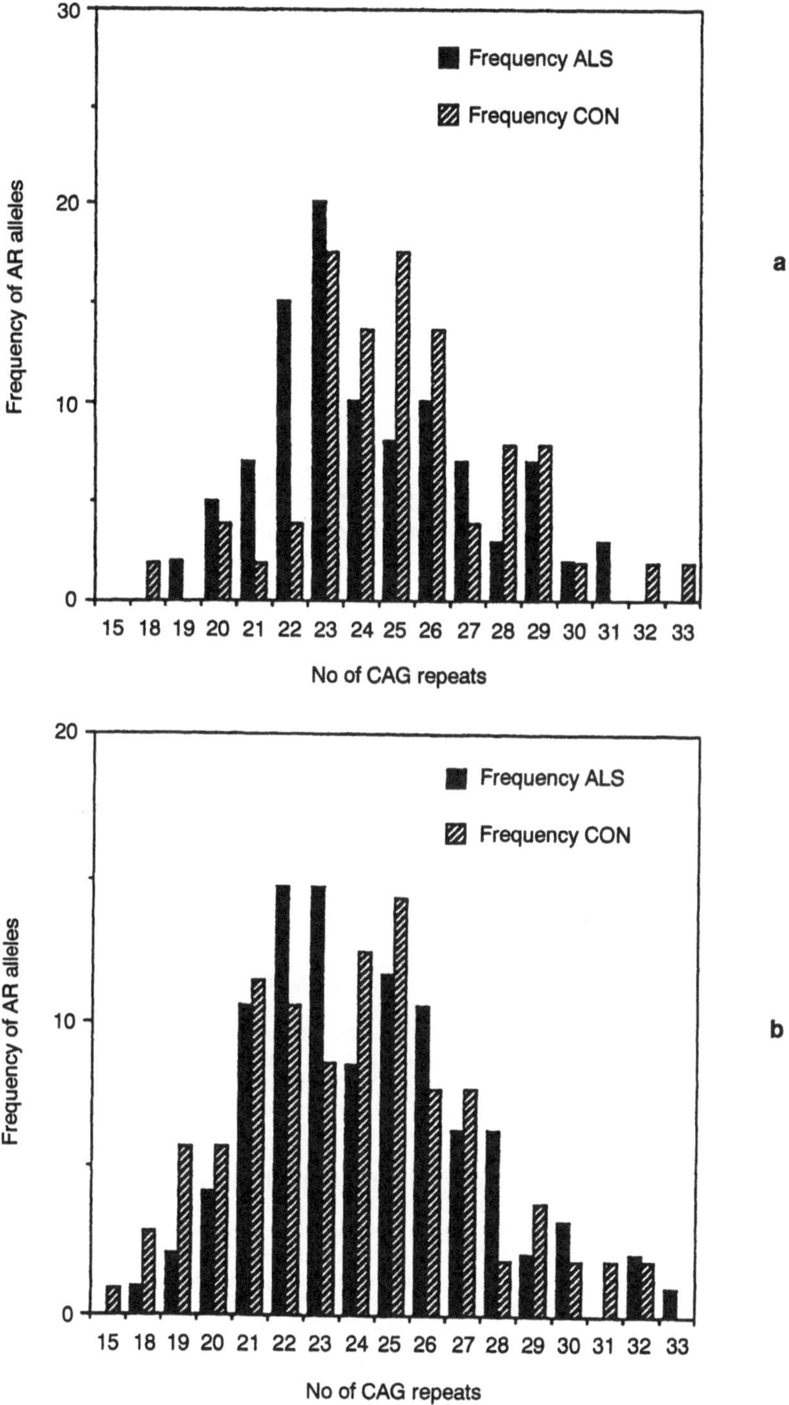

Fig. 3.3. Relative androgen receptor (AR) allele frequency distributions in the ALS and control male and female populations.

Summary and Future Prospects

The identification of mutations in Kennedy's syndrome, and particularly in FMND, has introduced new possibilities for understanding pathogenic mechanisms of motor neuron degeneration. Fortuitously, in both situations, a great deal was already known about both the genes and their products. This has had the salutary effect of concentrating research into specific areas of molecular cell biology, although much remains to be done. Although free radicals are likely to be involved in some way in patients with the Cu,Zn SOD mutations, the precise mechanisms by which free radicals cause relatively selective motor neuron cell death is unknown, and the majority of familial cases are not caused by such mutations. We do not know yet how heterogeneous the non-chromosome 21-linked families will turn out to be. There may be several other genes causing autosomal-dominant FMND. The candidate gene approach may prove fruitful, but there may be many surprises, and some of the mutations may be in as yet unknown genes, in which case reconstructing the mechanisms of pathogenesis may take many years. Nevertheless, the new insights will undoubtedly advance our understanding of the molecular mechanisms of motor neuron death, and may be helpful in identifying rational new treatments.

Large areas of the genome have been excluded in non-chromosome 21-linked families, and a number of candidate genes have been tested and excluded, including the genes for glutamate dehydrogenase and the GluR-5 receptor subunit. Collaborative studies will still be necessary to characterise other genes in FMND, and co-operation between the centres involved in this research is still of crucial importance for further advance.

References

Appelbaum JS, Roos RP, Salazar-Grueso EF et al. (1992) Intrafamilial heterogeneity in hereditary motor neuron disease. Neurology 42:1488–1492

Arbizu T, Santamaria J, Gomez JM et al. (1983) A family with adult spinal and bulbar muscular atrophy, X-linked inheritance and associated testicular failure. J Neurol Sci 59:371–382

Beckman JS, Carson M, Smith CD (1993) ALS, SOD and peroxynitrite. Nature 364:584

Bowling AC, Schulz JB, Brown RH Jr et al. (1993) Superoxide dismutase activity, oxidative damage, and mitochondrial energy metabolism in familial and sporadic amyotrophic lateral sclerosis. J Neurochem 61:2322–2325

Brown RH (1989) Gene analysis in familial amyotrophic lateral aclerosis: the problems and the prospects. Int ALS/MND Update 4:13–15

Brzustowicz LM, Lehner T, Castilla LH et al. (1990) Genetic mapping of chronic childhood-onset spinal muscular atrophy to chromosome 5q11.2–13.3. Nature 344:540–541

Camu W, Billiard M (1993) Coexistence of amyotrophic lateral sclerosis and Werdnig–Hoffmann disease within a family. Muscle Nerve May 1993:569–570

Chancellor AM, Warlow CP (1992) Adult onset motor neuron disease: worldwide mortality, incidence and distribution since 1950. J Neurol Neurosurg Psychiatry 55:1106–1115

Chancellor AM, Swingler RJ, Fraser H et al. (1992) A prospective study of adult onset motor neuron disease in Scotland. Methodology, demography and clinical features of incident cases in 1989. J Neurol Neurosurg Psychiatry 55:536–551

Chio A, Brignolio F, Meineri P et al. (1987) Phenotypic and genotypic heterogeneity of dominantly inherited amyotrophic lateral sclerosis. Acta Neurol Scand 75:277–282

Choi W-T, MacLean HE, Chu S (1993) Kennedy's disease: genetic diagnosis of an inherited form of motor neuron disease. Aust NZ J Med 23:187–192

Constantinidis J, Richard J, Tissot R (1974) Pick's disease: histological and clinical correlations. Eur Neurol 11:208–217

Coté F, Collard JF, Julien P (1993) Progressive neuronopathy in transgenic mice expressing the human neurofilament heavy gene: a mouse model of amyotrophic lateral sclerosis. Cell 73:35–46

Coyle JT, Puttfarcken P (1993) Oxidative stress, glutamate, and neurodegenerative disorders. Science 262:689–695

Deng H-X, Hentati A, Tainer JA et al. (1993) Amyotrophic lateral sclerosis and structural defects in Cu,Zn superoxide dismutase. Science 261:1047–1051

Doyu M, Sobue G, Mukai E et al. (1992) Severity of X-linked recessive bulbospinal neuronopathy correlates with size of the tandem CAG repeat in androgen receptor gene. Ann Neurol 32:707–710

Emery A, Holloway S (1982) Familial motor neurone disease. In: Rowland LP (ed.) Human motor neurone diseases. Raven Press, New York, pp 139–147

Engel WK, Kurland LT, Latzo I (1959) An inherited disease similar to amyotrophic lateral sclerosis with a pattern of posterior column involvement: an intermediate form? Brain 82:203–303

Finlayson MH, Guberman A, Martin JB (1973) Cerebral lesions in familial amyotrophic lateral sclerosis and dementia. Acta Neuropathol (Berlin) 26:237–246

Garofalo O, Figlewicz DA, Leigh PN et al. (1993) Androgen receptor gene polymorphisms in amyotrophic lateral sclerosis. Neuromusc Disord 3:195–199

Gilliam TC, Brzustowicz LM, Castilla LH et al. (1990) Genetic homogeneity between acute and chronic forms of spinal muscular atrophy. Nature 345:823–825

Gunarsson L-G, Dahlbom K, Strandman E (1991) Motor neuron disease and dementia reported among 13 members of a single family. Acta Neurol Scand 84:429–433

Halliwell B, Gutteridge JMC, Cross CE (1992) Free radicals, antioxidants, and human disease: where are we now? J Lab Clin Med 598–620

Harding AE, Bradbury PG, Murray NMF (1983) Chronic asymmetrical spinal muscular atrophy. J Neurol Sci 59:69–83

Harding AE, Thomas PK, Baraitser M et al. (1982) X-linked recessive bulbospinal neuronopathy: a report of ten cases. J Neurol Neurosurg Psychiatry 45:1012–1019

Hawkes CH, Cavanagh JB, Mowbray S (1984) Familial motor neurone disease: report of a family with 5 post-mortem studies. In: Clifford Rose F, (ed), Research progress in motor neuron disease. Pitman, London, pp 70–98

Hayashi H, Kato S (1989) Total manifestations of amyotrophic lateral sclerosis (ALS) in the totally locked-in state. J Neurol Sci 93:19–35

Hirano A, Kurland LT, Sayre GP (1967) Familial amyotrophic lateral sclerosis. Arch Neurol 16:232–243

Hirano A (1991) Cytopathology of amyotrophic lateral sclerosis. In: Rowland LP (ed.) Amyotrophic lateral sclerosis and other motor neuron diseases. Raven Press, New York, pp 91–101 (Advances in neurology 56)

Horton WA, Eldridge R, Brody JA (1976) Familial motor neuron disease. Neurology 26:460–465

Hudson AJ (1981) Amyotrophic lateral sclerosis and its association with dementia, parkinsonism, and other neurological disorders: a review. Brain 104:217–253

Igarashi S, Tanno Y, Onodera O, et al. (1992) Strong correlation between the number of CAG repeats in androgen receptor genes and the clinical onset of features of spinal and bulbar muscular atrophy. Neurology 42:2300–2303

Jeffreys AJ, Brookfield JFY, Semeonoff R (1985) Positive identification of an immigration test-case using human DNA fingerprinting. Nature 317:818–819

Jokelainen M (1977) Amyotrophic lateral sclerosis in Finland. I. An epidemiological study. Acta Neurol Scand 56:185–193

Kennedy W, Alter M, Sung K (1968) Progressive proximal spinal and bulbar muscular atrophy of late onset: a sex-linked recessive trait. Neurology 18:671–680

Kew J, Leigh N (1992) Dementia with motor neuron disease. In: Rossor MN (ed.) Unusual dementias, Baillière Tindall, London (Baillières clinical neurology 1(3))

King A, de Belleroche J (1991) Molecular genetics of familial motor neurone disease: studies in the UK. Cytogenet Cell Genet. 58:2100–2101

King, A, Houlden H, Hardy J, Lane R, Chancellor A, de Belleroche J (1993) Absence of linkage between chromosome 21 loci familial amyotrophic lateral sclerosis. J Med Genet 30:318

King A, Orrell R, Lane R, de Belleroche J (1994) More than one locus for familial amyotrophic lateral sclerosis: absence of linkage to the SMA locus. Biochem Soc Trans 22:1495

Kurland LT, Mulder DW (1955) Epidemiological investigations of amyotrophic lateral sclerosis. 2.

Familial aggregrations indicative of dominant inheritance, parts I and II. Neurology 5:182–196, 249–268

Kusaka H, Imai T, Hashimoto T et al. (1988) Ultrastructural study of chromatolytic neurons in an adult-onset sporadic case of amyotrophic lateral sclerosis. Acta Neuropathol (Berlin) 75:523–528

La Spada AR, Wilson EM, Lubahn EB et al. (1991) Androgen receptor gene mutations in X-linked spinal and bulbar muscular atrophy. Nature 352:77–79

Leigh PN, Whitwell H, Garofalo O et al. (1991) Ubiquitin-immunoreactive intraneuronal inclusions in amyotrophic lateral sclerosis: morphology, distribution and specificity. Brain 114:775–788

Li T-M, Alberman E, Swash M (1988) Comparison of sporadic and familial disease amongst 580 cases of motor neuron disease. J Neurol Neurosurg Psychiatry 51:778–784

Lien LY, Boyce FM, Kleyn P et al. (1991) Mapping of human microtubule-associated protein 1B in proximity to the spinal muscular atrophy locus at 5q13. Proc Natl Acad Sci USA 88:7873–7876

McNamara JO, Fridovich I (1993) Did radicals strike Lou Gehrig? Nature 362:20–21

Melki J, Sheth P, Abdelhak S et al. (1990a) Mapping of acute (type 1) spinal muscular atrophy to chromosome 5q12–q14. Lancet 336:271–273

Melki J, Abdelhak S, Sheth P et al. (1990b) Gene for chronic proximal spinal muscular atrophies maps to chromosome 5q. Nature 344:767–768

Mhatre AN, Trifiro MA, Kaufman M et al. (1993) Reduced transcriptional regulatory competence of the androgen receptor in X-linked spinal and bulbar muscular atrophy. Nature Genet 5:184–188

Mitchell JD, Jackson MJ (1992) Free radicals, amyotrophic lateral sclerosis, and neurodegenerative disease. In: Smith RA (ed.) Handbook of amyotrophic lateral sclerosis. Marcel Dekker, New York, pp 533–541

Mizutani T, Sakamaki S, Tsuchiya N et al. (1992) Amyotrophic lateral sclerosis with ophthalmoplegia and multisystem degeneration in patients on long-term use of respirators. Acta Neuropathol 84:372–377

Mulder DW, Kurland LT, Otford KP (1986) Familial adult motor neuron disease: amyotrophic lateral sclerosis. Neurology 36:511–517

Murayama S, Okawa Y, Mori H et al (1989) Immunocytochemical and ultrastructural study of Lewy body-like hyaline inclusions in familial amyotrophic lateral sclerosis. Acta Neuropathol 78:143–152

Murros K, Fogelholm R (1983) Amyotrophic lateral sclerosis in middle Finland: an epidemiological study. Acta Neurol Scand 67:41–47

Nakamura Y, Leppert M, O'Connell P et al. (1987) Variable number of tandem repeat (VNTR) markers for human genetic mapping. Science 235:1616–1622

Neary D, Snowden JS, Mann DMA et al. (1990) Frontal lobe dementia and motor neuron disease. J Neurol Neurosurg Psychiatry 53:23–32

Okamoto K, Murakami N, Kusaka H et al. (1992) Ubiquitin-positive intraneuronal inclusions in the extramotor cortices of presenile dementia patients with motor neuron disease. J Neurol 239:426–430

Orr HT, Chung M, Banfi S et al. (1993) Expansion of an unstable trinucleotide CAG repeat in spinocerebellar ataxia type 1. Nature Genet 4:221–226

Richards RI, Sutherland GR (1992) Dynamic mutations: a new class of mutations causing human disease. Cell 709–712

Robertson EE (1953) Progressive bulbar paralysis showing heredofamilial incidence and intellectual impairment. Arch Neurol Psychiatry 69:197–207

Rosen DR, Siddique T, Patterson D et al. (1993) Mutations in Cu/Zn superoxide dismutase gene are associated with familial amyotrophic lateral sclerosis. Nature 362:59–62

Ross CA, McInnis MG, Margolis RL, Li S-H (1993) Genes with triplet repeats: candidate mediators of neuropsychiatric disorders. TINS 7:254–260

Rothstein JD, Jin L, Dykes-Hoberg M et al. (1993) Chronic inhibition of glutamate uptake produces a model of slow neurotoxicity. Proc Natl Acad Sci USA 90:6591–6595

Shaw PJ, Ince PG, Goodship J (1992) Adult-onset motor neuron disease and infantile Werdnig–Hoffmann disease (spinal muscular atrophy type 1) in the same family. Neurology 42:1477–1480

Siddique T, Pericak-Vance MA, Brook BR, et al. (1989) Linkage analysis in familial amyotrophic lateral sclerosis. Neurology 39:919–926

Siddique T, Figlewicz DA, Pericak-Vance MA et al. (1991) Linkage of a gene causing familial amyotrophic lateral sclerosis to chromosome 21 and evidence of genetic-locus heterogeneity. N Engl J Med 324:1381–1384

Sobue G, Hashizuma Y, Mukai E et al. (1989) X-linked bulbospinal neuronopathy. A clinicopathological study. Brain 112:209–232

Swash M, Leigh N (1992) Workshop report. Criteria for diagnosis of familial amyotrophic lateral sclerosis. Neuromusc Disorders 2:7–9

Swash M, Scholtz CL, Vowles G et al. (1988) Selective and asymmetric vulnerability of corticospinal and spinocerebellar tracts in motor neuron disease. J Neurol Neurosurg Psychiatry 51:785–789

Ting-Ming L, Alberman E, Swash M (1988) Comparison of sporadic and familial disease amongst 580 cases of motor neuron disease. J Neurol Neurosurg Psychiatry 51:778–784

Troost J (1991) Spinal muscular atrophy of infantile and juvenile onset, due to metabolic derangement. In: de Jong JMBV (ed.) Diseases of the motor system. Elsevier, Amsterdam, pp 97–105 (Handbook of clinical neurology 59)

Troy CM, Muma NA, Greene LA et al. (1990) Regulation of peripherin and neurofilament expression in regenerating rat motor neurons. Brain Res 529:232–238

Veltema AN, Roos RAC, Bruyn GW (1990) Autosomal dominant adult amyotrophic lateral sclerosis. A six generation Dutch family. J Neurol Sci 97:93–115

Warner CL, Griffin JE, Wilson JD (1992) X-linked spinomuscular atrophy: a kindred with associated abnormal androgen receptor binding. Neurology 42:2181–2184

Weber JL, May PE (1989) Abundant class of human DNA polymorphisms which can be typed using the polymerase chain reaction. Am J Hum Genet 44:388–396

Wightman G, Anderson VER, Martin J et al. (1992) Hippocampal and neocortical ubiquitin-immunoreactive inclusions in ALS with dementia. Neurosci Lett 139:269–274

Williams C, Kozlowski MA, Hinton DR et al. (1990) Degeneration of spinocerebellar neurons in amyotrophic lateral sclerosis. Ann Neurol 27:215–225

Williams DB, Floate A, Leicester J (1988) Familial motor neuron disease: differing penetrance in large pedigrees. J Neurol Sci 86:215–230

Willis CL, Meldrum BS, Nunn PB et al. (1993) Neuronal damage induced by β-*N*-oxalylamino-L-alanine (BOAA) in rat hippocampus can be prevented by a non-NMDA antagonist, 2,3-dihydroxy-6-nitro-7-sulfamoyl-benzo(F)quinoxaline (NBQX). Brain Res 627:55–62

Xu Z, Cork LC, Griffin JW et al. (1993) Increased expression of neurofilament subunit NF-L produces morphological alterations that resemble the pathology of human motor neuron disease. Cell 73:23–33

4 Pathology of Motor System Disorder

Samuel M. Chou

Introduction

Motor neuron disease (MND) or amyotrophic lateral sclerosis (ALS), the commonest form of motor system degeneration in man, occurs sporadically in middle or late adult life and can be divided into three clinical subtypes irrespective of bulbar or spinal predominance. This classification, convenient for conceptualization of the pathology germane to MND, includes: (1) lower motor neuron type, or primary muscular atrophy (PMA) (Aran's disease 1850); (2) upper motor neuron type, or primary lateral sclerosis (PLS) (Erb's disease 1891); and (3) upper-lower motor neuron type, or classic ALS. On the basis of some 20 clinical and 5 autopsy cases, Charcot (1869, 1885) coined the term "la sclérose laterale amyotrophique", recognizing that the disease started in the lateral columns of the spinal cord, hence "lateral sclerosis", propagated to the bulbar gray and anterior horns, and secondarily produced muscle atrophy as shown by the term "amyotrophic". Many atypical forms of MND, all with a high familial incidence, have been recently added. They include: (a) familial ALS, (b) Guamanian ALS, (c) Kii-peninsula ALS in Japan, (d) ALS in Auyu and Jaki people of West New Guinea, and (e) familial juvenile ALS. Gowers (1888) believed that the degeneration in both upper and lower motor neurons might occur simultaneously or independently and, following this concept, pure lower motor neuron diseases have been regarded as separate entities. With this notion, one is obliged to compare the pathology of ALS with other MNDs such as Werdnig–Hoffmann disease, Fazio–Londe disease, Kugelberg–Welander disease, familial spastic paraplegia of Strümpell and the neuronal type of Charcot–Marie–Tooth disease. Strümpell disease is chosen for comparison in this chapter.

The aim of this chapter is to provide an overview of ALS pathology based on a number of reviews (Bertrand and Van Bogaert 1925; Brownell et al. 1970; Castaigne et al. 1972; Comant & Marie1958; Lawyer and Netsky 1953; Hirano and Iwata 1979; Chou 1979) and the data from more recent studies on the pathology of ALS/MND.

Physiological Anatomy of the Motor System

Whether the disease process of ALS involves only the motor system remains to be seen. However, there is little doubt that the motor system is the cardinal target of the pathologic insult. The organisation of the motor system in the CNS is complex. The systematized concept of Kuypers (1973) is particularly useful in explaining the clinical and pathologic features of MND. Kuypers emphasized that in order to gain a functionally meaningful understanding of the motor system, the destinations of the various descending pathways to the spinal motoneurons and to the interneurons should be defined. In the monkey, and therefore probably in humans, three major termination patterns of descending motor pathways have been recognized. Their modularity roles in motor functions in terms of their final common pathways after directly or indirectly synapsing with lower motor neurons are summarized in Fig. 4.1. In addition to the well-known direct corticospinal (or pyramidal) tract arising from the upper motor neurons (Fig. 4.1a), there are two groups of brainstem fibers which terminate in the intermediate zone of the spinal gray matter and indirectly synapse with motoneurons, via interneurons. These are the anteromedial (Fig. 4.1b) and lateral brainstem pathways (Fig. 4.1c); both receive projections from premotor and precentral cortical fibers. The lateral pathway originates mainly in the contralateral red nucleus, descends in parallel to the direct corticospinal tract, and is termed the indirect corticospinal tract. Its fibers terminate on interneurons in the contralateral posterolateral intermediate zone of the spinal gray matter, whereas the direct corticospinal tract fibers terminate monosynaptically at the contralateral antero-posterolateral and bilateral anteromedial intermediate zones of the spinal gray matter, as shown in Fig. 4.1. The anteromedial brainstem pathway originates in many brainstem nuclei including the interstitial nucleus of Cajal, superior colliculus, vestibular complex and bulbar medial reticular formation; it projects to the anteromedial intermediate zones of the spinal gray, bilaterally (Fig. 4.1b). It is important to recognize that the nerve fibers of the anteromedial pathway distribute rather diffusely throughout the anterolateral funiculi of the spinal cord before they synapse either monosynaptically, or less directly, with α motoneurons. This anatomical concept is crucial to understanding the diffuse degeneration of anterolateral funiculi commonly seen in the ALS spinal cord (Fig. 4.2).

Lower motoneurons in the anterior horns throughout the spinal cord as well as the brainstem (Cranial nerves III–VII, IX–XIII) separately innervate corresponding skeletal muscles. Each α motoneuron innervates a number of extrafusal striated muscle fibers, forming the motor unit. The major part of synaptic input to α and other motoneurons is contributed by interneurons of the intermediate zone in the spinal cord rather than by direct corticospinal projections.

Axonal Transport in the Motor System

Pertinent to the discussion of pathology in ALS is the concept of axonal transport and primary impairment of axonal transport or "axostasis" (Chou and Hartmann 1964, 1965). The motoneurons, both Betz cells and anterior spinal neurons, are among the largest cells in the human body. The reason for the large cell body is evident, for the motoneuron has to maintain the integrity of its dendritic branches, which often extend more than 1 cm, and its axon which may project more than 100 cm from the cell body. The cell body communicates with the periphery by its electrical activity, and also by different types of axaplasmic flow. The latter is a mechanism by which structural and functional proteins are transported to the periphery along the axon and by which the cell body receives chemical feedback signals (retrograde flow). Two major types of *orthograde axonal transport* include: (1) the fast (400 mm/day) bidirectional flow which transports membrane-associated

A) Descending Corticospinal Pathway (DCP)

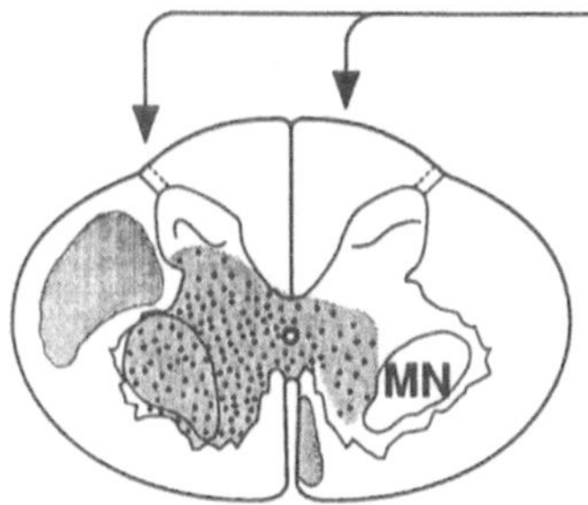

From: Motor & premotor cortex

Via: Direct Corticospinal tract

Contralaterally to:
-posterior lateral interneurons
-motoneurons (MN)

Bilaterally to:
-anteromedial interneurons

Control: Capacity for independent use of distal extremities and fingers

B) Anteromedial Brainstem Pathway (ABP)

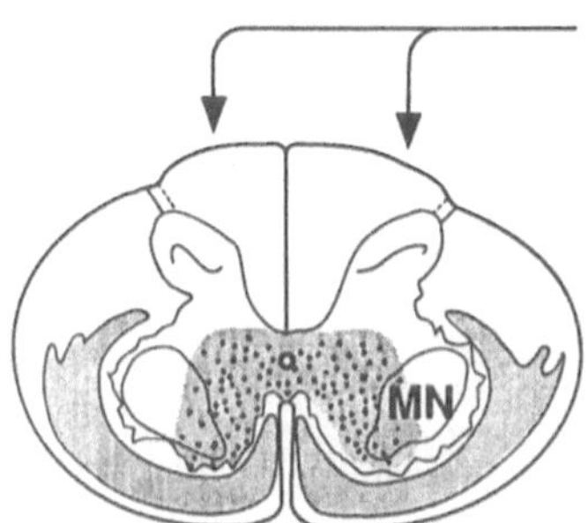

From: Interstitial N of Cajal: Med. vest. N: Sup. collic.: Bulbar med. RF

Via: Anteriolateral columns

Bilaterally to: Anteromedial interneurons & MNs

Control: Axial and proximal extremity muscles: Head and trunk: to maintain erect posture

C) Lateral Brainstem Pathway (LBP)

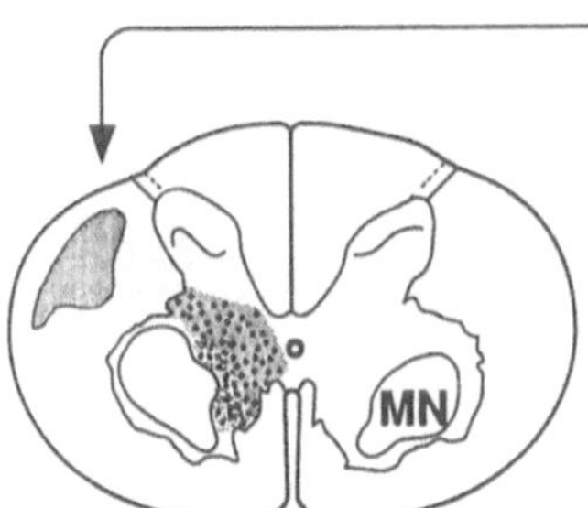

From: Red nucleus: Bulbar lat. RF

Via: Indirect Corticospinal tract

Contralaterally to: posterolateral interneurons

Control: Flexion of distal extermity muscles and distal reflex movements

Fig. 4.1. Schematic representation of three motor pathways in primates.

proteins and glycoproteins and (2) the slow (a few mm/day) flow which transports a network of interconnected microfilaments, microtubules and neurofilaments as component "a" (0.1–2 mm/day) and a large complex of soluble proteins as component "b" (2–4 mm/day) (Lasek 1986). In addition, retrograde axonal transport carries both endogenous (e.g. amino acids, nerve growth factor) and exogenous substances (e.g. tetanus toxin, polio virus, herpes-simplex virus, rabies virus, horseradish peroxidase, lectins, etc.) from the terminal axons to the cell body at a rate of more than 75 mm/day. Pertinent to the possible etiology of ALS is the concept of "suicide transport". This concept implies that a neurotoxic factor as a

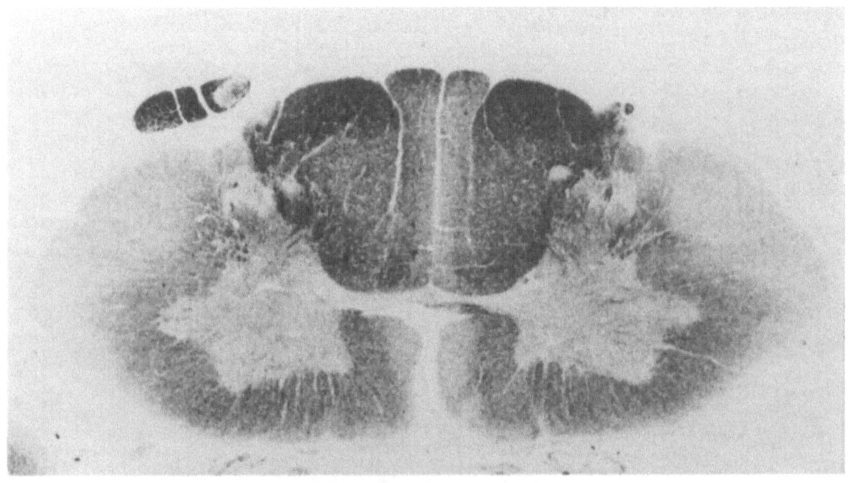

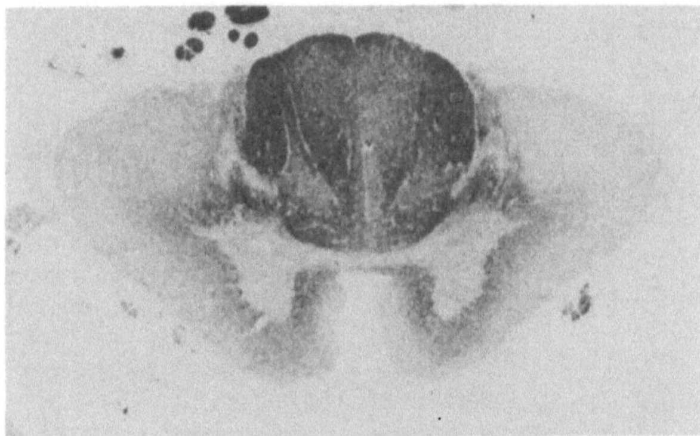

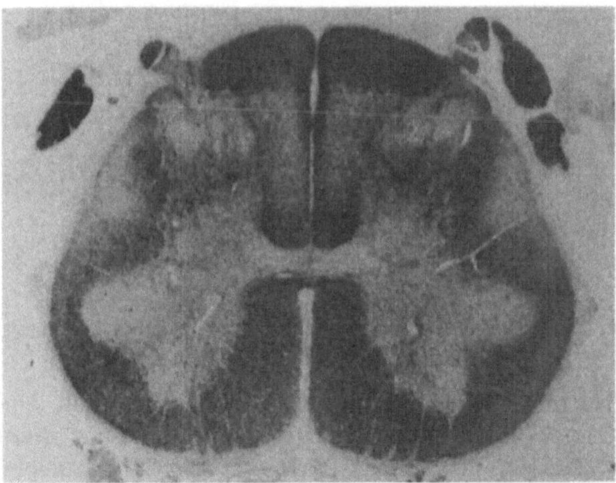

Fig. 4.2. "Total" tract degeneration in the cervical, dorsal and lumbar cords from a sporadic ALS case survived beyond the usual terminal point, involving the gracile, anterolateral, posterior spinocerebellar, corticospinal tracts. Note the involvement of Flechsig's middle root zone in the posterior column of the dorsal and lumbar cords and atrophic anterior roots.

virus can be transported retrogradely to the cell body and so selectively kill those neurons that transport it (Wiley et al. 1982).

Selective alterations in axonal transport induce predictable patterns of axonal pathology. Impairment of slow transport in axons leads to neurofilament accumulation and proximal axonal swelling (Chou and Hartmann 1964; Troncoso et al. 1982; Gajdusek 1985). The resulting proximal axonopathy may induce distal axonal atrophy as well as secondary demyelination (Griffin et al. 1978; Griffin and Watson 1988) consistent with *central-distal axonopathy* (Thomas et al. 1984) or "dying back" degeneration (Cavanagh 1979), as depicted in Fig. 4.3. There is ample evidence suggesting that the basic properties of axonal transport in the CNS differ from those of large peripheral sensory or motor axons (Reh et al. 1987; Oblinger 1988). Accordingly, certain cytopathological differences should exist between the upper and lower motor neurons in ALS (Chou 1991). Transneuronal degeneration from the upper to lower motor neurons may be expected, especially if they are monosynaptically linked, from impairment in axonal transport analogous to denervation atrophy occurring in skeletal muscles (Chou & Norris 1993).

Macroscopic Pathology of ALS

In many cases, both the brain and spinal cord will appear normal except for the alterations related to the normal aging processes. In some cases, selective atrophy of the precentral gyri (Fig. 4.4), first described by Kahler and Pick in 1879, may be conspicuous. Generalized atrophy of the spinal cord may be noted only in very chronic forms, but atrophic and grayish anterior spinal nerve roots, in contrast to the normal-appearing posterior spinal nerve roots are common features. Sclerotic discoloration and shrinkage of lateral corticospinal tracts may be detected on the cut surfaces of chronic ALS spinal cords. Perhaps most striking is the atrophied, distal skeletal muscles which are usually shrunken, pale and fibrotic.

Microscopic Pathology of ALS

The cardinal and characteristic histopathological features of ALS include:

1. Loss of large motoneurons with focal astrogliosis
2. "Senescent changes" with lipofuscin pigment atrophy
3. Various cytoplasmic inclusions with chromatolysis
4. Proximal and distal axonopathy with axonal spheroids
5. Tract degenerations
6. Degeneration of motor nerve fibers, motor end-plates and muscle atrophy (see Chapter 7).

Motoneuron Loss

Motoneuron loss in the cortex, brainstem and spinal cord varies from case to case, and the literature on this concept is inconsistent. One way to explain this incon-

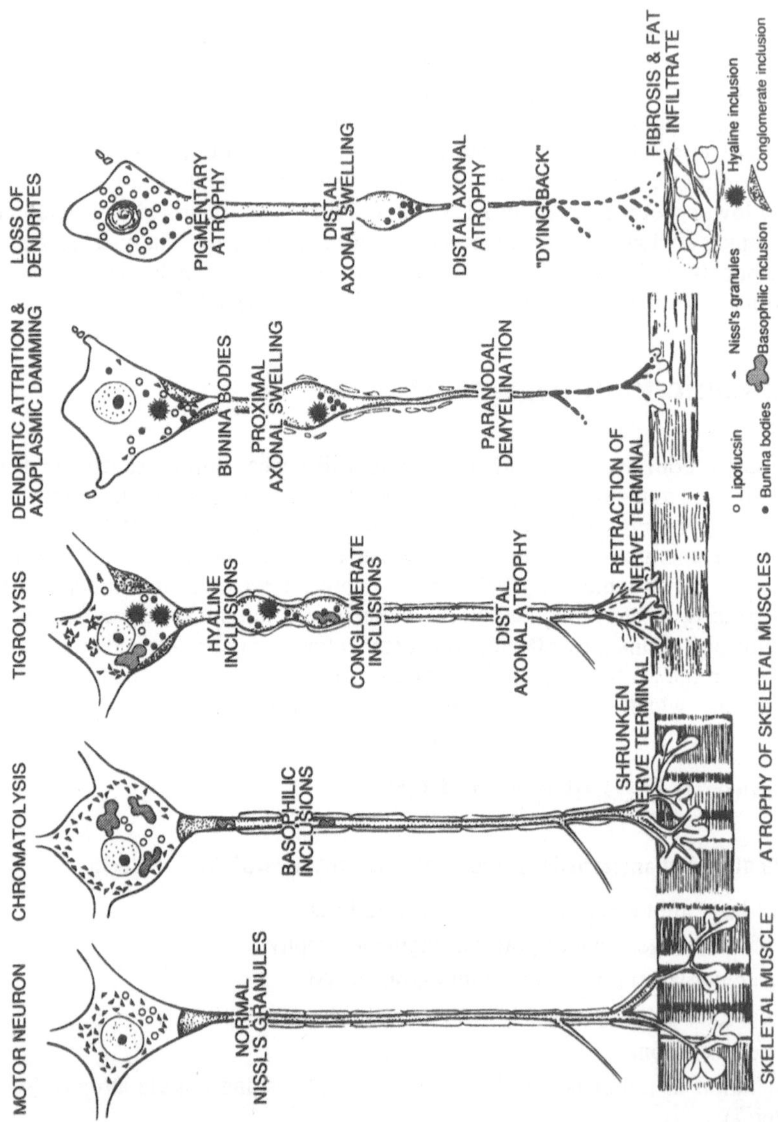

Fig. 4.3. Scheme of hypothetical evolution of ALS as a disease of primary "axostasis".

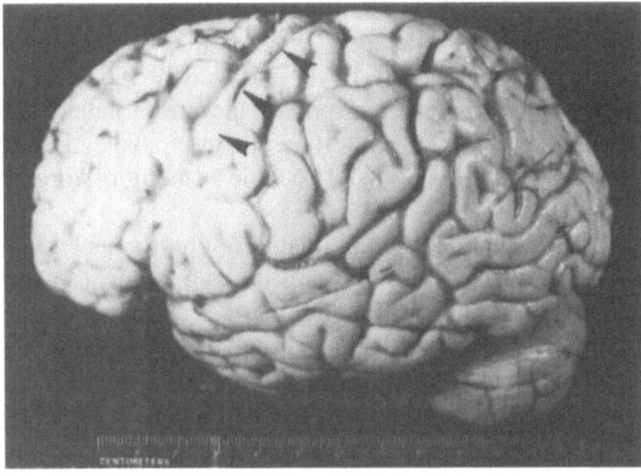

Fig. 4.4. Lateral view of the brain from a sporadic ALS case of 3-year duration, showing atrophic and sunken motor strip (arrow heads).

sistency is to consider that the disease process in ALS may start in any upper or lower motoneuron group, whether in the motor cortex, bulbar motor nuclei or spinal cord, and may spread eventually to other motoneuron groups at different organisational levels with diverse intensity. Premature death of motoneurons is in keeping with the concept of "abiotrophy" (Cavanagh 1979). However, motoneuron loss in ALS is more likely to constitute a selective loss with pathogenetic significance.

Loss of Betz cells in the motor cortex was first described by Charcot and Marie in 1885. They demonstrated degeneration of corticospinal tracts from the motor cortex through the internal capsule, cerebral peduncles, pons, medulla oblongata and spinal cord.

Scattered foci of patchy astrocytosis were described in the second to third layers of the motor cortex in 8 of 11 ALS cases by (Kamo et al. 1983, 1987). We were unable to verify this finding, but noted marked astrocytosis confined to the motor subcortex, deep from the Betz cell layer and intensifying along the gray-white junction, especially in a group of relatively young ALS patients, aged 32 through 52 years. Foci of patchy astrocytosis were found, in our experience, only in association with early or late senile degeneration. The reason for this focal distribution of reactive astrocytosis is unclear. However, as will be discussed below, the astrocytosis may be related to distal axonal swelling, and to degeneration of cortical Betz cells. Myelin sheaths are relatively well preserved in the white matter of the motor subcortex in ALS, supporting the concept of "dying back" axonopathy in the disease.

It has been questioned whether the motor cortex is singularly involved or if a more diffuse involvement of the entire frontal cortex occurs (Bertrand and van Bogaert 1925). We have seen cases of ALS in which Betz cells of the motor cortex are well preserved and others in which anterior spinal motoneurons are relatively well preserved. This variation may be due to tissue sampling since the motor system involvement in ALS is seldom uniform and symmetrical (Swash et al. 1986, 1988) In the motor cortex, Davison (1941) found involvement of the Betz cells with complete and continuous degeneration of the pyramidal tracts in only 12 of 42 ALS patients; Friedman and Freedman (1950) in only one third of 50 patients.

Of interest is the observation that among motor neuron groups at the same level, certain groups are more susceptible or more resistant to the disease process. In the brainstem, the hypoglossal nucleus, the nucleus ambiguus, and the trigeminal and facial motor nuclei are most susceptible, whereas the oculomotor, trochlear and abducent nuclei are rarely involved (Lawyer and Netsky 1953). In the spinal cord, there is a tendency to involve motoneurons of the posterolateral group but cases with predominant involvement of the anteromedial group have been described. Here again, the different termination patterns of each descending motor pathway must be taken into consideration.

Selective lack of involvement, often dramatic, of Onuf's nuclei in the sacral (S2) cord (Fig. 4.5a) is correlated with the well-known clinical feature of lack of abnormality of vesicorectal sphincter function in ALS patients (Mannen et al. 1977). It is intriguing that there is similar lack of involvement of Onuf's nuclei in both acute and chronic cases of poliomyelitis (Kojima et al. 1989). While one may infer that polio viral-specific receptor protein may exist on certain motoneuron groups, a recent immunohistochemical study has indicated that Onuf's neurons in some respects resemble autonomic neurons more than motoneurons (Katagiri 1988). Similarly, the well-preserved extraocular muscles in ALS patients are consistent with sparing of the corresponding bulbar nuclei. Although there are rare cases of ALS with ophthalmoplegia, as well as abnormalities in the oculomotor nuclei in some cases (Takahata et al. 1976; Harvey et al. 1979; Akiymama et al. 1987) these findings are exceptional. This selective sparing of certain motor nuclei has been speculatively explained as due to the inherent absence of androgen receptors in neurons of those motor nuclei (Weiner 1980).

"Senescent Changes" in ALS

The most consistent and readily discernible microscopic finding of aging is "accumulation" of lipofuscin granules in atrophied perikarya, the so-called pigment atrophy. This change can be misinterpreted as a "senescent change" since these wear-and-tear pigments are characteristically present in neurons, particularly neurons of the aged. Lipofuscin granules also ostensibly increase in quantity in many atrophic mammalian cells, for example, in myofibres in brown atrophy and in adrenal cortex cells in brown degeneration. Hence, an apparent increase of lipofuscin granules at the expense of lost cytoplasmic components including Nissl granules in neuronal perikarya must not be construed as a phenomenon necessarily closely related to senescence. In the motoneurons in ALS, it is the perikaryal atrophy which produces an apparently increased amount of perikaryal lipofuscin. Although a relative increase in the incidence rate of ALS with increasing age has been found in population-based studies (Juergens et al. 1980; Mulder 1987), there is no unequivocal pathological evidence indicating that the pathogenic process of ALS is related in any way to that of aging.

Similarly, the presence of neurofibrillary tangles (NFT) in both brain and spinal cord, unassociated with senile plaques, described by Hirano (1973) in all 70 Guamanian ALS patients, suggested that an aging process might be a contributory factor in the high prevalence of ALS in Guam. Similar findings were reported in ALS occurring in the Kii peninsula in Japan (Shiraki and Yase 1975) and in the Guam parkinsonism-dementia (PD) complex (Hirano 1973). NFT may be a characteristic finding in the general Guam population and therefore a characteristic feature of Guamanian ALS and PD. Indeed, a control study of 69 members of the Guamanian Chamorro without dementia, ALS, or parkinsonism (Anderson et al.

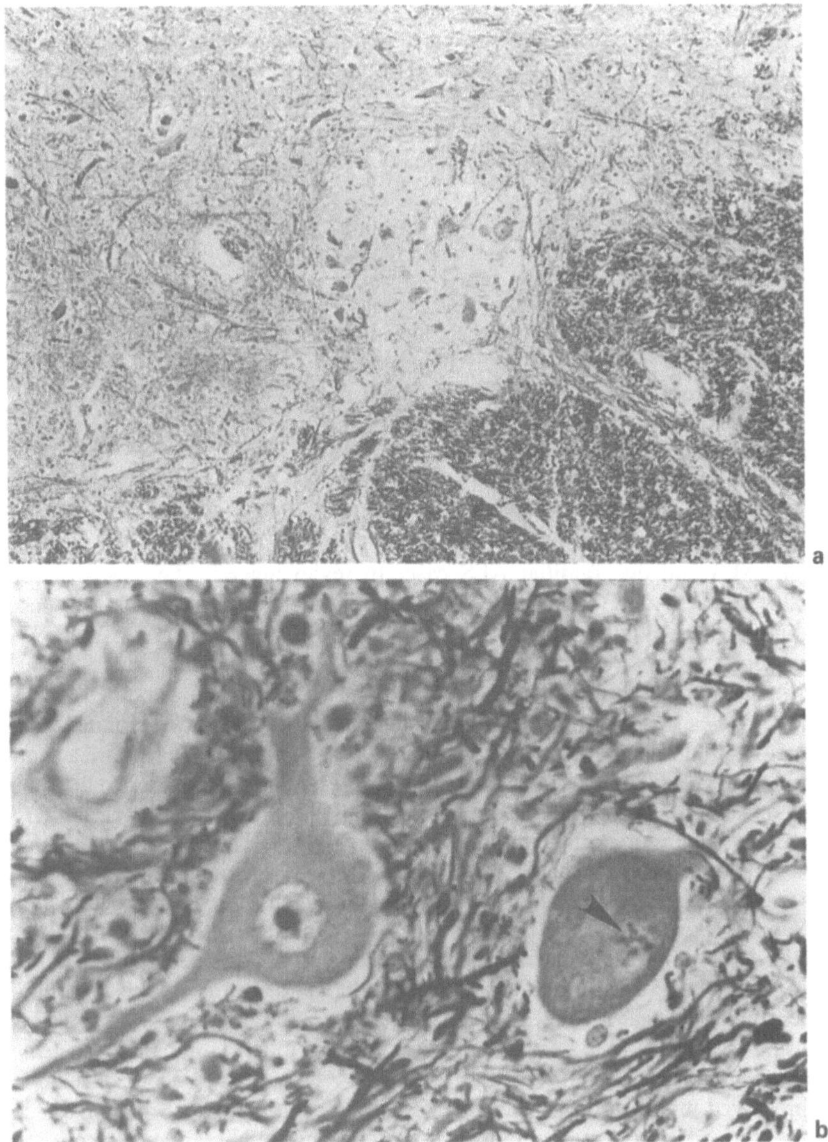

Fig. 4.5. **a** Well-preserved small neurons of Onuf's nucleus; all the large alpha-motonerons in the anterior horn disappeared. Klüver-Barrera stain. ×150. **b** Rounded spinal motoneuron soma with attrition of dendrites and Bunina bodies (arrow head) in the cytoplasm in comparison to the relatively preserved motoneuron above. (Bielschowsky stain. ×900.)

1979) disclosed a very high rate of NFT at an early age in this population. Furthermore, it has been well established that NFT by itself is not an aging process (Mandybur et al. 1977), although this change may promote the pathogenic process of ALS by contributing to the phenomenon of disturbed axonal flow.

Although non-specific neuronal alterations including abnormal Nissl granules, pyknosis, ischaemic change, neuronophagia, satellitosis, pigment atrophy, etc. were described in ALS Betz cells by Davison (1941), no specific alteration has been described. Golgi-stained Betz cells revealed dendritic fragmentation with numerous irregularities and loss of dendritic spines (Hammer et al. 1979), with encroaching gliosis related to the soma. These findings were considered similar to those found in Betz cells in normal aging. Such preterminal dendritic alterations (Fig. 4.5b) have been described in various clinical situations (Carpenter et al. 1988) but appear more frequently in ALS.

Cytoplasmic Inclusions in ALS

Among the wide variety of neuronal changes described in the ALS literature, most intriguing is the presence of intracytoplasmic inclusions of variegated tinctorial characteristics and shapes (Chou 1979). These inclusions include (a) eosinophilic inclusions, or Bunina bodies, (b) basophilic inclusions, (c) hyaline inclusions, and (d) conglomerate inclusions.

Bunina bodies are small (2–7 μm diameter), intracytoplasmic, eosinophilic, refractile inclusions (Fig. 4.6a,b). They were first described by Bunina in 1962 in two cases of familial ALS and later by Hirano (1965) in two Guamanian ALS, two sporadic ALS and one familial ALS case. In reviewing 82 ALS cases of three clinical subtypes, we found these eosinophilic inclusions in 67% (55 cases) including all three clinical subtypes PLS (1), PMA (22), and ALS (28) (Chou 1979). Several reports (Hart et al. 1977; Tomonaga et al. 1978; Okamoto et al. 1980) have described the ultrastructure appearance of Bunina bodies as honeycombed, densely granular bodies resembling the autophagosomes that are often found among lipofuscin granules (Fig. 4.7a,b). Within honeycombed vacuolar spaces are accumulations of trapped filaments or other granular or vesicular organelles. Small Bunina bodies are seldom ubiquitinated (Murayama et al. 1990), but those appearing larger and reticular are more often ubiquitinated (Fig. 4.6c). The presence of Bunina bodies in motoneuron somas in the vicinity of large proximal axonal spheroids raises a possible causal relationship (Fig. 4.6d) between these two structures.

Basophilic inclusions were first described by Wohlfart and Swank in 1941 as "peculiar-shaped intracellular bodies in a few ventral horn cells of the lumbar region. They varied in size and shape, were homogeneous and weakly basophilic. Frequently they appear like coagulated fluid". Basophilic inclusions vary greatly in shape and size (Fig. 4.8a) may be globoid, and are much larger (4 to 16 μm in average diameter) than Bunina bodies (Fig. 4.8b). They tend to occur in relatively well-preserved motoneurons in juvenile-onset or young adult ALS patients. Occasionally, they appear in motoneurons that show changes in Nissl granules consisting either of central chromatolysis or tigrolysis (Fig. 4.8d). They were described in detail by Nelson and Prensky (1972) as RNA-rich inclusions (Fig. 4.8c) and are probably identical to those described in sporadic juvenile ALS cases by others (Wohlfart and Swank 1941; Berry et al. 1969; Tsujihata et al. 1980; Oda et al. 1978). We described large basophilic inclusions in two familial ALS patients (father and son) who died, both at the age of 33 (Chou 1979) and in an adult male of 31 years of age from Australia (Chou et al. 1988). Rarely, both eosinophilic and basophilic inclusions coexist in the cytoplasm of the same neurons. Of great interest is the description in the original article by Bunina (1962) of the dimension of the inclusions measuring 5 to 20 μm in diameter, an observation which hints at the possibility of a relationship with hyaline inclusions (see below). Electron microscopy of basophilic inclusions (Oda et al. 1978; Chou 1979) revealed aggregates of fuzzy

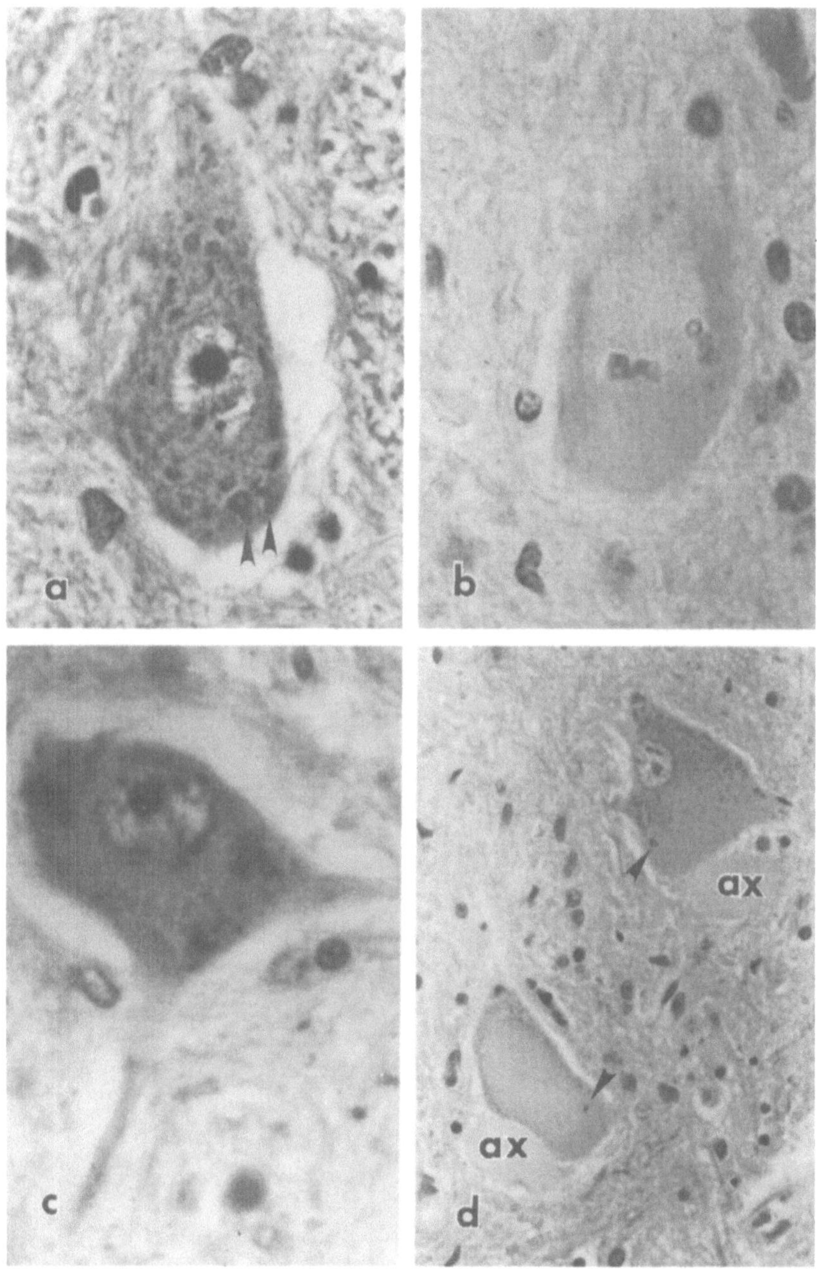

Fig. 4.6. **a** Two aggregates of eosinophilic refractile Bunina bodies (arrow heads) in a relatively well-preserved Betz cell. (H&E stain. ×1700.) **b** Spinal motoneuron from a sporadic ALS case, showing "pigment atrophy" of the perikaryon with abundant lipofuscin accumulation in the center with two aggregates of Bunina bodies. (H&E stain. ×1700.) **c** Ubiquitinated granules of irregular and reticular shape apparently related to Bunina body formation in a large spinal motoneuron soma. (AEC-hematoxylin counterstain. ×1700.) **d** Two chromatolytic motoneurons with a Bunina body in each perikaryon (arrow head) with a large axonal spheroid each (ax) neighbouring the perikaryon. (H&E stain. ×900.)

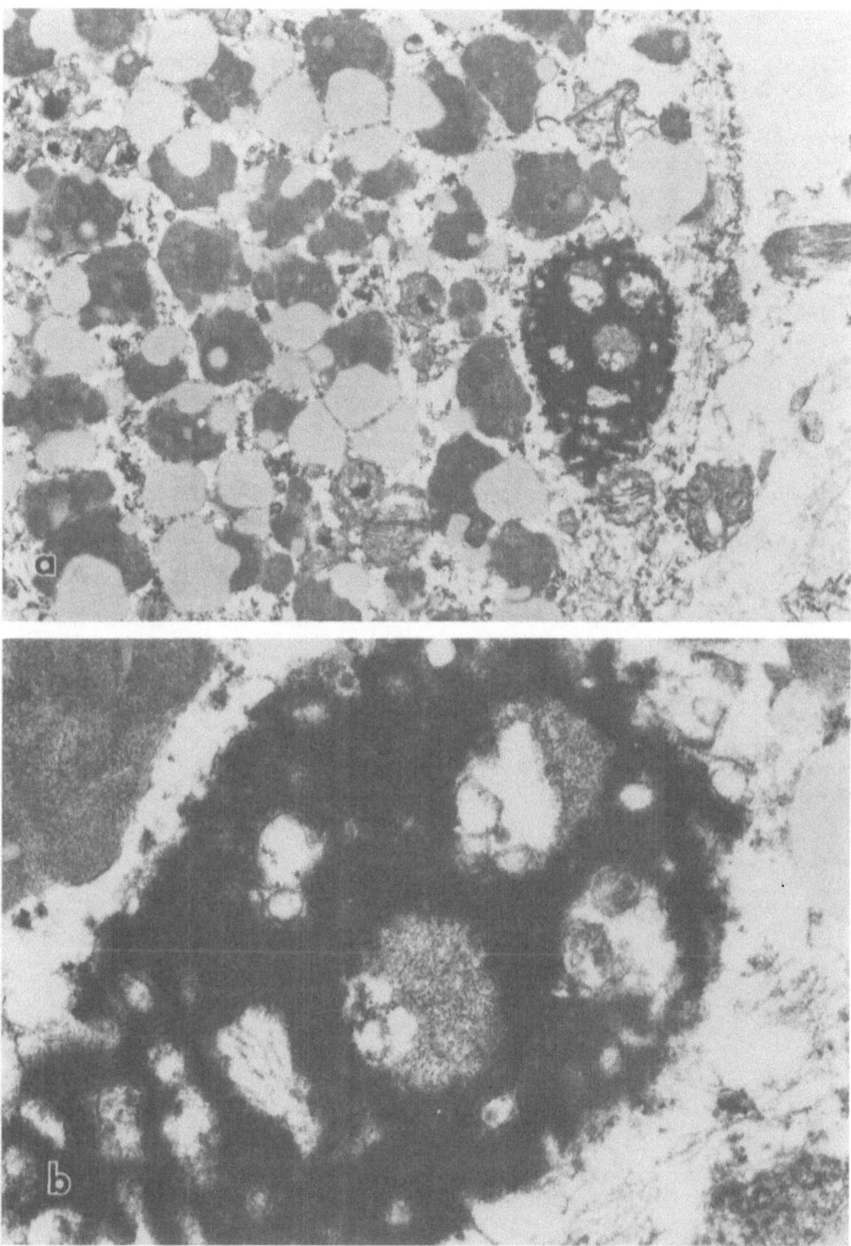

Fig. 4.7. a A single Bunina body in a lake of lipofuscin granules in a motoneuron soma, showing its characteristic high electron density and multivacuolation. (×15 000.) **b** Higher magnification of the Bunina body showing multiple vacuoles containing variegated structures including filaments, granules and vesicles, suggestive of autophagosomal nature. (×45 000.)

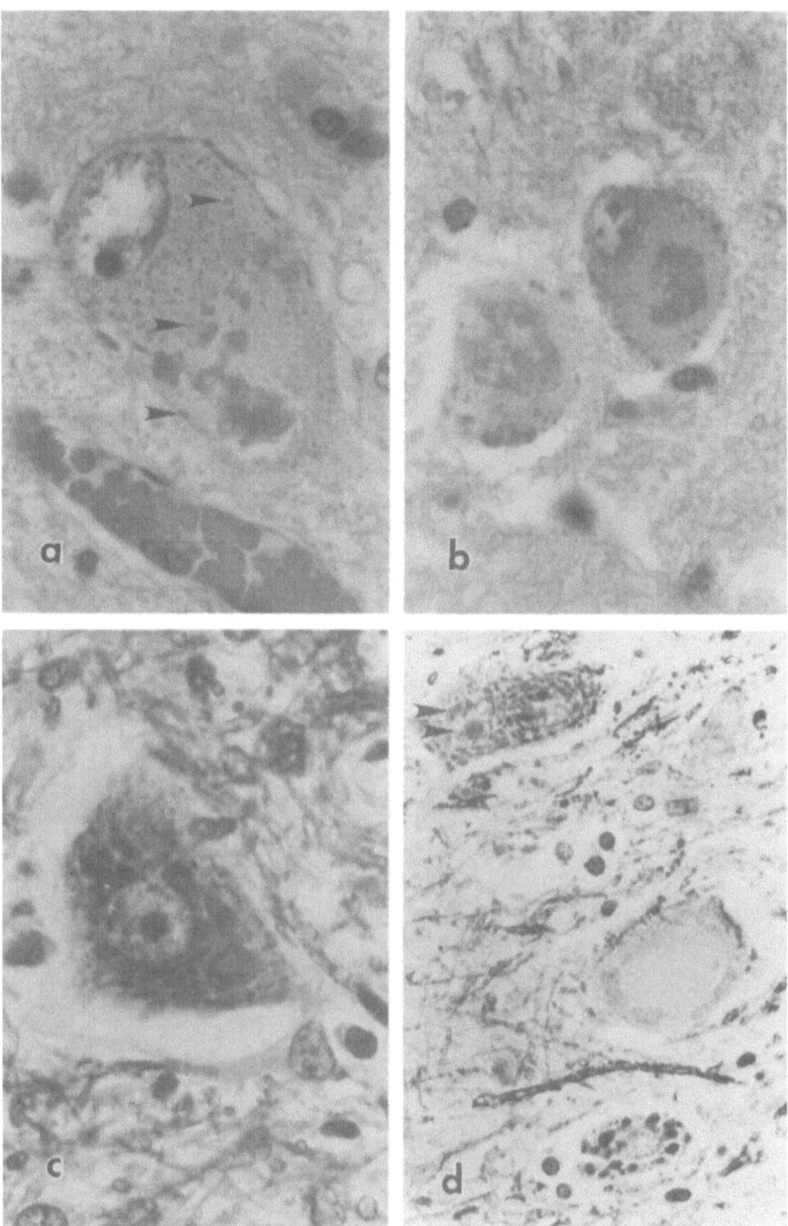

Fig. 4.8. **a** Chromatolytic motoneuron soma containing "peculiar-shaped" basophilic inclusions admixed with Bunina bodies (arrow heads). (H&E stain. ×1700.) **b** Two bulbar reticular neurons: one containing reticulated or filigree-shaped inclusion and another, more solid spheroidal inclusion. (H&E stain. ×900.) **c** Strongly Nissl-stained positive basophilic globose inclusions suggesting possible presence of Nissl granules or ribosomes in the inclusions. (Klüver–Barrera stain. ×900.) **d** Three Betz cells with a top Betz cell containing two basophilic inclusions; marked central chromatolysis in the middle showing faintly basophilic inclusions at the periphery; and the bottom with condensation of Nissl's granules ("tigrolysis"). (Klüver–Barrera stain. ×900.)

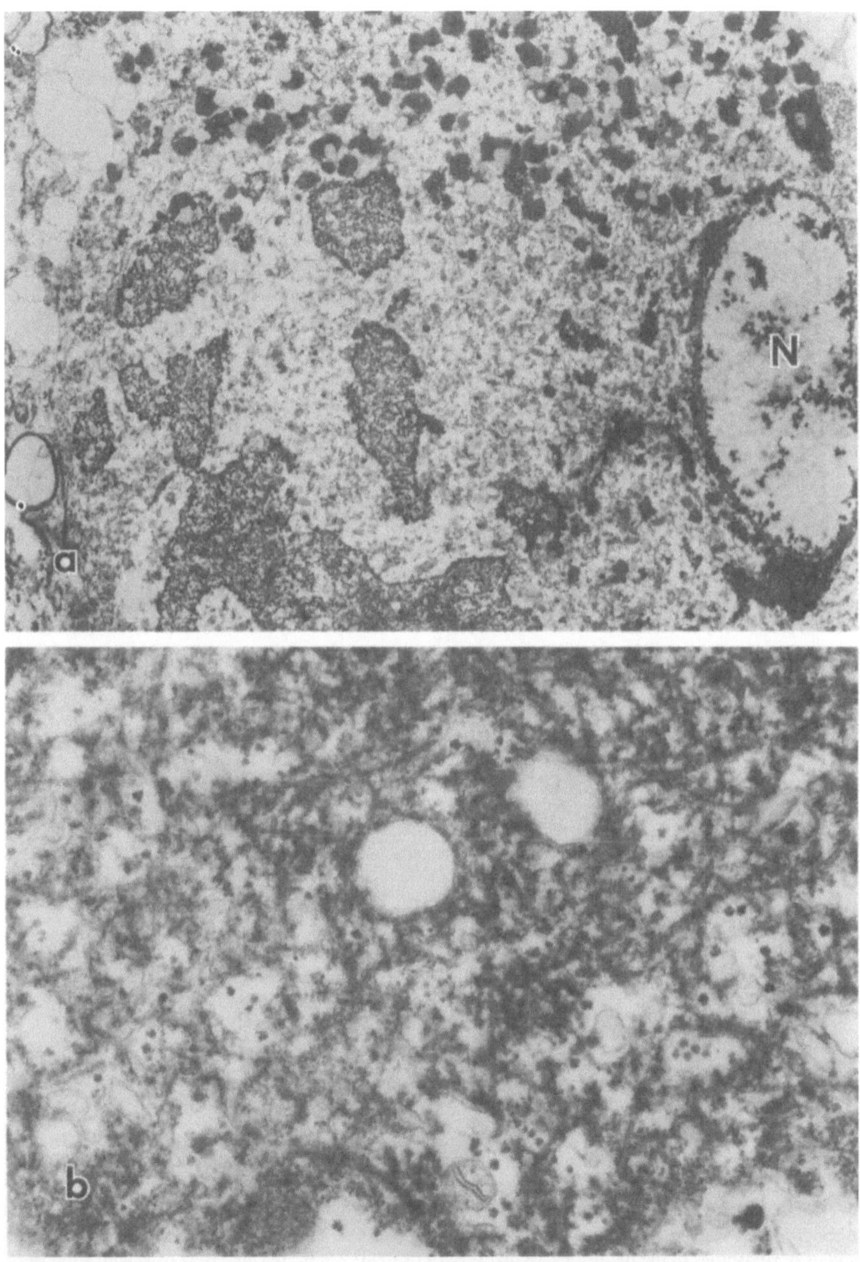

Fig. 4.9. a Chromatolytic motoneuron containing multiple filigree-shaped inclusions at the center of the perikarya pushing the nucleus (N) and lipofuscin granules to the periphery. (×6500.) **b** Higher magnification of the basophilic inclusion showing granule (ribosome and/or glycogen) coated microtubules (15 nm in diameter) and a few vesicles. (×75 000.)

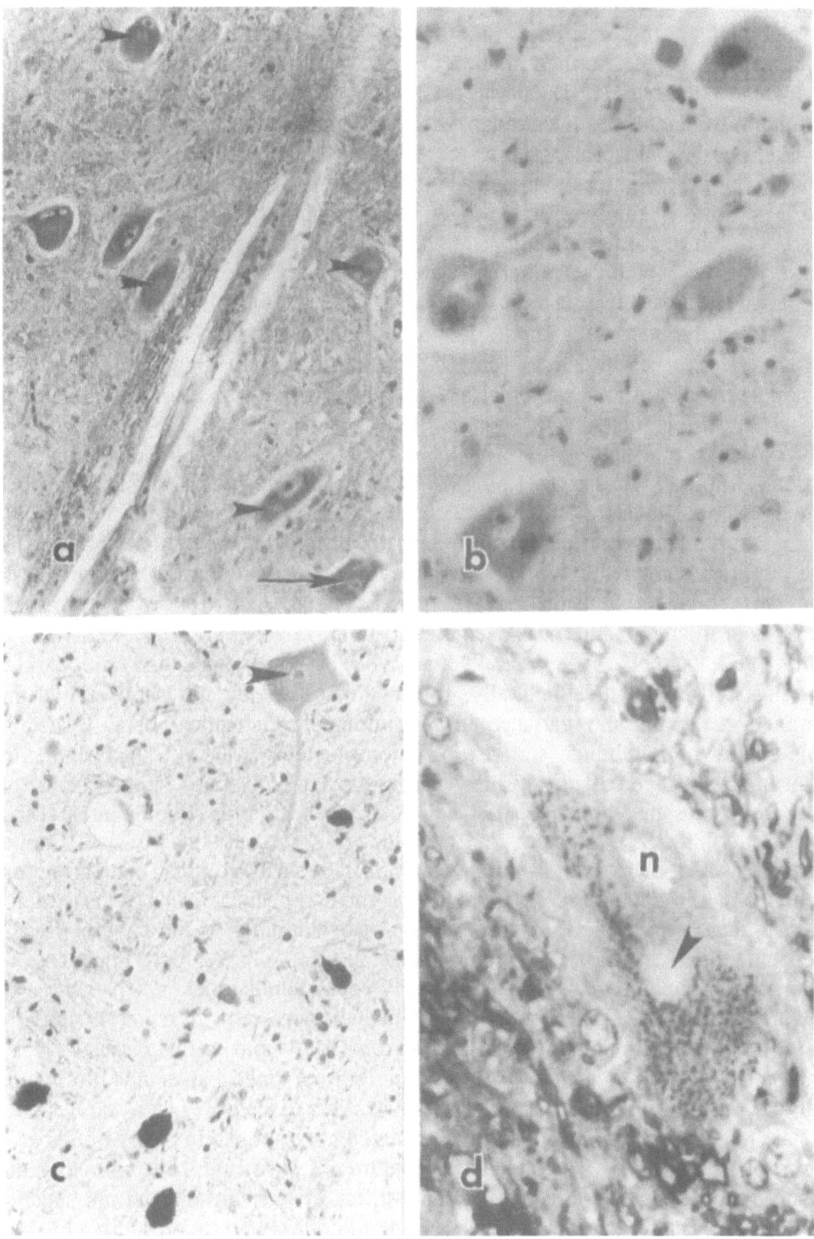

Fig. 4.10. **a** Anterior horn with relatively populated motoneurons from a subacute ALS case showing at least five motoneurons containing hyaline inclusions (arrow heads) and one of them with Bunina body in the center (arrow). (H&E stain. ×150.) **b** Intensely ubiquitinated hyaline inclusion in each of three motoneuron somas. AEC-hematoxylin counter-stain. (×300.) **c** Immunostained anterior horn for phosphorylated neurofilament (250 kD) showing intensely stained large axonal spheroids but not in the perikaryon nor the hyaline inclusion (arrow head). (AEC-hematoxylin. ×200.) **d** Semi-thin resin section showing a hyaline inclusion with a central Bunina body (arrow) encroaching on lipofuscin aggregate in a motoneuron soma. (Toluidine blue stain. ×1200.)

microtubules, with an outer diameter of 12–15 nm, closely associated with ribosomes and rough endoplasmic reticulum (Fig. 4.9a,b), in reticular or filigree patterns.

Hyaline inclusion, described first by Hirano et al. (1967) and later by others (Metcalf and Hirano 1971; Takahashi et al. 1972; Tanaka et al. 1980, 1984) was at first thought to occur only in familial ALS. They consist of a hyaline-like, or Lewy body-like, poorly stainable substance, occasionally with a central core which stains basophilic but is seldom concentrically laminated (Fig. 4.10a). They measure up to 3 or 5 μm in diameter, rarely resemble Lewy bodies, and are surrounded by a clear zone or halo. Twelve of 82 cases demonstrated this inclusion including 3 PMA cases, 7 sporadic ALS cases and 2 familial ALS cases (Chou 1979). In 10 of these, hyaline inclusions were seen in association with Bunina bodies. Hyaline inclusions have also been observed frequently in sporadic ALS (Delisle and Carpenter 1984; Munoz et al. 1988; Kuroda et al. 1986; Kato et al. 1988). It has also been suggested that hyaline inclusions along with large proximal axonal spheroids may be more prominent and abundant in ALS patients with shorter clinical courses (Chou 1987a). Many histochemical characteristics of these inclusions overlap with those of other neuronal inclusions (Chou 1987b; Leigh et al. 1989) and an inescapable impression is gained that these three inclusions are interrelated, i.e., that both basophilic and hyaline inclusions may undergo an evolutionary change terminating as Bunina bodies. This course might be comparable to the development of autophagic vacuoles. Immunohistochemically, hyaline inclusions are strongly decorated by ubiquitin but not necessarily by neurofilament antibodies (Leigh et al. 1988; Lowe et al. 1988; Murayama et al. 1989, 1990; Kato et al. 1989; Sasaki et al. 1988) both in sporadic and familial ALS cases (Fig. 4.10b,c). Ultrastructurally, hyaline inclusions do not show a uniform appearance, reflecting the variance in the immunohistochemical data. Basically, granule-coated microtubules, sheaves of microfilaments, vacuoles and miniature Bunina bodies all coexist in hyaline inclusions (Fig. 4.11a,b). Immunoelectron microscopy shows that granule-associated filaments are highly ubiquitinated (Murayama et al. 1989, 1990; Lowe et al. 1988). Indeed, the concurrence of both Bunina body and hyaline inclusions is not uncommon (Fig. 4.10a,d). By comparing the histochemical, immunohistochemical and ultrastructural characteristics of these inclusions, one may propose the following scheme of evolution as . . . basophilic > hyaline > eosinophilic (Bunina) . . . as summarized in Fig. 4.3. A high frequency of occurrence of basophilic inclusions in unusually young adults or juvenile patients with ALS is in keeping with this scheme, in that this inclusion seems to represent an early stage in the evolution of intracytoplasmic neuronal inclusions in this disease.

Hyaline conglomerate inclusions, first described by Schochet et al (1969) in a sporadic ALS case and later also in a sporadic ALS case by Hughes and Jerrome (1971) are relatively rare. This inclusion (Fig. 4.12a,b) was found in only 2 of 82 cases reviewed by Chou (1979). The clinical features in these two cases, one with sporadic ALS and the other with familial PMA, suggest that these inclusions are not specific for any particular motor syndrome. Electron microscopic studies indicate that hyaline conglomerate inclusions consist mainly of neurofilaments and mitochondria, and that they may thus be interpreted not as cytoplasmic inclusions in a strict sense, but as focal intracytoplasmic accumulations of axoplasmic components (Kusaka et al. 1988). Such an intracytoplasmic accumulation has been described not only in motoneurons but also in other neurons including those in the pontine reticular formation of a sporadic ALS case (Kondo et al. 1986). While it is possible that such a neurofilamentous aggregate is secondary to over production of overexpression of neurofilament subunits it is more likely that it is secondary to the retrograde "axoplasmic damming" due to an impairment in slow anterograde axonal transport. This would explain why phosphory-

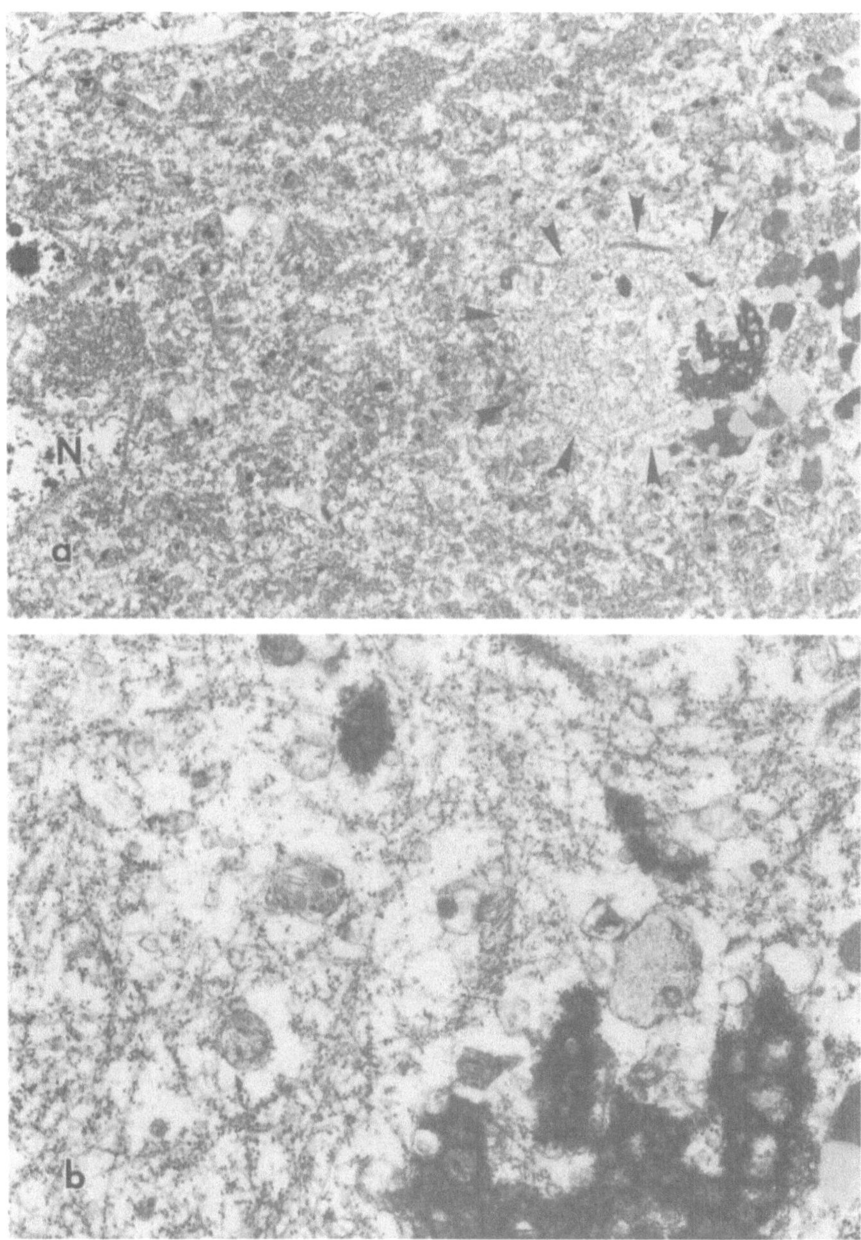

Fig. 4.11. **a** Early hyaline inclusion (arrow heads) in the relatively intact perikaryon positioned between nucleus (N) and lipofuscin granules. (×9500.) **b** Higher magnification of the hyaline inclusion consisting of small Bunina body, sheaves of filaments, ribosome-coated microtubules (12–15 nm), and a few vesicles. (×28 500.)

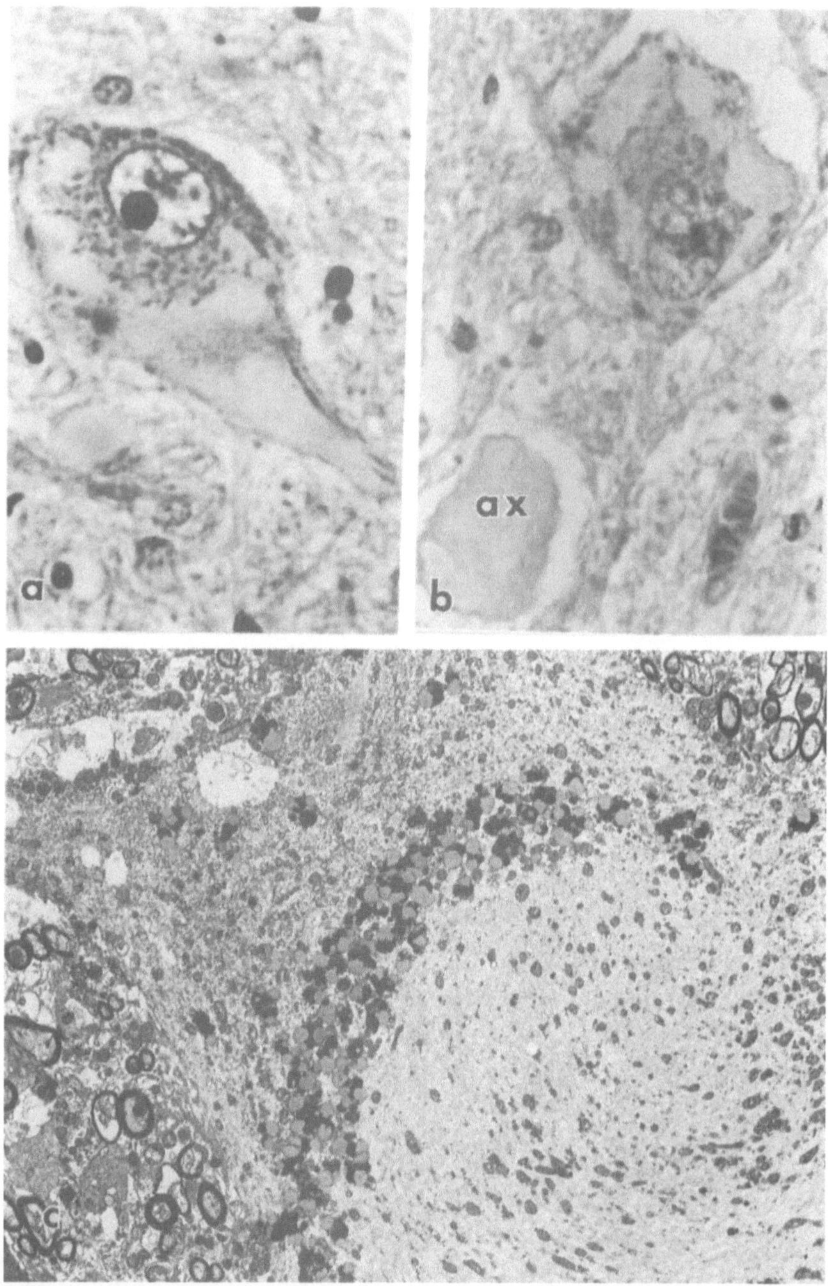

Fig. 4.12. **a** Spinal motoneuron filled with conglomerated hyaline inclusions at the periphery of the perikaryon and the proximal axon. Note a Bunina body along the inclusion. (H&E stain. ×2000.) **b** Conglomerate hyaline inclusions in a large motoneuron soma accompanied by a large proximal axonal spheroid (ax). (×1800). **c** Ultrastructure of intracytoplasmic axoplasmic damming or regurgitation of neurofilaments pushing lipofuscin granules to the periphery of a motoneuron cytoplasm. (×8200.)

lated neurofilament epitopes (of distal axon type) tend to accumulate in the perikarya of ALS motoneurons (Munoz et al. 1988; Chou 1987a,b; Leigh et al. 1989; Kurisaki et al. 1983; Mannetto et al. 1988; Xu et al. 1993; Coté et al. 1993). However, this type of retrograde intracytoplasmic damming of phosphorylated neurofilaments is not specific for ALS, since it is found also in control anterior horn cells (Leigh et al. 1989). Similar, if not identical, intracytoplasmic conglomerate inclusions have been induced in the upper motor neurons of IDPN-intoxicated monkeys; these experimentally-induced inclusions are composed of whorled neurofilaments (Fig. 4.12c) associated with mitochondria (Chou 1983).

Proximal and Distal Axonopathy in ALS

Large, proximal axonal swellings were first described by Carpenter (1968) in 11 cases of subacute ALS (with clinical involvement of 10 months duration or less) and, later, in our laboratory in two ALS patients who died 4 and 11 months after onset (Chou et al. 1970), and by others (Inoue and Hirano 1979). A direct connection between the proximal axonal swelling and the perikarya has been unequivocally demonstrated (Fig. 4.13a–d) (Sasaki et al. 1989a,b). Similar axonal swellings packed with whirled neurofilaments and mitochondria (Fig. 4.13e) can be induced by a single injection of the neurotoxin IDPN (β-β'-iminodipropionitrile) a toxic dimer crystallized from *Lathyrus odoratus* that causes neurolathyrism. An intra-axonal or primary impairment of axonal transport has been thought to be important in the pathogenesis of the resulting axonopathy, and the concept of "primary axostasis" was proposed (Chou and Hartmann 1964, 1965). Indeed, a selective impairment of axoplasmic flow affecting the slow component of the orthograde axonal transport system (Griffin et al. 1978, 1988; Chou and Klein 1972) without affecting the fast component or the retrograde transport (Griffin et al. 1978; Kuzuhara and Chou 1981) has been demonstrated. It is tempting to correlate the role of intracytoplasmic and intra-axonal inclusions in inducing the impairment of the axonal flow with a "dying back" degeneration of both upper and lower motor nerve fibers. Alternatively, a primary insult resulting in axostasis may predispose to the formation of intracytoplasmic inclusions. Immunohistochemical studies of various neurofilament subunits (Fig. 4.10c) in axonal spheroids indicate the presence of neurofilament proteins of axonal or phosphorylated type and fails to suggest any post-translational abnormalities (Chou 1987b; Kurisaki et al. 1983; Schmidt et al. 1987; Dickson et al. 1986). This may be due to the lack of more specific antibodies against neurofilament subunits. The ultrastructure of axonal spheroids is characterized by densely packed neurofilaments with a few mitochondria, but with rare microtubules (Fig. 4.13e).

Tract Degeneration

Corticospinal Tract Degeneration

Tract degeneration is most commonly found in the corticospinal tracts of the ALS spinal cord. The extent of this degeneration, however, is not consistently correlated with the clinical features of pyramidal signs. The involvement may be asymmetrical (Fig. 4.14) (Swash et al. 1988). The atrophy and pyramidal tract signs occur in the absence of corticospinal tract involvement at autopsy; the reverse has also been reported (Lawyer and Netsky 1953; Friedman and Freedman 1950). A few authors considered that corticospinal degeneration starts in the internal capsule and that the

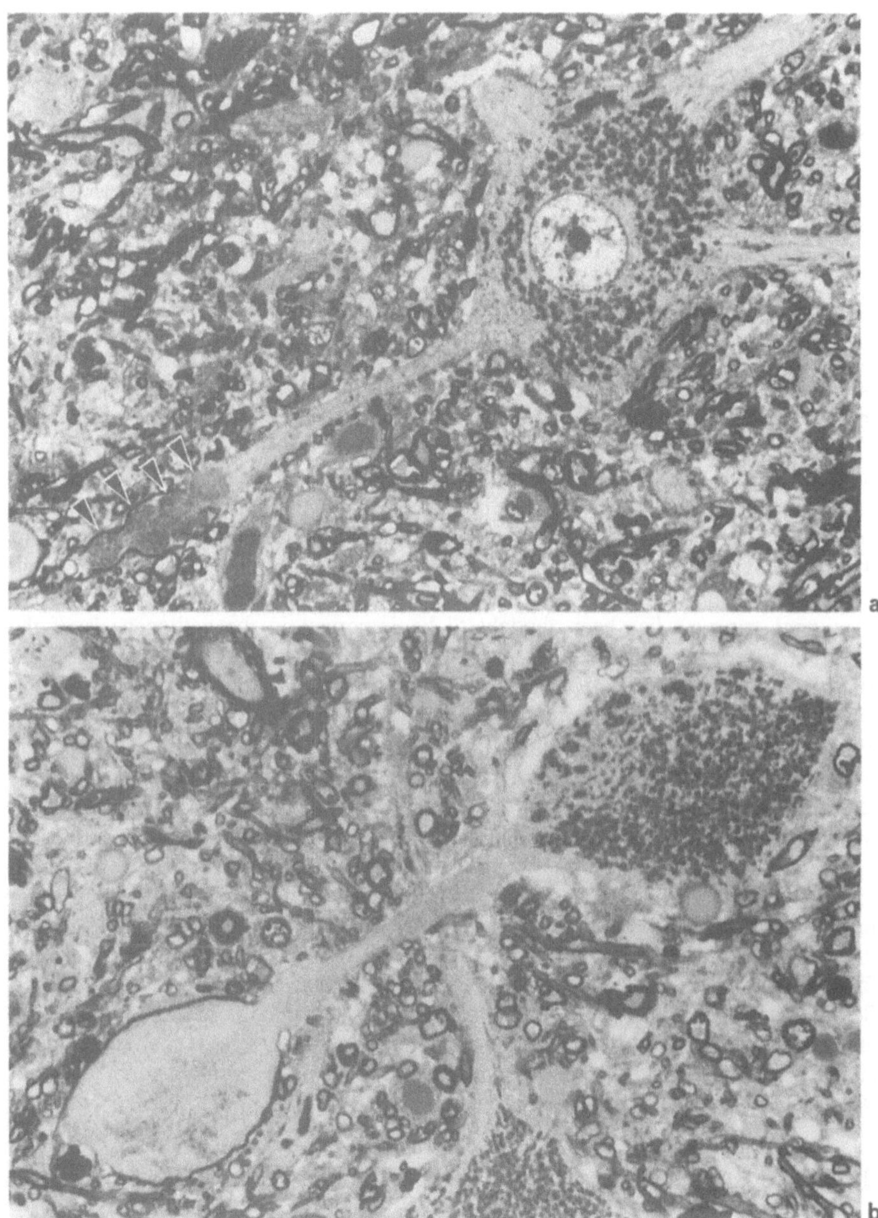

Fig. 4.13. **a** Proximal axon of a motor neuron from an MND patient, showing intra-axonal inclusions at the junction beginning of myelinated segment (arrow heads); they are stained more densely with toluidine blue (×1400). **b** Anterior motor neurons from an MND patient showing a globular swelling of the proximal axon surrounded by a thin myelin sheath in direct continuation to the perikaryon. (Toluidine blue. ×1400.)

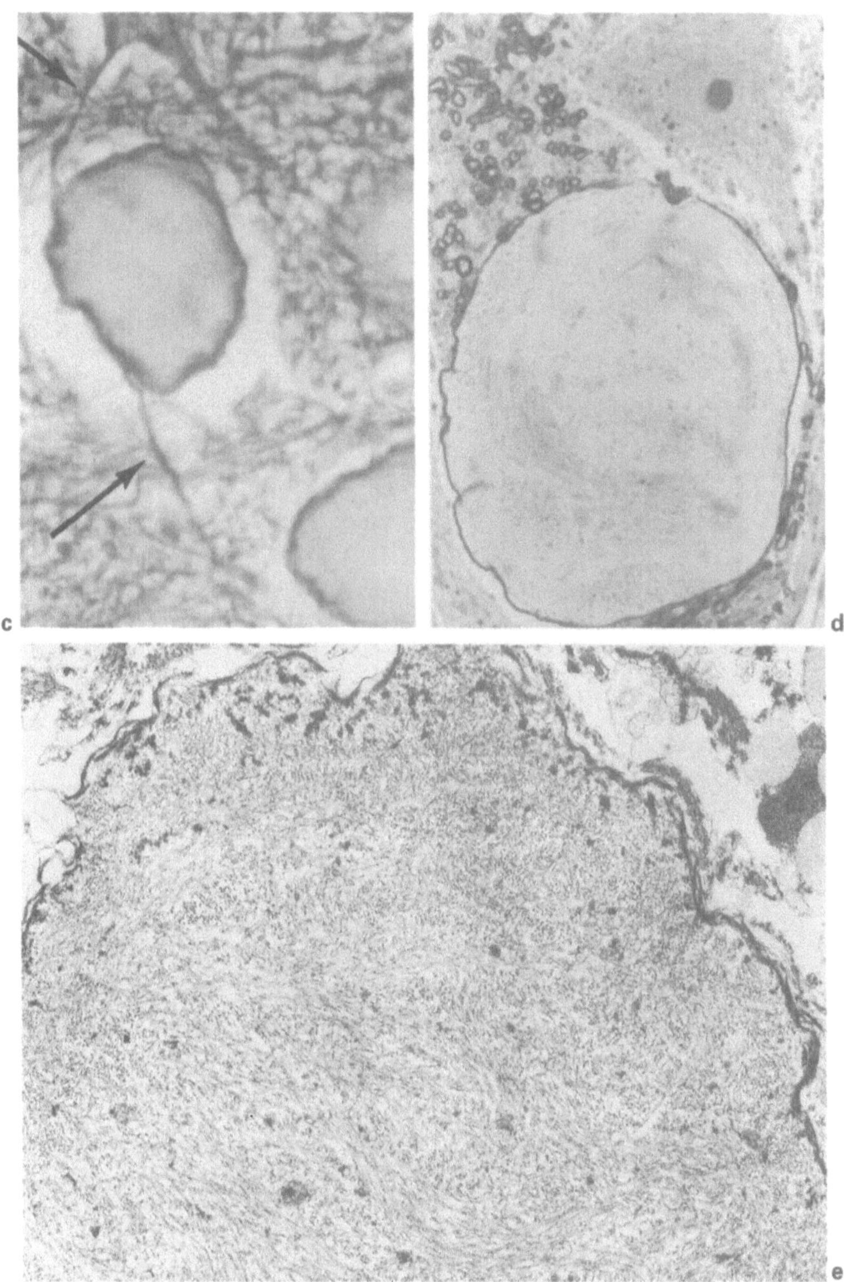

Fig. 4.13. (*continued*) **c** A large proximal axonal spheroid linked by a proximal axonal segment (arrow) to the perikaryon and to the distal segment (arrow) accompanied by two large axonal spheroids. Bodian stain. (×2000.) **d** Resin-embedded giant axonal spheroid nearby a spinal motoneuron, showing a thin myelin sheath surrounding the spheroid and whorls of mitochondria within the dammed axoplasm. (Toluidine blue stain. ×2200.) **e** Ultrastructure of thinly myelinated axonal spheroid consisting of densely packed neurofilaments, without microtubules and a few mitochondria. (×15 000)

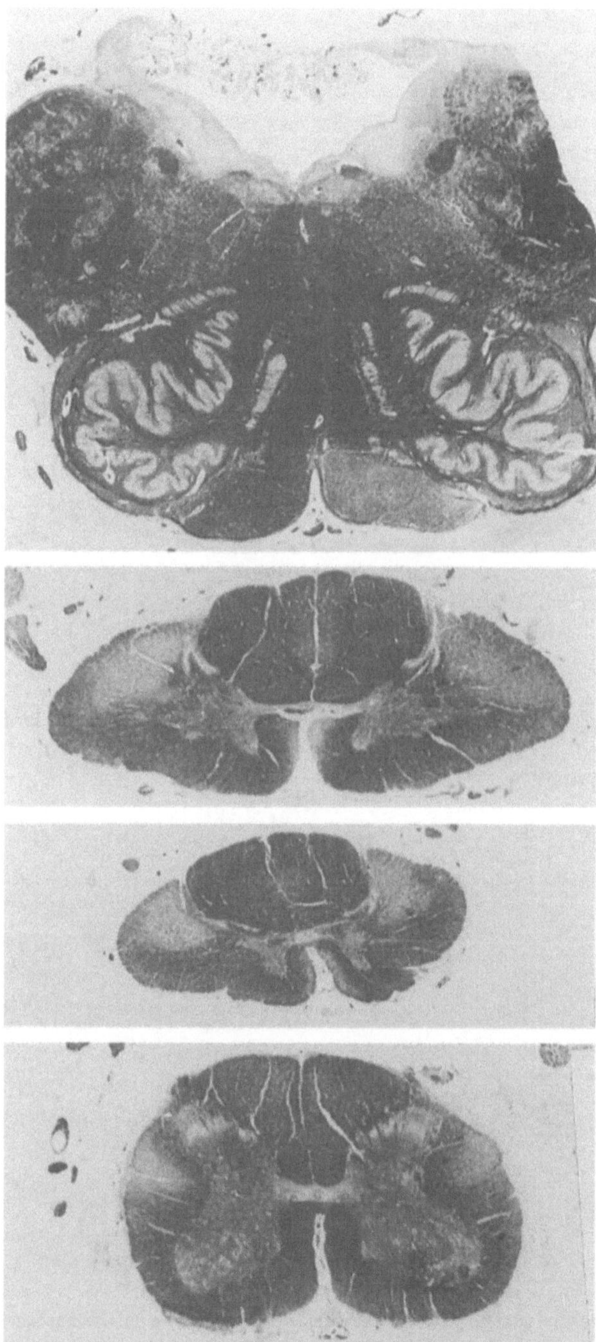

Fig. 4.14. Asymmetrical right corticospinal tract degeneration at the bulbar pyramids and bilateral degeneration with more severely involved left corticospinal tract in the spinal cord after decussation. Note the well-preserved spinocerebellar tracts.

cortical lesions are attributable to *ascending* degeneration. The credit for the first description of the entire pyramidal tract degeneration is given to Kojwenikoff (1883), who described degeneration occupying the anterior third quarter of the posterior limb of the internal capsule in horizontal sections (Fig. 4.15a). Later, the presence of this restricted area of pyramidal tract degeneration was confirmed by Hirayama et al. (1962), who corrected the traditional misconception for the topographic location of pyramidal tracts that largely originated from Dejerine's description (1901). In general, corticospinal tract involvement is uneven at various levels, as stressed by Bertrand and Van Bogaert (1925). Of 37 ALS cases studied by Davison in 1941, 12 showed total pyramidal degeneration from the motor cortex to the spinal cord, 2 from the cerebral peduncle, 7 from the pons, 12 from the medulla, and in 4 cases, the degeneration was limited to the spinal cord. Of 45 cases examined by Brownell et al. (1970), less than half demonstrated complete pyramidal degeneration from the cortex and in 10 corticospinal tract degeneration was undetectable. Among 36 "classical" ALS cases, the same authors noted several patterns of corticospinal tract degeneration: no pyramidal degeneration in 8, degeneration in pyramidal tracts alone in 8, both pyramidal and anterolateral column involvement in 17, and both pyramidal and anterolateral column involvement, with some posterior column degeneration. This variation in involvement of the corticospinal tracts, and of the anteromedial brainstem pathway can be explained, in part, by the variable pattern of involvement of different motoneuron groups especially of cortical, bulbar and spinal motoneurons, according to the functional organizational scheme elucidated by Kuypers (1973). Furthermore, it is of critical importance to recognize the clinicopathologic substrates of indirect and direct corticospinal tract degeneration since these two tracts may be independently involved. Such a clinicopathologic differentiation can better be understood by comparing ALS with Strümpell's disease.

However, corticospinal tract degeneration limited to different levels of the CNS cannot be readily explained. In this regard, the finding of giant axonal spheroids collected segmentally along the corticospinal tracts in internal capsules (Fig. 4.15b), cerebral peduncles and medullary pyramids in two ALS cases from Guam and in two sporadic cases with clinical courses ranging from 7 months to 2 years may provide a plausible explanation (Chou et al. 1980, 1988). Those giant axonal spheroids, which are immunoreactive with antibodies to neurofilaments and weakly so for ubiquitin antibodies (Fig. 4.15c) provide additional evidence suggesting an impairment in both distal and proximal axonal transport in ALS. Because the axonal transport profiles in the CNS are different from those in lower motoneurons, the process of formation of these axonal spheroids in the distal part of the upper corticospinal tracts, although pathogenetically similar, shows differences from spheroids formed in the proximal axons of spinal motoneurons. Indeed, proximally located axonal spheroids have never been observed, for example in or near Betz cells. However, spheroids in the upper corticospinal tracts might have been derived from either Betz cells or upper brainstem neuron somas, carried distally, and aggregated and conglomerated until they could be transported no further, as strongly indicated by "tadpole" formation of spheroids in the segments of corticospinal tracts showing a unidirectional conformation (Fig. 4.15d) with the "tadpole" tails pointing proximally.

Selective involvement and loss of large myelinated motor fibers has been reported by several authors at anterior spinal roots (Tsukagoshi et al. 1979; Sobue et al. 1983) or peripheral motor axons (Dyck et al. 1975) and at various levels of the corticospinal tract pathway (Hirayama et al. 1962). This selective involvement of large motoneuron axons implies that a certain physiobiochemical property unique to

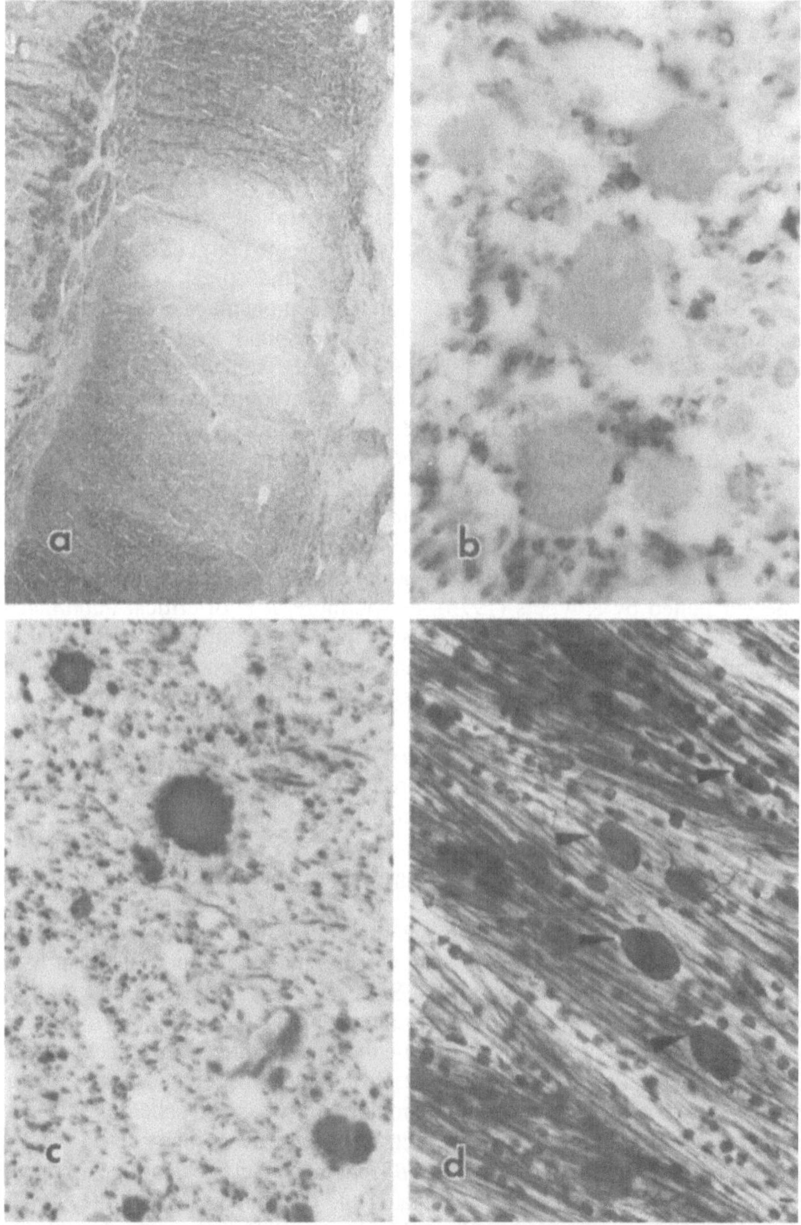

Fig. 4.15 **a** Corticospinal tract degeneration in the internal capsule localized at the third quadrant of the posterior limb characterized by myelin loss and vacuolation. (Klüver–Barrera stain. ×10.) **b** Large axonal spheroids (70–80 μm in diameter) trapped at the internal capsule admixed with relatively preserved myelinated axons. (Klüver-Barrera stain. ×250.) **c** Axonal spheroids in the internal capsule showing a strong immunoreactivity for phosphorylated neurofilaments (200 kD). (×200.) **d** Wet formalin-fixed internal capsule sectioned along the axonal axis with a vibratome and stained with Bodian stain, showing "tadpole" shaped spheroids with their tails (arrow heads) uniformly pointing towards the proximal. (×150.)

large motoneurons, which is potentially shared by some large sensory neurons, renders them susceptible to such an axonopathy. This hypothesis may be applicable to certain large neurons such as those in Clarke's columns or large reticular neurons of the brainstem and may further explain involvement of these systems in ALS, as described below. However, gamma motoneurons are also affected, as shown by illustrations of the motor innervation of muscle spindles in ALS (Swash and Fox 1974).

Anterolateral Column Degeneration

The relatively common observation of symmetrical and diffuse myelin pallor in the anterolateral columns in the ALS spinal cord, especially in the cervical and dorsal segment (Fig. 4.2), has puzzled neuropathologists for many years. This finding can now be explained on the basis of degeneration in the anteromedial brainstem pathway. The anteromedial pathway is poorly defined in man and the pathologic implication in ALS is uncertain. This pathway comprises many tracts including reticulospinal, vestibulospinal and interstitiospinal, all of which originate from brainstem reticular neurons.

The presence of pale basophilic globose cytoplasmic inclusions (Fig. 4.8a,b) was found to mark the distribution of certain motoneurons in the motor cortex, brainstem and spinal cord in three young male adults with ALS (age 31, 32 and 33 years) who died after relatively short durations of illness (7, 15 and 12 months respectively). The basophilic inclusions served as a convenient marker for the neurons which were selectively involved, presumably before their death and disappearance (Chou et al. 1988). The distribution of those inclusions in the brainstem reticular neurons was almost identical in all three of these cases. Among the median, medial and lateral groups of reticular formation in the brainstem, the median raphe or serotoninergic reticular neurons, were intact without basophilic inclusions, but both the medial and lateral reticular neurons displayed prominent basophilic inclusions readily detectable in H&E sections. Additionally, both the medial and lateral vestibular groups of neurons and, to a lesser extent, the neurons in the interstitial nucleus of Cajal, contained basophilic inclusions (Fig. 4.16). These ALS cases showed distinct myelin pallor in the anterolateral spinal funiculi, corresponding to the anteromedial pathway discussed above. Thus, both the anteromedial pathway and its neuronal origin appeared preferentially affected and provided us with a plausible explanation for the previously unexplained pallor of myelin in the anterolateral spinal funiculi in ALS. Neuronal loss in the brainstem reticular formation has seldom been mentioned in ALS. The reason for the absence of earlier recognition of degeneration of the brainstem reticular neurons may lie in the difficulty in quantitation and localization of reticular neurons. The correlation of neuronal degeneration in the reticular formation and degeneration of its tract becomes feasible when such affected neurons display the rather specific change of intracytoplasmic basophilic inclusions (Fig. 4.7a,b). Loss of reticular neurons and gliosis in the pontine reticular formation, as well as of the periaqueductal gray (interstitial nucleus of Cajal) has been noted in two sporadic ALS cases; these spinal cords showed severe and diffuse degeneration of the anterolateral funiculi (Hayashi et al. 1986). Hyaline conglomerate neurofilamentous inclusions have also been reported in bulbar reticular neurons from a case of sporadic ALS (Kondo et al. 1986). Similarly a marked loss of neurons and astrogliosis were described in the pontine reticular formation in a case of familial ALS by Tabuchi et al. (1983).

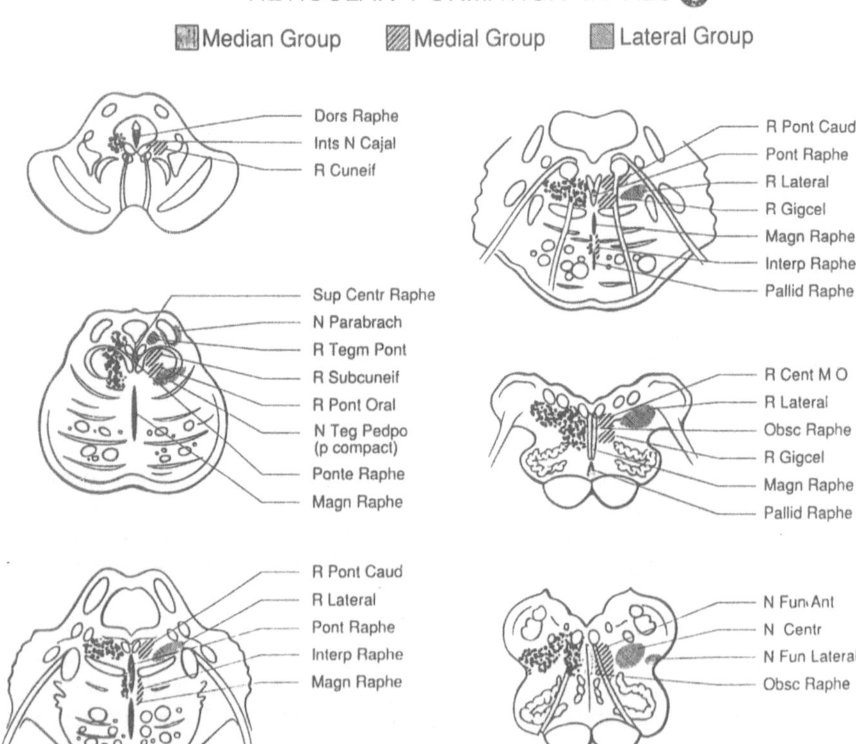

Fig. 4.16. Diagrammatic representation for distribution of basophilic globose inclusion (BGI) in bulbar reticular neurons in three ALS cases, sparing of the median group.

Spinocerebellar and Clarke's Column Degeneration

Degeneration of the spinocerebellar tracts and Clarke's columns has been described and perhaps overemphasized in the past as one of the characteristic pathological features unique to familial ALS. Earlier reports on the pathological findings in familial ALS (Hirano et al. 1967; Metcalf and Hirano 1971; Takahashi et al. 1972; Tanaka et al. 1980, 1984; Engel et al. 1959) contended that spinocerebellar tract, posterior column and motor system degeneration was a unique patho-anatomical characterization of familial ALS that did not occur in sporadic ALS. However, many cases of sporadic ALS with spinocerebellar degeneration have since been reported (Brownell et al. 1970; Hughes and Jerrome 1971; Swash et al. 1988; Page et al. 1977; Averback and Crocker 1981; 1982; Takasu et al. 1985), some with Clarke's column involvement (Hayashi et al. 1986; Okamoto et al. 1988). In 12 sporadic ALS cases, Averback and Crocker (1982) found loss of over one-third of the neurons in Clarke's column. A recent study with a surface marker selective for neurons in the Clarke's column has demonstrated marked loss of spinocerebellar neurons in five sporadic ALS cases (Williams et al. 1990).

Degeneration of the posterior spinocerebellar tracts has been related to a primary lesion of Clarke's column as best exemplified in Friedreich's ataxia. In ALS also, however, neurons in Clarke's columns are involved, and this probably accounts for the degeneration of the posterior spinocerebellar tract degeneration (Cavanagh 1979) found in many cases of ALS (Swash et al. 1988). Furthermore, many pathognomonic microscopic findings in lower motoneurons in ALS, e.g. proximal axonal swellings, axonal globules, Bunina bodies and neuronal loss, have also been reported in Clarke's column in sporadic ALS (Averback and Crocker 1981, Okamoto et al. 1988). Thus, it is fair to include spinocerebellar degeneration as a late concomitant of the motoneuron system degeneration in sporadic ALS. In familial ALS, the phenotypic expression of spinocerebellar degeneration may be more marked in the early stage (Hayashi and Kato 1989).

Posterior Column Degeneration

Sensory complaints in ALS patients are not uncommon (Mulder 1975), but are usually not accompanied by corresponding histopathologic findings in the sensory tracts. Posterior column degeneration, especially of the fasciculus gracilis, is common with aging. It has been estimated that approximately 5% of autopsied spinal cords from a population older than 65 years show posterior column degeneration. Posterior column degeneration in sporadic ALS has seldom been described in recent papers, probably because its specificity might be doubted, although many earlier reports noted the presence of such degeneration (Engel et al. 1959). The involvement of the middle root zone of Flechsig in the lumbar cord was once considered characteristic for posterior column involvement in familial ALS, but this finding simply means that the long and large nerve fibers for proprioception are not entirely spared in ALS. As expected, in many review articles on the pathology of ALS, one finds descriptions of posterior column degeneration (Lawyer and Netsky 1953; Brownell et al. 1970; Castaigne et al. 1972; Chou 1979; Malamud 1968). These findings, however, were dismissed as coincidental rather than unusual variants of ALS by some of those authors. A recent morphometric study tends to suggest otherwise. Kawamura et al. (1981) described a marked reduction in large cyton populations (54%) in lumbar (L5) spinal ganglion cells in five sporadic cases of ALS. The findings were interpreted as corresponding to the posterior column and spinocerebellar tract degeneration that has been reported in both sporadic and familial ALS. The organization and function of the dorsal columns has recently been redefined. According to that scheme, the dorsal columns have a major role in certain motor controls by transferring peripheral inputs to the motor cortex (Davidoff 1989). Hence, one can no longer completely dissociate the dorsal columns from the motoneuron system.

Basal Ganglia Degeneration

Mild diffuse degeneration is said to be common in the basal ganglia in ALS, including the thalamus and substantia nigra (Comant & Marie 1958). Castaigne et al. (1972) described basal ganglia involvement in 16 of 19 atypical ALS cases though extrapyramidal signs were absent clinically. In rare instances, basal ganglia lesions with corresponding extrapyramidal signs have been described in sporadic ALS cases. Bertrand and van Bogaert (1925) described diffuse involvement of both the cerebral cortex and basal ganglia. Involvement of thalamus has been occasionally mentioned (van Bogaert et al. 1965; Kosaka and Mehrein 1978) as well as substantia nigra (Castaigne et al. 1972; Bonduell 1975; Serratrice et al. 1983). With the Marchi technique, which is more sensitive in picking up degenerated fibers, Smith (1960) noted degenerated fibers in the thalamus, the subthalamus, the globus pallidus, the substantia nigra, the prerubal field, the semi-aqueductal gray, the superior colliculus, and the brainstem reticular formation. Thus, Marchi staining may serve as a useful tool for studying subtle involvement of the lateral and anteromedial brainstem pathways. In keeping with the concept of "total" ALS (Hayashi and Kato 1989), to be discussed below, it is of paramount importance to recognize degeneration of the neuronal groups from which the affected brainstem pathways arise. These rare ALS syndromes with both thalamic and substantia nigra involvement have sometimes been regarded as a separate disease from ALS (van Bogaert et al. 1965). The nosological and semiological position of ALS cases with involvement of basal ganglia and other systems has not yet been resolved. Those motoneuron diseases that occur in association with olivopontocerebellar, striatonigral or nigro-spino-dentate degenerations have, for the present, been conveniently classified among the group of multiple system diseases (Oppenheimer 1984).

Pathology of Atypical ALS and other MND

The characteristic pathologic features in each of the three major groups of atypical ALS (i.e. familial ALS, ALS in Guam or in the Kii peninsula of Japan, and Strümpell's familial spastic paraplegia) will be briefly discussed.

Familial ALS. The inheritance patterns of familial ALS have been suggestive of mainly autosomal-dominant transmission. Three different patterns of histopathological heterogeneity are known (Horton et al. 1976), but these are not at variance from those in sporadic ALS. They consist of involvement of: anterior horn cells with pyramidal tracts, of anterior horn cells without lesions in the pyramidal tracts, and of such motor system changes plus involvement of the spinocerebellar tracts, Clarke's columns and posterior columns.

In addition to the characteristic features of sporadic ALS, Engel et al. (1959) described a unique pathological feature in three members of an autosomal-dominant familial ALS syndrome. It was characterised by demyelination in the mid-root zones of the posterior spinal columns. The pattern of the degeneration was similar to that in tabes dorsalis except that no disturbance of proprioception was detected in those patients. The spinocerebellar tracts were also involved. This clinically silent, ascending degeneration is by no means the characteristic or essential pathological concomitant of familial ALS, since many familial ALS cases later reported did not show this ascending degeneration. It should again be emphasized that the posterior column degeneration may occur as a normal aging process and, also, as a common finding in patients with cachexia. Furthermore, in compression neuropathy of the posterior spinal roots, large calibre proprioceptive fibers are preferentially involved, producing mid-root zone degeneration. In addition, posterior column involvement has been described in many sporadic ALS cases (one case by Lawyer and Netsky 1953; three cases by Brownell et al. 1970, and three cases by Castaigne et al. 1972). Among our 82 cases, there were three sporadic ALS cases which showed mid-root zone posterior column degeneration; this was considered a complication rather than a patho-anatomic concomitant of ALS (Chou 1979). Combined posterior column and spinocerebellar tract degeneration has also been described in sporadic ALS cases (one case by Hughes and Jerrome 1971; two cases by Brownell et al. 1970; and one case by Page et al. 1977). These findings may be prominent in the absence of clear-cut pyramidal tract degeneration (Power et al. 1974). At any rate, the ascending tract degeneration in ALS, if it exists, should not be severe and should be considered a minor histopathological concomitant, otherwise it may be difficult to differentiate ALS from other heredofamilial motoneuron diseases such as spastic familial paraplegia, Charcot–Marie–Tooth disease or Friedreich's ataxia.

Western Pacific ALS. Neurofibrillary tangles (NFT) in both brain and spinal cord, unassociated with senile plaques, were described as characteristic features in 70 Guam ALS patients studied (Hirano 1973; Kurland and Brody 1975). These cells were compared with 90 sporadic ALS cases in New York City in which NFT were either completely absent, or present in negligible number. The neurons affected were not necessarily those of the motor system and included the hippocampal gyrus, amygdaloid nucleus, hypothalamus, substantia innominata, and locus coeruleus. Similar findings were reported in ALS in the Kii peninsula in Japan (Shiraki and Yase 1975). NFT unassociated with senile plaques was also the characteristic feature in the Guam parkinsonism-dementia (PD) complex (Hirano 1973). A control study of 69 members of the Guamanian Chamorro people without dementia, ALS or parkinsonism by Anderson et al. (1979) disclosed that NFT occurred at a very high rate and at an early age in this population and suggested that while a causal relationship between NFT and Guamanian ALS or PD exists, it may not be direct. By comparing histopathological findings in 20 ALS cases from Western Australia and those among 22 Guamanian ALS cases, Tan et al. (1984) concluded that NFTs are not a feature of ALS.

Familial Spastic Paraplegia (Strümpell's Disease). What are the clinicopathological characteristics which differentiate direct from the indirect corticospinal tract degeneration? This hypothetical question is asked in order to address the confusing issues and features sometimes encountered in ALS, e.g. primary lateral sclerosis without amyotrophy (Younger et al. 1988) or familial ALS without lateral sclerosis (Power et al. 1974). The issue is best discussed and explained by presenting a case of the "pure" form of familial spastic paraplegia (FSP) or Strümpell's disease and by corroborating these findings with the clinicopathological features of FSP previously reported.

Case Report

Clinical history. A 63-year-old woman who died after a one-week history of "flu", had first been hospitalized at the age of 45 years for evaluation of a slowly progressive decline in the strength and control of her lower legs beginning in her teens. By age 32 years, she was confined to a wheelchair but, despite the lower extremity disability, the use of her hands and arms was not impaired. She was able to play the piano, write letters, and knit and kept her cheerful disposition. In her early 40s, however, she became increasingly taciturn, withdrawn and irritable, probably because she developed increasing problems with faecal incontinence, became obsessed with "germs" and "contaminations", and developed compulsive hand washing. By this time, there was virtually complete paralysis of her legs, but she continued to use her hands and arms with remarkably good facility. Except for moderate dorsal kyphoscoliosis and distal muscle atrophy in her legs, the general examination was unremarkable. Neurological examination revealed normal cranial nerve functions, but slurred speech. She was unable to stand and the legs showed bilateral heel cord shortening and moderate equinovarus deformity. In contrast, strength and coordination of the upper extremities was intact. The plantar responses were bilaterally extensor; the deep tendon reflexes were hypoactive in the upper extremities, absent in the ankles and bilaterally hypoactive in the knees. Sensory examination revealed loss of vibration sense below the iliac crests, but sensation was otherwise intact. Her second hospitalization 11 years later, at the age of 56, was for drainage of large abscesses in her buttocks. At the time, in addition to markedly hyperactive knee jerks, urinary incontinence and spotty hypesthesia were recorded. The third hospitalization a year later was for recurrent dysphagia and, for the first time, moderate spasticity and hyperreflexia in both upper extremities were noted. Her final hospitalization at age 63, was for a week-long history of "flu" with lethargy and she died 2 hours after admission.

At least three other members including her brother, father and paternal uncle suffered from a similar disease with onset in the teens in three and in late middle age in the fourth. Each had a normal life span and died from unrelated illness.

Brain autopsy. The brain (1100 g) showed mild generalized cortical atrophy and ventricular dilatation. An old superficial cortical contusion (1 cm across) was noted over the left inferior temporal gyrus. The spinal cord appeared markedly thin and showed mild grayish discoloration of the spinal roots, especially of the posterior roots. Microscopically, a striking pattern of degeneration in corticospinal tracts and dorsal columns was seen (Fig. 4.17). Corticospinal tract degeneration was noticeable below the medulla and reached its maximum intensity at the lower thoracic level and lost its bulk at the lumbar levels, whereas, the dorsal column degeneration was slight at the lower lumbar level, pronounced at upper thoracic and cervical levels and involved mainly the fasciculus gracilis with relative sparing of the cuneatus. The dorsal column degeneration was slight at the lower lumbar level, pronounced at upper thoracic and cervical levels and involved mainly the fasciculus gracilis with relative sparing of the cuneatus. Posterior spinocerebellar tracts were equivocally degenerated but loss of neurons in Clarke's columns was rather distinct. In both corticospinal tracts and dorsal columns large axons were preferentially involved and lost. Anterior horn cells, especially of the lateral groups in the lumbar cord, were well preserved (Fig. 4.18a) despite more than a 40-year history of muscle weakness and atrophy of the lower limbs. Very large proximal axonal swellings up to 90 μm in diameter were seen near the anterior motoneurons and the latter showed central chromatolysis (Fig. 4.18b), however, cytoplasmic neuronal inclusions were not detected.

This case illustrates clinicopathological features typical of FSP but distinctively different from classic ALS in that there was extremely slow progress in the disease course; only the lower limbs were involved; there was severe spasticity; muscle weakness and atrophy was not marked; corticospinal tract involvement was limited to a level caudal to the lower brainstem; "dying back" or distal axonopathy of long and thick nerve fibers, including those of the funiculus gracilis, was a major feature (surprisingly) and there was good preservation of motoneurons despite distal arroploy and weakness. Over 200 cases of familial spastic paraplegia have been reported; of those only 11 autopsied cases have been accepted as a "pure" form of familial spastic paraplegia (Behan and Maia 1974; Harding 1983).

Strümpell (1880) first described two brothers presenting in middle age with muscle weakness and spasticity that preferentially involved the lower extremities. Pathological findings in one of the sibs and in another unrelated, but similar case were described by Strümpell (1886, 1904), showing degeneration confined to the corticospinal tracts and funiculus gracilis of the spinal cord with minor involvements of the spinocerebellar tracts at the thoracic and cervical levels. In contrast to familial ALS, neurons in the motor cortex and pyramidal tracts in the internal capsules, cerebral peduncles in the midbrain, and bulbar pyramids were spared. This unique pathological finding was soon substantiated by Schwartz and Liu (1956) who identified 7 similar cases in the literature, added 2 cases of their own, reviewed the pathological findings in all 9 cases, and concluded that the disease described by Strümpell was a different entity. This conclusion was strongly supported by Behan and Maia (1974) who added 2 cases of their own and summarized the pathologic features in FSP as consisting of corticospinal tract degeneration from the medulla pyramids caudally, the funiculus gracilis degeneration increasing cephalad, and autosomal inheritance, usually dominant. The two tracts

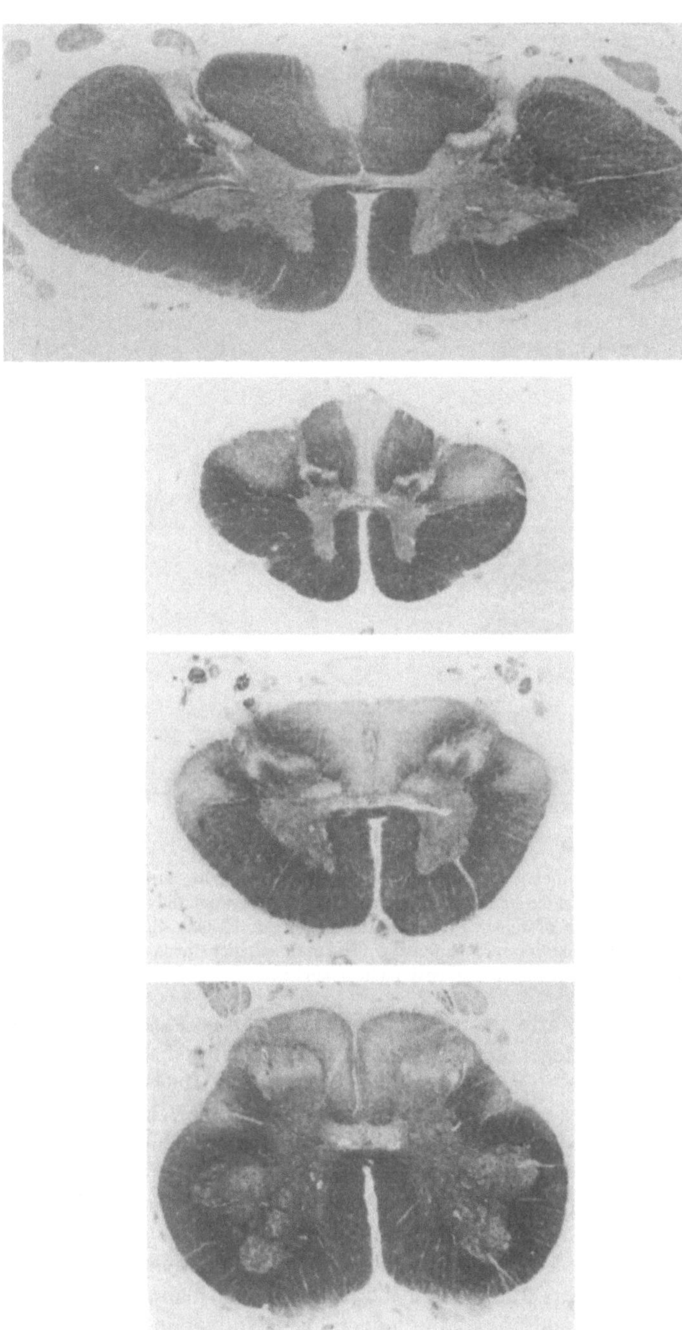

Fig. 4.17. Spinal cord sections from a patient with Strümpell's disease showing degeneration of gracile tract and indirect corticospinal tracts.

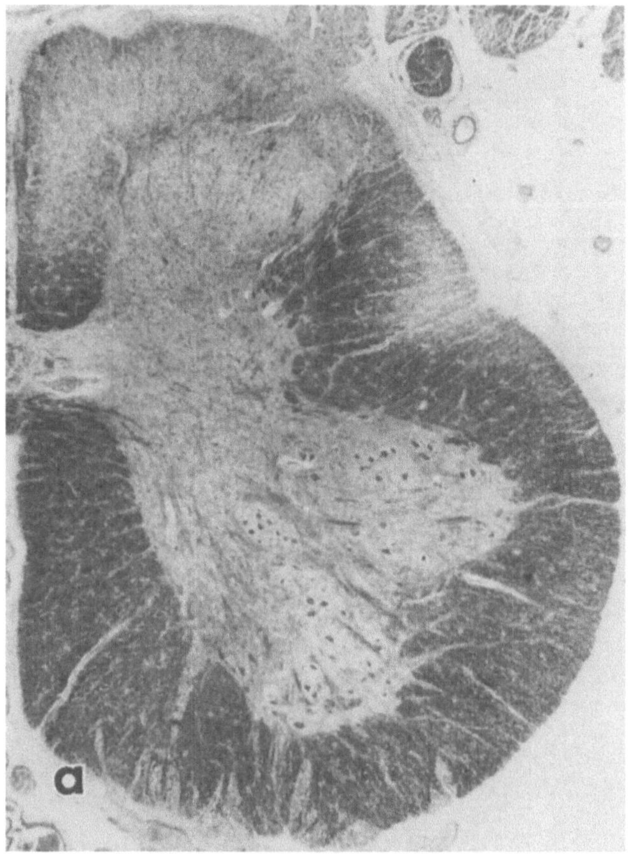

Fig. 4.18. **a** Higher magnification of the lumbar spinal cord showing well preserved anterior horn motoneurons, indicating that the indirect corticospinal tract degeneration did not affect the motoneurons despite the 15-year history of spasticity and muscle weakness. (Klüver–Barrera stain. ×20.) **b** (*see opposite*)Well-preserved anterior motoneurons with occasional giant axonal spheroids (arrow head) suggestive of primary "axostasis" as the early pathogenetic event. (×150.)

which were constantly affected in FSP represent the longest and thickest in the CNS and the pattern of degeneration fits best for a "dying back" or "entral-distal axonopathy syndrome" (Cavanagh 1979; Thomas et al. 1984). Involvement of the peripheral neuron does not occur in the classical syndrome. The clinicopathological distinction between ALS and FSP may be due to involvement of different cortico-spinal tracts, i.e., direct and indirect. As aforementioned, the indirect corticospinal tracts originate from large neurons of the red nuclei and the lateral bulbar reticular formation, and degeneration selective to this tract will show no abnormality in the direct corticospinal tracts, including those above the medulla. Since the indirect corticospinal tract fibers terminate at the posterolateral interneurons, degeneration selective to this tract will not readily induce transneuronal degeneration of anterior motoneurons, so that amyotrophy can ensue only as a result of disuse. Furthermore, since the indirect corticospinal tracts control reflex and flexion movements of distal muscles, especially of the legs, degeneration selective to this tract will cause spasticity and hyperreflexia with little muscle weakness and atrophy. However, the presence of large proximal axonal swellings and central chromatolysis in FSP

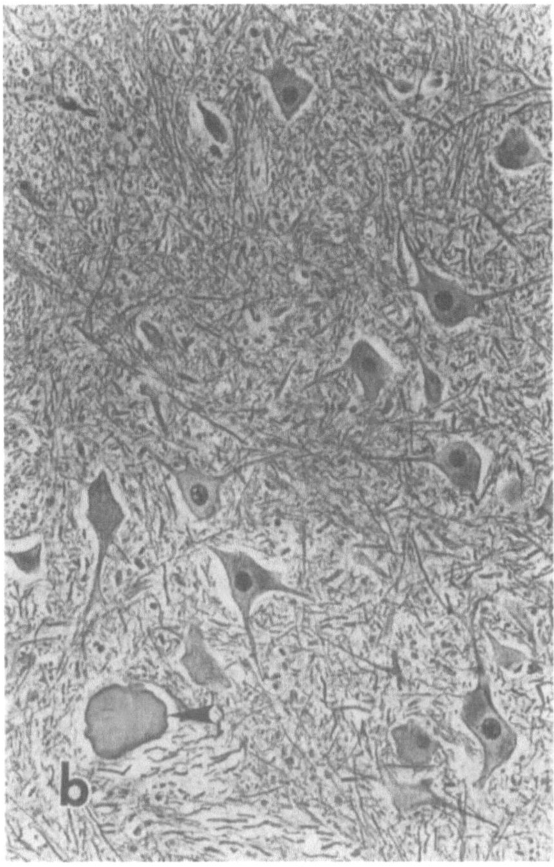

Fig. 4.18b

suggests "axostasis" as a common underlying pathogenic mechanism in both ALS and FSP, although intracytoplasmic inclusions have never been described in anterior horn motoneurons in FSP. Of interest in this regard is the description by Hirano et al. (1976) of the neuropathological findings in a case of neurolathyrism with more than 30-year history of spastic paraparesis after ingestion of *Lathyrus sativus*; degeneration of both pyramidal and gracile funiculi in the spinal cord and intracytoplasmic hyaline inclusions in relatively well-preserved motoneurons were described.

"Total" ALS Degeneration

In the vast majority of ALS patients, the terminal point is respiratory failure secondary to complete respiratory muscle paralysis. With the advent of modern technology which greatly improved the quality of the life-support apparatus including the respirator, ALS patients can now live beyond the previous terminal point of respiratory failure. Both the clinical and pathological features of those ALS patients who lived beyond the previous terminal point are unexpectedly different from the classical findings. Four patients with sporadic ALS whose lives were sustained with respirators far beyond the point of total voluntary muscle paralysis were described by Hayashi and Kato (1989). Clinically, these patients presented the "totally locked-in state" and were considered to represent the ultimate complex of clinical manifestations of ALS. In these patients, oculomotor muscles and sphincter muscles, previously considered to be spared in ALS, were markedly involved. The ophthalmoplegia was of supranuclear type and developed 9 months to 4 years after the development of respiratory failure. Autopsies of two patients who died 5 and 4.5 years after the onset of ALS revealed, in addition to the histopathological findings of classical motor system involvement, degeneration of the whole anterolateral funiculi in the spinal cord, including the spinocerebellar tracts with neuronal loss and gliosis in Clarke's columns, and of the middle root zones in the posterior columns, thus mimicking the changes of familial ALS. Of great interest, was selective degeneration of areas in the brainstem reticular formation with neuronal loss and gliosis in the periaqueductal gray, red nucleus, and of midbrain and pontine reticular neurons. These findings have seldom been described in typical ALS but are pertinent to the emerging concept of Kuypers' motor system as discussed above. The neuronal origins for both anteromedial brainstem pathway (i.e. periaqueductal neurons, interstitial nucleus of Cajal, and vestibular neurons) and the lateral brainstem pathway (i.e. red nucleus and brainstem reticular formation) were virtually destroyed. Furthermore, neuronal loss and gliosis were also marked in the globus pallidus, subthalamus, thalamus, substantia nigra, cerebellar dentate; unusual sites of involvement in typical ALS. By adding 3 previously reported cases of "total" ALS from Japan (Tabuchi et al. 1983; Takasu et al. 1985; Akiyama et al. 1987; Hayashi and Kato 1989), the authors have concluded that the previously unrecognized areas of involvement in the "total" ALS represent the "parapyramidal tract" (an archaic motor system) which were affected when the principal pyramidal pathways were completely destroyed in the advanced stage of ALS. This parapyramidal system is, according to Zülich (1975), situated in and arises from the mesencephalon and more caudal brainstem and may be simultaneously involved in the course of ALS (as in familial ALS) or subsequently involved, after the classical pyramidal systems are destroyed in the "totally" advanced stages of ALS.

Clinicopathologic correlations

In the foregoing discussion, an attempt was made to rationalize the validity of the classification into three clinical subtypes of ALS on a patho-anatomical basis. The involvement of different components and levels of motoneurons of the direct and indirect corticospinal tracts, and of the anteromedial brainstem pathway, as well as the lower motoneuron system, should lead to clinical variance in ALS. In 1963, MacKay reported a study of 70 ALS cases and clinically identified 23 cases of PMA, 11 cases of qualified PLS and 36 classic cases of ALS; at autopsy, those 70 cases were reclassified as 61 ALS, 8 PMA and 1 PLS cases. MacKay considered that the disorder represents a single degenerative disease with different clinical expressions. This frequency of distribution of the ALS subtypes was also reflected in the pathologically studied cases described by Lawyer and Netsky (1953) who included 51 ALS, 2 PMA and 1 probable PLS, and by Castaigne et al. (1972) who reviewed 42 typical ALS cases which included 5 chronic anterior poliomyelitis (i.e. PMA) and 2 PLS cases. Our pathological review of 82 cases disclosed 47 ALS, 29 PMA, 2 PLS, 2 Guamanian, and 3 familial ALS cases (Chou 1979).

Progressive lateral sclerosis (PLS) is characterized by involvement of corticospinal tracts and posterolateral interneurons with well-preserved spinal motoneurons, especially of the anterolateral group. This suggests involvement of the lateral brainstem pathway which terminates in the posterolateral part of the spinal intermediate zone and, unlike the direct corticospinal tracts, does not terminate monosynaptically in the anterior horn motoneurons. The latter neuron groups will be better preserved if the indirect corticospinal tracts are primarily involved as exemplified in Strümpell's disease, presenting with severe spasticity.

Progressive muscular atrophy (PMA) is characterized by severe loss of spinal motoneurons with intact or relatively well-preserved corticospinal tracts. The anteromedial brainstem pathway may be primarily involved in PMA, as in the bulbar form, with relative preservation of corticospinal tracts, and thus movements of the head, bulbar, axial and proximal limb muscles would be predominantly impaired.

If the direct corticospinal tracts are primarily involved, distal muscle weakness and atrophy, especially of the finger muscles and distal flexor muscles, should be more evident. Pathologically, the lateral motoneuron group should be more severely affected since many lateral anterior horn motoneurons receive monosynaptic terminations from direct corticospinal tract fiber projections. Thus, the clinical features in the subtypes of MND may be correlated with the degeneration patterns of both descending motor pathways.

Conclusions

While a certain personally biased view must inevitably be included in the foregoing discussion, emphasis has been placed on the following:

1. In the light of the recently elucidated anatomical organization of primate motor systems by Kuypers, the patho-anatomy of ALS/MND must be re-evaluated individually in ALS cases according to the clinical symptomatology referrable to the patterns of abnormality in the intermediate gray matter zones and anterior horns, as well as the patterns of descending tract degeneration.

2. Clinicopathologic correlation becomes feasible if one correlates the descending motor pathway degeneration with its clinical features as exemplified by the indirect corticospinal tract degeneration in Strümpell's disease or PLS.

3. Diffuse myelin pallor and degeneration of anterolateral funiculi in the spinal cord in ALS can be explained by involvement of the anteromedial brainstem pathway. Other tract degenerations are explained by the concept of "total" degeneration as found in ALS patients whose lives have been sustained beyond the natural "terminal" point.

4. The basic mechanism involved in MND may start in different neuronal groups (motor cortex, bulbar, and spinal) with their corresponding descending motor pathways, inducing both proximal and central-distal axonopathy in their related nerve fiber projections, comparable to "dying back" degeneration.

5. Such an axonopathy may be closely related to the formation of various intracytoplasmic and intra-axonal inclusions and the resultant "primary axostasis" in motoneurons of ALS patients of all clinical subtypes.

6. The three major types of neuronal inclusions (i.e. basophilic, hyaline and Bunina types) may represent different phases in the evolution of the ALS process. Conglomerated neurofilament inclusions are epiphenomena closely related to formation of those three inclusions.

7. The recent elucidation of "total" histopathological manifestations in long survivors maintained by sustained life-support strongly suggests that degeneration of both spinocerebellar tracts and dorsal column is a patho-anatomic concomitant of sporadic ALS. In familial ALS, the rate of ascending tract degeneration may be faster than that in sporadic ALS.

8. More data from immunohistochemistry and immunoelectron microscopy of the inclusion bodies in well-fixed specimens from ALS cases with relatively short clinical course, and short post-mortem delay, are needed to elucidate the pathogenesis of human ALS.

References

Akiyama KH, Tsutsumi H, Onoda N et al. (1987) An autopsy case of sporadic amyotrophic lateral sclerosis with sensory disturbances and ophthalmoplegia. Byori to Rinshou 5:921–927 (Japanese)

Anderson FH, Richardson EP Jr, Okazaki H et al. (1979) Neurofibrillary degeneration on Guam: frequency in Chamorros with no known neurological disease. Brain 102:65–77

Aran FA (1850) Recherches sur une Malade non encore décrite du système musculaire (atrophie musculaire progressive). Arch Gen Med 24:5–35, 172–214

Averback P, Crocker P (1981) Abnormal proximal axons of Clarke's columns in sporadic motor neuron disease. Can J Neurol Sci 8:173–175

Averback P, Crocker P (1982) Regular involvement of Clarke's nucleus in sporadic amyotrophic lateral sclerosis. Arch Neurol 39:155–156

Behan WMH, Maia M (1974) Strümpell's familial spastic paraplegia: genesis and neuropathology. J Neurol Neurosurg Psychiatr 37:8–20

Berry RG, Chambers RA, Duchett S, Terrers R (1969) Clinicopathologic study of juvenile amyotrophic lateral sclerosis. Neurology 19:312

Bertrand I, Van Bogaert L (1925) La sclérose latérale amyotrophique (anatomie pathologique). Rev Neurol 32:779–806

Bogaert L van, Martin L, Martin J (1965) Sclérose latérale amyotrophique avec dégénerescence spinocérébelleuse et délire épileptique. Acta Neurol Psychiatr Bel 65:845–872

Bonduell M (1975) Amyotrophic lateral sclerosis. In: Vinken RJ and Bruyn GW (eds) Handbook of clinical neurology 22(2):281–338

Brownell B, Oppenheimer DR, Hughes JT (1970) The central nervous system in motor neuron disease. J Neurol Neurosurg Psychiatr 33:338–357

Bunina TL (1962) On intracellular inclusions in familial amyotrophic lateral sclerosis. AH Neuropatol Psikhit Korsakov 62:1293–1299

Capenter S (1968) Proximal axonal enlargement in motor neuron diseases. Neurology 18:841–851

Carpenter S, Karpati G, Durham H (1988) Dendritic attrition precedes motor neuron death in amyotrophic lateral sclerosis (ALS) Neurology 38 (Suppl 1):252

Castaigne P, Lhermitte G, Cambier J et al. (1972) Étude neuropathologique de 61 observations de sclérose latérale amyotrophique: Discussion nosologique. Rev Neurol 127:401–414

Cavanagh JB (1979) The "dying back" process. Arch Pathol Lab Med 103:659–664

Charcot JM, Joffroy A (1869) Deux cas d'atrophie musculaire progressive avec lésions de la substance grise et des faisceaux antérolatéraux de la moelle épinière. Arch Physiol (Paris) 2:354–367, 629–649, 744–760

Charcot JM, Marie P (1885) Deux nouveaux cas de la sclérose latérale amyotrophique suivis d'autopsie. Arch Neurol 10:1–35, 168–186

Chou SM, Hartmann HA (1964) Axonal lesions and waltzing syndrome after IDPN administration in rats. With a concept: "Axostasis". Acta Neuropathol 3:428–450

Chou SM, Hartmann, HA (1965) Electron microscopy of focal neuroaxonal lesions produced by β-β' iminodipropionitrile (IDPN) in rats. Acta Neuropathol 4:590–603

Chou SM, Martin JD, Gutrecht JA et al. (1970) Axonal balloons in subacute motor neuron disease. J Neuropathol Exp Neurol 29:141–142

Chou SM, Klein R (1972) Autoradiographic studies of protein turnover in motor neurons of IDPN-treated rats. Acta Neuropathol 22:183–189

Chou SM (1979) Pathognomy of intraneuronal inclusions in ALS: In: Tsubaki T, Toyokura Y (eds) Amyotrophic lateral sclerosis. University of Tokyo Press, Tokyo, pp 135–176

Chou SM, Kuzuhara S, Gibbs CJ Jr, Gajdusek DC (1980) Giant axonal spheroids along corticospinal tracts in a case of Guamanian ALS. J Neuropathol Exp Neurol 39:345
Chou SM (1983) Selective involvement of both upper and lower motoneurons in (β-β' iminodipropionitrile) intoxicated monkeys. J Neuropathol Exp Neurol 42:309
Chou SM (1987a) Immunocytochemical characterization of inclusions in motoneurons of subacute ALS. In: Tsubaki T, Yase Y (eds) International conference of amyotrophic lateral sclerosis, p 232
Chou SM (1987b) Immunoreactivity of neurofilament epitopes in motor neurons of subacute ALS. J Neuropathol Exp Neurol 46:375
Chou SM (1991) Neuropathology of upper motor neurons in ALS. Proc XIth Int Cong Neuropathol, 595–598
Chou SM, Huang TE (1988) Giant axonal spheroids in internal capsules of amyotrophic lateral sclerosis brains revisited. Ann Neurol 24:168
Chou SM, Norris FH (1993) Amyotrophic lateral sclerosis: lower motor neuron disease spreading to upper motor neurons. Muscle Nerve 16:864–869
Chou SM, Tan N, Kakulas BA (1988) Involvement of anteromedial brainstem pathway in ALS: its neuropathologic implication. Neurology 36(Suppl 1):252
Comant JM, Marie P (1958) Die Myatrophische Lateralsclerose. Handbuch Spec Path Anat Histol (Henke-Lubarsch), Springer, Berlin Gottingen Heidelberg, 13:2624–2692
Coté F, Collard J-F, Julian J-P (1993) Progressive neuronopathy in transgenic mice expressing human neurofilament heavy gene: a mouse model of amyotrophic lateral sclerosis. Cell 73:35–46
Davidoff RA (1989) The dorsal columns. Neurology 39:1377–1385
Davison C (1941) Amyotrophic lateral sclerosis: origin and extent of the upper motor neuron lesion. Arch Neurol Psychiatr 46:1039–1056
Dejerine J (1901) Anatomie des centres nerveux. Paris Rueff II(1): 128–137, 232–235
Delisle MB, Carpenter S (1984) Neurofibrillary axonal swellings in amyotrophic lateral sclerosis. J Neurol Sci 63:241–250
Dickson DW, Yen SH, Suzuki KI, Davies P, Garcia JH, Hirarro A (1986) Ballooned neurons in selective neurodegenerative disease contain phosphorylated neurofilament epitopes. Acta Neuropathol 71:216–223
Dyck RJ, Stevens JC, Mulder DW et al. (1975) Frequency of nerve fiber degeneration of peripheral motor and sensory neurons in amyotrophic lateral sclerosis: morphometry of deep and superficial peroneal nerve. Neurology 25:781–785
Engel WK, Kurland LT, Klatzo I (1959) An inherited disease similar to amyotrophic lateral sclerosis with a pattern of posterior column involvement. An intermediate form? Brain 82:203–220
Friedman AP, Freedman D (1950) Amyotrophic lateral sclerosis. J Neurol Ment Dis 111:1–11
Gajdusek DC (1985) Hypothesis on interference with axonal transport of neurofilament as a common pathogenetic mechanism in certain diseases of central nervous system. N Engl J Med 312:714–719
Gowers WR (1888) A manual of diseases of the nervous system, vol. 1. Churchill, London
Griffin JW, Hoffman PN, Clark AW et al. (1978) Slow axonal transport of neurofilament proteins: Impairment by β-β'-iminodipropionitrile administration. Science 202:633–635
Griffin JW, Watson DF (1988) Axonal transport in neurological disease. Ann Neurol 23:3–13
Hammer RP Jr, Tomiyasu U, Scheibel AB (1979) Degeneration of the human Betz cell due to amyotrophic lateral sclerosis. Exp Neurol 63:336–346
Harding AE (1983) Classification of the hereditary ataxias and paraplegias. Lancet ii:1151–1155
Hart MN, Cancilla PA, Frommes S, Hirano A (1977) Anterior horn cell degeneration and Bunina-type inclusions associated with dementia. Acta Neuropathol 38:225–228
Harvey DG, Torack RM, Rosenbaum HE (1979) Amyotrophic lateral sclerosis with ophthalmoplegia: a clinicopathologic study. Arch Neurol 36:615–617
Hayashi H, Nagashima K, Urano Y, Iwata M (1986) Spinocerebellar degeneration with prominent involvement of the motor neuron system: autopsy report of a sporadic case. Acta Neuropathol 70:82–85
Hayashi H, Kato S (1989) Total manifestations of amyotrophic lateral sclerosis: ALS in the totally locked-in state. J Neurol Sci 93:19–35
Hirano A (1965) Pathology of amyotrophic lateral sclerosis. In: Gajdusek DC, Gibbs CJ (eds) Slow, latent and temperate virus infection. NINDB, Washington DC, pp 23–37
Hirano A, Kurland LT, Sayre GP (1967) Familial amyotrophic lateral sclerosis. Arch Neurol 16:232–243
Hirano A (1973) Progress in the pathology of motor neuron diseases. In: Zimmerman H (ed) Progress in neuropathology, vol. 2. Grune and Stratton, New York, pp 181–215
Hirano A, Llena JF, Streifler M, Cohn DF (1976) Anterior horn cell changes in a case of neurolathyrism. Acta Neuropathol 35:277–283

Hirano A, Iwata M (1979) Pathology of motor neurons with special reference to amyotrophic lateral sclerosis and related diseases. In: Tsubaki T, Toyokura Y (eds) Amyotrophic lateral sclerosis. University of Tokyo Press, Tokyo, pp 107–133

Hirayama K, Tsubaki T, Toyokura Y, Okinaka S (1962) The representation of the pyramidal tract in the internal capsule and basis pedunculi. Neurology 12:337–342

Horton WA, Eldridge R, Brody JA (1976) Familial motor neuron disease: Evidence for at least three different types. Neurology 26:460–465

Hughes JT, Jerrome D (1971) Ultrastructure of anterior motor neurons in the Hirano–Kurland–Sayre type of combined system degeneration. J Neurol Sci 13:389–399

Inoue K, Hirano A (1979) Early pathological changes of amyotrophic lateral sclerosis: autopsy findings of a case of 10 months duration. Neurol Med (Tokyo) 11:448–455

Juergens SM, Kurland LT, Okazaki H, Mulder DW (1980) ALS in Rochester, Minnesota 1925–1977. Neurology 30:463–470

Kahler O, Pick L (1879) Über die progressiven Spinaler Amyotrophien. Atschr f Nervenh 5:169

Kamo H, Haebara H, Akiguchi I, Kameyama M, Kimura H (1983) Peculiar patchy astrocytosis in the precentral cortex of amyotrophic lateral sclerosis. Clin Neurol 23:974–981

Kamo H, Haebara H, Akiguchi I, Kameyama M, Kimura H, McGeer P (1987) A distinctive distribution of reactive astroglia in the precentral cortex in amyotrophic lateral sclerosis. Acta Neuropathol 74:33–38

Katagiri T, Kuzikai T, Nihei K, Honda K, Sasaki H, Polak M (1988) Immunocytochemical study of Onuf's nucleus in amyotrophic lateral sclerosis. Jpn J Med 17:23–28

Kato T, Katagiri T, Hirano A, Sasaki H, Arai S (1988) Sporadic lower motor neuron disease with Lewy body-like inclusions: a new subgroup? Acta Neuropathol 76:208–211

Kato T, Katagiri T, Hirano A, Kawanami T, Sasaki H (1989) Lewy body-like hyaline inclusions in sporadic motor neuron disease are ubiquitinated. Acta Neuropathol 77:391–396

Kawamura Y, Dyck PJ, Shimono M et al. (1981) Morphometric comparison of the vulnerability of peripheral motor and sensory neurons in amyotrophic lateral sclerosis. J Neuropathol Exp Neurol 40:667–675

Kojima H, Furuta Y, Fujita M, Fujioka Y, Nagashima K (1989) Onuf's motoneuron is resistant to poliovirus. J Neurol Sci 93:85–92

Kojwenikoff A (1883) Cas de sclérose amyotrophique. La dégénérescence de faisceaux pyramidaux se propegeant à travers tout l'éncéphale. Arch Neurol Sci 6:357

Kondo A, Iwaki T, Tateishi J, Kirimotok, Morimoto T, Oomura I (1986) Accumulation of neurofilaments in a sporadic case of amyotrophic lateral sclerosis. Jpn J Psychiat Neurol 40:677–684

Kosaka K, Mehrein P (1978) Myatrophische Lateralsklerose kombiniert mit Degeneration im Thalamus und der Substantia nigra. Acta Neuropathol 44:241–244

Kurisaki H, Ihara Y, Nukina N, Toyokura Y (1983) Immunocytochemical study of ALS spinal cords using an antiserum to 200 k peptide of neurofilament and antibodies to tubulin. Clin Neurol 23:1013

Kurland LT, Brody JA (1975) Amyotrophic lateral sclerosis Guam type. In: Vinken PJ, Bruyn GW (eds) Handbook of clinical neurology 22(2):339–351

Kuroda S, Kuyama K, Morioka E, Ohtsuki S, Nanba R (1986) Sporadic amyotrophic lateral sclerosis with intracytoplasmic eosinophilic inclusions. A case closely akin to familial ALS. Neurol Chir (Tokyo) 24:31–37

Kusaka H, Imdi T, Hashimoto S, Yamamoto T, Maya K, Yamasaki M (1988) Ultrastructural study of chromatolytic neurons in adult-onset sporadic cases of amyotrophic lateral sclerosis. Acta Neuropathol 75:533–528

Kuypers HGJM (1973) The anatomical organization of the descending pathways and their contribution to motor control especially in the primates. In: Desmedt JE (ed) New developments in electromyography and clinical neurophysiology. 3:38–68

Kuzuhara S, Chou SM (1981) Retrograde axonal transport of HRP in IDPN-induced axonopathy. J Neuropathol Exp Neurol 40:300

Lasek RJ (1986) Polymer sliding in axons. J Cell Sci (Suppl 5):161–179

Lawyer T Jr, Netsky MG (1953) Amyotrophic lateral sclerosis: a clinico-anatomic study of 53 cases. Arch Neurol 69:171–192

Leigh PN, Anderton BH, Dodson A, Gallo JM, Swash M , Power DM (1988) Ubiquitin deposits in anterior horn cells in motor neuron disease. Neurol Lett 93:197–203

Leigh PN, Dodson A, Swash M, Brion JP, Anderton BH (1989) Cytoskeletal abnormalities in motor neuron disease. Brain 112:521–535

Lowe J, Lennox G, Jefferson D et al. (1988) A filamentous inclusion body within anterior horn neurons in motor neuron disease defined by immunocytochemical localization of ubiquitin. Neurosci Lett 94:203–210

McHolm GB, Aguilar MJ, Norris FH (1984) Lipofuscin in amyotrophic lateral sclerosis. Arch Neurol 41:1187–1188
MacKay RP (1963) Course and prognosis in amyotrophic lateral sclerosis. Arch Neurol 8:117–127
Malamud N (1968) Amyotrophic lateral sclerosis. In: Minckley J (ed) Neuromuscular disease: pathology of the nervous system, vol. 1. pp 712–725
Mandybur TI, Nagpaul AS, Pappas Z, Niklowitz WJ (1977) Alzheimer neurofibrillary changes in subacute sclerosing encephalitis. Ann Neurol 1:103–107
Mannen T, Iwata M, Toyokura Y et al. (1977) Preservation of a certain motoneuron group of the sacral cord in amyotrophic lateral sclerosis: its clinical significance. J Neurol Neurosurg Psychiatr, 40:464–469
Mannetto V, Sternberger NH, Perry G, Sternberger LA, Gambetti P (1988) Phosphorylation of neurofilaments is altered in amyotrophic lateral sclerosis. J Neuropathol Exp Neurol 47:642–653
Metcalf CW, Hirano A (1971) Clinicopathological studies of a family with amyotrophic lateral sclerosis. Arch Neurol 24:518–523
Moss TH, Campbell MJ (1987) Atypical motor neuron disease with features of a multisystem degeneration: a non-familial case with prominent sensory involvement. Clin Neuropathol 6:55–60
Mulder DW (1975) Motor neuron disease. In: Dyck PJ et al. (eds) Peripheral neuropathy. Saunders, Philadelphia, pp 709–770
Mulder DW (1987) The clinical syndrome: what does it tell us of etiology? International Congress ALS, Kyoto Meeting, Abstract, p 33
Munoz DG, Greene C, Perl DP, Selkol DJ (1988) Accumulation of phosphorylated neurofilaments in anterior horn motoneurons of amyotrophic lateral sclerosis patients. J Neuropathol Exp Neurol 47:9–18
Murayama S, Ookawa Y, Nori H et al. (1989) Immunocytochemical and ultrastructural study of Lewy body-like hyaline inclusions in familial ALS. Acta Neuropathol 78:142–152
Murayama S, Mori H, Ihara Y, Bouldin TW, Suzuki K, Tomonaga M (1990) Immunocytochemical and ultrastructural studies of lower motor neurons in amyotrophic lateral sclerosis. Ann Neurol 27:137–148
Nelson JS, Prensky AL (1972) Sporadic juvenile amyotrophic lateral sclerosis. Arch Neurol 27:300–306
Oblinger MM (1988) Biochemical composition and dynamics of the axonal cytoskeleton in the corticospinal system of the adult hamster. Metabol Brain Dis 3:49–64
Oda M, Akagawa N, Tabuchi Y, Tanabe H (1978) A sporadic juvenile case of the amyotrophic lateral sclerosis with neuronal intracytoplasmic inclusions. Acta Neuropathol 44:211–216
Okamoto K, Hirai S, Marimatsu M, Ishida Y (1980) The Bunina bodies in amyotrophic lateral sclerosis. Neurol Med (Toyko) 13:133–141
Okamoto K, Yamazaki T, Yamaguchi H, Shooji M, Hirai S (1988) Pathology of Clarke's nucleus in sporadic amyotrophic lateral sclerosis. Clin Neurol 28:536–542.
Oppenheimer DR (1984) Diseases of the basal ganglia, cerebellum, and motor neurons. In: Adams JH, Corsellis JAN, Duchen LW (eds) Greenfield's neuropathology, 4th ed. Wiley, New York, pp 699–747
Page RW, Moskowicz RW, Nash RE, Roesmann U (1977) Lower motor neuron disease with spinocerebellar degeneration. Ann Neurol 2:524–527
Power JM, Horoupian DS, Shaumburg HH (1974) Documentation of a neurological disease in a Vermont family 90 years later. J Can Sci Neurol May:139–140
Reh TA, Redshaw JD, Bisby MA (1987) Axons of the pyramidal tract do not increase their transport of growth-associated proteins after axotomy. Mol Brain Res 2:1–6
Sasaki S, Kamei H, Yamane K, Murayama S (1988) Swellings of neuronal processes in motor neuron disease. Neurology 38:1114–1118
Sasaki S, Murayama S, Yamane K, Sakuma H, Takeishi M (1989a) Swellings of proximal axons in a case of motor neuron disease. Ann Neurol 25:520–522
Sasaki S, Yamane K, Sakuma H, Murayama S (1989b) Sporadic motor neuron disease with Lewy body-like hyaline inclusions. Acta Neuropathol 78:555–560
Schmidt ML, Carden MJ, Lee V M-Y, Trojanowski JQ (1987) Phosphate dependent and independent neurofilament epitopes in the axonal swelling of patients with motor neuron disease and controls. Lab Invest 56:282–294
Schochet SS, Hardman JM, Ludwig PP, Earle KM (1969) Intraneuronal conglomerates in sporadic motor neuron disease: light and electron microscopy. Arch Neurol 20:548–553
Schwartz GA, Liu C-N (1956) Hereditary (familial) spastic paraplegia. Arch Neurol Psychiatr 75:144–162
Serratrice GT, Toga M, Pellisier JF (1983) Chronic spinal muscular atrophy and pallidonigral degeneration: Report of a case. Neurology 33:306–310

Shiraki H, Yase Y (1975) Amyotrophic lateral sclerosis in Japan. In: Vinken PJ, Bruyn GW (eds) Handbook of clinical neurology. 22(2):353–419

Smith M (1960) Nerve fiber degeneration in the brain in amyotrophic lateral sclerosis. J Neurol Neurosurg Psychiatr 23:269–282

Sobue G, Sahashi K, Takahashi A et al. (1983) Degenerating compartment and functioning compartment of motor neuron loss. Neurology 33:654–657

Strümpell A (1880) Beiträge zur Pathologie der Ruckermarks. Arch Psychiatr Nervenkrank 10:676–717

Strümpell A (1886) Über eine bestimmte form der primaren kombinierten Systemerkrank ung des Rüchenmarks. Arch Psychiatr Nervenkrank 17:227–238

Strümpell A (1904) Die primare Seitenstrangsklerose (spastische Spinal-paralyse) Dtsch Z Nervenheilk 27:291–339

Swash M, Fox KP (1974) The pathology of the muscle spindle: effect of denervation J Neuro Sci 22:1–24

Swash M, Leader M, Brown A, Swetherlam KW (1986) Focal loss of anterior horn cells in the spinal cord in motor neuron disease. Brain 104: 939–952

Swash M, Scholtz CL, Vowles G, Aingram D (1988) Selective and asymmetric vulnerability of corticospinal and spinocerebellar tracts in motor neuron disease. J Neurol Neurosurg Psychiatr, 51:785–789

Tabuchi TK, Takahashi K, Tananka J (1983) Familal ALS with ophthalmoplegia. Clin Neurol 23:278–287

Takahashi K, Nakamura H, Okada E (1972) Hereditary amyotrophic lateral sclerosis. Arch Neurol 27:292–299

Takahata N, Yamaouchi T, Fukatsu R et al. (1976) Brain stem gliosis in a case with clinical manifestations of amyotrophic lateral sclerosis. Folio Psychiatr Neurol (Japan) 30:41–48

Takasu T, Mizutani T, Sakamaki S, Tsuchiya N, Kamei S, Kohzu H (1985) An autopsy case of sporadic amyotrophic lateral sclerosis with degeneration of the spinocerebellar tracts and the posterior column. Ann Rep Neurodeg Dis Res Committee in Japan, pp 94–100

Tan N, Kakulas BA, Masters CL et al. (1984) Observation on the clinical presentation and the neuropathological findings of ALS in Australia and Guam. In: ALS in Asia and Oceania, Proc 6th Asia and Oceanian Congress of Neurology, ALS Workshop, pp 31–40

Tanaka S, Yase Y, Yoshimasu H (1980) Familial amyotrophic lateral sclerosis. Ultrastructural study of intraneuronal hyaline inclusion material. Adv Neurol Sci 24:386–387

Tanaka S, Nakamura H, Tabuchi Y, Takahashi K (1984) Familial amyotrophic lateral sclerosis: features of multisystem degeneration. Acta Neuropathol 64:22–29

Thomas PK, Shaumburg HH, Spencer PS et al. (1984) Central distal axonopathy syndrome: newly recognized models of naturally occurring human degenerative diseases. Ann Neurol 15:313–314

Tomonaga M, Saito M, Yoshimura M, Shimada H, Tohgi H (1978) Ultrastructure of the Bunina bodies in anterior horn cells of amyotrophic lateral sclerosis. Acta Neuropathol 42:81–86

Troncoso JC, Price DL, Griffin JW, Parhad IM (1982) Neurofibrillary axonal pathology in aluminum intoxication. Ann Neurol 12:278–283

Tsujihata M, Hazama R, Ishii N et al. (1980) Ultrastructural localization of acetylcholine receptor at the motor endplate: myasthenia gravis and other neuromuscular diseases. Neurology 30:1203–1211

Tsukagoshi H, Yanagisawa N, Oguchi K et al. (1979) Morphometric quantification of the cervical limb motor cells in controls and in amyotrophic lateral sclerosis. J Neurol Sci 41:287–297

Weiner LP (1980) Possible role of androgen receptors in amyotrophic lateral sclerosis. Arch Neurol 37:129–131

Wiley RG, Blessing WW, Reiss DJ (1982) Suicide transport: destruction of neurons by retrograde transport of ricin, abrin, and modeccin. Science 216:889–890

Williams C, Kozlowski MA, Hinton DR, Miller CA (1990) Degeneration of spinocerebellar neurons in amyotrophic lateral sclerosis. Ann Neurol 27:215–225

Wohlfart G, Swank RL (1941) Pathology of amyotrophic lateral sclerosis: fiber analysis of the ventral roots and pyramidal tracts of the spinal cord. Arch Neurol Psychiatr 46:783–799

Xu Z, Cork LG, Griffin JW, Cleveland DW (1993) Increased expression of neurofilament subunit NF-L produces morphological alterations that resemble the pathology of human motor neuron disease. Cell 73:23–33

Younger DS, Chou SM, Hays AP et al. (1988) Primary lateral sclerosis: a clinical diagnosis re-emerges. Arch Neurol 45:1304

Zülich KJ (1975) Pyramidal and parapyramidal systems in man. In: Zülich KJH, Creutzfeldt O, Galbraith GC (eds) Cerebral localization. An Otfrid Foerster Symposium. Springer, Heidelberg New York, pp 32–47

5 The Pathology of Motor Neuron Disease

J.E. Martin and M. Swash

Introduction

In the descriptions of progressive muscular atrophy by Aran (1850) and Duchenne (1853) and of progressive bulbar palsy by Duchenne (1860) involvement of the muscular and not the nervous system was implicated. This issue was hotly debated (Cruveilhier 1853; Luys 1860), and an abnormality of the nervous system came to be regarded as the prime pathogenetic mechanism. Consistent central nervous system involvement in these disorders was described by Charcot and Joffroy (1869) and Charcot (1870). The term amyotrophic lateral sclerosis was introduced by Charcot (1874) following his demonstration of pyramidal tract lesions in cases of progressive muscular atrophy. Dejerine (1883) championed the now generally accepted view that progressive muscular atrophy, progressive bulbar palsy and amyotrophic lateral sclerosis are clinical variants of the same disorder: motor neuron disease (MND).

The cardinal pathological features of MND are loss of anterior horn cells, and of motor cells in the lower cranial nerve nuclei, and degeneration of the crossed and uncrossed corticospinal tracts. The classical descriptions of the pathology, based on autopsied cases, represent the end-stage of the disease (Holmes 1909; Bertrand and Van Bogaert 1925; Lawyer and Netsky 1953; Brownell et al. 1970; Castaigne et al. 1972). In addition to involvement of the motor system, there is loss of neurons in Clarke's column (Holmes 1909; Averback and Crocker 1982) and this is associated with degeneration of the spinocerebellar tracts (Averback and Crocker 1982; Williams et al. 1990). Although myelin is generally pale in the anterior and lateral parts of the cord, the posterior column myelin is usually normal, apart from a minor degree of pallor in the gracile columns. However, in familial cases, pallor of the posterior columns may be more prominent (Iwata and Hirano 1979). Several variant forms of MND are well recognised including an association with dementia and with parkinsonism (Feller et al. 1966; Boudouresques et al. 1967; Farmer and Allen 1969; Mitsuyama and Takamatsu 1971; Reed and Brody 1975; Horton et al. 1976; Hudson 1981; Mitsuyama 1984).

In most major series the presence of familial forms of MND and of atypical cases, either in clinical picture or neuropathological findings, raises the question of the relationship of classical MND to other neuronal degenerations (Lawyer and Netsky 1953; Brownell et al. 1970). Recent studies of the genetics of spinal muscular atrophies and MND, together with advances in the pathological study of MND and other disorders using immunohistochemical techniques, may help to clarify some of these questions.

General Autopsy and Neuropathological Findings

At autopsy the striking feature is often of profound muscular atrophy, this feature being the basis of Aran's and Duchenne's belief in a myopathic process being responsible for the condition (Aran 1850; Duchenne 1860). In earlier series the muscular atrophy was compounded by cachexia, particularly in cases of bulbar palsy, but this is less common with the advent of assisted feeding by nasogastric tube or by gastrostomy where appropriate. Death often results from respiratory failure, complicated in the majority of cases by the presence of bronchopneumonia (Charcot and Joffroy 1869; Puscariu and Lambrior 1906; Dagnelie and Cambier 1933; Lawyer and Netsky 1953; Carpenter 1968; Brownell et al. 1970; Hughes 1982; Averback and Crocker 1982). Aspiration of food or gastric contents into the respiratory tract is perceived as a clinical problem in the care of patients with MND, and whilst aspiration of small quantities of food may contribute to pulmonary infection or infarction (Lawyer and Netsky 1953), this is considered to result from recumbency (Hughes 1982). In fact, aspiration of food or gastric contents does not appear to be a direct cause of death in any major autopsy series. In two of the most completely reported general autopsy series additional findings included cerebral haemorrhage, atherosclerosis, thyroid adenomata, aortic aneurysm, trichinosis, duodenal ulcer, carcinoma of the colon, carcinoma of the kidney and myelomatosis (Lawyer and Netsky 1953; Brownell et al. 1970). These findings are consistent with coincidental pathological findings in an autopsy series in this age group (Hughes 1982, Henson and Urich 1982).

There is a striking absence of bedsores in patients dying with MND, despite the profound degree of incapacity prior to death (Charcot 1874; Forrester 1976; Fukukawa and Tokoyura 1978). It has been suggested that bedsores develop in patients with neurological disease due to vasomotor paralysis related to sympathetic nervous system dysfunction and that sympathetic vasomotor activity is spared in MND (Forrester 1976; Fukukawa and Tokoyura 1978). Preservation of sympathetic vasomotor tone corresponds with preservation of sympathetic neurons in the lateral horn of the spinal cord (Iwata and Hirano 1979).

Changes in skin collagen structure and in collagen and mucopolysaccharide content have been reported in MND (Fullmer et al. 1960; Ono et al. 1986). Other "systemic" structural alterations in MND include the association of abnormal liver function tests with the presence in liver biopsy specimens of swollen mitochondria and intramitochondrial inclusion bodies with a high copper content (Masui et al. 1985; Nakano et al. 1987).

Macroscopic abnormalities of the nervous system are well reported in MND. Cruveihier (1853) noted thinning of the anterior roots in MND in the celebrated case of Prosper Laconte, studied by Duchenne (described by Dejerine, 1883). Charcot and Joffroy (1869) and Dejerine himself (1883) confirmed anterior root atrophy occurring in cases of MND with both a clinical picture of progressive muscular atrophy and of bulbar palsy. In both these studies the thinning of the anterior roots is noted to be more pronounced in the cervical cord than in the lumbar region. Most authors agree that posterior roots are unaffected (Holmes 1909). However, there is some disagreement as to the appearance of peripheral nerves, some suggesting that there is no gross alteration (Lawyer and Netsky 1953; Hughes 1982), others claiming to observe atrophy (Dejerine 1883; Bertrand and Van Bogaert 1925). It is well recognised, however, that considerable loss of fibres may occur before atrophy becomes apparent to the naked eye (Wohlfart and Swank 1941).

The spinal cord in MND has been described as normal, showing generalised atrophy, or a chalkiness or greyness of the pyramidal tracts (Kojewnikoff 1883; Lawyer and Netsky 1953). Macroscopic changes in the texture of the white matter of the spinal cord are more readily appreciated after fixation. Some series report haemorrhages in the anterior horns of the spinal cord, but these are regarded as agonal (Lawyer and Netsky 1953). Flattening or shrinkage of the pyramids is well described (Lawyer and Netsky 1953).

Visible atrophy of the cerebral cortex has been reported by several authors, notably Kahler and Pick (1884), Bertrand and Van Bogaert (1925), Lawyer and Netsky (1953) and Brownell et al. (1970). The atrophy has been described as predominantly affecting the motor cortex, although generalised atrophy of the cerebral hemispheres has also been reported (Brownell et al. 1970).

A consistent feature of many autopsy series of patients dying with a clinical history of MND is the presence of cases found to have features of other disorders, particularly disseminated sclerosis, or of atypical forms on neuropathological examination (Lawyer and Netsky 1953; Brownell et al. 1970).

Motor Cortex

Loss of the giant pyramidal cells of Betz is accepted as a major pathological feature of MND (Fig. 5.1) and was taken as evidence linking motor cortex with pyramidal tracts in anatomical tracing studies (Kojewnikoff 1883; Holmes 1909; Hughes 1982), but the consistency of this finding has proved to be the subject of some debate in the literature. Charcot and Marie (1885) are credited with the first description of degeneration of large pyramidal cells in MND in the pre- (and post) central gyrus, and particularly the paracentral lobule. In single case reports and in larger series many authors describe the loss of giant pyramidal cells from lamina 5 of the cerebral cortex of the precentral gyrus (Rossi and Roussy 1906; Holmes 1909; Bertrand and Van Bogaert 1925; Lawyer and Netsky 1953; Smith 1960; Hughes 1982). Other authors suggest that there may not always be evidence of loss of Betz cells in typical cases with perhaps a shorter duration of illness than cases showing a loss of such cells (Puscariu and Lambrior 1906; Brownell et al. 1970). Davison (1941) however, stated that the length of illness or the age of the patient had no influence on the severity or extent of the pathological process. It has been suggested that changes in the motor cortex may be overlooked unless appropriate methods are employed for staining, especially when seeking supporting evidence for upper motor neuron degeneration such as subcortical white matter degeneration (Smith 1960; Hughes 1982). The presence or absence of Betz cell degeneration is relevant to discussions of the relative susceptibility of proximal or distal parts of the upper motor neuron (Davison 1941; Smith 1960; Brownell et al. 1970).

The presence of degeneration in areas of cerebral cortex other than the precentral gyrus has provided the opportunity for speculation as to the origin of the pyramidal tract and the true extent of the motor cortex in man (Holmes 1909; Smith 1960). In particular, degeneration of pyramidal cells of the postcentral gyrus has been noted

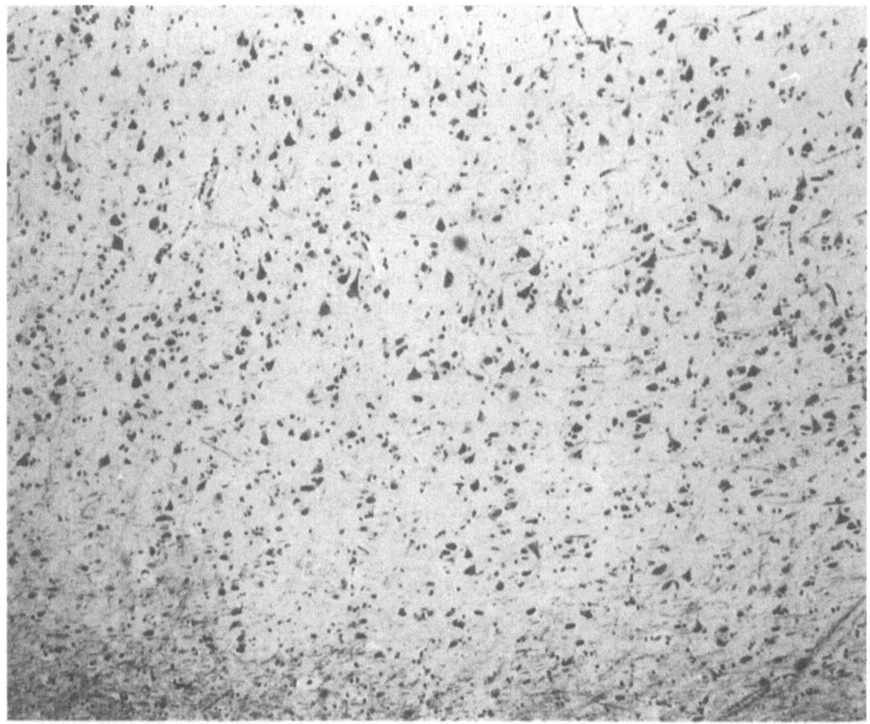

Fig. 5.1. Motor cortex from a patient dying with sporadic ALS showing loss of Betz cells and mild gliosis. (Luxol fast blue ×100.)

(Charcot and Marie 1885; Holmes 1909). However, it is generally accepted that the motor area may extend anteriorly and posteriorly of the precentral gyrus or motor cortex (Smith 1960), and changes have also been noted in the "pre motor" frontal cortex, and opercular areas (Bertrand and Van Bogaert 1925; Davison 1941; Friedman and Freedman 1950).

Reports of Betz cell degeneration in MND also comment on accompanying gliosis (Holmes 1909; Hughes 1982). A specific plaque-like distribution of reactive astrocytes in laminae 2, 3, 4 and 5 of the motor cortex has been described in MND, but not in other neurological disorders (Kamo et al. 1987). Takahata and colleagues (1976) have also described a case in which there was pronounced gliosis of the brain stem in a patient presenting with MND. These studies have suggested that there may be a primary abnormality of astroglia in MND.

Neuronal changes described include disappearance of the giant pyramidal cells, or shrinkage of these cells, pyknosis, displacement of the nucleus, incrustations, accumulation of pigment (lipofuscin), loss of dendrites, satellitosis and vacuolation with fat deposition (Holmes 1909; Bertrand and Van Bogaert 1925; Davison 1941; Lawyer and Netsky 1953; Castaigne et al. 1972). Chromatolysis, axonal change and neuronophagia are also claimed in some reports to be common features (Davison 1941; Lawyer and Netsky 1953), but other authors have stated that true chromatolysis is rare and may be related to agonal changes (Bertrand and Van Bogaert 1925) and that neuronophagia is unusual (Holmes 1909; Hughes 1982).

Other layers of the motor cortex may be affected; Bertrand and Van Bogaert (1925), Davison (1941) and Lawyer and Netsky (1953) described loss of small pyramidal cells in the motor cortex, as did Holmes (1909) who, in addition, commented on the preservation of neurons in the multiform layer (lamina 6).

Subcortical White Matter and Cortical Pyramidal Tracts

Following the demonstration of degeneration of the spinal pyramidal tracts in MND by Charcot (1874), several authors, including Kojewnikoff (1883), Charcot and Marie (1885), Probst (1898, 1903), Rossi and Roussy (1906) and Marie et al. (1923) showed degeneration of the pyramidal tract extending into the cerebral cortex, up to the subcortical white matter adjacent to the origin of upper motor neurons (Holmes 1909).

In several pathological studies, even in the presence of MND with predominantly lower motor neuron clinical signs ("progressive muscular atrophy"), there is usually evidence of pyramidal tract degeneration (Lawyer and Netsky 1953; Brownell et al. 1970).

Hirayama et al. (1962) described loss of large calibre fibres with associated demyelination of the corticospinal tract in the third quarter of the posterior limb of the internal capsule in MND. These latter findings are, however, at variance with the classical description of the pyramidal tract as being present solely in the anterior two thirds of the posterior limb of the internal capsule (Hirayama 1978). This finding, of degeneration in the posterior part of the posterior limb of the internal capsule, has been described in other large studies (Rossi and Roussy 1906; Bertrand and Van Bogaert 1925; Brownell et al. 1970).

Many studies have noted degeneration of fibres in the corpus callosum, chiefly in the middle ½–⅓. (Probst 1898, 1903; Spiller 1900, 1905; Holmes 1909; Smith 1960; Brownell et al. 1970; Hughes 1982). This fibre degeneration is thought to be related to degeneration of commissural fibres linking right and left motor areas (Brownell et al. 1970).

Corticospinal tract degeneration can be traced through the middle third of the crus in the midbrain, the corticospinal regions of the pons and the pyramids of the medulla into the spinal cord (Holmes 1909; Davison 1941; Brownell et al. 1970; Hughes 1982).

Changes have been noted in the white matter of the temporal lobe and in the long tracts of the occipitotemporal and frontothalamic region, suggesting to some authors that MND may be, whilst mainly affecting frontal and motor areas, a diffuse cortical disorder (Bertrand and Van Bogaert 1925). Smith (1960) and Hughes (1982), also noted degenerating fibres in frontal gyri, parietal, temporal and cingulate gyri.

Central Grey Matter and Basal Ganglia

Degenerating fibres and reactive gliosis have been described in several areas of the cerebrum other than those generally associated with the pyramidal tracts. Areas reported to show such degenerative features include the thalamus (Holmes 1909; Smith 1960; Brownell et al. 1970; Hughes 1982), globus pallidus (Smith 1960; Brownell et al. 1970; Hughes 1982), ansa and fasciculus lenticularis (Bertrand and Van Bogaert 1925; Smith 1960; Hughes 1982), field of Forel (Smith 1960; Hughes 1982) and hypothalamus (Smith 1960; Hughes 1982). Smith (1960) suggested that many of these degenerating fibres may originate in the globus pallidus and that the pallidal degeneration may contribute to the severity of the muscular paralysis in MND.

Neuronal changes including atrophy, chromatolysis, vacuolar or pigmentary degeneration have been documented in the basal ganglia, with astrocytosis or products of degeneration present in the globus pallidus, putamen and caudate nuclei (Bertrand and Van Bogaert 1925; Brownell et al. 1970). Other reports state that degeneration of the caudate or lenticular nuclei is not a feature of MND (Holmes 1909).

Degenerating fibres have been reported in the substantia nigra, superior colliculus and tegmentum in a small minority of cases of sporadic MND (Smith 1960; Brownell et al. 1970; Hughes 1982), whereas other series note little pathology in the substantia nigra (Kojewnikoff 1883; Bertrand and Van Bogaert 1925). Major degenerative features are seen in the substantia nigra and globus pallidus in the high incidence foci of MND on the island of Guam (Hirano et al. 1966). This is discussed more fully later.

Brainstem and Cranial Nerve Nuclei

Degeneration of the motor nuclei of the 5th, 7th, 9th, 10th, 11th and 12th cranial nerves is well documented in MND (Bertrand and Van Bogaert 1925; Davison 1941; Lawyer and Netsky 1953; Malamud 1968; Hughes 1982; Hartmann et al. 1989). In the facial nucleus the ventral and lateral parts of the nucleus are said to be more affected than the dorsal region, correlating with the relative sparing of the upper branch of the facial nerve (Holmes 1909). Degeneration of the nucleus ambiguus, which supplies motor fibres to cranial nerve nuclei 9, 10 and 11 to innervate the larynx, also occurs in up to 80% of patients with MND (Lawyer and Netsky 1953; Malamud 1968; Brownell et al. 1970; Hartmann et al. 1989). Degeneration of the paired nuclei may be symmetrical or asymmetrical (Lawyer and Netsky 1953; Brownell et al. 1970).

Cellular changes described in the motor nuclei of the brainstem include cell loss, gliosis, neuronal atrophy, cytoplasmic pallor, vacuolation and neuronophagia (Brownell et al. 1970; Hirano and Iwata 1979; Chou 1979) (Fig. 5.2).

The 3rd, 4th and 6th cranial nerve nuclei are described as unaffected in classical MND, even in later stages of the disease, so that ocular movements are usually spared (Holmes 1909; Bertrand and Van Bogaert 1925; Malamud 1968; Brownell et al. 1970; Castaigne et al. 1972; Le Bigot 1972; Iwata and Hirano 1979). Pathological involvement of the oculomotor nuclei has been described in MND without loss of oculomotor function (Carpenter 1968; Hughes 1982; Mann and Yates 1974), but morphometric analysis of fibre density in the spinal roots of cranial nerves 3, 4 and 6 have shown minor alterations only (Sobue et al. 1981a). Ophthalmoplegia in

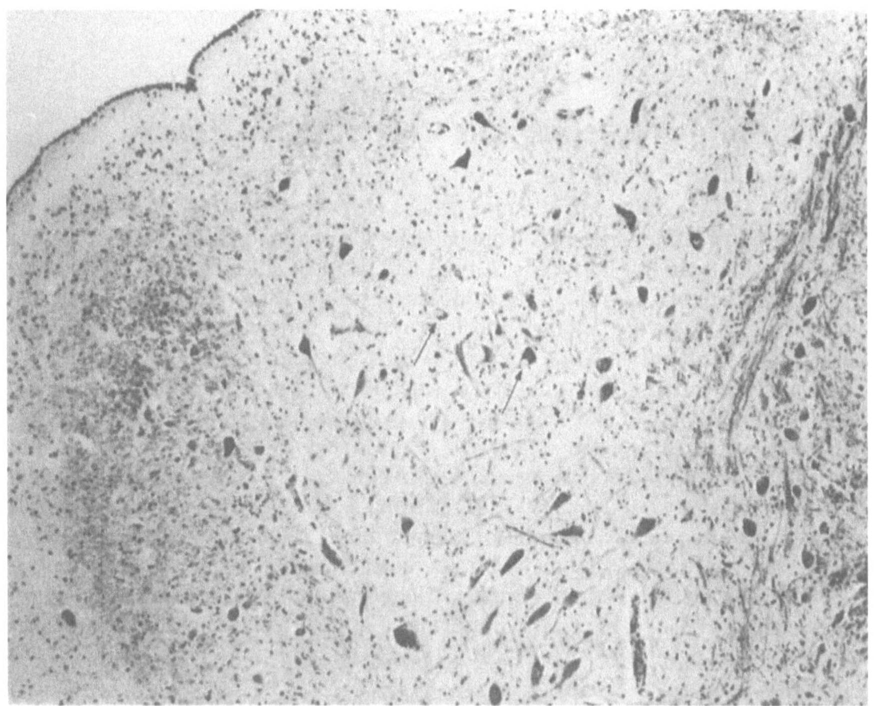

Fig. 5.2. Hypoglossal nucleus showing neuronal degeneration in ALS, including two vacuolated neurons (arrows). (Luxol fast blue ×150.)

association with loss of oculomotor neurons (Bertrand and Van Bogaert 1925; Harvey et al. 1979) has, however, been reported in MND. Reports of ocular changes detectable on electro-oculography (Leveille et al. 1982) and in patients maintained on respiratory support (Hayashi and Kato 1989) suggest that, whilst the motor neurons subserving oculomotor function are initially relatively resistant to the disease process, if the patient survives for a longer period these neurons may also become involved.

Sensory nuclei have received less attention than the motor nuclei in most pathological studies; however, Bertrand and Van Bogaert (1925) state that involvement of the sensory nuclei is as common as that of the motor. Holmes (1909) comments, however, that the sensory nuclei are spared.

The tractus solitarius is described as being spared in MND (Holmes 1909; Bertrand and Van Bogaert 1925), but the associated nucleus is reported to show chromatolytic changes in neurons (Bertrand and Van Bogaert 1925).

Degeneration is described in the vestibular nuclei and in the fibres of the medial longitudinal fasciculus (Holmes 1909; Bertrand and Van Bogaert 1925; Smith 1960). Deafness has been reported in association with juvenile MND, but not in the sporadic adult form of the disease (Wadia et al. 1987). Nystagmus has also been reported in MND (Kushner et al. 1984).

Degenerating fibres have been demonstrated in the reticular formation of the brainstem (Bertrand and Van Bogaert 1925; Smith 1960). The pontine nuclei and olives are spared in MND or show minimal changes only (Holmes 1909; Bertrand

and Van Bogaert 1925). Lawyer and Netsky (1953) describe degeneration in the external arcuate nucleus in 2 of 53 cases of MND.

Lower Motor Neurons and Corticospinal Tracts

A variety of lesions of the lower motor neuron have been described in MND, from atrophy with loss of dendrites (Cruveihier 1853; Charcot and Joffroy 1869) to total disappearance of anterior horn cells (Luys 1860). Loss of anterior horn cells is most prominent in the cervical region, in comparison to lumbar changes which may be minimal or equally severe (Charcot and Joffroy 1869; Kojewnikoff 1883; Rossi and Roussy 1906; Bertrand and Van Bogaert 1925; Castaigne et al. 1972) (Fig. 5.3).

Despite some variation in the degree to which levels of the cord are affected, most groups of motor neurons show degeneration, with the exception of the nucleus of Onufrowicz at the second sacral segment (Onuf 1889, 1890; Mannen et al. 1977; Iwata and Hirano 1978; Schroeder and Reske-Nielsen 1984; Hudson and Kiernan 1988). Kawamura and colleagues (1981) showed a selective vulnerability of alpha motor neurons with relative sparing of gamma motor neurons in MND. However, other studies suggest that gamma motor neurons are also lost in MND (Tsukagoshi et al. 1979). Clinical, electromyographic (Schwartz and Swash 1982) and pathological studies (Tsukagoshi et al. 1980; Swash 1980) show that involvement of the

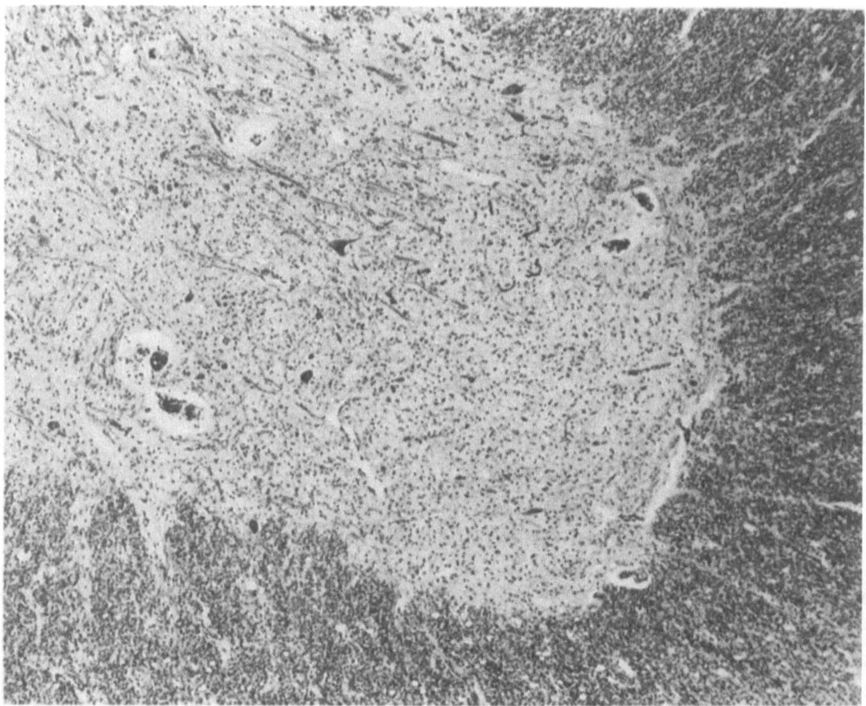

Fig. 5.3. Lumbar anterior horn in ALS showing loss of alpha motor neurons and gliosis. (Luxol fast blue ×60.)

corticospinal tracts is asymmetrical, and that loss of anterior horn cells may be diffuse or focal within individual spinal cord segments. In addition, the disease tends to involve anterior horn cells that innervate distal muscles in the extremities rather than those supplying proximal muscles. The loss of anterior horn cells is thought to proceed in a random chronological manner (Sobue et al. 1983; Swash et al. 1986) producing progressive functional deficit, despite the occurrence of axonal sprouting tending to provide some degree of compensation for the denervation atrophy (Wohlfart 1957; Sobue et al. 1983; Swash et al. 1986).

Cellular changes in lower motor neurons in MND vary, as stated above, from atrophy with loss of dendrites to total loss of anterior horn cells (Cruveihier 1853; Luys 1860; Nakano and Hirano 1987). In addition, the presence of increased amounts of lipofuscin is a feature of many reports (Puscariu and Lambrior 1906; Holmes 1909; Bertrand and Van Bogaert 1925; Lawyer and Netsky 1953; Hirano and Iwata 1979). The presence of chromatolysis is described in MND (Puscariu and Lambrior 1906; Holmes 1909; Bertrand and Van Bogaert 1925; Castaigne et al. 1972), but is regarded as a relatively unusual feature (Hirano and Iwata 1979). Several different histological types of intracellular inclusion have been described in the anterior horn in MND (Hirano et al. 1968; Hirano and Iwata 1979; Inoue and Hirano 1979; Hirano and Inoue 1980; Sasaki et al. 1982; Sasaki and Hirano 1983; Leigh et al. 1988) (Figs. 5.4–5.7). Eosinophilic cytoplasmic, intraneuronal inclusions – Bunina bodies, have been reported in familial, sporadic and Guam-type MND (Bunina 1962; Hirano 1965; Hirano and Iwata 1979). Hirano bodies are eosinophilic rod-shaped inclusions in the perikarya, dendrites or axons of neurons and whilst found in MND, they are not specific to this disorder (Chou 1979). Hyaline bodies consist of eosinophilic accumulations of neurofilaments, 10 nm in diameter, in the cell bodies or processes of motor neurons (Schochet et al. 1969; Hirano and Iwata 1979; Hirano et al. 1984a,b). Granulovacuolar cytoplasmic degeneration has been noted in motor neurons in MND, especially in the Guamanian form (Hirano et al. 1966). Lewy-like bodies have also been reported in familial MND (Takahashi et al. 1972). Neuronophagia and vacuolation are described in lower motor neurons in MND, but are unusual (Hirano and Iwata 1979). Although it has been suggested (Munoz et al. 1988; Manetto et al. 1988) that phosphorylated neurofilament epitopes were present in increased concentration in anterior horn cells in MND, Leigh et al. (1989) could not confirm this.

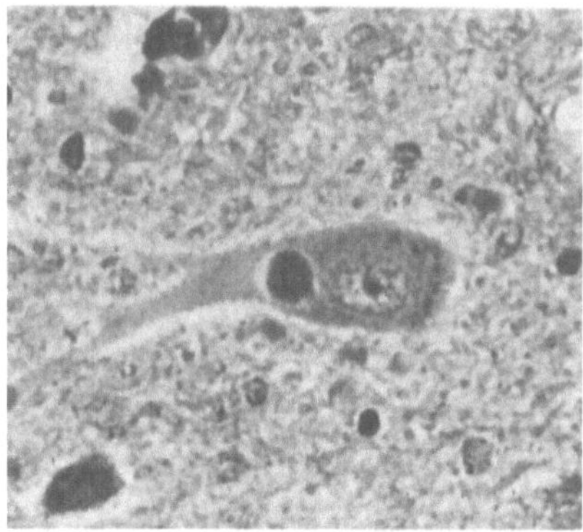

Fig. 5.4. Anterior horn cell in ALS showing a Lewy-like inclusion. (Haematoxylin and eosin ×150.)

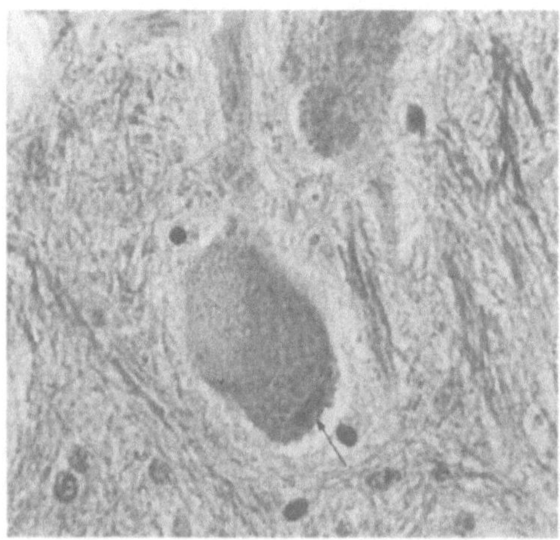

Fig. 5.5. Anterior horn cell in ALS showing a Bunina body (arrow). (Haematoxylin and eosin ×200.)

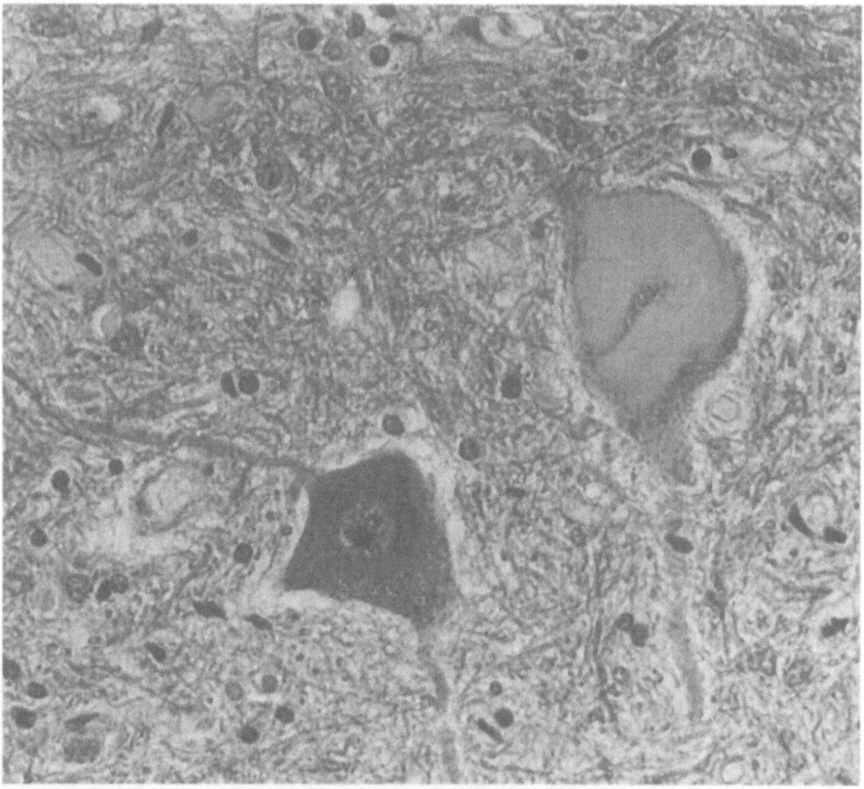

Fig. 5.6. Anterior horn cells in ALS, one showing increased lipofuscin, the other a hyaline inclusion. (Haematoxylin and eosin ×200.)

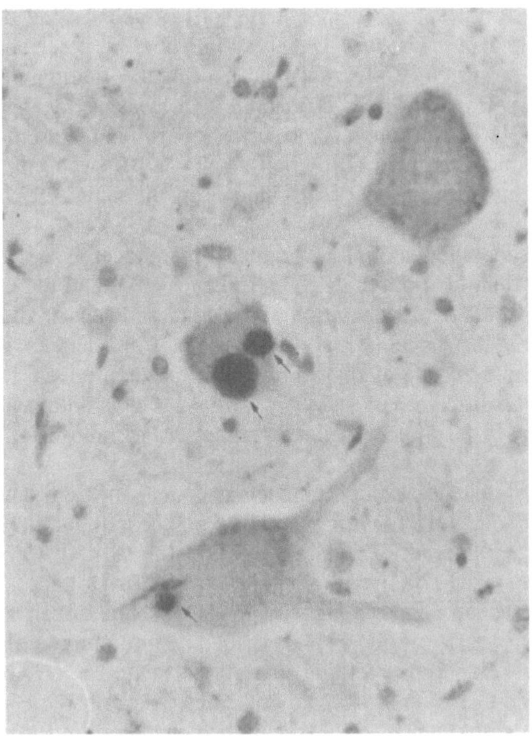

Fig. 5.7. Anterior horn cells in ALS showing ubiquitinated inclusions (arrows). (Immunoperoxidase ×200.)

Accumulations of 10 nm neurofilaments occur in proximal axonal swellings (spheroids); the latter are a common feature of MND and are found in the anterior horn of the spinal cord and in the somatic motor nuclei of the brainstem (Wohlfart 1959; Carpenter 1968; Chou et al. 1970; Hirano et al. 1984a; Schmidt et al. 1987; Sasaki et al. 1988, 1989). The neurofilamentous material in spheroids consists of filaments arranged individually, or in small bundles with side arms. Other cellular organelles are found in association with these neurofilamentous bundles. Axonal swellings are, however, a common normal finding in the spinal cords of ageing humans and other primates (Clark et al. 1984).

Other reported abnormalities of the spinal cord include the presence of aberrant myelinated neurites (Troost et al. 1989), heterotopic neurons (Kozlowski et al. 1989) and the presence of viral particles within motor neurons in a single case (Pena 1977). Small perivascular lymphocytic infiltrates are also reported commonly in MND spinal cord (Lawyer and Netsky 1953; Bunina 1962; Engelhardt et al. 1993).

Spinocerebellar Pathways

In 1909 Holmes reviewed the evidence for degeneration of the spinocerebellar tracts in sporadic MND, having demonstrated such changes in his material. Holmes (1909), and other contemporary authors, whilst showing spinocerebellar tract degeneration, were not able to demonstrate loss of cells in the dorsal nucleus of Clarke (Rossi and Roussy 1906). Degeneration of the dorsal and ventral spinocerebellar tracts is currently regarded as an accepted finding in sporadic MND (Bertrand and Van Bogaert 1925; Davison 1941; Swash et al. 1988; Hayashi and Kato 1989; Leigh 1990), although some series report this feature only in a minority of cases with atypical features (Brownell et al. 1970). Iwata and Hirano (1979), however, comment on the absence of changes in the spinocerebellar tracts in 37 cases of MND.

There is a well documented higher incidence of spinocerebellar tract involvement in familial MND than in sporadic MND (Hudson, 1981) and the tract involvement is usually accompanied by degeneration of the dorsal nucleus of Clarke (Engel et al. 1959; Hirano et al. 1967; Kubo et al. 1967; Metcalf and Hirano 1971; Igisu et al. 1974). Pathological evidence of spinocerebellar degeneration and loss of neurons in the nucleus dorsalis of Clarke in cases of familial MND is not usually accompanied by clinical features of spinocerebellar dysfunction (Hirano et al. 1967; Hudson, 1981).

The incidence of degeneration of the nucleus dorsalis of Clarke was thought to be very low in sporadic MND, with many reports describing no cellular loss or degeneration in this disorder (Friedman and Freedman 1950; Lawyer and Netsky 1953; Brownell et al. 1970, Iwata and Hirano 1979). Holmes (1909) described possible degeneration in Clarke's nucleus, in the absence of cell loss, in one case, and Bertrand and Van Bogaert (1925) described central chromatolysis, loss of cell processes, cell loss and other cellular alterations in up to 30% of their cases of MND. Averback and Crocker (1982), however, have described the loss of up to 30% of neurons in the dorsal nucleus of Clarke in the spinal cord in sporadic MND using morphometric methods, with all cases showing a reduction in numbers of neurons in the nucleus. These findings have been supported by Swash et al. (1988) and by Williams and colleagues (1990) who suggest in addition that degeneration of the spinocerebellar system in MND may be accompanied by loss or internalisation of specific spinocerebellar cell surface antigens.

Whilst degeneration is described in the fibres of the cerebellar peduncles in sporadic MND (Holmes 1909; Bertrand and Van Bogaert 1925), the same reports document normal cerebellar morphology (Holmes 1909; Bertrand and Van Bogaert 1925). Cerebellar changes including the loss of Purkinje cells and the presence of hyalinised inclusion bodies in the vermis have, however, been reported in familial MND (Hirano et al. 1967).

Sensory Systems

Whilst clinical evidence of sensory impairment is relatively unusual in sporadic MND (Cruveilhier 1853; Davison and Wechsler 1936; Lawyer and Netsky 1953; Mulder 1957; Mulder 1982; Toyokura 1979), clinical, physiological and morphological evidence of sensory impairment are well reported in the disorder (Wechsler et al. 1929; Dagnelie and Cambier 1933; Davison and Wechsler 1936; Friedman and Freedman 1950; Lawyer and Netsky 1953; Dyck et al. 1975; Kawamura et al. 1981; Bosch et al. 1985; Jamal et al. 1985; Di Trapani et al. 1986; Ben Hamida et al. 1987).

Demyelination of the posterior columns is seen occasionally in sporadic MND (Charcot and Marie 1885; Bertrand and Van Bogaert 1925; Lawyer and Netsky 1953; Castaigne et al. 1972). This feature may be seen in conjunction with clinical evidence of sensory disturbance (Dagnelie and Cambier 1933; Davison and Wechsler 1936) or without clinical evidence of sensory abnormalities (Holmes 1909; Brownell et al. 1970). Some authors have not reported alterations in the posterior columns in sporadic MND (Iwata and Hirano 1979) or have regarded single incidences of demyelination in the posterior columns as an incidental finding (Luys 1860; Lawyer and Netsky 1953).

Posterior column involvement is, however, more frequent in familial MND, occurring in up to 70% of cases (Engel et al. 1959; Hirano et al. 1967; Iwata and Hirano 1979; Hudson 1981). The degeneration of the posterior columns is usually seen in the middle root zone (Iwata and Hirano 1979; Hudson 1981). Despite the striking incidence of posterior column involvement in familial MND, sensory symptoms are unusual (Hirano et al. 1967; Ohta 1975; Iwata and Hirano 1979). It has been suggested that the posterior column lesions seen in MND are secondary to degeneration of target neurons in the dorsal nucleus of Clarke, and that sensory deficits do not develop since only proprioceptive neurons are involved (Ikuta et al. 1975). Not all cases of posterior column degeneration in MND affect the middle root zone, however, and the cuneate and gracile fasciculi may be involved (Iwata and Hirano 1979). In addition, cases of familial MND with posterior column involvement do not always show evidence of degeneration in the dorsal nucleus of Clarke (Takahashi et al. 1972).

Despite pathological observations of preservation of dorsal root ganglia (Iwata and Hirano 1979), in morphometric studies cells of the posterior root ganglia are consistently affected in sporadic MND, showing shrinkage of the cell body, but to a lesser degree than alpha motor neurons (Kawamura et al. 1981). Posterior root fibres and sensory peripheral nerves show morphological evidence of involvement in sporadic MND (Dyck et al. 1975; Bradley et al. 1983).

The cells of the posterior horn of the spinal cord are generally regarded as being well preserved in MND (Luys 1860), but some reports detail the loss of cells in the dorsal horn of the spinal cord in MND in association with sensory loss (Charcot and Joffroy 1869; Wechsler et al. 1929) or with no such loss (Holmes 1909).

Involvement of the ascending spinothalamic tracts has also been reported in MND (Bertrand and Van Bogaert 1925; Wechsler et al. 1929; Dagnelie and Cambier 1933; Davison 1941; Iwata and Hirano 1979). Degenerative features have been reported in neurons and subcortical white matter related to the post-central gyrus, but interpretation of such changes is confused by uncertainty as to the area of origin of the motor system (Charcot and Marie 1885; Holmes 1909; Smith 1960).

Thalamic gliosis in the absence of posterior column involvement was noted in a case of sporadic MND with sensory disturbance reported by Brownell and colleagues (1970). Thalamic fibre degeneration is commonly reported in sporadic MND, although usually in the absence of sensory abnormality (Holmes 1909; Smith 1960; Brownell et al. 1970; Hughes, 1982). Parietal lobe fibre degeneration in the absence of sensory problems is also well described in sporadic MND (Davison 1941), but the significance of this finding is also unclear.

In general, despite the often striking changes in the posterior columns or the spinothalamic tracts, there is little correlation between the pathological findings of sensory system degeneration in sporadic or familial MND and the presence of clinically apparent sensory deficit.

Autonomic Nervous System

Accounts of MND tend to emphasise preservation of autonomic function as typical of the disorder (Toyokura 1979). Abnormalities of visceral function, including gut, bladder and heart have, however, been reported in association with MND (Leri 1902; Oppenheim 1911; De Jong 1950; Bondin et al. 1954; Smith et al. 1957; Tsubaki 1979; Chida et al. 1989). Preservation of the secretory function of the gut may be related to sparing of the cells of the dorsal motor nucleus of the vagus (Holmes 1909, Malamud 1968; Kerr and Preshaw 1969; Iwata and Hirano 1979). Other reports, however, document degeneration in these neurons (Bertrand and Van Bogaert 1925; Lawyer and Netsky 1953; Castaigne et al. 1972; Le Bigot 1972). A certain degree of degeneration in this nucleus may be insufficient to cause clinical problems or measurable abnormalities of gastrointestinal function.

The absence of bedsores in patients with MND is well established (Charcot 1874; Forrester 1976; Fukukawa and Tokoyura 1978). It has been suggested that bedsores develop in patients with neurological disease due to vasomotor paralysis related to sympathetic nervous system dysfunction and that sympathetic vasomotor activity is spared in MND (Forrester 1976; Fukukawa and Tokoyura 1978). Preservation of sympathetic vasomotor tone corresponds with preservation of neurons in the intermediolateral column of the spinal cord (Holmes 1909; Bertrand and Van Bogaert 1925; Brownell et al. 1970; Castaigne et al. 1972; Le Bigot 1972; Iwata and Hirano 1979). However, Probst (1898, 1903) and Kennedy and Duchen (1985) reported a decrease in the numbers of cells in the intermediolateral column in MND.

The preservation of bladder and anal sphincter function in MND is also a distinct clinical feature of the disorder (Sakuta et al. 1978; Toyokura 1979) and appears to be related to preservation of the nucleus of Onufrowitz in the second sacral segment of the spinal cord (Mannen et al. 1977). The nature of the neurons of this nucleus is disputed, however, with disagreement as to the classification of these neurons as motor neurons or parasympathetic neurons. Further anatomical studies are awaited.

Peripheral Nerve

The macroscopic observations of pronounced anterior root atrophy (Cruveihier 1853; Luys 1860; Charcot and Joffroy 1869; Dejerine 1883) have their origin in the loss of large diameter fibres with relative sparing of smaller diameter fibres (Wohlfart and Swank, 1941) which in turn has been shown to be directly related to the loss of anterior horn cells in the spinal cord with relative sparing of gamma motor neurons (Tsukagoshi et al. 1979).

Other peripheral nerves have been described as of normal appearance or showing simple atrophy (Kojewnikoff 1883). Axonal loss of greater or lesser degree has been shown in studies of MND phrenic, sural, superficial and deep peroneal nerves as well as cervical and lumbar ventral roots (Bradley et al. 1983; Dyck et al. 1975; Hanyu et al. 1982; Sobue et al. 1981a,b). Demyelination was not prominent in these studies. The changes reported tend to preferentially affect larger myelinated fibres, with distal changes being little different from those seen proximally (Sobue et al.

1981b; Bradley et al. 1983), supporting the concept of MND as a neuronopathy (Thomas 1991).

Studies of peripheral nerve biopsies with a prominent motor component have been conducted using the motor branch of the median nerve in the hand (Breuer et al. 1987). These studies have shown loss of the normal fascicular structure, axonal loss and an increase in numbers of Renaut bodies (Breuer et al. 1987; Asbury 1973). The study by Breuer and colleagues (1987) also studied fast axonal transport in the nerve biopsies using videomicroscopy and showed an increase in speed of anterograde organelle transport and a decrease in retrograde transport. Studies of microtubular proteins in intercostal nerves of patients with MND suggest that, prior to evidence of morphological damage there may be changes in β-tubulin and tau proteins (Binet and Meininger 1988).

Muscle

Initial reports of MND described the marked muscular atrophy typical of the disorder (Aran 1850). Histological studies of muscle in MND have described the typical features of fatty infiltration of muscle (Cruveihier 1853; Lawyer and Netsky 1953) and the features of denervation atrophy, with groups of atrophic angular muscle fibres (Charcot and Joffroy 1869; Kojewnikoff 1883).

The presence of atrophic fibres has been directly correlated with the loss of alpha motor neurons in the spinal cord (Tsukagoshi et al. 1979). Occasional hypertrophic fibres may be seen (Tsukagoshi et al. 1979; Adams 1975) as are occasional dystrophic fibres (Lawyer and Netsky 1953).

Striated muscle biopsies show fibre type grouping in enzyme histochemical preparations (Hughes 1982), and biopsies demonstrate axonal sprouting from the residual intramuscular nerve fibres, secondary to denervation (Wohlfart 1957).

Studies of the motor end plate show increased segmentation and an increase in area, but the primary significance of these changes in end plate morphology, if any, is not clear (Bjornskov et al. 1975). Motor end plate changes are reviewed in detail in Chapter 6.

Intellect, the Frontal Lobe and Parkinsonism

Cruveihier (1853), in his description of the clinicopathological features of progressive muscular atrophy to nerve cell degeneration, specifically commented on the preservation of intellect in this disorder. Dementia is still regarded as an uncommon manifestation of sporadic MND, although it may be more frequently observed in familial MND (Jokelainen 1977; Toyokura 1979; Hudson 1981; Mulder 1982; Mitsumoto et al. 1988; Leigh 1990). The clinical features of the dementia associated with sporadic MND differ from those described in Alzheimer's disease and suggest a dementia of predominantly frontal lobe type (Neary et al. 1990). This is supported by studies which suggest that patients with classical MND may show subclinical frontal lobe dysfunction on neuropsychological testing, in the absence of dementia (David and Gillham 1986). The dementia associated with MND appears not to be transmissible to primates, in contrast to the amyotrophic form of Creutzfeldt–Jakob disease (Salazar et al. 1983).

Parkinsonism, or a combination of dementia and parkinsonism, is less common than dementia alone in association with sporadic or familial MND (Hudson 1981). Much debate has centred on the relationship of subgroups of patients with atypical features to the classical form of MND and whether those patients with dementia or parkinsonism in addition to MND are victims of coincidental disorders, novel disorders or represent part of the phenotypical expression of MND (Wechsler and Davison 1932; Brownell et al. 1970; Mitsuyama and Takamiya 1979; Salazar et al. 1983; Neary et al. 1990). The sex ratio, age of onset and duration of illness of sporadic MND does not differ from the subgroups with dementia or parkinsonism, or both, suggesting that the disorders are part of the same process (Hudson 1981).

Pathological findings in sporadic (and familial) MND with dementia include typical findings of lower motor neuron and pyramidal tract degeneration (Wechsler and Davison 1932; Hudson 1981). In addition, cases with dementia have been reported to show macroscopic atrophy of the frontal and/or temporal lobes (Wechsler and Davison 1932; Finlayson et al. 1973; Hudson 1981; Wikstrom et al. 1982; Neary et al. 1990). Histologically, most reports document neuronal loss and gliosis in layers 2 and 3, and occasionally laminae 5 and 6, of the frontal temporal lobes. Spongy degeneration of the superficial layers of the cortex has also been reported. In addition, changes are commonly reported in the globus pallidus, caudate nucleus, putamen, thalamus and substantia nigra in the absence of clinical features related to such degeneration (Hudson 1981; Morita et al. 1987; Neary et al. 1990). Hippocampal degeneration with neurofibrillary tangle formation is not a feature of the dementia associated with MND, although an occasional, presumably incidental, senile plaque may be found (Hudson 1981; Morita et al. 1987; Neary et al. 1990).

Histological findings in sporadic MND associated with dementia and parkinsonism are similar to those seen in MND with dementia alone, but with a higher (90%) incidence of degeneration in the substantial nigra or globus pallidus, but no Lewy body formation (Hudson 1981; Salazar et al. 1983). Greenfield and Matthews (1954) reported a case of sporadic MND with parkinsonism, but no dementia, in which there were degenerative changes in the substantia nigra which included the presence of neurofibrillary tangles. Neurofibrillary tangles with cell loss have also been reported in the substantia nigra in association with lower motor neuron loss and in more extensive areas of the cerebrum in association with pyramidal tract degeneration in two cases of post-encephalitic MND and parkinsonism (Hirano and Zimmerman 1962; McMenemy et al. 1967).

High incidence foci of MND occur in the Western Pacific region, in the Auyu and Jakai people of West New Guinea, the Kii peninsula of Japan and the Chamorros of the Mariana Islands (Kurland and Mulder 1954; Hirano et al. 1966; Hudson 1981; Garruto and Yase 1986). Clinical features of the disease in these foci are the same as for classical descriptions of the sporadic disorder elsewhere (Garruto and Yase 1986). The Guam population of Chamorro people has been most extensively documented. In the same population with a high incidence of MND, there is a high incidence of parkinsonism and dementia, with a tendency to a combination of the two disorders. Approximately 7% of patients with MND, dementia or parkinsonism have a combination of all three features (Stanhope et al. 1972). Studies of the sex ratio, age of onset, duration of onset and overlap of symptoms have suggested that these disorders may also be variants of the same disease process (Hirano et al. 1961a, 1966).

Neuropathological studies of the Guamanian form of MND show loss of anterior horn cells and pyramidal tract degeneration. There is sparing of the posterior columns and of the spinocerebellar tracts (Hirano et al. 1966). In addition to these features, Hirano and colleagues (Malamud et al. 1961; Hirano et al. 1966) and Anderson and colleagues (1979) described neurofibrillary changes of varying degree in the substantia nigra, locus ceruleus, substantia innominata, other basal nuclei, cerebral cortex and other areas. These features were present in all 34 cases of MND occurring in the Chamorro population in one such study (Hirano et al. 1966). Occasional anterior horn cells also showed neurofibrillary changes in 13 of 34 cases (Hirano et al. 1966). Granulovacuolar degeneration of hippocampal neurons was also a constant finding (Malamud et al. 1961; Hirano 1965; Hirano et al. 1966). Cases of Guamanian parkinsonism-dementia complex show frontotemporal atrophy, with marked cell loss in the substantia nigra, locus ceruleus, substantia innominata and globus pallidus. Neurofibrillary change is widespread and hippocampal neurons also show granulovacuolar degeneration. The presence of Lewy bodies in the substantia nigra is not a feature of Guamanian parkinsonism-demential complex (Hirano et al. 1961b, 1966; Hirano 1965). An important feature of the neuropathological findings in Guamanian MND and parkinsonism-dementia is the considerable degree of overlap in the cellular pathology of these disorders (Hirano 1965; Hirano et al. 1966; Hudson 1981).

Relation to Other Neuronal Degenerations

The diagnosis of MND, or of its clinical variants, requires the fulfilment of certain clinical criteria, discussed elsewhere in this book, but also the exclusion of certain conditions known to mimic MND. Such conditions include lead and mercury intoxication, motor neuron degeneration associated with plasma cell dyscrasias or with paraneoplastic syndromes, motor neuropathies, hexosaminidase deficiencies, poliovirus infection (Fig. 5.8), the amyotrophic form of Creutzfeldt–Jakob disease, hypoglycaemia, disorders of calcium metabolism, thyroid disorders, ischaemia and syphilis (Roos et al. 1980; Johnson 1981; Rowland 1982; Salazar et al. 1983; Mitsumoto et al. 1985, 1988; Bradley et al. 1991; Thomas 1991). In the determination of other causes of motor neuron degeneration clinical findings and history are of paramount importance. Lawyer and Netsky (1953) specifically comment that the findings in the spinal cord of a patient dying with old poliomyelitis may be similar to those of progressive muscular atrophy, but that the clinical differences are obvious. Neuropathological validation of a clinical diagnosis may be valuable, particularly in cases with atypical features, for example pathological studies of MND have commented that cases with a clinical diagnosis of MND may show evidence of other pathological conditions, such as multiple sclerosis or multisystem atrophy (Lawyer and Netsky 1953; Brownell et al. 1970; Tanaka et al. 1984).

The presence of inherited forms of MND, atypical forms, and of the high incidence foci in the Western Pacific with neuropathological features that differ

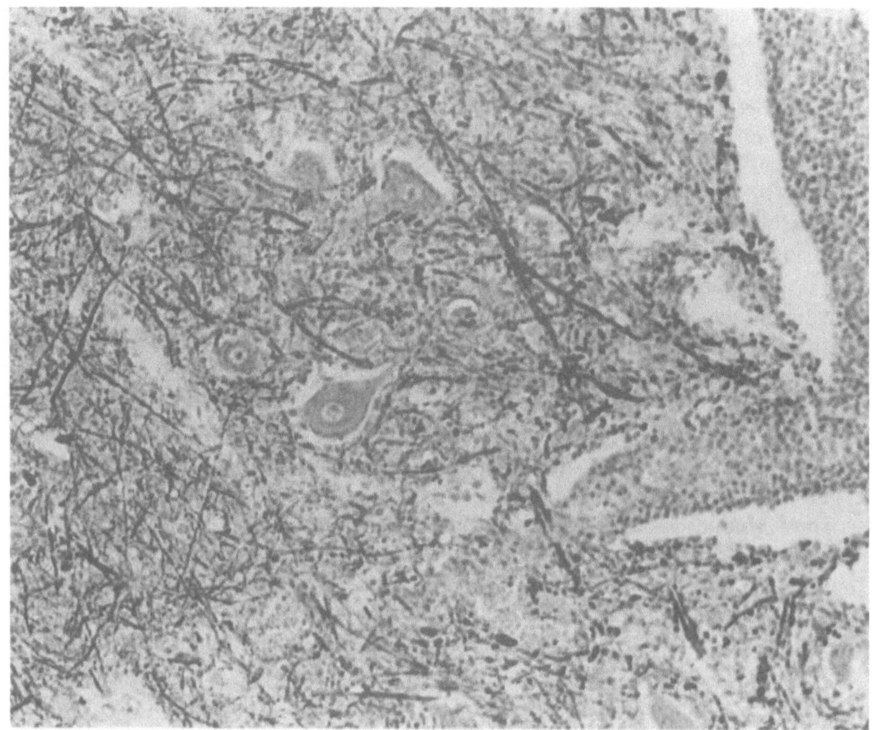

Fig. 5.8. Anterior horn in acute poliomyelitis showing intense perivascular lymphocytic infiltrate. (Luxol fast blue ×150.)

from the classical sporadic form of MND provides problems with the classification of MND and thus with hypotheses of pathogenesis (Rowland, 1991). Advances in the understanding of the genetic basis of the β-hexosaminidase deficiencies (Neote et al. 1991), spinal muscular atrophy (Gilliam et al. 1990), X-linked spinal and bulbar muscular atrophy (La Spada et al. 1991) and possibly familial MND (Conneally 1991; Şiddique et al. 1991) may help with the understanding of the relationships between these disorders at a genetic level. As the molecular mechanisms consequent on gene defects become understood it may be possible to relate such defects to the neuropathological findings in MND.

The apparent heterogeneity of MND in clinical presentation and neuropathological findings has provided much speculation regarding the overlap of forms of the disease, and whether they truly represent clinical variants of one disease, or are different disorders (Hirano et al. 1966; Hudson 1981). Immunohistochemical techniques have provided opportunities to study the expression of cytoskeletal and stress protein expression in MND, MND variant forms and related motor neuronal degenerations (Munoz et al. 1988; Leigh et al. 1988, 1989a, b; Leigh and Swash 1991). It has become apparent that, despite the presence of other differing pathological features, the presence of ubiquitinated inclusion bodies within anterior horn cells of the spinal cord is constant in sporadic, familial and Guamanian forms of MND (Leigh et al. 1988; Lowe et al. 1988a). Such distinctive anterior horn cell inclusions (Fig. 5.7) are not seen in other neurodegenerative diseases, poliomyelitis, spinal muscular atrophy and a range of other disorders, although other inclusions such as neurofibrillary tangles and Lewy bodies are also stained with anti-ubiquitin antibodies (Lowe et al, 1988b; Leigh et al. 1989b).

The apparently diffuse involvement of motor neurons in MND differs from the neuronal changes seen in the lipidoses such as Hurler's syndrome (Hirano and Iwata 1979), in that affected groups of neurons in MND may show individual neurons which appear normal, implying a random process or a process of differential susceptibility among an at-risk population (Hirano and Iwata 1979; Sobue et al. 1983). Alzheimer's disease shows a similar pattern of individual cell loss and preservation within a susceptible neuronal population (Hirano and Iwata 1979), as does Werdnig–Hoffmann disease – infantile spinal muscular atrophy (Hirano and Iwata 1979). Such pathology, with a specific at-risk population and individual cell death, resembles few toxic or infective processes. In spinal muscular atrophy it has been suggested that anterior horn cells may die due to skeletal muscle apoptosis and subsequent lack of neurotrophic support to the related motor neurons (Fidzianka et al. 1990). This hypothesis requires further investigation, and integration with evidence of the gene defect in this disorder (Gilliam et al. 1990).

Recent interest in adult-onset neurodegeneration has been stimulated by the discovery of a family of mutations in the genome of *Caenorhabditis elegans*, a small nematode, which code for late-onset neuronal degeneration in specific neuronal groups (Driscoll and Chalfie 1991). The morphology of the cell death caused by mutations in this *mec*-4 gene family is that of neuronal swelling and vacuolation, in contrast to the single cell or apoptotic cell death seen in programmed cell death (Chalfie and Wolinsky 1990). Whilst the exact mechanisms underlying programmed cell death are not yet clear, some of the genes controlling the process in *Caenorhabditis elegans* have been identified. Mutations in these *ced*-3 and *ced*-4 genes can prevent programmed cell death from occurring, but do not prevent the cell death caused by *mec*-4 mutations which appear to operate through a transmembrane protein abnormality (Yuan and Horvitz 1990; Bargmann 1991). Proteins associated with this neurodegenerative process have been termed degenerins, and homologous proteins have been identified in other species. Neuronal degeneration resembling programmed cell death or apoptosis can be induced in vitro by the withdrawal of neurotrophic factor support (Martin et al. 1988). Neurodegeneration caused by a recapitulation of a programmed cell death mechanism has attractions as an hypothesis for the pathogenesis of neurodegenerative diseases such as MND since it affects specific neuronal populations, affects single cells within these specific populations, and can be induced by a variety of stimuli such as trophic factor withdrawal or genetic mutations (Appel 1981; Martin et al. 1988; Bargmann 1991). Thus, like the occurrence of MND in sporadic, hereditary, juvenile and variant or overlap forms, a common end-point of selective neuronal degeneration may be reached by a variety of pathogenetic mechanisms involving a fundamental cellular process.

Acknowledgements. JEM is a Wellcome Trust Research Fellow and holds the Gillson Scholarship in Pathology of the Worshipful Society of Apothecaries. Our work is supported by the Motor Neurone Disease Association of the United Kingdom.

References

Adams RD (1975) Diseases of muscle, 3rd edn. Harper and Row, Hagerstown, pp 425–430

Anderson FH, Richardson EP Jr, Okazaki H, Brody JA (1979) Neurofibrillary degeneration on Guam. Brain 102:65–77

Appel SH (1981) A unifying hypothesis for the cause of amyotrophic lateral sclerosis, parkinsonism and Alzheimer disease. Ann Neurol 10:499–505

Aran FA (1850) Recherches sur une maladie non encore décrite du système musculaire (atrophie musculaire progressive). Arch Gen Med 24:5–35, 172–214

Asbury AL (1973) Renaut bodies: a forgotten endoneurial structure. J Neuropathol Exp Neurol 32:334–343

Averback P, Crocker P (1982) Regular involvement of Clarke's nucleus in sporadic amyotrophic lateral sclerosis. Arch Neurol 39: 155–156

Bargmann CL (1991) Death from natural and unnatural causes. Curr Biol 1:388–390

Ben Hamida M, Letaief F, Hentati F, Ben Hamida C (1987) Morphometric study of the sensory nerve in classical (or Charcot disease) and juvenile amyotrophic lateral sclerosis. J Neurol Sci 78:313–329

Bertrand I, van Bogaert L (1925) Rapport sur la sclérose latérale amyotrophique (anatomie pathologique). Rev Neurol 32:779–806

Binet S, Meininger V (1988) Modifications of microtubule proteins in ALS nerve precede detectable histologic and ultrastructural changes. Neurology 38:1596–1600

Bjornskov EK, Dekker NP, Norris FH, Stuart ME (1975) End-plate morphology in amyotrophic lateral sclerosis. Arch Neurol 32:711–712

Bondin G, Barbizet J, Hillemand B, Lote J (1954) Atonie et dilatation gastro-oesophagienne au cours de la sclérose latérale amyotrophique. Bull Mem Soc Med Hop Paris 2:641–648

Bosch EP, Yamada T, Kimura J (1985) Somatosensory evoked potentials in motor neuron disease. Muscle Nerve 8:556–562

Boudouresques J, Toga M, Roger J et al. (1967) Etat dementiel, sclérose latérale amyotrophique, syndrome extrapyramidal. Etude anatomique. Discussion nosologique. Rev Neurol 116:693–704

Bradley WG, Good P, Rasool CG, Adelman LS (1983) Morphometric and biochemical studies of peripheral nerves in amyotrophic lateral sclerosis. Ann Neurol 14:267–277

Bradley WG, Robison SH, Tanden R, Besser D (1991) Post-radiation motor neuron syndromes. In: Rowland LP (ed) Amyotrophic lateral sclerosis and other motor neuron diseases. Advances in neurology, vol 56. Raven Press, New York pp 341–355

Breuer AC, Lynn MP, Atkinson MB et al. (1987) Fast axonal transport in amyotrophic lateral sclerosis: An intra-axonal traffic analysis. Neurology 37:738–748

Brownell B, Oppenheimer DR, Hughes JT (1970) The central nervous system in motor neuron disease. J Neurol Neurosurg Psychiatr 33:338–357

Bunina TL (1962) On intracellular inclusions in familial amyotrophic lateral sclerosis. Ah Nerropat Psikhit Korsakov 62:1293–1299

Carpenter S (1968) Proximal axonal enlargement in motor neuron disease. Neurology 18:842–851

Castaigne P, Lhermitte F, Cambier J, Escourolle R, Le Bigot P (1972) Etude neuropathologique de 61 observations de sclérose latérale amyotrophique. Discussion nosologique. Rev Neurol 127:401–414

Chalfie M, Wolinsky E (1990) The identification and suppression of inherited neurodegeneration in *Caenorhabditis elegans*. Nature 345:410–416

Charcot JM (1870) Note sur un cas de paralysie glosso-laryngée suivi d'autopsie. Arch Physiol (Paris) 3:247–260

Charcot JM (1874) De la sclérose latérale amyotrophique. Prog Med 2:325–327, 341–342, 453–455

Charcot JM, Joffroy A (1869) Deux cas d'atrophie musculaire progressive avec lésions de la substance grise et des faisceaux antérolateraux de la moelle épinière. Arch Physiol (Paris) 2:354–367, 629–644, 744–760

Charcot JM, Marie P (1885) Deux nouveaux cas de sclérose amyotrophique suivis d'autopsie. Arch Neurol 10:1–35

Chida K, Sakamaki S, Takusu T (1989) Alteration in autonomic function and cardiovascular regulation in amyotrophic lateral sclerosis. J Neurol 232:127–130

Chou SM (1979) Pathognomy of intraneuronal inclusions in ALS. In: Tsubaki T, Tokoyura Y (eds) Amyotrophic lateral sclerosis. University Park Press, Baltimore, pp 135–176

Chou SM, Martin JD, Gutrecht JA, Thompson HG Jr (1970) Axonal balloons in subacute motor neuron disease. J Neuropathol Exp Neurol 29:141–142

Clark AW, Parhad IM, Griffin JW, Price DL (1984) Neurofilamentous axonal swellings as a normal finding in the spinal anterior horn of man and other primates. J Neuropathol Exp Neurol 43:253–262

Conneally PM (1991) A first step toward a molecular genetic analysis of amyotrophic lateral sclerosis. N Engl J Med 324:1430–1432

Cruveihier J (1853) Sur la paralysie musculaire, progressive, atrophique. Bull Acad Med (Paris) 18:546–583

Dagnelie J, Cambier P (1933) Contribution anatomo-clinique à l'étude de la sclérose latérale amyotrophique. Rev Neurol 40 (2):25–36

David AS, Gillham RA (1986) Neuropsychological study of motor neurone disease. Psychosomatics 27:441–445

Davison C (1941) Amyotrophic lateral sclerosis. Origin and extent of the upper motor neuron lesion. Arch Neurol Psychiatr 46:1036–1056

Davison C, Wechsler IS (1936) Amyotrophic lateral sclerosis with involvement of posterior column and sensory disturbances. A clinicopathologic study. Arch Neurol Psychiatr 35:229–239

Dejerine J (1883) Etude anatomique et clinique sur la paralysie labio-glosso-laryngée. Arch Physiol Normale Serie 3(2):180–227

De Jong RN (1950) The neurologic examination. Paul B Hoeber, New York p 330

Di Trapani G, David P, La Cara A, Servidei S, Tonali P (1986) Morphological studies of sural nerve biopsies in the pseudopolyneuropathic form of amyotrophic lateral sclerosis. Clin Neuropathol 5:134–138

Driscoll M, Chalfie M (1991) The *mec*-4 gene is a member of a family of *Caenorhabditis elegans* genes which can mutate to induce neuronal degeneration. Nature 349:588–593

Duchenne GBA (1853) Etude comparée des lesions anatomiques dans l'atrophie musculaire progressive et dans la paralysie générale. Union Med 7:246–247

Duchenne GBA (1860) Paralysie musculaire progressive de la langue, du voile du palais et des levres. Arch Gen Med 16:283–296, 431–445

Dyck PJ, Stevens JC, Mulder DW, Espinosa RE (1975) Frequency of nerve fiber degeneration of peripheral and sensory neurons in amyotrophic lateral sclerosis. Morphometry of deep and superficial peroneal nerves. Neurology 25:781–785

Engel WK, Kurland LT, Klatzo I (1959) An inherited disease similar to amyotrophic lateral sclerosis with a pattern of posterior column involvement. An intermediate form? Brain 82:203–220

Engelhardt JI, Tatji J, Appel SH (1993) Lymphocytic infiltrates in the spinal cord in amyotrophic lateral sclerosis. Arch Neurol 50:30–36

Farmer TW, Allen JN (1969) Hereditary proximal amyotrophic lateral sclerosis. Trans Am Neurol Assoc 94:140–144

Feller TG, Jones RE, Netsky MG (1966) Amyotrophic lateral sclerosis and sensory changes. Virginia Med Monthly 93:328–335

Fidzianka A, Goebel HH, Warlo I (1990) Acute infantile spinal muscular atrophy. Muscle apoptosis as a proposed pathogenetic mechanism. Brain 113:443–445

Finlayson MH, Guberman A, Martin JB (1973) Cerebral lesions in familial amyotrophic lateral sclerosis and dementia. Acta Neuropathol 26:237–246

Forrester JM (1976) Amyotrophic lateral sclerosis and bedsores. Lancet i:970

Friedman AP, Freedman DA (1950) Amyotrophic lateral sclerosis. J Nerv Mental Dis 111:1–18

Fukukawa T, Tokoyura Y (1978) Amyotrophic lateral sclerosis and bedsores. Plethysmographic analysis. Lancet i:159

Fullmer HM, Siedler HD, Krooth RS, Kurland LT (1960) A cutaneous disorder of connective tissue in amyotrophic lateral sclerosis: a histochemical study. Neurology 10:717–724

Garruto RM, Yase Y (1986) Neurodegenerative disorders of the Western Pacific: the search for mechanisms of pathogenesis. Trends Neurosci 9:368–374

Gilliam TC, Brzustowicz LM, Castilla LH et al. (1990) Genetic homogeneity between acute and chronic forms of spinal muscular atrophy. Nature 345:823–825

Greenfield JG, Matthews WB (1954) Post-encephalitic parkinsonism with amyotrophy. J Neurol Neurosurg Psychiatr 17:50–56

Hanyu N, Oguchi K, Yangisaua N, Tsukagoshi H (1982) Degeneration and regeneration of ventral root motor fibers in amyotrophic lateral sclerosis. J Neurol Sci 55:99–115

Hartmann HA, McMahon S, Sun DY, Abbs JH, Uemura E (1989) Neuronal RNA in nucleus ambiguus and nucleus hypoglossus of patients with amyotrophic lateral sclerosis. J Neuropathol Exp Neurol 48:669–673

Harvey DG, Torack RM, Rosenbaum HE (1979) Amyotrophic lateral sclerosis with ophthalmoplegia. A clinicopathologic study. Arch Neurol 36:615–617

Hayashi H, Kato S (1989) Total manifestations of amyotrophic lateral sclerosis (ALS) in the totally locked-in state. J Neurol Sci 93:19–35

Henson RA, Urich RA (1982) Cancer and the nervous system: the neurological manifestations of systemic malignant disease. Blackwell Scientific Publications, Oxford, pp 441–445

Hirano A (1965) Pathology of amyotrophic lateral sclerosis. In: Gajdusek DC, Gibbs CJ (eds) Slow, latent and temperate infections. NINDB Monograph No. 2. National Institutes of Health, Washington, DC, pp 23–37

Hirano A, Inoue K (1980) Early pathological changes in amyotrophic lateral sclerosis: electron microscopic studies of chromatolysis, spheroids and Bunina bodies. Neurol Med (Tokyo) 13:148–160

Hirano A, Iwata M (1979) Pathology of motor neurons with special reference to amyotrophic lateral sclerosis and related diseases. In: Tsubaki T, Tokoyura Y (eds) Amyotrophic lateral sclerosis. University Park Press, Baltimore, pp 107–133

Hirano A, Kurland LT, Krooth RS, Lessell S (1961a) Parkinsonism-dementia complex and endemic disease on the island of Guam. I. Clinical features. Brain 84:642–661

Hirano A, Malamud N, Kurland LT (1961b) Parkinsonism-dementia complex, an endemic disease on the island of Guam. II. Pathological features. Brain 84:662–679

Hirano A, Zimmerman HM (1962) Alzheimer's neurofibrillary changes. A topographic study. Arch Neurol 7:227–242

Hirano A, Malamud N, Elizan TS, Kurland LT (1966) Amyotrophic lateral sclerosis and parkinsonism-dementia complex on Guam. Further pathologic studies. Neurology 15:35–51

Hirano A, Kurland LT, Sayre GP (1967) Familial amyotrophic lateral sclerosis. A subgroup characterised by posterior and spinocerebellar tract involvement and hyaline inclusions in the anterior horn cells. Arch Neurol 16:232–243

Hirano A, Dembitzer HM, Kurland LT, Zimmerman HM (1968) The fine structure of some intra-ganglionic alterations. J Neuropathol Exp Neurol 27:167–182

Hirano A, Donnenfeld H, Sasaki S, Nakano I (1984a) Fine structural observations of neurofilamentous changes in amyotrophic lateral sclerosis. J Neuropathol Exp Neurol 43:461–470

Hirano A, Hakano I, Kurland LT, Mulder DW, Holley PW, Saccomanno G (1984b) Fine structural study of neurofibrillary changes in a family with amyotrophic lateral sclerosis. J Neuropathol Exp Neurol 43:471–480

Hirayama K (1978) The representation of the tract in the cerebrum in cases of ALS. In: Tsubaki T, Tokoyura Y (eds) Amyotrophic lateral sclerosis. University Park Press, Batimore, pp 209–219

Hirayama K, Tsubaki T, Tokoyura Y, Okinaka S (1962) The representation of the pyramidal tract in the internal capsule and basis pedunculi: a study based on three cases of amyotrophic lateral sclerosis. Neurology 12:337–342

Holmes G (1909) The pathology of amyotrophic lateral sclerosis. Rev Neurol Psychiatr 7:693–725

Horton WA, Eldrige R, Brody WA (1976) Familial motor neuron disease. Evidence for at least three different types. Neurology 26:460–465

Hudson AJ (1981) Amyotrophic lateral sclerosis and its association with dementia, parkinsonism and other neurological disorders: a review. Brain 104:217–247

Hudson AJ, Kiernan JN (1988) Preservation of certain voluntary muscles in motoneuron disease. Lancet i:652–653

Hughes JT (1982) Pathology of amyotrophic lateral sclerosis. In: Rowland LP (ed) Human motor neuron diseases. Raven Press, New York, pp 61–74

Igisu K, Ohta M, Yamashita M, Kuroiwa Y (1974) Familial motor neuron disease. Jpn Hum Gene 19:108–109

Ikuta F, Makibuchi T, Ohama E (1975) Neuropathological study of the spinal posterior column lesion in familial amyotrophic lateral sclerosis. ALS Progress Report of the Japanese Ministry of Welfare, 70

Inoue K, Hirano A (1979) Early pathological changes in amyotrophic lateral sclerosis: autopsy findings in a case of ten months duration. Neurol Med (Tokyo) 11:448–455

Iwata M, Hirano A (1978) Sparing of the Onufrowicz nucleus in sacral anterior horn lesions. Ann Neurol 4:245–249

Iwata M, Hirano A (1979) Current problems in the pathology of amyotrophic lateral sclerosis. In: Zimmerman HM (ed) Progress in neuropathology, vol 4. Raven Press, New York, pp 277–298

Jamal GA, Weir AL, Hansen S, Ballantyne JP (1985) Sensory involvement in motor neuron disease: further evidence from automated thermal threshold determination. J Neurol Neurosurg Psychiatr 48:906–910

Johnson WG (1981) The clinical spectrum of hexosaminidase deficiency diseases. Neurology 31:1453–1456

Jokelainen M (1977) Amyotrophic lateral sclerosis in Finland. II. Clinical characteristics. Acta Neurol Scand 56:194–204

Kahler O, Pick L (1884) Beitrage zur Pathologie und pathologische Anatomie des zentralen Nervensystems. Z Nervenh 5:169

Kamo H, Haebara H, Akiguchi I, Kameyama M, Kimura H, McGeer PL (1987) A distinctive distribution of reactive astroglia in the precentral cortex in amyotrophic lateral sclerosis. Acta Neuropathol 74:33–38

Kawamura Y, Dyck PJ, Shimono M, Okazaki H, Tateishi J, Doi H (1981) Morphometric comparison of the vulnerability of peripheral motor and sensory neurons in amyotrophic lateral sclerosis. J Neuropathol Exp Neurol 40:667–675

Kennedy PGE, Duchen LW (1985) A quantitative study of intermediolateral column cells in motor neurone disease and the Shy-Drager syndrome. J Neurol Neurosurg Psychiatr 48:1103–1106

Kerr PWL, Preshaw RM (1969) Secretomotor function of the dorsal motor nucleus of the vagus. J Physiol 205:405–415

Kojewnikoff A (1883) Cas de sclérose latérale amyotrophique, la dégénérescence des faisceaux pyramidaux se propageant à travers tout l'éncéphale. Arch Neurol 6:356–376

Kozlowski MA, Williams C, Hinton DR, Miller CA (1989) Heterotopic neurons in ALS spinal cord. Neurology 39:644–648

Kubo H, Ikuta F, Tsubaki T (1967) An autopsy case with a history of familial amyotrophic lateral sclerosis and posterior column involvement. Clin Neurol 7:45–50

Kurland LT, Mulder DW (1954) Epidemiologic investigations of amyotrophic lateral sclerosis. I. Preliminary report on geographic distribution, with special reference to the Mariana Islands, including clinical and pathological observations. Neurology 4:355–378, 438–448

Kushner MJ, Parrish M, Burke A, Behrens M, Hays AP, Frame B, Rowland LP (1984) Nystagmus in motor neuron disease: clinicopathological study of two cases. Ann Neurol 16:1–77

La Spada AR, Wilson EM, Lubahn DB, Harding AE, Fishbeck KH (1991) Androgen receptor gene mutations in X-linked spinal and bulbar muscular atrophy. Nature 352:77–79

Lawyer T Jnr, Netsky MG (1953) Amyotrophic lateral sclerosis: a clinicopathological study of 53 cases. Arch Neurol Psychiatr 69:171–192

Le Bigot P (1972) Contribution à l'étude neuropathologique de la sclérose latérale amyotrophique et au probléme de ses limites nosologiques (à propos de 56 observations anatomo-cliniques). Thse de Paris.

Leigh PN (1990) Amyotrophic lateral sclerosis and other motor neurone diseases. Curr Opin Neurol Neurosurg 3:567–575

Leigh PN, Anderton BH, Dodson A, Gallo J-M, Swash M, Power DM (1988) Ubiquitin deposits in anterior horn cells in motor neurone disease. Neurosci Lett 93:197–203

Leigh PN, Dodson A, Swash M, Brion J-P, Anderton BH (1989a) Cytoskeletal abnormalities in motor neuron disease. An immunocytochemical study. Brain 112:521–535

Leigh PN, Probst A, Dale GE, Power DP, Brion J-P, Dodson A, Anderton BH (1989b) New aspects of the pathology of neurodegenerative disorders as revealed by ubiquitin antibodies. Acta Neuropathol 79:61–72

Leigh PN, Swash M (1991) Cytoskeletal pathology in motor neuron diseases. In: Rowland LP (ed) Amyotrophic lateral sclerosis and other motor neuron diseases. Advances in neurology, vol 56. Raven Press, New York, pp 115–124

Leri A (1902) Atrophie généralisée de la musculature de tous les visceres dans une amyotrophie progressive type Aran–Duchenne. Rev Neurol 10:394–401

Leveille A, Kiernan J, Goodwin JA, Antel J (1982) Eye movements in amyotrophic lateral sclerosis. Arch Neurol 39:684–686

Lowe J, Lennox G, Jefferson D et al. (1988a) A filamentous inclusion body within anterior horn neurons in motor neuron disease defined by immunocytochemical localisation of ubiquitin. Neurosci Lett 94:203–210

Lowe J, Blanchard A, Morrell K et al. (1988b) Ubiquitin is a common factor in intermediate filament inclusion bodies of diverse type in man, including those of Parkinson's disease, Pick's disease, and Alzheimer's disease, as well as Rosenthal fibres in cerebellar astrocytomas, cytoplasmic bodies in muscle and Mallory bodies in alcoholic liver disease. J Pathol 155:9–15

Luys J (1860) Atrophie musculaire progressive. Lésions histologiques de la substance grise de la moelle épinière. Gazette Med Paris 15:505

Malamud N (1968) Neuromuscular diseases. In: Mincler J (ed) Pathology of the nervous system, vol 1. McGraw-Hill, New York, pp 712–725

Malamud N, Hirano A, Kurland LT (1961) Pathoanatomic changes in amyotrophic lateral sclerosis on Guam: special reference to occurrence of neurofibrillary changes. Arch Neurol 5:301–415

Manetto V, Sternberger NH, Perry G et al. (1988) Phosphorylation of neurofilaments is altered in amyotrophic lateral sclerosis. J Neuropathol Exp Neurol 47:642–653

Mann DMA, Yates PO (1974) Motor neuron disease: the nature of the pathogenic mechanisms. J Neurol Neurosurg Psychiatr 37:1036–1046

Mannen T, Iwata M, Toyokura M, Nagashima K (1977) Preservation of a certain motoneurone group of the sacral cord in amyotrophic lateral sclerosis; its clinical significance. J Neurol Neurosurg Psychiatr 40:464–469

Marie P, Bouttier H, Bertrand I (1923) Etude anatomique d'un cas de sclérose latérale amyotrophique à préponderance hemiplégique. Bull Mem Soc Med Hop Paris 47:481–485

Martin DP, Schmidt RE, Di Stafano PS, Lowry OH, Carter JG, Johnson EM (1988) Inhibitors of protein synthesis and RNA synthesis prevent neuronal death caused by nerve growth factor deprivation. J Cell Biol 106:829–844

Masui Y, Mozai T, Kakeh K (1985) Functional and morphometric study of the liver in motor neuron disease. J Neurol 232:15–19

McMenemy WH, Barnard RO, Jellinek EH (1967) Spinal amyotrophy. A late sequel of epidemic encephalitis (von Economo). Rev Roumaine Neurol 4:251–259

Metcalf CW, Hirano A (1971) Amyotrophic lateral sclerosis. Clinicopathological studies of a family. Arch Neurol 24:518–523

Mitsumoto H, Sliman RJ, Schafer IA et al. (1985) Motor neuron disease and adult hexosaminidase A deficiency in two families: evidence for multisystem degeneration. Ann Neurol 17:378–385

Mitsumoto H, Hanson MR, Chad DA (1988) Amyotrophic lateral sclerosis. Recent advances in pathogenesis and therapeutic trials. Arch Neurol 45:189–202

Mitsuyama Y (1984) Presenile dementia with motor neurone disease in Japan: clinico-pathological review of 26 cases. J Neurol Neurosurg Psychiatr 47:953–959

Mitsuyama Y, Takamatsu I (1971) An autopsy case of presenile dementia with motor neurone disease. Brain Nerve 23:409–416

Mitsuyama Y, Takamiya S (1979) Presenile dementia with motor neuron disease in Japan. A new entity? Arch Neurol 36:592–593

Morita K, Kaiya H, Ikeda T, Namba M (1987) Presenile dementia combined with amyotrophy: a review of 34 Japanese cases. Arch Gerontol Geriatr 6:263–277

Mulder DW (1957) The clinical syndrome of amyotrophic lateral sclerosis. Proc Mayo Clin 32 (17):427–436

Mulder DW (1982) Clinical limits of amyotrophic lateral sclerosis. In: Rowland LP (ed) Human motor neuron diseases. Advances in neurology, vol 36. Raven Press, New York, pp 15–22

Munoz DG, Greene C, Perl DP, Selkoe DJ (1988) Accumulation of phosphorylated neurofilaments in anterior horn motoneurons of amyotrophic lateral sclerosis patients. J Neuropathol Exp Neurol 47:9–18

Nakano I, Hirano A (1987) Atrophic cell processes of large motor neurons in the anterior horn in amyotrophic lateral sclerosis: observation with silver impregnation method. J Neuropathol Exp Neurol 46:40–49

Nakano Y, Hirayama K, Terao K (1987) Hepatic ultrastructural changes and liver dysfunction in amyotrophic lateral sclerosis. Arch Neurol 44:103–106

Neary D, Snowden JS, Mann DMA, Northen B, Goulding PJ, Macdermott N (1990) Frontal lobe dementia and motor neuron disease. J Neurol Neurosurg Psychiatr 53:23–32

Neote K, Mahuran DJ, Gravel RA (1991) Molecular genetics of β-hexosaminidase deficiencies. In: Rowland LP (ed) Amyotrophic lateral sclerosis and other motor neuron diseases. Advances in neurology, vol 56. Raven Press, New York, pp 189–207

Ohta M (1975) Familial amyotrophic lateral sclerosis. Neurol Med (Tokyo) 2:33–39

Ono S, Tokoyura Y, Manner T et al. (1986) Amyotrophic lateral sclerosis: Histologic, histochemical and ultrastructural abnormalities of skin. Neurology 36:948–956

Onuf (Onufrowitz) B (1889) Notes on the arrangement and function of the cell groups of the sacral region of the spinal cord. J Nerv Mental Dis 26: 498–504

Onuf (Onufrowtiz) B (1890) On the arrangement and function of the cell groups of the sacral region of the spinal cord in man. Arch Neurol Psychopathol 3:387–411

Oppenheim H (1911) Textbook of nervous diseases for physicians and students, 5th ed, vol 1. TN Foulis, London, p 220

Pena CE (1977) Virus-like particles in amyotrophic lateral sclerosis. Electron microscopical study of a case. Ann Neurol 1:290–297

Probst M (1898) Zu den fortschreitenden Erkrankungen der motorischen Leitungsbahnen. Arch Psychiatr Nervenkr 30:766–844

Probst M (1903) Zur Kenntnis der amyotrophischen Lateralsklerose. Sitsungsberichte der Akademie der Wissenschaften in Wien 112:683–824

Puscariu E, Lambrior AA (1906) Un cas de sclérose latérale amyotrophique. Rev Neurol 14:789–799

Reed DM, Brody JA (1975) amyotrophic lateral sclerosis and parkinsonism-dementia on Guam, 1945–1972. I. Descriptive epidemiology. Am J Epidemiol 101:287–301

Roos RP, Viola MV, Wollmann R, Hatch MH, Antel JP (1980) Amyotrophic lateral sclerosis with antecedent poliomyelitis. 37:312–313

Rossi I, Roussy G (1906) Un cas de sclérose latérale amyotrophique avec dégénération de la voie pyramidal suivie au Marchi de la moelle jusqu'au cortex. Rev Neurol 14:393–406

Rowland LP (1982) Diverse forms of motor neuron diseases. In: Rowland LP (ed) Human motor neuron diseases. Advances in neurology, vol 36. Raven Press, New York, pp 1–14

Rowland LP (1991) Ten central themes in a decade of ALS research. In: Rowland LP (ed) Amyotrophic lateral sclerosis and other motor neuron diseases. Advances in neurology, vol 56. New York: Raven Press pp 3–23

Sakuta M, Nakanishi T, Toyokura Y (1978) Anal muscle electromyograms differ in amyotrophic lateral sclerosis and Shy–Drager syndrome. Neurology 28:1289–1293

Salazar AM, Masters CL, Gajdusek DC, Gibbs CJ (1983) Syndromes of amyotrophic lateral sclerosis and dementia: relation to transmissible Creutzfeldt–Jakob disease. Ann Neurol 14:17–26

Sasaki S, Hirano A (1983) A study of small argyrophilic bodies in motor neuron disease. Neurol Med (Tokyo) 18:334–340

Sasaki S, Okamoto K, Hirano A (1982) An electron microscopic study of small argyrophilic bodies in the human spinal cord. Neurol Med (Tokyo) 17:570–576

Sasaki S, Kamei H, Yamane K, Maruyama S (1988) Swelling of neuronal processes in motor neuron disease. Neurology 38:1114–1118

Sasaki S, Maruyama S, Yamane K, Sakuma H, Takeishi M (1989) Swellings of proximal axons in a case of motor neuron disease. Ann Neurol 25:520–522

Schmidt ML, Carden MJ, Lee VM-Y, Trojanowski JQ (1987) Phosphate-dependent and independent neurofilament epitopes in the axonal swellings of patients with motor neuron disease and controls. Lab Invest 56:282–294

Schochet SS, Hardman JM, Ladewig PP, Earle KM (1969) Intraneuronal conglomerates in sporadic motor neuron disease. Arch Neurol 20:548–553

Schroeder HD, Reske-Nielsen E (1984) Preservation of the nucleus X-pelvic floor motosystem in amyotrophic lateral sclerosis. Clin Neuropathol 3(5):210–216

Schwartz MS, Swash M (1982) Pattern of involvement in the cervical segments in the early stage of motor neuron disease. A single fibre EMG study. Acta Neurol Scand 65:424–431

Siddique T, Figlewicz DA, Pericak-Vance MA et al. (1991) Linkage of a gene causing familial amyotrophic lateral sclerosis to chromosome 21 and evidence of genetic-locus heterogeneity. N Engl J Med 324:1381–1384

Smith AWM, Mulder DW, Code CF (1957) Esophageal motility in amyotrophic lateral sclerosis. Proc Mayo Clin 32:438–440

Smith MC (1960) Nerve fibre degeneration in the brain in amyotrophic lateral sclerosis. J Neurol Neurosurg Psychiatr 23:269–282

Sobue G, Matsuoka Y, Mukai E, Takayanagi T, Sobue I (1981a) Spinal and cranial motor nerve roots in amyotrophic lateral sclerosis and X-linked recessive bulbospinal muscular atrophy: morphometric and teased-fiber study. Acta Neuropathol 55:227–235

Sobue G, Matsuoka Y, Mukai E, Takayanagi T, Sobue I (1981b) Pathology of myelinated fibers in cervical and lumbar ventral spinal roots in amyotrophic lateral sclerosis. J Neurol Sci 50:413–421

Sobue G, Sahashi K, Takahashi A, Matsuoka Y, Muroga T, Sobue I (1983) Degenerating compartment and functioning compartment of motor neurons in ALS: possible process of motor neuron loss. Neurology 33:654–657

Spiller WG (1900) Contributions from the William Pepper Laboratories of Clinical Medicine, Philadelphia, p 63

Spiller WG (1905) Primary degeneration of the pyramidal tracts. Univ Pennsylvania Med Bull 18:390

Stanhope JM, Brody JA, Morris CE (1972) Epidemiological features of amyotrophic lateral sclerosis and parkinsonism-dementia in Guam, Marianas Islands. Int J Epidemiol 1: 199–210

Swash M (1980) Vulnerability of lower brachial myotomes in motor neuron disease. J Neurol Sci 47:59–68

Swash M, Leader M, Brown A, Swettenham KW (1986) Focal loss of anterior horn cells in the cervical cord in motor neuron disease. Brain 109:939–952

Swash M, Scoltz CL, Vowles G, Ingram DA (1988) Selective and asymmetrical vulnerability of corticospinal and spinocerebellar tracts in motor neuron disease. J Neurol Neurosurg Psychiatr 51:785–789

Takahashi K, Nakamura H, Okada E (1972) Hereditary amyotrophic lateral sclerosis. Histochemical and electronmicroscopic study of hyaline inclusions in motor neuron disease. Arch Neurol 27:292–299

Takahata N, Yamanouchi N, Fukatsu R (1976) Brainstem gliosis in a case with clinical manifestations of amyotrophic lateral sclerosis. Folia Psychiatr Neurol Jpn 30:41–48

Tanaka J, Nakamura H, Tabuchi Y, Takahashi K (1984) Familial amyotrophic lateral sclerosis: features of multisystem degeneration. Acta Neuropathol 64:22–29

Thomas PK (1991) Separating motor neuron diseases from pure motor neuropathies. In: Rowland LP (ed) Amyotrophic lateral sclerosis and other motor neuron diseases. Advances in neurology, vol 56. Raven Press, New York, pp 381–384

Toyokura Y (1979) Negative features in ALS. In: Tsubaki T, Tokoyura Y (eds) Amyotrophic lateral sclerosis. University Park Press, Baltimore, pp 53–58

Troost D, Louwerse ES, De Jong JMBV, Van Leersum, Van Raalte JA (1989) Aberrant myelinated neurites in the anterior horns of a patient with amyotrophic lateral sclerosis. Clin Neuropathol 8:152–155

Tsubaki T (1979) Review of studies of the Ministry of Health and Welfare motor neuron disease research committee, Japan. In: Tsubaki T, Tokoyura Y (eds) Amyotrophic lateral sclerosis. University Park Press, Baltimore, pp 43–51

Tsukagoshi H, Yanagisawa N, Oguchi K, Nagashima K, Murakami T (1979) Morphometric quantification of the cervical limb motor cells in controls and in amyotrophic lateral sclerosis. J Neurol Sci 41:287–297

Tsukagoshi H, Yanagisawa N, Oguchi K (1980) Morphometric quantification of the cervical limb motor cells in various neuromuscular diseases. J Neurol 47:463–472

Wadia PN, Bhatt MH, Misra VP (1987) Clinical neurophysiological examination of deafness associated with juvenile motor neuron disease. J Neurol Sci 78:29–33

Wechsler IS, Davison C (1932) Amyotrophic lateral sclerosis with mental symptoms. A clinicopathological study. Arch Neurol Psychiatr 27:857–880

Wechsler IS, Brock S, Weil A (1929) Amyotrophic lateral sclerosis with objective and subjective (neuritic) sensory disturbances. A clinical and pathological report. Arch Neurol Psychiatr 21:299–310

Wikstrom J, Paetau A, Paol J, Sulkava R, Haltia M (1982) Classic amyotrophic lateral sclerosis with dementia. Arch Neurol 39:681–683

Williams C, Kozlowski MA, Hinton DR, Miller CA (1990) Degeneration of spinocerebellar neurons in amyotrophic lateral sclerosis. Ann Neurol 27:215–225

Wohlfart G (1957) Collateral regeneration from residual motor nerve fibres in amyotrophic lateral sclerosis. Neurology 7:124–134

Wohlfart G (1959) Degenerative and regenerative axonal changes in the ventral horns, brainstem and cerebral cortex in amyotrophic lateral sclerosis. Acta Univ Lundensis 56:1–13

Wohlfart G, Swank RL (1941) Pathology of amyotrophic lateral sclerosis: fiber analysis of the ventral roots and pyramidal tracts of the spinal cord. Arch Neurol Psychiatr 46:783–799

Yuan J, Horvitz HR (1990) The Caenorhabditis genes *ced*-3 and *ced*-4 act cell-autonomously to cause programmed cell death. Dev Biol 138:33–41

6 Alternative Approaches to the Pathology of Motor Neuron Disease

J. E. Martin and M. Swash

The pathology of motor neuron disease (MND) is well recognised, but poses many problems of interpretation and understanding. There is, as yet, no insight into the pathobiological mechanisms that underlie the development of the specific neurodegenerative changes that must characterise the disease. In order to try to understand this problem a number of other approaches have been tried. These include:

1. A search for naturally occurring and experimental models of MND
2. Studies of the motor end plate including axonal sprouting
3. Studies of axonal transport in the peripheral nervous system in MND
4. Evaluation of the mechanism of selective neuronal death, and of selective resistance of Onuf's nucleus.

These alternative approaches to understanding of the pathology of MND are reviewed in this chapter.

Animal Models in the Study of MND

Two basic groups of experimental approaches have been used in this study and development of animal models designed to mimic human motor neuron disease. The two groups comprise naturally occurring animal models and experimentally induced MND-like disorders. At this time however, there is no animal model of MND that is generally agreed to adequately mimic the range of clinical and pathological features of the disease seen in man.

Naturally Occurring Animal Models of MND

The most extensively studied naturally occurring putative animal models of MND occur in the Brittany spaniel and the wobbler mouse. An autosomal dominant form of spinal muscular atrophy – hereditary canine spinal muscular atrophy (HCSMA) may develop in the Brittany spaniel. This disorder resembles the progressive muscular atrophy form of MND (Cork et al. 1979). In this model the affected dogs develop a rapidly progressive, predominantly proximal weakness if homozygous for the trait, whereas heterozygotes have milder or more chronic disease (Sack et al. 1984). The pathology of the canine disease is notable for the prominence of axonal spheroids in the anterior horn of the spinal cord (Cork et al. 1982), a feature also seen in human motor neuron disease (Carpenter 1968; Sasaki et al. 1988). In HCSMA there is thought to be a disturbance in the slow components of axonal transport (Griffin et al. 1982) and neurofilaments accumulate in neuronal perikarya. The model is useful in the study of genetic influence in human MND, and the presence of axonal swellings in anterior horns suggests some relationship to the pathology of human MND. It is well recognised, however, that in the commonest form of human MND, i.e. sporadic MND, there is no simple relationship between genetic history and disease (Mulder et al. 1986; Kurtzke and Beebe 1980). Similarly, the pathological findings in motor neuron disease, whilst including axonal swellings (Carpenter 1968), also include other features such as Bunina bodies (Bunina 1962) and Hirano bodies (Hirano 1965), not found in HCSMA. In addition, there is no loss of motor neurons in HCSMA.

Hirano bodies associated with degeneration of cortical and hippocampal neurons may be found in a natural murine mutant, the brindled mouse, characterised by a deficiency in copper metabolism (Nagara et al. 1980). The clinical and pathological features of the neuronal degeneration in the brindled mouse are however, otherwise unlike those found in human MND. A spontaneous lower motor neuron disease that has been suggested as a possible model for the human disorder may also develop in rabbits and pigs and an hereditary amyotrophic disorder in pointer dogs has been described, although preliminary neuropathological studies failed to demonstrate neuropathological abnormalities of anterior horn cells (Yamaguchi et al. 1978).

The wobbler mouse is another naturally occurring disorder which has been used as a model of human MND (Duchen et al. 1968; Andrews et al. 1974). In this model affected mice develop weakness, atrophy and contractures chiefly affecting forelimb muscles, and the pathological changes are seen chiefly in the cervical cord (Duchen et al. 1968; Andrews 1975). The disorder is inherited as an autosomal recessive trait (Duchen et al. 1968). Affected mice are small for their age. The pathological features consist of vacuolation and degeneration of anterior horn cells, although only a small proportion of anterior horn cells are affected at any given clinical stage (Mitsumoto and Bradley, 1982). Occasionally dorsal root ganglion cells may also degenerate. There is abnormal protein synthesis in the anterior horn cells (Murakami et al. 1980) with accumulation of neurofilaments. Further, there is reduced regenerative capacity of these motor neurons after nerve crush (Mitsumoto 1985). Slow axonal transport has been found to be reduced in the distal nerve segments (Bird et al., 1971, Mitsumoto and Gambetti, 1983) representing data acquired from the mixed sensorimotor axons found in limb nerves. These abnormalities in the wobbler mouse have been ascribed to a primary neuronopathy according to the classification described by Spencer and Schaumberg (1981); that is, the degeneration starts in the neuronal perikaryon and then spreads to the proximal axon (Mitsumoto and Bradley 1982). This animal model is useful because the rapid timecourse of the degeneration permits many cycles to be studied, but it does not provide an accurate model of human motor neuron disease from the pathological or genetic point of view. However, it is useful in the study of neuronal degeneration and regeneration.

Another murine mutant, the "wasted" mouse, has been described by Shultz et al. (1982) and by Lutsep and Rodriguez (1989). The mutants, mice homozygous for the "wasted" recessive gene, develop tremor, ataxia, weakness of the hind limbs and weight loss (Shultz et al. 1982). Neuropathological studies of this mutant show vacuolar degeneration of anterior horn cells in the spinal cord and of motor neurons in the brainstem motor nuclei (Lutsep and Rodriguez, 1989). There is no loss of motor neurons, but prominent staining for phosphorylated 200 kD neurofilament epitopes was demonstrated (Lutsep and Rodriguez 1989), a feature that together with neuronal vacuolation may also be seen in human MND (Manetto et al. 1988; Munoz et al. 1988), although not all workers have found such changes (Leigh et al. 1989; Fig. 6.1).

A murine mutant *Mnd* has been described by Messer and Flaherty (1986) and subsequently by Messer et al. (1987) and Mazurkiewitz et al. (1988, 1990). In mice homozygous for this dominant trait weakness and wasting of limbs develops leading to paralysis, particularly of the hind limbs. degeneration of upper and lower motor neurons occurs, and this neuronal degeneration is accompanied by intraneuronal inclusion body formation (Fig. 6.2) and increased expression of ubiquitin within anterior horn cells (Mazurkiewitz 1990). The pathological appearances of neuronal degeneration in this mutant appeared to be similar to the human disorder in early studies, but further studies have shown that there is a widespread accumulation of Luxophilic inclusions of possible lysosomal origin within both sensory and motor neurons in the spinal cord and cranial nerve nuclei (Mazurkiewitz et al.; Green and Martin, data in press). The underlying defect in the *Mnd* mouse is unknown, but there may be an abnormality of lysosomal processing within neurons.

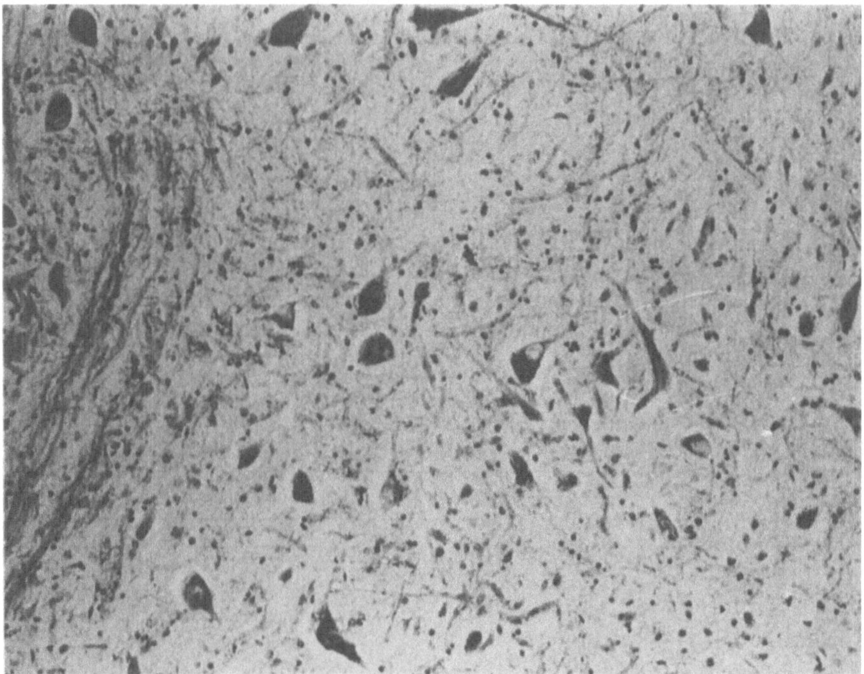

Fig. 6.1. Vacuolated neurons from the hypoglossal nucleus in a patient dying with MND (Luxol-fast blue/cresyl violet).

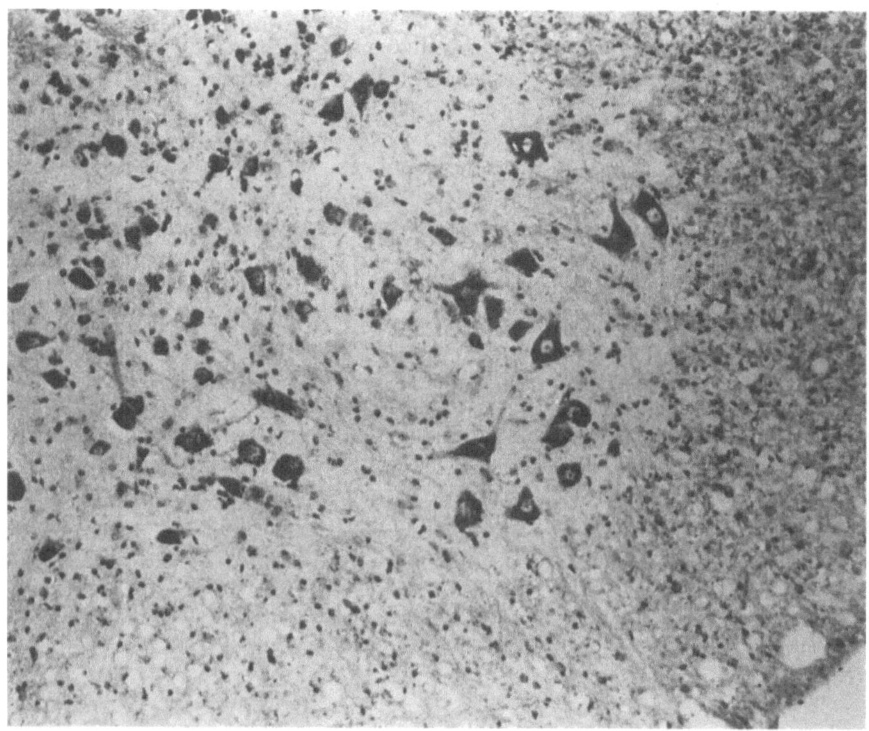

Fig. 6.2. Anterior horn cells in the *Mnd* mouse mutant containing multiple Luxophilic inclusion bodies (Luxol-fast blue/cresyl violet).

The progressive motor neuronopathy (*pmn*) mouse is another recently described spontaneous mutant which develops paralysis of the hindlimbs in the early weeks of life, followed by forelimb weakness and death by the age of six to seven weeks after birth (Schmalbruch et al. 1991). This disorder is autosomal recessive and heterozygotes are unaffected. There is progressive axonal degeneration, starting at the motor end plates and progressing proximally, but with only mild neuronal changes until late in the disease (Schmalbruch et al. 1991; Sendtner et al. 1992b). This

Table 6.1. Possible animal models of MND

Natural	Neurotoxic
Brittany spaniel (HCSMA)	Doxorubicin
Wobbler mouse	Vitamin C deficiency (guinea pig)
Brindled mouse	Poliovirus (with cyclophosphamide)
Wasted mouse	Mouse neurotropic retrovirus
Mnd mouse	Bovine akabane virus
pmn mouse	Excitotoxins
Rabbit MND	
Pig MND	
Pointer amyotrophy	

model, together with the *Mnd* mouse, has been used in studies of the rescue of motor neuronal function by ciliary neurotrophic factor (Sendtner et al. 1992a,b).

Neurotoxic Substances

Experimental animal models of MND chiefly involve the use of neurotoxic agents, whether chemical or viral. A variety of chemical agents have been used to produce neurological diseases in animal models that, in some way, resemble human disorders. A recent example in this field has been the discovery of an acute parkinsonian syndrome in drug addicts using heroin contaminated with 1-methyl-4-phenyl-1,2,3,6-tetrahydropyridine-MPTP (Ballard et al. 1985). This human neurotoxic model has been reproduced in experimental animals with production of parkinsonism and histological changes suggestive of Lewy body formation (Forno et al. 1986). This neurotoxicological approach promises to lead to greater understanding of the pathogenesis of Parkinson's disease (Gibb 1989). Many neurotoxic models of MND and of other neuronal disorders depend on the experimental production of abnormalities of neuronal metabolism. For example, doxorubicin is transported retrogradely along axons in the CNS and produces translational defects in protein synthesis in anterior horn cells in the rat (Yamamoto et al. 1984). Deficiency of vitamin C in the guinea pig may cause death of anterior horn cells and corticospinal tract degeneration, sometimes in the presence of a bleeding disorder (Den Hartog Jager 1985).

Neurotropic Viruses

Certain viruses may produce selective neuronal degeneration. Poliovirus is an RNA virus with specific neurotropism to anterior horn cells, and studies of this infection are therefore of particular interest in relation to MND (Martyn 1990). It has been shown that the virus attaches to anterior horn cells via a glycoprotein receptor expressed solely on cells capable of infection by poliovirus (Holland et al. 1959). In addition, infection with active replication of poliovirus results in intracellular degradation of specific polypeptides (Urzukainqui and Carrasco 1990). Although poliomyelitis is typically an acute illness a chronic poliovirus infection leading to paralysis may be induced in mice using cyclophosphamide pretreatment (Jubelt and Meagher 1984): the latency of the disorder is inversely related to the infective dose of virus. Histological abnormalities in anterior horn cells are seen only in symptomatic animals. The pathological picture of acute or remote polio in humans, however, whilst showing loss of anterior horn cells, does not generally include the presence of inclusion bodies such as those described by Bunina and Hirano, and the ubiquitinated inclusions virtually specific for MND described in anterior horn cells by Leigh et al. (1988) and by Lowe et al. (1988).

Mouse neurotropic (leukaemia) retrovirus infection may produce lower motor neuron degeneration and axonal swellings (Gardner et al. 1976) and may also prove useful as a model of neuronal degeneration with similarities to that found in human MND. Spongiform change and glial cell involvement has been described in this infection (Brooks et al. 1980), features not seen in the typical human disorder. Bovine akabane disease results from akabane arbovirus infection. Transplacental infection of calves may occur and affected animals develop arthrogryposis associated with degeneration and loss of anterior horn cells in the spinal cord. Wallerian degeneration is found in the descending motor tracts and there is resultant neurogenic muscular atrophy (Hartley et al. 1977). In bovine akabane disease there is a loss of muscarinic cholinergic, glycine/strychnine and central-type benzodiazepine receptors in the anterior horn (Grundlach et al. 1990) reflecting neurotransmitter changes similar to those associated with anterior horn cell loss seen in human MND (Whitehouse et al. 1983).

Excitotoxic models of anterior horn cell degeneration are reviewed in Chapter 10.

In conclusion, there is no entirely satisfactory animal model of sporadic human motor neuron disease. Naturally occurring animal models of anterior horn cell degeneration and acquired neurotoxic models fail to mimic the highly specific features of MND. It is likely that further understanding of the pathogenesis of these models in relation to the natural human form of MND will be required in order to allow a rational approach to the development of experimental systems to study the mechanisms of anterior horn cell death, and thus of possible therapeutic approaches to human MND.

Axonal Sprouting and the Motor End Plate

The phenomenon of the fibre type grouping seen in muscle biopsies is the result of denervation (Fig. 6.3) followed by reinnervation from axonal sprouting derived from the terminal innervation or motor end plate of parts of neighbouring motor units (Edstrom and Kugelberg 1968; Dubowitz and Brooke 1973). Sprouting occurs as a normal feature of intramuscular nerve development, and the process is typical of the potential effects of neuronal plasticity (Cuppini et al. 1990). At early stages in development there is an excess of nerve endings per muscle fibre, resulting in multiple innervation (Brown et al. 1981; Gorio et al. 1983). Numerous sprouts then degenerate to yield the adult nerve fibre–muscle fibre ratio (Brown et al. 1981, Gorio et al. 1983). Sprouting occurs if a normal nerve fibre lies close to degenerating nerve fibre. The degenerating nerve induces nodal sprouting in surviving neurons, the normal nerve fibre sending out numerous collateral sprouts along its course. The outgrowing fibres are tipped by growth-clones which are seen as terminal expansion of the fibre (Wohlfart 1957). Several sprouts enter the tubular Bungner bands formed by Schwann cells, grow along these paths and enter the motor endplate region (Wohlfart 1957).

The molecular mechanisms underlying the processes of axonal sprouting are not clear. Some authors have considered that axonal regeneration is mediated by release of a stimulatory factor from degenerating neurons (Brown et al. 1981; Hoffman and Springell 1951), and others have postulated a reaction to a decrease in the retrograde transport of neurotrophic factors from the end-organ (see Schwab 1990).

When sprouting is a result of axonal severance, the regenerative process is accompanied by nerve cell body alterations which comprise the "axonal reaction". This is seen in histological preparations as chromatolysis, with dispersion of granular endoplasmic reticulum, perikaryal swelling and nuclear displacement. Concomitant with these morphological changes there is a decrease in neurofilament synthesis and an increase in microtubule synthesis. These alterations are thought to

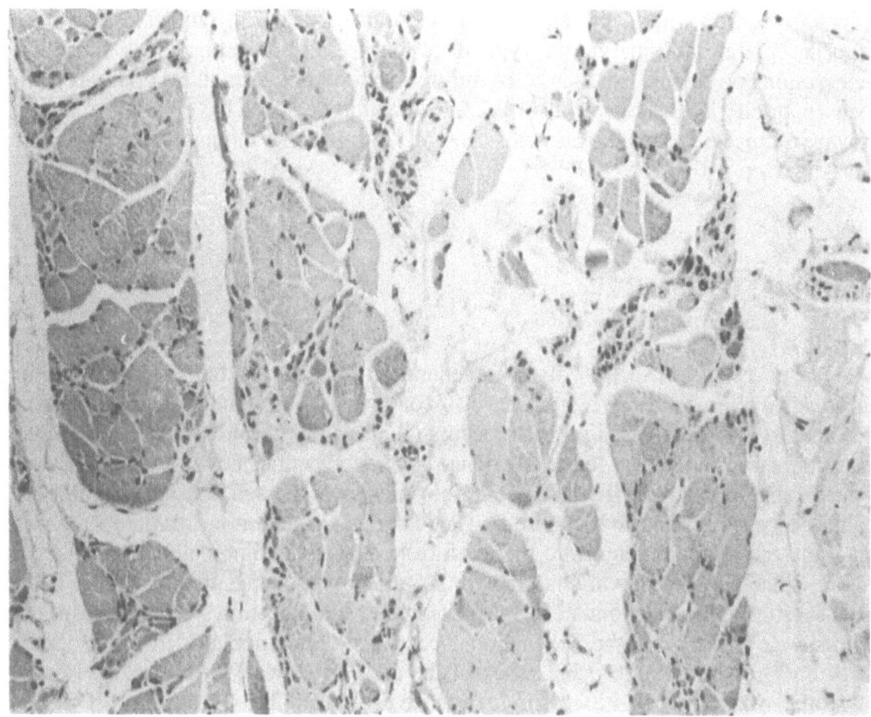

Fig. 6.3. Section of tongue muscle from a patient dying with MND, showing denervation atrophy, with clusters of small angular fibres.

reflect the diversion of cellular metabolism from maintenance of axonal calibre to extension of axons via growth cone and neurite extension. From cell culture models it has been shown that the growth cone of axons contains mainly actin microfilaments in the motile filopodial tips, whereas the microtubules form the basis of the points at which neurite extension occurs, neurofilaments appearing once axonal regeneration is established (Yamada et al. 1970, 1971; Spiegelman et al. 1979; Jockusch and Jockusch 1981). Evidence of reinnervation by sprouting in humans with MND has been demonstrated both by electrophysiological and morphological methods (Buchthal & Pinelli 1953; Wohlfart 1957). Gurney et al. (1984) suggested, using botulinum-induced sprouting models, that an anti-sprouting antibody may be present in the serum of patients with MND, but this has not been confirmed.

The morphology of the end plate and axonal branches has been studied using a variety of methods including double-labelling techniques to demonstrate acetylcholinesterase and axons (Pestronk and Drachman 1978 ; Bjornskov et al. 1975; 1982; Alderson et al. 1989). The end plate in MND shows a number of abnormalities, especially segmentation and elongation (Wohlfart 157; Bjornskov et al. 1982). A "sprouting index" has been used to measure axonal sprouting in human and experimental lower motor neuron diseases (Coers et al. 1973a; Pestronk and Drachman 1978; Bjornskov et al. 1975; 1982). This index is derived from the mean number of branch points, or number of branches of the terminal axon, multiplied by the mean end plate length. It is thought that in denervation the axonal sprouts reinnervate end plates, whereas in myopathies there is expansion of the end plate or of its arborisation without collateral sprouts (Coers et al. 1973b). In MND however, both the collateral innervation typical of progressive neuronal loss implying denervation and reinnervation and the multiple innervation seen in myopathies have been reported (Bjornskov et al. 1982). Myopathic features have however, been shown in muscle from patients with a rapidly progressive clinical course (Achari and Anderson 1974).

Axonal Transport

When a nerve is transected, distal degeneration occurs, but the proximal stump survives (Waller 1852). When a nerve is compressed, accumulation of axoplasm proximal to the compressed segment occurs (Weiss and Hiscoe 1948). These single observations provide the basic evidence for axoplasmic flow. It has become apparent that axoplasmic flow of macromolecules from the cell body is essential for maintenance of axonal architecture and metabolism, since only a limited range of materials can be transferred into the axon from its surrounding cells (Gainer et al. 1977). Since early observations on axoplasmic flow, several methods have been used to study this phenomenon: morphological studies with specialised optics, radioisotope tracing and the use of fluorescent-labelled compounds in living tissue (Droz and Leblond 1962; Ochs 1982; Okabe and Hirokawa 1990).

Axonal transport can be classified into two phases: the slow transport phase (0.1–5 mm/day) and the fast transport phase (200–400 mm/day). The slow component of axonal transport (SC) carries the bulk of the cytoplasmic cytoskeletal proteins from the cell body into the axon (Ochs 1982: Lasek et al. 1984). Fast transport (retrograde, FCr, and also anterograde, FCa: Aquino et al. 1987) carries small vesicles and membranous organelles thought to be related to neuronal transmission (Tsukita and Ishikawa 1981; Ochs 1982).

Slow Axonal Transport

All major structural proteins of the axon (tubulin, actin, neurofilament, clathrin, fodrin), the proteins regulating their polymerisation (tau factors and calmodulin) and enzymes concerned with glycolysis (neuron-specific enolase, creatine kinase, aldolase, pyruvate kinase and lactic dehydrogenase) are carried by the slow component of axonal transport (SC) transport. SC axonal transport moves material down the axon, i.e. in the anterograde direction. It consists of subcomponents classified on the basis of transport rates and type of molecules carried, although some variability between species and nerve types is seen (McQuarrie et al. 1986). The slower component of slow axonal transport (SCa) moves at 0.1–2 mm/day and carries proteins including the three neurofilament subunits (200 kD, 160 kD and 70 kD), tau and tubulin (Hoffman and Lasek 1975; Willard et al. 1979; Black and Lasek 1980; Lasek et al. 1984; Tytell et al. 1984). The faster component of slow axonal transport (SCb) moves at 2 to 5 mm/day and carries actin, myosin, clathrin, calmodulin and glycolytic enzymes (Hoffman and Lasek 1975; Willard et al. 1979; Black and Lasek 1980; Lasek et al. 1984). Both SCa and SCb carry fodrin (Lasek et al. 1984) and spectrin (Mangeat and Burridge 1984).

The mechanism of slow axonal transport is uncertain. Some experiments suggest that slow axonal transport results from translocation of the cytoskeleton in relation to the axolemma (Grafstein and Forman 1980; Lasek 1982). This translocation is thought to result from sliding polymers, associated in particular with microtubules, (Lasek 1986), and the SCa and SCb rates may reflect the speed at which sets of polymers slide past each other (Filliatreau et al. 1988). However, experiments with cells in culture including photobleaching of microinjected fluorescent-labelled tubulin and actin suggest that assembly and disassembly of tubulin and actin may give a false impression of axonal transport (Bambourg et al. 1986; Okabe and Hirokawa 1990) and that these molecules are stationary. Slow axonal transport has not been studied in MND since many techniques for such studies involve invasive procedures unsuitable for human investigation.

Fast Axonal Transport

The fast component of axonal transport (FC) can be visualised by direct microscopy in living cells as a movement of organelles (Breuer et al. 1987; Vallee et al. 1989) and the molecular mechanisms underlying this process are better understood than those of slow axonal transport (Sheetz et al. 1989; Vallee et al. 1989). As well as vesicles and membranous organelles FC axonal transport carries glycolipids, glycoproteins and gangliosides (Tsukita and Ishikawa 1981; Ochs 1982; Aquino et al. 1987). This process has been shown to be dependent on oxidative phosphorylation (Ochs and Hollingworth 1971). In addition, much evidence suggests that microtubules are an essential part of FC transport, since there is spatial association between organelles and microtubules (Tsukita and Ishikawa 1981) and agents that prevent tubulin polymerisation or which disassemble microtubules impair FC axonal transport (Kreutzberg 1969; Schlaepfer 1971; Dahlstrom 1971; Ghetti and Ochs 1978; Griffin et al. 1983).

The purification of microtubules for the study of FC axonal transport produces a mixture of tubulin and "microtubule-associated proteins" (MAPs). Two such MAPs have been purified and are thought to have a role as motors for fast axonal transport. One such MAP, kinesin, is a mechano-chemical ATPase activated by microtubules that was shown to promote gliding of microtubules across a glass coverslip (Brady 1985; Vale et al. 1985; Porter et al. 1987). Kinesin is associated with membrane bound organelles in vivo (Pfister et al. 1989), and whilst initially thought to be a motor for anterograde axonal transport (Sheetz et al. 1989), antibodies to kinesin have been shown to inhibit both FCa and FCr axonal transport (Brady et al. 1990). Dynein (MAP1C), another MAP, is also an ATPase that will promote microtubule gliding in vitro (Paschal et al. 1987), and is thought to be the prime mover in retrograde axonal transport (Vallee et al. 1989, Sheetz et al. 1989). Brain dynein is a cytoplasmic form of the ciliary and flagellar ATPase of the same name (Vallee et al. 1989).

Many neurodegenerative and neuropathic disorders have been considered as disorders of axonal transport (Gajdusek, 1985). Experimental models of chemical modification of neurofilaments have produced axonal swellings similar to those seen in MND, thought to arise from interruption of normal mechanisms of axonal transport of these proteins (Sayre et al. 1985). Retrograde axonal transport has also been implicated as the mechanism for access to the central nervous system of viruses such as polio and rabies (Jubelt et al. 1980; Tsaing 1979), and of heavy metals such as lead and mercury (Baruah et al. 1981; Arvidson 1987). Axonal transport has itself been useful in the study of the anatomy of the central nervous system since retrograde axonal transport will carry horseradish peroxidase (Matsushita and Hosoya 1979) and monoclonal antibodies are both anterogradely and retrogradely transported (Ritchie et al. 1986).

Axonal Transport in MND

Abnormalities of axonal transport have been inferred in MND since Carpenter's description of axonal spheroids (1968). Hirano and colleagues (1980, 1984a,b) and Inoue and Hirano (1979) have described accumulations of 10 mm neurofilaments within axonal swellings and within swollen anterior horn cells. These changes have been used as evidence to support the hypothesis that neurofilament transport (i.e. slow axonal transport, SC) is abnormal in MND (Gajdusek 1985). Direct studies using video-enhanced contrast microscopy to study the motion of organelles (i.e. fast axonal transport) in peripheral nerve from patients with MND have shown increased speed of anterograde transport (FCa) and diminished speed of retrograde transport (FCr) (Breuer et al. 1987). A previous study of fast axonal transport in peripheral nerve from patients with MND failed to distinguish anterograde and retrograde components of fast transport, and is therefore difficult to interpret (Norris 1979). Abnormalities of fast axonal transport in peripheral nerve in MND imply that there is an abnormality of the microtubular system (Breuer et al. 1987) or, perhaps, of the molecular motor, or attachments to the microtubules, i.e. the microtubule associated proteins (MAPs). No studies have been undertaken, as yet, on the biological role of MAPs and their relation to axonal transport in MND. Abnormalities of microtubular proteins have, however, been described in peripheral nerve in MND, and are thought to precede histological and ultrastructural evidence of nerve degeneration (Binet and Meininger 1988).

A primary abnormality in the microtubular system with disruption of fast, and subsequently, slow axonal transport is an attractive hypothesis for a fundamental pathogenetic mechanism to explain ultrastructural and axonal transport abnormalities seen in MND. Neuronal degeneration in the presence of microtubular abnormalities may also be directly consequent upon axonal transport failure if a motor nerve growth factor exists which is retrogradely transported from a target organ (muscle) to promote or stimulate the function and survival of the parent cell body (Slack et al. 1983), analogous to nerve growth factor in the sympathetic and sensory nervous networks (Hendry et al. 1974).

Selective Neuronal Vulnerability in MND: Onuf's Nucleus and other Nuclei

In MND muscle weakness and wasting is consequent on degeneration of anterior horn cells in the spinal cord, bulbar palsy on loss of lower cranial nerve motor neurons, and hyperreflexia on degeneration of the upper motor neurons and pyramidal tracts. Sensory systems are not usually affected. There is also preservation of certain motor functions until late in the natural history of the disease, notably eye movement and bladder and bowel function including the voluntary sphincter musculature (Charcot and Joffroy 1869; Toyokura 1979; Iwata and Hirano 1979). Sparing of eye movements is related to preservation of the 3rd, 4th and 5th cranial nerve nuclei (Iwata and Hirano, 1979). Preservation of bladder and bowel function in MND appears to be related to resistance of neurons in Onuf's nucleus in the sacral spinal cord to degeneration (Mannen et al. 1977; Schrøder and Reske-Nielsen 1984).

Onuf's nucleus is a small, round cluster of neurons in the anterior horn of the sacral spinal cord in man first described in 1889 and 1890 by Onufrowicz. It was initially called nucleus X and was found to be present only in segments S2–S3. In animals the homologue of this nucleus has been found to innervate the striated external anal and periurethral sphincters (Sato et al. 1978; De Aranjo et al. 1982; De Groat et al. 1981). A somatic origin for the nucleus is supported by the motor functions of this nucleus, by the fact that pelvic floor muscles innervated by projections from Onuf's nucleus are under voluntary control (Holstege and Tan 1987) and by the morphology of the neurons. Previously, it was thought that this nucleus might contain autonomic neurons because it showed selective sparing of the neurons in MND (Mannen et al. 1977; Iwata and Hirano 1978), in Werdnig–Hoffmann disease and in acute poliomyelitis (Iwata and Hirano 1978; Sung 1982) in which autonomic functions are preserved. Conversely, in Shy–Drager syndrome, mannosidosis and Hurler's syndrome the anterior horn cells are spared and Onuf's nucleus shows degeneration (Sung et al. 1979; Mannen et al. 1977). In addition, a study by Yamamoto et al. (1978) has suggested that the Onuf nucleus may innervate smooth muscle as well as the striated sphincters.

The neurons of Onuf's nucleus are reported to be smaller than typical motor neurons although, in man, Onuf's nucleus may contain a mixed population of neurons of different sizes (Mannen et al. 1977) and neurons in Onuf's nucleus in the cat and macaque average over 30 microns in diameter. There is a high density of serotonin, substance P, encephalin, C-flanking peptide of neuropeptide Y and somatostatin immunoreactive fibres in the Onuf's nucleus (Gibson et al. 1988, Tashiro et al. 1989). The pattern of peptide immunoreactivity in Onuf's nucleus is similar to that of the sacral autonomic nucleus (Gibson et al. 1988; Katagiri et al. 1986). This evidence, together with electrophysiological evidence has led to the alternative suggestion that the nucleus may be an extension of the intermediolateral cell column (Sung et al. 1979; Holstege et al. 1986). A cellular bridge linking Onuf's nucleus and the sacral intermediolateral cell group has been described by Rexed (1954), Petras and Cummings (1972) and Yamamoto et al. (1978), suggesting coordination of function or control. There is however, supraspinal control of the Onuf's motor neurons (Holstege and Tan 1987), more suggestive of somatic origin (Swift 1989). In conclusion, current evidence tends to support the concept of Onuf's nucleus with somatic origin, but a close association with the autonomic system.

The problems posed by selective neuronal degeneration in MND with sparing of specific neuronal groups such as Onuf's nucleus require a number of experimental approaches. Studies of the anatomy and physiology of neurons resistant to a disease may reveal features unique to these cells that protect them from degeneration. Another approach involves the postulation of a hypothetical mechanism of disease and the testing of such hypotheses.

In the study of features that may be unique to groups of disease resistant neurons, both classical anatomical and newer experimental methods may be used. In relation to the study of Onuf's nucleus, classical methods of anatomical study have established interconnection of neurons of Onuf's nucleus and the intermediolateral column in animals and humans (Rexed 1954; Petras and Cummings 1972; Yamamoto et al. 1978), and clinicopathological and electrophysiological studies indicate the functional interrelation of the nucleus and the striated sphincters (Mannen et al. 1977; Iwata and Hirano 1979; Sung et al. 1979). Anatomical tracing methods in animals such as the cat have also played a useful role in establishing connectivity (De Aranjo et al. 1982; De Groat et al. 1981). Recent advances in the tracing of neurons in whole mount thick-slice preparations of formalin-fixed material by microinjection of fluorescent dyes may also have uses in establishing single cell and anatomical connections in humans in health and disease (Einstein, 1988). The characterisation of neurotransmitters aids in establishing cell origin and function, for example the studies of Gibson et al. (1988), and others on neuropeptide expression in the spinal cord (Tashiro et al. 1989). Other methods that may help to classify neuronal types are studies of the ultrastructural features of synapses. Conradi (1969) showed that C-type synapses uniquely characterise somatic alpha-motoneurons. Using this criterion, Pullen (1988) demonstrated C-type synapses on cytochemically identified sphincteric motoneurons in Onuf's nucleus of the cat, thus unequivocally characterising them as somatic. However, ultrastructural studies on human central nervous system are hampered by the requirement for fresh material. Recently, C-type synapses have been identified on human motor neurons by Pullen et al. (1992), using short delay post-mortem spinal cord. C-type synapses were also identified on neurons in Onuf's nucleus (Pullen et al. 1992).

Reverse genetic studies may also help elucidate neuronal characteristics specific to a group or subset of neurons. For example, the identification of genes responsible for the late-onset degeneration of a subset of neurons (*ced*) (*c*ell *d*eath) in a particular nucleus in the nematode *Caenorhabditis elegans*, and the suppression of such degeneration by a mutation in a permissive gene (*mec*-6) (Chalfie and Wolinsky 1990; Hengartner et al. (1992) may lead to the identification of metabolic pathways specific to this subset of neurons and, by extrapolation, to a specific target for degenerative or toxic damage. The identification of the degenerin family of proteins responsible for such neuronal death is another advance that may lead to the identification of related motor neuronal proteins. Similar metabolic or structural specificities in motor neurons may be revealed when the gene for acute and chronic spinal muscular atrophy (SMA), linked to chromosome 5q (Gilliam et al. 1990), is cloned, sequenced and the gene product identified. This product, already at risk in mutations affecting 5q, may prove a target for other toxic, viral or degenerative processes in selective motor neuron degeneration.

Studies involving the postulation of causes or mechanisms for selective vulnerability also provide an approach to the problem. Such suggestions include trans-synaptic spread of toxins or other factors to account for involvement of neurons known to lie in direct anatomical continuity (Hudson and Kiernan 1988). It has been shown that diphtheria, tetanus, botulinum and a variety of other toxins are all spread from cell to cell across synapses with retrograde transport (Pullen 1990; Wiley et al. 1982). Any possible such toxin is, however, as yet unidentified in MND. Neurotrophic viruses have been implicated as a cause of MND (Kennedy 1990) and poliovirus has long been regarded as a candidate virus, since it causes a motor neuron-specific pattern of damage (Martyn et al. 1988; Martyn 1990). However, the presence of persistent or latent poliovirus infection in patients with MND has not been demonstrated using virological, immunohistochemical or in situ hybridisation methods (Kennedy 1990). The technique of polymerase chain DNA amplification from the RNA poliovirus genome is a more sensitive method for viral detection that is well suited to the resolution of such problems (Kennedy 1990).

Further advances in the study of selective neuronal vulnerability are likely to come only with better understanding of neurons and their interactions with neurons and with glial cells, whether close or distant from the neuronal cell body.

Conclusion

There is, as yet, no insight into the pathobiological mechanisms that underlie the development of specific neurodegenerative changes that characterise MND. Research into the mechanisms of neuronal degeneration in MND has also been hampered by the absence of a suitable or representative animal model; further studies of the *Mnd* mutant may redress this deficiency. Whilst an animal model is awaited, there are however, novel approaches that may lead to such insight in the near future. Such approaches include the development of molecular genetic approaches to the study of mechanisms of selective degeneration and molecular pathological approaches to the study of biochemical and protein chemical changes in neuronal degeneration.

Acknowledgements. J.E.M. is a Wellcome Trust Fellow and holds the Gillson Scholarship in Pathology of the Worshipful Society of Apothecaries. Our work is supported by the Motor Neurone Disease Association of the United Kingdom. We thank Dr A. H. Pullen for kind help and advice.

References

Achari AN, Anderson S (1974) Myopathic changes in amyotrophic lateral sclerosis. Neurology 24:477–481

Alderson K, Pestronk A, Yee W-C, Drachman DB (1989) Silver cholinesterase immunocytochemistry: a new neuromuscular junction stain. Muscle Nerve 12:9–14

Andrews JM (1975) The fine structure of the cervical spinal cord, ventral root and brachial nerves in the wobbler (wr) mouse. J Neuropathol Exp Neurol 34:12–27

Andrews JM, Gardner MB, Wolfgram FJ et al. (1974) Studies on a murine form of spontaneous lower motor neuron degeneration – the wobbler (wr) mouse. Am J Pathol 76:63–78

Aquino DA, Bisby MA, Ledeen RW (1987) Bidirectional transport of gangliosides, glycoproteins and neutral glycosphingolipids in the sensory neurons of rat sciatic nerve. Neuroscience 20:1023–1029

Arvidson B (1987) Retrograde axonal transport of mercury. Exp Neurol 98:198–203

Ballard PA, Tetrud JW, Langston JW (1985) Permanent human parkinsonism due to 1-methyl-4-phenyl-1,2,3,6-tetrahydropyridine (MPTP): seven cases. Neurology 35:949–956

Bambourg JR, Bray D, Chapman K (1986) Assembly of microtubules at the tip of growing axons. Nature 231:788–790

Baruah JK, Chandri GR, Bradley WG, Munsat TL (1981) Retrograde transport of lead in rat sciatic nerve. Neurology 31:612–616

Binet S, Meininger V (1988) Modifications of microtubule proteins in ALS nerve precede detectable histologic and ultrastructural changes. Neurology 38:1596–1600

Bird MT, Shuttleworth E Jnr, Koestner A et al. (1971) The wobbler mouse mutant: an animal model of hereditary motor system disease. Acta Neuropathol 19:39–50

Bjornskov EK, Dekker NP, Norris FH et al. (1975) End-plate morphology in amyotrophic lateral sclerosis. Arch Neurol 32:711–712

Bjornskov EK, Norris FH, Mower-Kuby J (1982) Histochemical staining of the acetylcholine receptor, acetylcholinesterase and the axon terminal. Muscle Nerve 5:140–142

Black MM, Lasek RJ (1980) Slow components of axonal transport; two cytoskeletal networks. J Cell Biol 86:616–623

Brady ST (1985) A novel brain ATPase with properties expected for the fast axonal transport motor. Nature 317:73–75

Brady ST, Lasek RJ (1982) Axonal transport; a cell-biological method for studying proteins that associate with the cytoskeleton. Methods Cell Biol 25(b):365–398

Brady ST, Pfister KK, Bloom GS (1990) A monoclonal antibody against kinesin inhibits both anterograde and retrograde fast axonal transport in squid axoplasm. Proc Natl Acad Sci USA 87:1061–1065

Breuer AC, Lynn MP, Atkinson MB et al. (1987) Fast axonal transport in amyotrophic lateral sclerosis: An intra-axonal organelle traffic analysis. Neurology 37:738–748

Brooks BR, Swarz JR, Johnson RT (1980) Spongiform polioencephalomyelopathy caused by a murine retrovirus. I. Pathogenesis of infection in newborn mice. Lab Invest 43:480–486

Brown MC, Holland RL Hopkins WG (1981) Motor nerve sprouting. Ann Rev Neurosci 4:17–42

Buchthal F, Pinelli L (1953) Action potentials in muscular atrophy of neurogenic origin. Neurology 3:591–603

Bunina TL (1962) On intracellular inclusions in familial amyotrophic lateral sclerosis. An Neuropat Psikhit Korsakov 62:1293–1299

Carpenter S (1968) Proximal axonal enlargement in motor neuron disease. Neurology 18:842–851

Chalfie M, Wolinsky E (1990) The identification and suppression of inherited neurodegeneration in *Caenorhabditis elegans*. Nature 345:410–416

Charcot JM, Joffroy A (1869) Deux cas d'atrophie musculaire progressive avec lésions de la substance grise et faisceaux antéro-latéraux de la moelle épinière. Arch Physiol Norm Pathol 2:354, 629, 744

Coers C, Teleman-Toppet N, Gerard JM (1973a) Terminal innervation ratio in neuromuscular disease. I. Methods and controls. Arch Neurol 49:210–214

Coers C, Telerman-Toppet N, Gerard JM (1973b) Terminal innervation ratio in neuromuscular diseases. II. Disorders of lower motor neuron, peripheral nerve and muscle. Arch Neurol 49:215–222

Conradi S (1969) Ultrastructure and distribution of neuronal and glial elements on the motoneurone surface in the lumbosacral spinal cord of the adult cat. Acta Physiol Scand (Suppl) 332:5–48

Cork LC, Griffin JW, Munnell JF et al. (1979) Hereditary canine spinal muscular atrophy. J Neuropathol Exp Neurol 38:209–222

Cork LC, Griffin JW, Choy C et al. (1982) Pathology of motor neurons in accelerated hereditary canine spinal muscular atrophy. Lab Invest 46:89–99

Cuppini R, Cecchini T, Cuppini C, Ciaroni S, Del Grande P (1990) Time course of sprouting during muscle reinnervation in vitamin E-deficient rats. Muscle Nerve 13:1027–1031

Dahlstrom A (1971) Effects of vinblastine and colchicine on monoamine containing neurons of the rat, with special regard to the axonal transport of amine granules. Acta Neuropathol Suppl 5:226–235

De Aranjo CG, Schmidt RA, Tanaglio RA (1982) Neural pathways to lower urinary tract identified by retrograde axonal transport of horseradish peroxidase. Urology 19:290–295

De Groat WC, Nadelhaft I, Miline RJ et al. (1981) Organisation of the sacral parasympathetic reflex pathways to the urinary bladder and large intestine. J Auton Nerv Syst 3:135–160

Den Hartog Jager WA (1985) Experimental motor neuron disease in the guinea-pig. J Neurol Sci 67:133–142

Droz B, Leblond CP (1962) Migration of proteins along the axons of the sciatic nerve. Science 137:1047–1048

Dubowitz V, Brooke MH (1973) Muscle biopsy: a modern approach. Saunders, Philadelphia

Duchen LW, Strich SJ, Falconer DS (1968) An hereditary motor neuron disease with progressive denervation of muscle in the mouse: the mutant 'wobbler'. J Neurol Neurosurg Psychiatr 31:535–542

Edstrom L, Kugelberg E (1968) Histochemical composition, distribution of fibres and fatiguability of single motor units. J Neurol Neurosurg Psychiatr 31:424–433

Einstein G (1988) Intracellular injection of Lucifer yellow into cortical neurons in lightly fixed sections and its application to human autopsy material. J Neurosci Methods 26:95–103

Filliatreau G, Denoulet P, de Nechaud B, Di Giamberardino L (1988) Stable and metastable cytoskeletal polymers carried by slow axonal transport. J Neurosci 8:2227–2233

Forno LS, Langston JW, DeLanney LE, Irwin I, Ricaurte GA (1986) Locus coeruleus lesions and eosinophilic inclusions in MPTP-treated monkeys. Ann Neurol 20:449–455

Gajdusek DC (1985) Hypothesis: interference with axonal transport of neurofilament as a common pathogenetic mechanism in certain diseases of the central nervous system. N Engl J Med 312:714–719

Gainer H, Tasaki I, Lasek RJ (1977) Evidence for the glia-neuron transfer hypothesis from intracellular perfusion studies of squid giant axons. J Cell Biol 74:524–530

Gardner MB, Rasheed S, Klement V et al. (1976) Lower motor neuron disease in wild mice caused by indigenous type C virus and search for a similar etiology in human amyotrophic lateral sclerosis. In: Andrews JM, Johnson RT, Brazier MAB (eds) Amyotrophic lateral sclerosis. Academic Press, Orlando, pp 217–234

Ghetti B, Ochs S (1978) On the relation between microtubule density and axoplasmic transport in nerves treated with maytansine. In: Canal N, Pozza G (eds); Peripheral neuropathies, developments in neurology. Elsevier-North Holland, Amsterdam.

Gibb WRG (1989) Neuropathology of Parkinson's disease. The Parkinson's Disease Society, London (The Parkinson Papers 4)

Gibson SL, Polak JM, Katagiri T et al. (1988) A comparison of the distribution of eight peptides in spinal cord from normal controls and cases of motor neuron disease, with special reference to Onuf's nucleus. Brain Res 474:255–278

Gilliam TC, Brzustowicz LM, Castilla LH et al. (1990) Genetic homogeneity between acute and chronic forms of spinal muscular atrophy. Nature 345:823–825

Gorio A, Carmignoto G, Finesso M, Polata P, Nunci MG (1983) Muscle reinnervation. II. Sprouting, synapse formation and repression. Neuroscience 8:403–416

Grafstein B, Forman DS (1980) Intracellular transport in neurons. Physiol Rev 60:1167–1283

Griffin JW, Cork LC, Adams RJ et al. (1982) Axonal transport in hereditary canine spinal muscular atrophy. J Neuropathol Exp Neurol 41:370

Grundlach AL, Grabara CSG, Johnston GAR et al. (1990) Receptor alterations associated with spinal motoneuron degeneration in bovine akabane disease. Ann Neurol 27:513–519

Gurney ME (1984) Suppression of sprouting at the neuromuscular junction by immune sera. Nature 307:546–548

Gurney ME, Belton AC, Cashman N et al. (1984) Inhibition of terminal axonal sprouting by serum from patients with amyotrophic lateral sclerosis. N Engl J Med 311:933–939

Hartley WJ, De Saram WG, Della-Porta AJ et al. (1977) Pathology of congenital bovine epizootic arthrogryposis and hydranencephaly and its relationship to Akabane virus. Aust Vet J 53:319–325

Hendry IA, Stoeckel K, Thoenen H, Iversen LL (1974) The retrograde axonal transport of nerve growth factor. Brain Res 68:103–121

Hengartner MO, Elis RE, Horvitz HR (1992) *Caenorhabditis elegans* gene *ced*-9 protects cells from programmed cell death. Nature 356:494–499

Hirano A (1965) Pathology of amyotrophic lateral sclerosis. In: Gajdusek DC, Gibbs CJ (eds) Slow, latent and temperate infections. NINDB Monograph No 2. National Institutes of Health, Washington, DC, pp 3–37

Hirano A, Inoue K (1980) Early pathological changes in amyotrophic lateral sclerosis: electron microscopic studies of chromatolysis, spheroids and Bunina bodies. Neurol Med (Tokyo) 13:148–160

Hirano A, Donnenfeld H, Sasaki S, Nakano I (1984a) Fine structural observations of neurofilamentous changes in amyotrophic lateral sclerosis. J Neuropathol Exp Neurol 43:461–470

Hirano A, Nakano I, Kurland LT, Mulder DW, Holley PW, Saccomanno G (1984b) Fine structural study of neurofibrillary changes in a family with amyotrophic lateral sclerosis. J Neuropathol Exp Neurol 43:471–480

Hoffman H, Springell PH (1951) An attempt at the chemical identification of "neurocletin" (the substance evoking axon sprouts). Aust J Exp Biol Med 29:417–424

Hoffman PH, Lasek RJ (1975) The slow component of axonal transport; identification of major structural polypeptides of the axon and their generality among mammalian neurons. J Cell Biol 66:351–366

Holland JJ, McLaren LC, Syverton JT (1959) Mammalian cell-virus relationship. III. Poliovirus production by non-primate cells exposed to poliovirus ribonucleic acid. Proc Soc Exp Biol 100:843–845

Holstege G, Tan J (1987) Supraspinal control of motoneurons innervating the striated muscles of the pelvic floor including urethral and anal sphincters in the cat. Brain 110:1323–1344

Holstege G, Griffiths D, Wall HD, Dalm E (1986) Anatomical and physiological observations on supraspinal control of bladder and urethral sphincter muscles in the cat. J Comp Neurol 250:449–461

Hudson AJ, Kiernan JN (1988) Preservation of certain voluntary muscles in motoneuron disease. Lancet i:652–653

Inoue K, Hirano A (1979) Early pathological changes in amyotrophic lateral sclerosis: autopsy findings in a case of 10 months duration. Neurol Med (Tokyo) 11:448–455

Iwata M, Hirano A (1978) Sparing of the Onufrowicz nucleus in sacral anterior horn lesions. Ann Neurol 4:245–249

Iwata M, Hirano A (1979) Current problems in the pathology of amyotrophic lateral sclerosis. In: Zimmerman HM (ed) Progress in neuropathology. Raven Press, New York, pp 277–298

Jockusch H, Jockusch BM (1981) Structural proteins in the growth cone of cultured spinal cord neurons. J Cell Biol 131:345–352

Jubelt M, Meagher JB (1984) Poliovirus infection of cyclophosphamide-treated mice results in persistent and late paralysis. I. Clinical, pathologic and immunologic studies. Neurology 34:486–493

Jubelt B, Narayan O, Johnson RT (1980) Pathogenesis of human poliovirus infection in mice. II. Age-dependency of paralysis. J Neuropathol Exp Neurol 39:149–159

Katagiri T, Gibson SJ, Su HC, Polak JM (1986) Composition and central projections of the pudendal nerve in the rat investigated by combined peptide immunocytochemistry and retrograde fluorescent labelling. Brain Res 372:313–322

Kennedy PGE (1990) On the possible role of viruses in the aetiology of motor neuron disease: a review. J R Soc Med 83:784–787

Kreutzberg GW (1969) Neuronal dynamics and axonal flow. IV. Blockage of intra-axonal enzyme transport by colchicine. Proc Natl Acad Sci USA 62:722–728

Kurtzke JF, Beebe GW (1980) Epidemiology of amyotrophic lateral sclerosis. I. A case–control comparison based on ALS deaths. Neurology 30:453–462

Lasek RJ (1982) Translocation of the neuronal cytoskeleton and axonal locomotion. Philos Trans R Soc 299:313–327

Lasek RJ (1986) Polymer sliding in axons. J Cell Sci (Suppl 5):161–179

Lasek RJ, Garner JA, Brady ST (1984) Axonal transport of the cytoskeletal matrix. J Cell Biol 99:212s–221s

Leigh PN, Anderton BH, Dodson A, Gallo J-M, Swash M, Power DM (1988) Ubiquitin deposits in anterior horn cells in motor neuron disease. Neurosci Lett 93:197–203

Leigh PN, Dodson A, Swash M, Brion J-P, Anderton BH (1989) Cytoskeletal abnormalities in motor neuron disease. An immunocytochemical study. Brain 112:521–535

Lowe J, Lennox G, Jefferson D et al. (1988) A filamentous inclusion body within anterior horn neurons in motor neuron disease defined by immunocytochemical localisation of ubiqutin. Neurosci Lett 94:204–210

Lutsep HL, Rodriguez M (1989) Ultrastructural, morphometric and immunocytochemical study of anterior horn cells in mice with 'wasted' mutation. J Neuropath Exp Neurol 48:519–533

Manetto V, Sternberger NH, Perry G et al. (1988) Phosphorylation of neurofilaments is altered in amyotrophic lateral sclerosis. J Neuropathol Exp Neurol 47:642–653

Mangeat PH, Burridge K (1984) Immunoprecipitation of non-erythrocyte spectrin within live cells following microinjection of specific antibodies: relation to cytoskeletal structures. J Cell Biol 98:1363–1377

Mannen T, Iwata M, Toyokura M et al. (1977) Preservation of a certain motoneurone group of the sacral cord in amyotrophic lateral sclerosis; its clinical significance. J Neurol Neurosurg Psychiatr 40:464–469

Martyn CN (1990) Poliovirus and motor neuron disease. J Neurol 237:336–338

Martyn CN, Barker DJP, Osmond C (1988) Motoneuron disease and post poliomyelitis in England and Wales. Lancet i:1319–1322

Matsushita M, Hosoya Y (1979) Cells of origin of the spinocerebellar tract in the rat, studied with the method of retrograde transport of horseradish peroxidase. Brain Res 173:185–200

Mazurkiewitz JE, Callahan LM, Messer A (1988) Distribution of neurofilament epitopes in spinal motoneurons in the normal and motor neuron degeneration mutant (*Mnd*) mouse. J Cell Biol 107:725a

Mazurkiewitz JE (1990) Ubiquitin deposits in spinal motoneurons of the *Mnd* (motor neuron degeneration) mouse. J Neurol Sci (Suppl) 98:349

McQuarrie IG, Brady ST, Lasek RJ (1986) Diversity in the axonal transport of structural proteins: major differences in between optic and spinal axons in the rat. J Neurosci 6:1593–1605

Messer A, Flaherty L (1986) Autosomal dominance in a late-onset motor neuron disease in the mouse. J Neurogenet 3:345–355

Messer A, Strominger NL, Mazurkiewitz JE (1987) Histopathology of the late-onset motor neuron degeneration (*Mnd*) mutant in the mouse. J Neurogenet 4:201–213

Mitsumoto H (1985) Axonal regeneration in wobbler motor neuron disease: quantitative histologic and axonal transport studies. Muscle Nerve 8:44–51

Mitsumoto H, Bradley WG (1982) Murine motor neuron disease (the wobbler mouse). Degeneration and regeneration of the lower motor neurons. Brain 105:811–834

Mitsumoto H, Gambetti P (1983) Slow axonal transport in the wobbler mouse (Murine motor neuron disease). Soc Neurosci 9:151

Mulder DW, Kurland LT, Offord KP et al. (1986) Familial adult motor neuron disease: amyotrophic lateral sclerosis. Neurology 36:511–517

Munoz DG, Greene C, Perl DP, Selkoe DJ (1988) Accumulation of phosphorylated neurofilaments in anterior horn motoneurons of amyotrophic lateral sclerosis patients. J Neuropathol Exp Neurol 47:9–18

Murakami T, Mastaglia L, Bradley WG (1980) Reduced protein synthesis in spinal anterior horn neurons in wobbler mouse mutant. Exp Neurol 67:423–432

Nagara H, Yakajima K, Suzuki K (1980) An ultrastructural study of the cerebellum of the brindled mouse. Acta Neuropathol 52:41–50

Norris FH (1979) Moving axon particles of intercostal nerve terminals in benign and malignant ALS. Proceedings of the International Symposium on Amyotrophic Lateral Sclerosis. University of Tokyo Press, Tokyo, pp 375–385

Ochs S (1982) Axoplasmic transport and its relation to other nerve functions. Wiley, New York

Ochs S, Hollingworth D (1971) Dependence of fast axonal transport in nerve on oxidative phosphorylation. J Neurochem 18:107–114

Okabe S, Hirokawa N (1990) Turnover of fluorescently labelled tubulin and actin in the axon. Nature 343:479–482

Onuf (Onufrowitz) B (1889) Notes on the arrangement and function of the cell groups of the sacral region of the spinal cord. J Nerve Mental Dis 26:498–504

Onuf (Onufrowitz) B (1890) On the arrangement and function of the cell groups of the sacral region of the spinal cord in man. Arch Neurol Psychopathol 3:387–411

Paschal BM, Shpetner HS, Vallee RB (1987) MAP1C is a microtubule-activated ATPase which translocates microtubules in vitro and has dynein-like properties. J Cell Biol 105: 1273–1282

Pestronk A, Drachman D (1978) Motor nerve sprouting and acetylcholine receptors. Science 199:1223–1225

Petras JM, Cummings JF (1972) Autonomic neurons in the spinal cord of the rhesus monkey. A correlation of the findings of cyto-architechtonics and sympathectomy with fiber degeneration following dorsal rhizotomy. J Comp Neurol 146:189–218

Pfister KK, Wagner MC, Stenoien DS, Brady ST, Bloom GS (1989) Monoclonal antibodies to kinesin heavy and light chains vesicle-like structures, but not microtubules, in cultured cells. J Cell Biol 108:1453–1463

Porter ME, Scholey JM, Stemple DL, Vigers GPA, Sheetz MP, McIntosh JR (1987) Characterisation of the microtubule movement produced by sea urchin egg kinesin. J Biol Chem 262:2794–2802

Pullen AH (1988) Quantitative synaptology of feline motoneurons to external anal sphincter muscle. J Comp Neurol 269:414–424

Pullen AH (1990) Morphometric evidence from C-synapses for phased Nissl body response in α-motoneurons retrogradely intoxicated with diphtheria toxin. Brain Res 509:8–16

Pullen AH, Martin JE, Swash M (1992) Ultrastructure of presynaptic input to motor neurones in Onuf's nucleus; controls and motor neuron disease. Neuropathol Appl Neurobiol 18:213–231

Rexed BA (1954) A cytoarchitechtonic atlas of the spinal cord in the cat. J Comp Neurol 100:297–379

Ritchie TC, Fabian RH, Choate JVA, Coulter JD (1986) Axonal transport of monoclonal antibodies. J Neurosci 6:1177–1184

Sack GH Jnr, Cork LC, Morris JM et al. (1984) Autosomal dominant inheritance of hereditary canine spinal muscular atrophy. Ann Neurol 15:369–373

Sasaki S, Kamei H, Yamane K et al. (1988) Swelling of neuronal processes in motor neuron disease. Neurology 38:1114–1118

Sato M, Mizuno M, Konishi A (1978) Localisation of motoneurones innervating peroneal muscles: a HRP study in cat. Brain Res 140:149–154

Sayre LM, Autilio-Gambetti L, Gambetti P (1985) Pathogenesis of experimental giant neurofilamentous axonopathies: A unified hypothesis based on chemical modification of neurofilaments. Brain Res Rev 10:69–83

Schlaepfer WW (1971) Vincristine-induced axonal alterations in rat peripheral nerve. J Neuropathol Exp Neurol 30:488–505

Schmalbruch H, Jensen, H-JS, Bjaerg M, Kamieniecka Z, Kurland L (1991) A new mouse mutant with progressive motor neuronopathy. J Neuropathol Exp Neurol 50:192–204

Schrøder HD, Reske-Nielsen E (1984) Preservation of the nucleus X-pelvic floor motosystem in amyotrophic lateral sclerosis. Clin Neuropathol 3(5):210–216

Schwab ME (1990) Myelin-associated inhibitors of neurite growth and regeneration in the CNS. Trends Neurosci 13:452–456

Sendtner M, Schmalbruch H, Stockli KA, Carroll P, Kreutzberg GW, Thoenen H (1992a) Ciliary neurotrophic factor prevents degeneration of motor neurons in mouse mutant progressive motor neuronopathy. Nature 358:502–504

Sendtner M, Stocki KA, Carroll P, Kreutzberg GW, Thoenen H, Schmalbruch H (1992b) More on motor neurons. Nature 360:541–542

Sheetz MP, Steuer ER, Schroer TA (1989) The mechanism and regulation of fast axonal transport. Trends Neurosci 12:474–478

Shultz LD, Sweet HO, Davisson MT et al. (1982) 'Wasted' a new mutant of the mouse with abnormalities characteristic of ataxia telangiectasia. Nature 297:402–404

Slack JR, Hopkins WG, Pockett S (1983) Evidence for a motor nerve growth factor. Muscle Nerve 6:243–252

Spencer PS, Schaumberg HH (1981) Classification of neurotoxic disease: a morphological approach. In: Spencer PS, Schaumberg HH (eds) Experimental and clinical neurotoxicology. Williams and Wilkins, Baltimore, pp 92–101

Spiegelman BM, Lopata MA, Kirschner MW (1979) Aggregation of microtubule initiation sites preceding neurite outgrowth in mouse neuroblastoma cells. Cell 16:253–263

Sung JH (1982) Autonomic neurons of the sacral spinal cord in amyotrophic lateral sclerosis, anterior poliomyelitis and "neuronal intranuclear hyaline inclusion disease". Distribution of sacral autonomic neurons. Acta Neuropathol 56:233–237

Sung JH, Mastri AR, Segal E (1979) Pathology of Shy–Drager syndrome. J Neuropathol Exp Neurol 38:353–368

Swift TR (1989) Neurons in Onuf's nucleus. Arch Neurol 46:606–607

Tashiro T, Sadota T, Matsushima R et al (1989) Convergence of serotonin, enkephalin- and substance P-like immunoreactive afferent fibres on single pudendal motoneurones in Onuf's nucleus of the cat: a light microscope study combining the triple immunocytochemical staining technique with the retrograde HRP-tracing method. Brain Res 481:392–398

Toyokura Y (1979) Negative features in ALS. In: Tsubaki T, Toyokura Y (eds) Amyotrophic lateral sclerosis. University Park Press, Baltimore, pp 53–58

Tsaing H (1979) Evidence for an intra-axonal transport of fixed and street rabies virus. J Neuropathol Exp Neurol 38:286–296

Tsukita S, Ishikawa H (1981) The cytoskeleton in myelinated axons: Serial section study. Biomed Res 2:424–437

Tytell M, Brady ST, Lasek RL (1984) Axonal transport of a subclass of tau proteins: evidence for the regional differentiation of microtubules in neurons. Proc Natl Acad Sci USA 81:1570–1574

Urzukainqui A, Carrasco L (1990) Degradation of cellular proteins during poliovirus infection: studies by two-dimensional electrophoresis. J Virol 63:4729–4735

Vale R, Reese T, Sheetz M (1985) Identification of a novel force-generating protein, kinesin, involved in microtubule-based motility. Cell 42:39–50

Vallee RB, Shpetner HS, Paschal BM (1989) The role of dynein in retrograde axonal transport. Trends Neurosci 12:66–70

Waller AV (1852) A new method for the study of the nervous system. Lond J Med 43:609–625

Weiss P, Hiscoe HB (1948) Experiments on the mechanism of nerve growth. J Exp Zool 107:315–395

Whitehouse PJ, Walmsley JK, Zarbin MA et al. (1983) Amyotrophic lateral sclerosis: alterations in neurotransmitter receptors. Ann Neurol 14:8–16

Wiley RG, Blessing WW, Reis DJ (1982) Suicide transport: destruction of neurons by retrograde transport of ricin, abrin, and moddecin. Science 216:889–890

Willard M, Wiseman M, Levine J, Skene P (1979) Axonal transport of actin in rabbit retinal ganglion cells. J Cell Biol 81:581–591

Wohlfart G (1957) Collateral regeneration from residual motor nerve fibres in amyotrophic lateral sclerosis. Neurology 7:124–134

Yamada KM, Spooner RS, Wessels NK (1970) Axon growth: role of microfilaments and microtubules. Proc Natl Acad Sci USA 66:1206–1212

Yamada KM, Spooner RS, Wessels NK (1971) Ultrastructure and function of growth cones and axons of cultured nerve cells. J Cell Biol 49:614–635

Yamaguchi C, Ineda S, Sakamoto H et al. (1978) Progressive hereditary neurogenic muscular atrophy in dogs of the pointer breed. Exper Animals 27:202–204

Yamamoto T, Iwasaki Y, Konno H (1984) Retrograde axoplasmic transport of Adriamycin: An experimental form of motor neuron disease? Neurology 34:1299–1304

Yamamoto T, Satomi H, Ise H, Takatama H, Takahashi K (1978) Sacral spinal innervations of the rectal and vesical smooth muscles and the sphincteric muscles as demonstrated by the horseradish peroxidase method. Neurosci Lett 7:41–47

7 The Molecular Pathology of Motor Neuron Disease

P. N. Leigh and O. Garofalo

Introduction

The importance of unravelling molecular changes associated with neuronal degeneration in motor neuron disease has been underlined by the identification of the probable molecular basis of several neurodegenerative disorders. These advances have arisen directly from studies on molecular pathology. Specifically, point mutations of the ß-amyloid precursor protein (APP) gene have been identified in some families with early onset Alzheimer's disease and mutations of the prion gene have been implicated as the cause of familial forms of spongiform encephalopathy (Hsiao et al. 1989; Collinge et al. 1989; Owen et al. 1989; Brown et al. 1991). In these disorders, abnormal proteins are deposited or accumulate in the extracellular matrix of the brain, and probably contribute to neuronal degeneration, although the precise mechanisms by which this occurs are as yet unknown. In addition to the extracellular deposit of amyloid in Alzheimer's disease and the prion disorders, many neurodegenerative disorders are associated with characteristic intraneuronal inclusions which are usually composed of cytoskeletal and other neuronal protein (Tables 7.1 and 7.2). The accumulation of intraneuronal inclusions may determine cell viability by interfering with such vital functions as axonal transport. It is noteworthy that in Alzheimer's disease the formation of neurofibrillary tangles (NFTs) is associated with disappearance of the normal cytoskeleton.

While there is no evidence that amyloid or other abnormal proteins accumulate in the extracellular matrix in MND, there are indications that there are altered proteins at present within vulnerable neurons, in the form of characteristic inclusions. It is thus possible that the identification of the protein constituents of inclusions might point towards the biochemical processes contributing to neuronal death. There are several ways in which this approach could be helpful. Firstly, if a particular molecule (be it a cytoskeletal protein or some other molecule important for motor neuron function) is identified as a component of inclusions in MND, the gene controlling the synthesis of that molecule would become a candidate in the search for mutations in familial MND. Secondly, analysis of post-translational changes of such proteins (e.g. phosphorylation, ubiquitination) may be important in reconstructing the biochemical events leading to selective neuronal death, and could even suggest new treatments designed to slow or halt the degenerative process. Finally, if particular proteins are implicated in the pathogenesis of MND, the new approaches using transfection in vitro, or the creation of transgenic animals, will allow new cellular and animal models to be developed, as has already happened in Alzheimer's disease and the prion disorders (Brown et al. 1991).

Table 7.1. Cytoskeletal proteins. The principal proteins that comprise the cytoskeleton, including some of the better characterised proteins that are associated with or bind to the structural proteins of the three filament types

Filament type	Main structural protein	M_r ($\times 10^{-3}$)	Associated proteins	M_r ($\times 10^{-3}$)	Source of associated proteins
Microtubules	α,β-Tubulin	50–55	MAP1A,1B IC	350	Brain
			MAP2A, 2B	270	Brain
			Tau	55–68	Brain
			Kinesin	110	Brain
Microfilaments	Actin	42	Myosin	200	Many cells
			Tropomyosine	29–39	Many cells
			Gelsolin	90	Many cells
			Villin	95	Epithelia
			Tibrin	68	Epithelia
			Spectrin	220, 235 240	Many cells
Intermediate filaments					
Keratin filaments	Keratin	40–70	α-B Crystallin		Epithelia
Vimentin filaments	Vimentin	55			Epithelia
Desmin filaments	Desmin	52			Mesenchyme Muscle
Glial filaments	Glial fibrillary acidic protein	50			Astrocytes
Neurofilaments	Neurofilament Triplet	68 (NFL) 150 (NFM) 200 NFH)			Neurons

The Cytoskeleton and Cytoskeletal Pathology in Neurodegenerative Disorders

The cellular pathology of MND has been described in detail in Chapters 4 and 5. The cytoskeleton (Table 7.1) comprises a group of filamentous and associated proteins with diverse functions, including maintenance of neuronal shape, axonal calibre, and axonal transport. Cytoskeletal proteins which are altered in neurodegenerative disorders include neurofilaments (NFs), and microtubule-associated proteins (MAPs). Some of the typical cellular lesions and cytoskeletal changes associated with neurodegenerative disorders are summarised in Table 7.2. Abnormalities of the microtubular system are prominent in disorders such as Alzheimer's disease, Pick's disease, progressive supranuclear palsy (PSP) and the Western Pacific ALS parkinsonism-dementia complex, which are associated with neurofibrillary degeneration. Neurofibrillary tangles, composed mainly of paired helical filaments in Alzheimer's disease and 15–25 nm straight filaments in Pick's disease and PSP, are identified by antibodies against the microtubule-associated protein tau and by anti-neurofilament (NF) antibodies (Anderton et al. 1982; Wood et al. 1986; Kosik et al. 1986). Lewy bodies and pale bodies typical of Parkinson's disease (PD)

Table 7.2. Major central nervous system disorders involving the neuronal cytoskeleton

Neuropathological lesion	Ultrastructure	Disease	Protein constituents
Alzheimer neurofibrillary tangle	Mainly PHF but with some 10–15 nm straight filaments	Alzheimer's disease Down's syndrome, Postencephalitic Parkinson's, ALS Parkinson-dementia of Guam, Dementia pugilistica, Subacute sclerosing panencephalitis, Tuberous sclerosis Hallervorden–Spatz disease, Ceroid lipofuscinosis,	Antigens shared with tau, neurofilaments, MAP2 and PHF Ubiquitin
Pick body	1–20 nm straight	Pick's disease	Antigens shared with tau, neurofilaments and PHF Ubiquitin
Globose tangle	15 nm straight	Progressive supranuclear palsy (PSP); corticobasal degeneration	Antigens shared with tau, neurofilaments and PHF Ubiquitin (+/−)
Lewy body (brain stem)	10 nm filaments and granular material	Parkinson's disease	Antigens shared with neurofilaments and tubulin, Ubiquitin
Lewy body (cortical)	10 nm filaments	Parkinson's disease	
Neurofilament accumulations (axonal, perikaryal)	10 nm neurofilaments	Motor neuron disease	Neurofilaments
		Spinal muscular atropy	Neurofilaments
		Giant axonal neuropathy	Neurofilaments
Perikaryal filamentous inclusions in MND	10–25 nm filaments	MND	Ubiquitin
Intra-oligodendroglial inclusions		Multiple system atrophy	Tubulin Ubiquitin Tau
Hippocampal dentate granule cell inclusions	20–25 nm filament	MND with dementia	Ubiquitin

are composed in part of straight 10–15 nm filaments and are decorated by antibodies against NFs, tubulin, and microtubule-associated proteins (MAPs 1 and 2) although tau is not a component of Lewy bodies (Kahn et al. 1985; Love et al. 1988; Galloway et al. 1988; Bancher et al. 1989a). Finally, there is increasing evidence that cytoskeletal and other neuronal proteins are altered in MND.

Neurofilaments and Neurofilamentous Pathology in MND

Most neurons express NF proteins, which are composed of three subunit polypeptides consisting of low molecular weight (NF-L, M_r 8 kDa), medium chain (NF-M,

M_r150 kDa), and heavy chain (NF-H, M_r200 kDa) subunits (Robinson and Anderton 1988). These subunits are assembled in the neuronal perikaryon and transported towards the axon. NFs are particularly abundant in large neurons with long axons, where they determine axonal calibre (Hoffman et al. 1984) and interact with other components of the cytoskeleton such as microtubules (Robinson and Anderton 1988). NFs are associated with the slowest phase of axonal transport (a rate of transport of about 1 mm per day) and NF-H tends to be more heavily phosphorylated than NF-M or NF-L; these phosphorylation sites are highly conserved (Schlaepfer 1987). NF polypeptide chains are translated and assembled in the neuronal perikaryon, and as they are transported towards the axon they become increasingly phosphorylated (Schlaepfer 1987). Since phosphorylation status plays a major part in determining the antigenic properties of NF subunits, many monoclonal and polyclonal antibodies raised against purified preparations of NF proteins selectively identify epitopes which are variably dependent on phosphorylation status. Such epitopes are termed P+, P++, or P+++ if they are partially (P+) or heavily (P++, P+++) dependent on phosphoryation status. The immunoreactivity of antibodies against phosphate-dependent epitopes is diminished or abolished by dephosphorylation with alkaline phosphatase, but some antibodies are directed against epitopes which are non-phosphorylated (P−) and the affinity of such antibodies increases after alkaline phosphatase treatment. Other epitopes (P-ind) are independent of phosphorylation status; their staining pattern on immunoblots or tissue sections does not change with alkaline phosphatase treatment. Antibodies against NF proteins may be directed against any of the NF subunits, and often cross-react with subunit proteins, most commonly NF-M and NF-H.

Antibodies directed against P++ and P+++ epitopes label axons more intensely than neuronal perikarya. Conversely, antibodies to P− epitopes identify perikarya and dendrites, but label axons weakly or not at all. Some NF phosphorylation occurs within neuronal perikarya (Schlaepfer 1987; Poltorak and Freed 1989), but phosphorylation increases as NF subunits are transported from the perikaryon to the axon (Sternberger and Sternberger 1983). Following neuronal damage or interruption of axoplasmic transport, perikaryal NFs may become more heavily phosphorylated, and thus phosphorylation detected by monoclonal antibodies (mabs) directed against phosphate-dependent epitopes is regarded as abnormal or "inappropriate". Such abnormal perikaryal NF phosphorylation can be detected following many types of neuronal injury, including nerve crush (Rosenfeld et al. 1987), exposure to neurotoxins such as aluminium, β,β'-iminodipropionitrile (IDPN) and 2,5-hexanedione (Gold 1987) and in hereditary disorders of animals, including canine spinal muscular atrophy, shaker calves, rabbit and pig motor neuron disease, swayback in goats and sheep, and zebra myelopathy (Cork et al. 1982a,b, 1988). Altered phosphorylation of neuron perikaryal NFs thus represents a non-specific response to neuronal injury. A common factor may be disruption of slow axonal transport.

Altered (and possibly abnormal) phosphorylation of cytoskeletal proteins is also a feature of other neurodegenerative disorders. Neurofibrillary tangles in Alzheimer's disease, Pick's disease and PSP, and Lewy bodies in Parkinson's disease (Table 7.2), are labelled by antibodies directed against phosphate-dependent NF epitopes (Kahn et al. 1985; Haugh et al. 1986; Love et al. 1988; Galloway et al. 1988; Bancher et al. 1989a); and microtubule-associated protein tau, which is a major component of neurofibrillary tangles, is abnormally phosphorylated (Wood et al. 1986; Kosik et al. 1986). Thus an abnormality of cytoskeletal protein phosphorylation may be a common feature of several disorders in which intracellular inclusions share NF and tau epitopes (Haugh et al. 1986). In Alzheimer's disease, abnormalities of the cytoskeleton are not restricted to neuronal inclusions, since abnormal tau is also present in dystrophic neurities within senile plaques (Wood et al. 1986; Grundke-Iqbal et al. 1986).

Cytoskeletal abnormalities were not recognised as a part of the pathology of MND until relatively recently. Wohlfart (1959), Carpenter (1968), and Chou (1979) noted that large axonal swelling (spheroids: Fig. 7.1a) within the gray matter of the anterior horn were a common finding in MND.

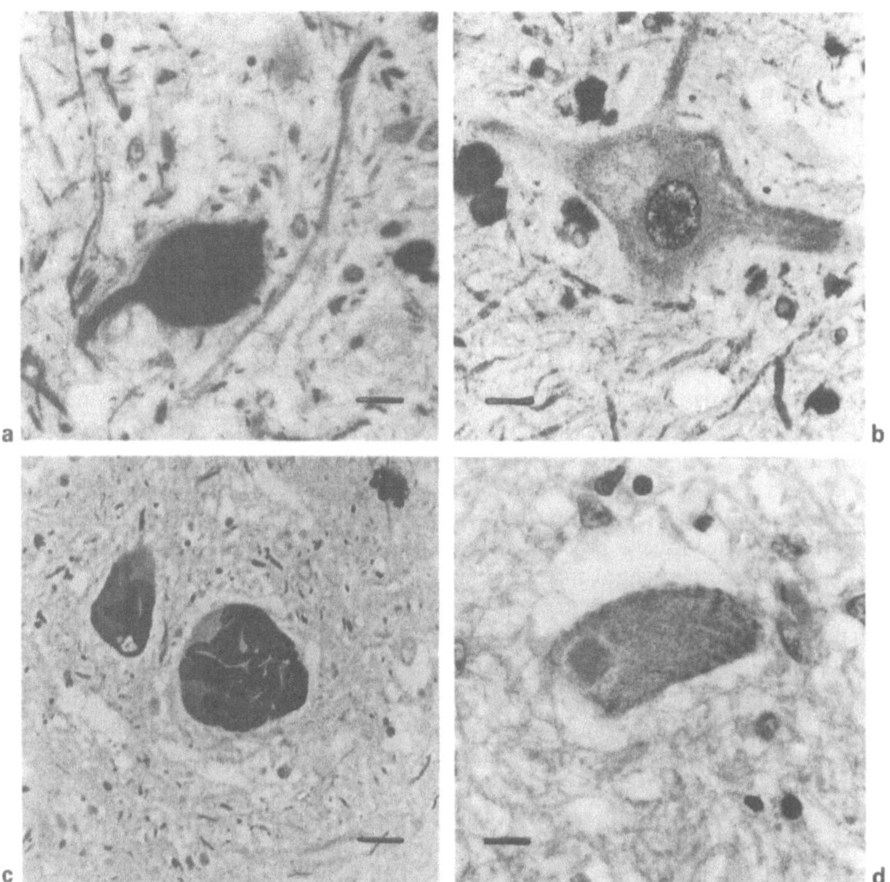

Fig. 7.1. a MND. Spinal cord, immunolabelled with a monoclonal antibody (MAB 147) against phosphorylated neurofilament protein. A large axonal swelling (spheroid) is present in the anterior horn. (Bar=5 μm. Reproduced from Leigh et al. 1989b with permission of the publishers). **b** MND. Lumbar spinal cord immunolabelled mab 147. There is intense labelling of small axonal swelling ("globules") and weak labelling of anterior horn cell cytoplasm. (Bar =5 μm.) **c** MND. Lumbar spinal cord labelled with an antibody against non-phosphorylated neurofilament heavy chain peptides, showing large neurofilamentous accumulations which were also identified by mabs such as BF10 which identify phosphorylated neurofilament epitopes. (Bar=25 μm. Reproduced with permission of the publishers from Leigh et al. 1989a). **d** Lewy body-like inclusion in spinal motor neuron. (H&E. Bar=5 μm.)

Spheroids are sometimes seen in motor neuron perikarya (Hirano et al. 1967; Carpenter, 1968; Schochet et al. 1969; Hughes and Jerrome, 1971; Delisle and Carpenter, 1984). These argyrophilic swellings are composed of disorganised 10 nm filaments typical of NFs. Small neurofilamentous swelling ("globules") are common in the spinal cords of neurologically normal adults (Fig. 7.1b), particularly in the ventral root exit zone, but large (greater than 25 μm diameter) proximal axonal swellings and tubular enlargements of proximal axons are characteristic of MND (Carpenter 1968; Delisle and Carpenter 1984; Clark et al. 1984; Leigh et al. 1989a). Spheroids are of limited diagnostic value since they are present in the spinal cord of about half the control cases, although they are more numerous and often very large in MND spinal cord (Delisle and Carpenter 1984). Spheroids are usually scarce in early onset forms of spinal muscular atrophy (SMA types I, II) and in X-linked bulbospinal neuronopathy (Sobue et al. 1990) but may be relatively abundant in type III SMA. Where there is severe loss of spinal motor neurons there are fewer spheroids, but the ratio of numbers of axonal swellings to surviving anterior horn cells is increased irrespective of whether there are many or few surviving neurons (Manetto et al. 1988; Leigh et al. 1989a; Sobue et al. 1990). Because there are often more surviving motor neurons in patients with short duration of illness, spheroids tend to be most abundant in such cases. Large spheroids are also found in motor nuclei of the brain stem (Carpenter, 1968; Delisle and Carpenter 1984) and in the internal capsule but they are seldom present in the motor cortex.

It is sometimes possible to show a connection between a spheroid and an anterior horn motor neuron (Carpenter 1968; Sasaki et al. 1988). Most spheroids are thought to be situated within the proximal axons of motor neurons, although it is possible that some may arise in dendrites (Kato et al. 1987; Sasaki et al. 1988) and some are situated more than 600 μm from the edge of the anterior horn. Sobue et al. (1990) noted that the distribution of spheroids and globules differed in MND and control subjects. In MND, spheroids were concentrated around the lateral nuclei of the anterior horn, an area containing large motor neurons which may be particularly vulnerable to the disease process. In contrast, "globules" in control patients are clustered around the ventral root exit zone.

It is difficult to know whether spheroids arise proximally and are transported distally, or vice versa – but the assumption has been that abnormalities of slow axonal transport (as occurs in IDPN or 2,5-hexanedione toxicity) lead to accumulation of proximal axonal or even perikaryal NFs which are heavily phosphorylated. Antibodies to P++ or P+++, but not P−, NF epitopes yield intense labelling of spheroids, indistinguishable from that of normal axons (Schmidt et al. 1987; Toyoshima et al. 1989). It seems likely that changes in the degree of phosphorylation of NFs in spheroids are secondary to abnormalities of axonal transport: anterograde, retrograde, or both. There remains the possibility that increased phosphorylation may further impair axonal function.

NF accumulations (Fig. 7.1c) within anterior horn motor neurons have been variously termed hyaline bodies or hyaline conglomerates (Schochet et al. 1969; Hirano et al. 1984a) although Lewy body-like inclusions (Fig. 7.1d) have also sometimes been referred to as hyaline inclusions (Hirano et al. 1967; Hirano et al. 1984b). The distinction is worth making, since the ultrastructural appearance and immunocytochemical features of these inclusions differ. Hyaline conglomerates as described by Hirano et al. (1967), Schochet et al. (1969), Chou (1979) and Kondo et al. (1986) consist of large bundles and aggregates of 10–15 nm filaments associated with thicker fibrils, mitochondria, lipofuscin, and granular material

(Hirano et al. 1984a). In some instances, honeycomb-like structures or linear densities associated with ribosome-like particles have been observed within groups of randomly arranged NFs (Hirano et al. 1984a). Neurofilamentous conglomerates may occupy the whole neuronal perikaryon, producing the appearance of chromatolysis or "achromasia" on H&E stains. These NF accumulations or conglomerates are not specific to MND and have been described in neurologically abnormal and normal controls (Leigh et al. 1989a; Sobue et al. 1990). Antibodies directed against both phosphate-dependent (P++/P+++) and P-ind NF epitopes label these accumulations (Leigh et al. 1989a).

Lewy body-like inclusions (sometimes termed hyaline inclusions) were initially thought to be associated with familial forms of MND (Hirano et al. 1967), although more recently they have been described in sporadic forms (Chou 1979), and in our own series were present in over 20% of patients, all sporadic cases (Leigh et al. 1991). Their appearance (Fig. 7.1d) is more variable than that of brainstem Lewy bodies in Parkinson's disease, but typically they consist of an achromasic or weakly eosinophilic core surrounded by a pale or basophilic halo, although lacking concentric lamination often present in Lewy bodies in PD. Lewy body-like inclusions may be surrounded by radially orientated filamentous structures and are often associated with neuronal achromasia. The ultrastructural appearances consist of randomly or radially arranged ill-defined linear densities intermixed with 10 nm filaments, larger filamentous structures 13–25 nm in diameter, dense granules and organelles such as mitochondria, rough endoplasmic reticulum, lipofuscin granules, and lysosomes (Hirano et al. 1984b; Sasaki et al. 1989). Sometimes the central core is associated with a condensation of these structures, and occasionally there is a more regular arrangement of radially orientated filaments, the structure appearing very similar to a typical Lewy body in PD (Kusaka et al. 1988). In view of the similarities to Lewy bodies of PD, which are labelled by antibodies against both P-ind and P++/+++ NF epitopes, it is not surprising that Lewy body-like inclusions in MND are labelled by anti-NF antibodies. Schmidt et al. (1987) described one 56-year-old patient in whom there were numerous Lewy body-like inclusions in the cervical spinal cord. The peripheral halo of the inclusions was strongly labelled by P− mabs against NF-M and NF-H, but only occasional inclusions were labelled by mabs against phosphate-dependent NF-H epitopes. Murayama et al. (1989), and Mizusawa et al. (1989) however noted labelling of the halo of Lewy body-like inclusions with anti-NF antibodies which are probably phosphate-dependent. It appears that some Lewy body-like inclusions do share epitopes with brain stem Lewy bodies in PD, although in our material antibodies that label typical Lewy bodies in PD have not labelled Lewy body-like inclusions. It is intriguing that diffuse Lewy body disease has been described in association with MND (Delisle et al. 1987; Gibb et al. 1989), although Lewy bodies in these cases have not been seen within anterior horn cells (Gibb et al. 1989) and furthermore Lewy bodies are rare in the substantia nigra of non-demented or demented MND patients, even in those cases in which there is striking degeneration of pigmented substantia nigra neurons (Neary et al. 1990).

The question arises as to whether there are abnormalities of perikaryal NF processing in MND, and several recent studies have suggested that perikaryal NFs may be abnormally phosphorylated. Munoz et al. (1988), using a monoclonal antibody (NF2F11) directed against a P+ NF epitope, noted that although P+ cells were more common in MND, the difference was not significant unless different categories of motor neuron abnormality were compared separately. Significantly more swollen diffusely immunoreactive neurons, and neurons with focal reactions for the antibody, were seen in MND cases. Leigh et al. (1989a), however, found no increase in perikaryal staining with antibodies against P+ mabs in MND using a panel of well-characterised mabs, some of which label neurofibrillary tangles in tissue sections and all of which strongly label chromatolytic neurons and axonal swellings. These mabs identify mainly P+ epitopes on the side arms of either NF-M or NF-H NF subunits (Miller et al. 1986; Haugh et al. 1986). Hyaline achromasic neurons and hyaline NF conglomerates in MND and controls and in infantile spinal muscular atrophy (SMA type 1) show marked immunoreactivity with these antibodies (Leigh et al. 1989a; Murayama et al. 1991). Thus our failure to show significantly increased levels of abnormal perikaryal labelling with antibodies directed against P+ NF epitopes is unlikely to be attributable solely to variables such as fixation. Nevertheless, Manetto et al. (1988) and Sobue et al. (1990) using different anti-NF mabs observed a significant increase in the number of spinal motor neurons in MND tissue which were labelled by P+ anti-NF mabs. The proportion of surviving anterior horn cells labelled by mab Ta-51 (which is directed against an epitope on NF-H) was increased in MND and in SMA type I, but not in multiple system atrophy or X-linked bulbospinal neuronopathy (Sobue et al. 1990).

Presumably the immunolabelling of Lewy body-like inclusions by antibodies to P-ind epitopes (and in the hypoglossal nucleus, antibodies to P+ NF epitopes) noted by Schmidt et al. (1987) and Munoz et al. (1988) reflects a different range of NF epitopes, as does the detection of "abnormal" phosphorylation in MND anterior horn motor neurons by some groups of antibodies but not others. None of these studies had revealed abnormal immunostaining of anterior horn cells by antibodies to other cytoskeletal proteins such as actin, tau and other microtubule associated proteins, tubulin, myosin, cytokeratin, peripherin, and vimentin (Manetto et al. 1988; Leigh et al. 1989a).

The balance of evidence thus supports a disturbance of NF phosphorylation in lower motor neurons in MND, although this only affects a proportion (about 10%–30%) of the surviving cells and is non-specific. While Lewy body-like inclusions have been identified by some P-ind and P+ NF antibodies, Bunina bodies have not yet been clearly linked to NF pathology.

Ubiquitin, the Stress Response and Neuronal Damage in MND

Ubiquitin, a 76-amino acid polypeptide which can be found in all eukaryotic cells, was identified as a component of neurofibrillary tangles in 1986 by Mori and colleagues, and subsequently many other groups have confirmed this observation. Ubiquitin is now known to be a component of Lewy bodies, Pick bodies, neurofibrillary tangles in many disorders, glial cell inclusions in multiple system atrophy, and inclusions in some muscle diseases (Table 7.2: Cole and Timiras, 1987; Perry et al. 1987; Kuzahara et al. 1988; Lowe et al. 1988a; Love et al. 1988; Bancher et al. 1989b; Manetto et al. 1989; Leigh et al. 1989b; Papp et al. 1989; Nakazato et al. 1990). The immunocytochemical localisaton of ubiquitin has become a useful tool in neuropathological diagnosis because it reveals new aspects of the molecular pathology of these disorders (Leigh et al. 1989b), and its presence has engendered speculation that it may represent a crucially important clue to the pathogenesis of neurodegenerative disorders. Thus it is relevant to survey some aspects of the biochemistry of this highly conserved protein.

Ubiquitin is present both free within the cytosol and conjugated to other proteins: this suggests that it plays a fundamental role in cellular processes. Ubiquitin is found covalently attached to nuclear histones H2A and H2B (Thorne et al. 1987; Nickel and Davie, 1989) and appears to be involved in regulation of gene expression. Intrinsic plasma membrane proteins have also been found to be ubiquitinated: one of these is the lymphocyte B homing receptor (Siegelman et al. 1986) involved in specific cell–cell recognition; another is the receptor for the platelet-derived growth factor (Yarden et al. 1986; Leung et al. 1987) and therefore, ubiquitination of these proteins may play a role in modification of receptor function. Recently, in vitro studies have suggested that the insertion of monoamine oxidase B into mitochondrial outer membranes may be mediated through a process involving ubiquitin (Zhaung and McCauley, 1989). Ubiquitin is also a member of the heat shock family of proteins (Bond and Schlesinger, 1985).

The non-lysosomal degradation of short-lived and abnormal proteins, however, is its best known role. This was first discovered by Hershko et al. (1980), while studying an ATP-dependent proteolytic system from reticulocytes. Ubiquitin was shown to be essential for the activity of this system, being conjugated to protein substrates in an ATP-requiring reaction. It was then proposed that conjugation is an indispensable process in protein breakdown and that conjugated proteins are then degraded by specific proteases (Ciechanover et al. 1980; Hershko et al. 1980).

However, Fried et al. (1987) have also suggested that ubiquitin has intrinsic proteolytic activity and that its conjugation to target proteins can convert these conjugates into ad hoc proteolytic enzymes.

The Ubiquitin Conjugation Pathway

The pathway leading to the conjugation of ubiquitin to the target proteins is illustrated in Fig. 7.2. The first step (1) is ubiquitin activation by a specific ubiquitin-activating enzyme, termed E1 (Ciechanover et al. 1981; Haas et al. 1982). E1 catalyses a two-step reaction in which ubiquitin adenylate is first formed in an ATP-dependent reaction; the activated ubiquitin is then transferred to a thiol site of the enzyme, with the release of AMP (Ciechanover et al. 1981; Haas et al. 1982). Hershko et al. (1981) have identified the activated amino acid residue of ubiquitin as the COOH-terminal glycine. The E1-ubiquitin thiol ester is then the donor for the formation of ubiquitin conjugates with proteins (Haas et al. 1982; Hershko et al. 1983) and two enzymes, E2 and E3, are involved in this process (Hershko et al. 1983).

E2 has a ubiquitin-carrier function (Fig. 7.2, step 2) and is the acceptor, by transacylation, of activated ubiquitin from E1, then transferring it to amino groups of proteins (Hershko et al. 1983). Pickart and Rose (1985) have shown that there is a family of five E2 proteins, which range in native molecular weight from 24 to 55 kDa, with most of them being homodimers. Only the smallest of these proteins is involved in E3-dependent ligation to substrates of the ubiquitin proteolytic pathway. Some E2 proteins have been found to transfer ubiquitin directly to histones (step 3 of Fig. 7.2). This reaction is ligase-independent and produces conjugates which usually have one molecule of ubiquitin per molecule of protein (Pickart and Rose,

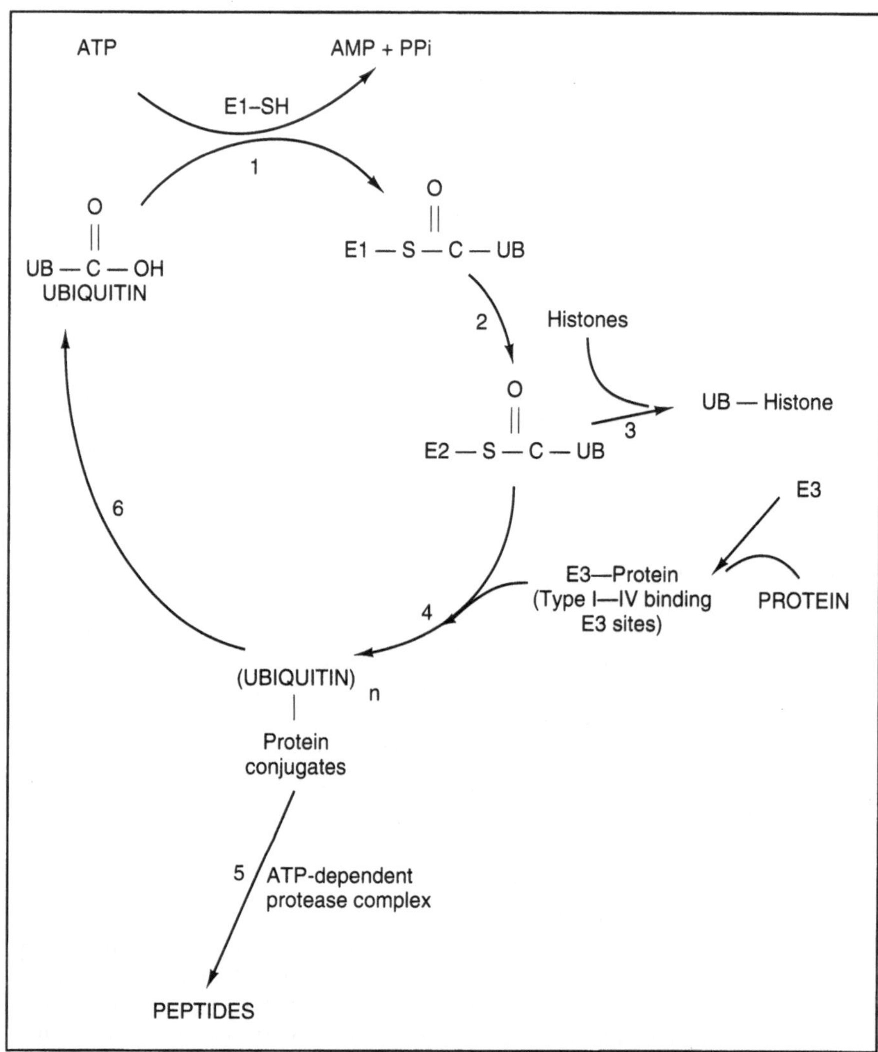

Fig. 7.2. The ubiquitin proteolytic system (see text for details).

1985; Pickart and Vella, 1988; Haas et al. 1988). Interestingly, the sequencing of the DNA encoding a 20 kDa ubiquitin carrier protein in yeast has shown it to be identical to the yeast DNA repair enzyme rad6 (Jentsch et al. 1987).

E3 catalyses the formation of the isopeptide bond (step 4 of Fig. 7.2) between ubiquitin and the target protein(s) and has a central role in selecting proteins suitable for degradation.

Selection of Substrates for Conjugation and Degradation

Structural determinants of the protein structure are recognised by the ubiquitin ligase. A free and exposed terminal amino group is an important signal for protein degradation by the ubiquitin system. This has been shown in experiments in which selective modification of the β-amino group of proteins inhibited their degradation in a ubiquitin-dependent system purified from reticulocytes. The addition of new β-amino groups, however, converted the proteins with blocked N-termini into efficient substrates (Hershko et al. 1984a; Chin et al. 1986; Bachmair et al. 1986). Four different types of substrates can be distinguished. Type I substrates have a basic N-terminal residue (His, Arg and Lys). Type II substrates are substrates with bulky hydrophobic N-termini (Leu, Trp, Phe and Tyr). More recently, a third group of substrates which display Ser, Ala or a Thr residue in the N-terminal position has been described (Gonda et al. 1989). Finally, proteins with N-terminal residues other than those cited above (Reiss et al. 1988) are recognised by a fourth type of protein-binding site on E3 (Reiss et al. 1988). However, some of these residues have to be post-translationally modified in order to be recognised by the protein binding sites on E3, and transfer RNA appears to participate in the covalent modification of selective proteolytic substrates, which makes them susceptible for recognition by the ubiquitin proteolytic system (Ciechanover et al. 1985; Ferber and Ciechanover 1986, 1987). The N-terminus of a protein is therefore a very important structural determinant for its recognition by the ubiquitin system. However, in vivo studies give evidence that important signals lie also outside the N-termini. Thus, cells incubated with puromycin which has no effect on the N-terminal residue, but terminates protein synthesis and releases puromycin peptides, renders cellular proteins susceptible to ubiquitin-dependent degradation (Ciechanover et al. 1984). In the same way, Hershko et al. (1982) observed that incubation of reticulocytes with amino acid analogues substituting residues distinct from the N-terminal residue of globin also renders the latter a substrate for ubiquitin-dependent protein degradation. More recently, Bachmair and Varshavsky (1989) have found that dihydrofolate reductase molecules with different N-terminal residues were metabolically stable. However, placing a fragment of 43 amino acid residues derived from the N-terminal region of β-galactosidase at the original N-terminus of the dihydrofolate reductase molecules made the chimaeric molecule susceptible to degradation. It was found that the β-galactosidase fragment contained two internal lysine residues proximal to the N-terminus, which are apparently necessary for lability of the protein. These lysine residues are bound to multiple ubiquitin residues which form polyubiquitin chains, labilizing the whole molecule (Chau et al. 1989).

Ubiquitin-Protein Ligase

E3 isoenzymes have been purified from rabbit reticulocytes and have been shown to contain the protein-binding sites (Hershko et al. 1986). These are remarkably specific (Reiss et al. 1988) and are capable of distinguishing between different N-terminal residues of the proteolytic substrates. Proteins with a suitable structure are bound to the specific binding sites on E3 and then ubiquitin is transferred from E2-ubiquitin (Fig. 7.2, step 2) to the substrate (Fig. 7.2, step 4). In ubiquitin-protein conjugates, many ubiquitin molecules are linked via their COOH-terminal glycine residue to -NH_2 groups of lysine residues of the protein(s) by isopeptide bonds (Ciechanover et al. 1980).

Protein Degradation and Regeneration of Free Ubiquitin

Proteins conjugated to ubiquitin chains through E3 are marked for degradation by an ATP-dependent large protease complex (1000 kDa), which has been identified and purified from reticulocytes (Hershko et al. 1984b; Hough et al. 1986, 1987; Ganoth et al. 1988). Three different factors, all of which are required for conjugate breakdown, combine to form a multienzyme complex (the multicatalytic proteinase complex, MCP), whose exact mode of action is still not known (Hershko, 1991; Rivett et al. 1991; Rivett 1989).

Free ubiquitin is finally recycled (Fig 7.2, step 6), by the action of specific hydrolases which cleave the peptide bond at its COOH-terminal glycine (Hershko et al. 1980). A family of such enzymes have been characterised from bovine thymus (Mayer and Wilkinson 1989). All of them appear to be thiol proteases. A cDNA for the isozyme which is predominant in bovine thymus has recently been cloned (Wilkinson et al. 1989) and analysis of its amino acid sequence has revealed a 54% homology with the neuron-specific protein PGP 9.5. Wilkinson et al. (1989) subsequently showed that PGP 9.5 actually possesses ubiquitin terminal hydrolase activity.

In summary, the ubiquitin proteolytic system is a complex pathway for degrading and modifying proteins (Rechsteiner, 1987; Hershko, 1991). Induction of ubiquitin synthesis occurs as a result of cell stress, including heat shock, and in some situations the ability to synthesise ubiquitin determines cell survival after heat shock. Failure to degrade abnormal proteins produced as a result of toxic damage by an environmental agent, or as the result of genetic defects, could lead to progressive cellular damage and the accumulation of insoluble, damaging material in the form of inclusions. While this is an attractive hypothesis, there is no direct evidence that neuronal inclusions are more than markers of cell damage. Neither is there evidence to suggest that a primary abnormality of the ubiquitin proteolytic system is responsible for any neurodegenerative process or disease. The association between ubiquitin and a wide range of morphologically different inclusions in many diverse diseases (neurological and non-neurological) argues against a pathogenic role for ubiquitin in neurodegeneration. There is a close but not invariable association between ubiquitin and altered intermediate filaments (Manetto et al. 1989; Lowe and Mayer 1990), but this by itself need not imply a defect in the ubiquitin system. Nevertheless, further investigation of the neurobiology of the ubiquitin and other cell stress systems may clarify aspects of the neurodegenerative process, and there can be no question that antibodies against ubiquitin have provided a powerful new tool in the pathological analysis of MND.

Ubiquitin and MND

Morphological Features of Ubiquitin-Immunoreactive Inclusions in MND

Following the discoveries discussed in the preceding section, it was natural to ask whether ubiquitin immunocytochemistry might reveal new aspects of the cellular pathology of MND, on the basis that antibodies might bind to ubiquitin conjugated to unknown or hitherto undetectable cellular proteins. This turned out to be the case, and Leigh et al. (1988) and Lowe et al. (1988b) independently reported the presence of characteristic ubiquitin-immunoreactive inclusions within anterior horn cells.

These observations have since been confirmed and extended in many studies from the USA, Europe and Japan.

Ubiquitin-immunoreactive (Ub-IR) inclusions in MND take several forms, and are present in anterior horn motor neurons, in brain stem motor neurons, and rarely in cortical motor neurons in patients dying with typical sporadic MND of Charcot type (ALS), in patients who have only lower motor neuron signs (progressive muscular atrophy), in patients with familial ALS, and in patients with MND and dementia (Leigh et al. 1988, 1991; Lowe et al. 1988b, 1989; Murayama et al. 1989; 1990a; Kato et al. 1989; Sasaki et al. 1989; Migheli et al. 1990; Schiffer et al. 1991). Filamentous inclusions (Fig. 7.3), originally described as skein-like inclusions, represent the most abundant type. These often form delicate interlacing bundles in the perikarya and sometimes in the dendrites of anterior horn cells or brain stem motor neurons.

Ultrastructurally they consist of bundles of filaments measuring 10–25 nm (and thus thicker than typical neurofilaments) arranged in the form of tubular structures,

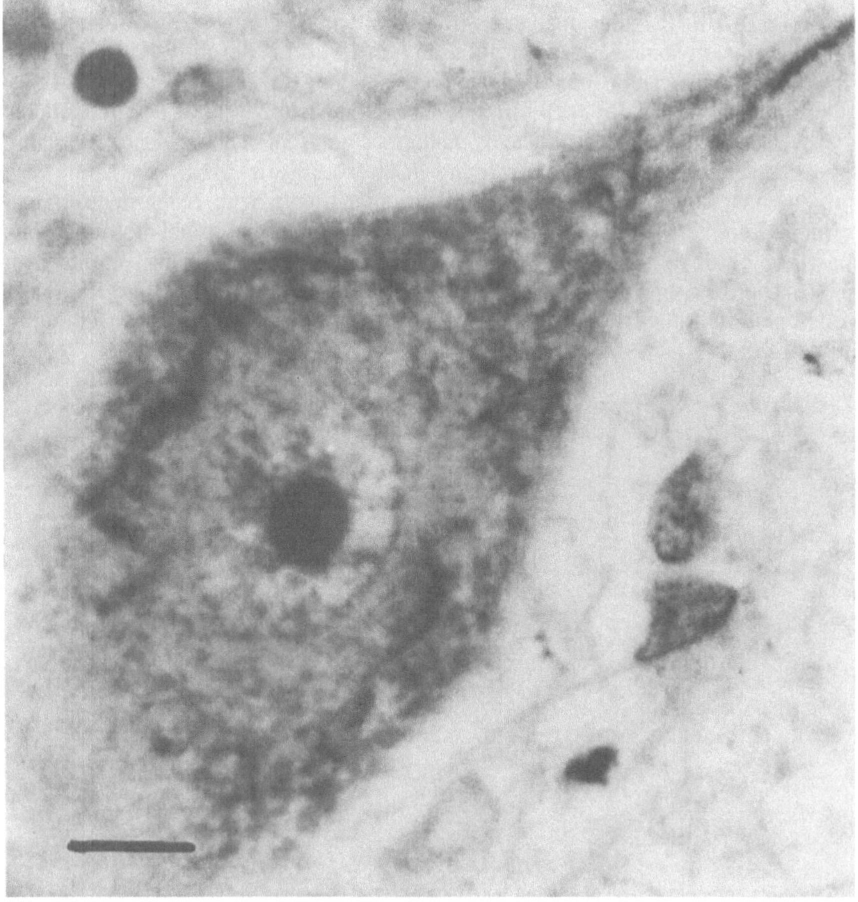

Fig. 7.3. MND, lumbar spinal cord labelled with a polyclonal antibody against ubiquitin and showing skein-like ubiquitin-IR inclusions (Bar = 5 μm.)

and associated with granular material (Fig. 7.4a,b) (Lowe et al. 1988b; Mizusawa et al. 1991; Schiffer et al. 1991).

The second most common type of Ub-IR inclusions is the dense accumulation or dense body (Fig 7.5a). In sections stained with H&E or other routine stains, dense bodies sometimes correspond to Lewy body-like inclusions or to poorly defined hyaline or vacuolar areas which are neither eosinophilic nor basophilic (Fig. 7.5b,c). Ultrastructurally, dense bodies of Lewy body-like type are composed of radially arranged filaments with a granulofilamentous core often containing lipofuscin granules (Kusaka et al. 1988; Lowe et al. 1988b; Murayama et al. 1989, 1990a; Schiffer et al. 1991). Some of the filaments are 10–15 nm diameter (and probably represent neurofilaments), while some are 15–25 nm in diameter and are thus slightly larger than typical neurofilaments.

Clinicopathological Correlations

Skeins usually outnumber Ub-IR dense bodies by about 2 : 1, but in some patients one may see only skeins or only dense bodies (Leigh et al. 1988, 1991). The prevalence of Ub-IR inclusions varies from less than 1% to over 30% of surviving motor neurons; in some cases a careful search in 10 or more sections must be made before an inclusion can be identified. As yet the diagnostic specificity and sensitivity of these inclusions has not been formally tested, and relatively few detailed studies have been reported on neurologically abnormal controls. The evidence available suggests that Ub-IR inclusions are found in 80%–100% of MND cases, but very rarely in neurologically normal controls or in patients with other neurological disorders (Leigh et al. 1991). Similar inclusions are not seen in spinal motor neurons following poliomyelitis (Lowe et al. 1989; Leigh et al. 1991), or in patients with type I or type II spinal muscular atrophy (SMA) (Murayama et al. 1991). In SMA, swollen neurons often show diffuse or concentric immunolabelling with ubiquitin antibodies.

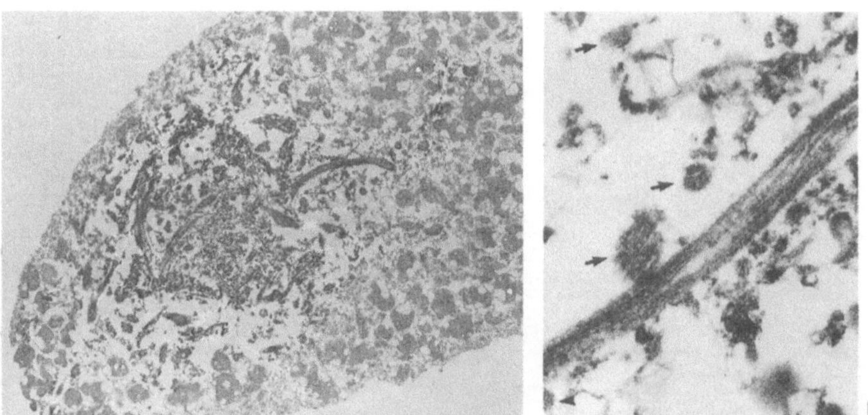

Fig. 7.4. **a** Electron micrograph of anterior horn motor neuron containing skein-like inclusion (×3300). **b** High-power view of filamentous structure in skein-like inclusion. The arrows indicate oblique and cross (arrowhead) sections of fibrils (×20 000). (Reproduced from Mizusawa et al. 1991, with kind permission of the authors and publishers.)

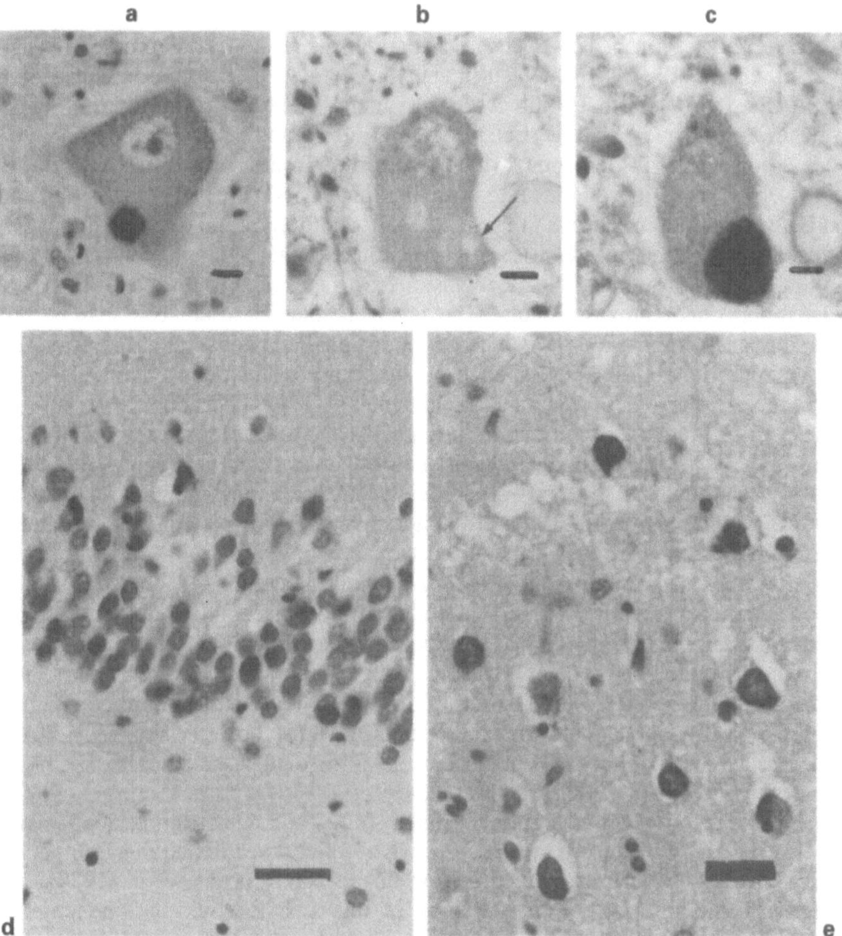

Fig. 7.5. **a** MND. Anterior horn motor neuron showing ubiquitin-immunoreactive dense body. (Bar=10 μm.) **b** MND. Anterior horn motor neuron showing poorly defined vacuolar area (arrow). Section labelled with mab 147 against phosphorylated neurofilament. The cell body is unlabelled. (Bar=10 μm.) **c** Same neuron in the adjacent section labelled with anti-ubiquitin antibody, showing large ubiquitin-immunoreactive inclusion corresponding to the vacuolar area arrowed in **b**. (Bar=10 μm.) **d** MND-dementia. Hippocampal dentate granule cells showing intraneuronal ubiquitin-immunoreactive dense bodies. (Bar=20 μm.) **e** MND-dementia. Medial frontal cortex, layers II and III, showing elongated, curved and rounded ubiquitin-immunoreactive inclusions in surviving neurons. (Bar=15 μm.) (**b** and **c** are reproduced from Leigh et al. 1988, with permission of the publishers, Elsevier.)

Ub-IR inclusions identical to those in typical MND are, however, present in the Western Pacific form of MND, the ALS parkinsonism-dementia complex (Matsumoto et al. 1990). This is not surprising since the pathology of motor system degeneration in Western Pacific ALS is virtually identical to that of MND-ALS elsewhere in the world. MND outside the Western Pacific endemic foci is sometimes associated with dementia, which is often of frontal lobe type (Neary et al. 1990). Ub-IR inclusions identical to those in typical MND are present in spinal cord and brain stem motor neurons in these cases. In addition, round or elongated Ub-IR inclusions are present in 5%–10% of hippocampal dentate granule cells (Fig. 7.5a) and filamentous or rounded Ub-IR inclusions are also seen in neurons of the frontal and temporal neocortex, particularly in layers II and III where cell loss is most marked in such cases (Fig. 7.5b). These hippocampal and neocortical inclusions are not argyrophilic, nor are they labelled by antibodies against neurofilaments or the microtubule-associated protein tau (Wightman et al. 1992). This contrasts with Pick bodies, which are composed of straight filaments and are strongly argyrophilic, and immunoreactive with antibodies against neurofilaments and tau (Murayama et al. 1990b; Wightman et al. 1992; Kew and Leigh 1992). Identical inclusions have been reported by Okamoto et al. (1991a) in non-demented MND patients and in patients with MND-dementia, suggesting that individuals with extra-motor cortex involvement may represent a subgroup of MND patients. It remains to be seen whether hippocampal inclusions are associated with cell loss, or cognitive abnormalities.

Several attempts have been made to correlate the clinical features of the disease with the distribution and type of Ub-IR inclusions. Although it is rare to find Ub-IR inclusions in the motor cortex (Leigh et al. 1991), they may be more common in patients with predominantly upper motor neuron signs (Lowe et al. 1989). Ub-IR inclusions in the motor cortex may take the form of dense bodies or skein-like inclusions (Lowe et al. 1989; Leigh et al. 1991).

The presence of Bunina bodies (Bunina 1962) and Ub-IR inclusions in the neurons of Onuf's nucleus in the sacral spinal cord has been linked to symptoms of urinary incontinence which sometimes occurs in MND (Lowe et al. 1989), and supports the notion that selective sparing of this area is a relative phenomenon (Kihira et al. 1991; Okamoto et al. 1991b). Similarly, although inclusions are very uncommon in the oculomotor nuclei, they are occasionally present (Leigh et al. 1991). The inclusions are said to be more abundant in cases with a relatively short clinical history (Schiffer et al. 1991), but this may be because there are more surviving neurons in such cases.

Do Ub-IR inclusions merely represent the end-stage of neuronal degeneration in MND? Skein-like inclusions are often present in anterior horn cells that appear relatively normal in other respects, although they are more frequently seen in shrunken achromasic neurons, and on average Ub-IR inclusions are associated with neurons lacking Nissl substance, and thus by inference more advanced in the degenerative process (Leigh et al. 1991).

Ub-IR inclusions, in the form of skeins, dense bodies or Lewy body-like inclusions, thus represent a characteristic feature of the cellular pathology of MND in its various forms. They are not found in some other types of lower motor neuron degeneration, such as SMA. They represent a pathological marker as important in the diagnosis of MND as the Lewy body or the neurofibrillary tangle are in the pathological diagnosis and classification of Parkinson's disease and Alzheimer's disease. Their biological significance is much less certain at present.

Pathogenic Significance of Ub-IR Inclusions in MND

Although it is has been assumed that the presence of Ub-IR material in motor and non-motor neurons in MND reflects the presence of ubiquitin-protein conjugates that are resistant to degradation via the ubiquitin-proteolytic pathway, there is no direct evidence in support of this notion. Similarly, the evidence that intermediate filaments are the target for ubiquitination in MND is scanty. Despite the presence of filamentous structures in skeins and Lewy body-like inclusions, there is no convincing evidence from many immunocytochemical studies for an association between ubiquitin and neurofilaments. In addition to the studies using different anti-neurofilament antibodies to which reference has been made earlier, a wide range of antibodies against other cytoskeletal proteins (Leigh et al. 1989a) have failed to identify any other component. Nevertheless, it is clear from ultrastructural studies that 10–15 nm filaments are sometimes associated with the ubiquitin deposits, and failure to detect NFs may reflect loss or unavailability of epitopes.

Attempts to isolate the Ub-IR inclusions in MND for biochemical analysis have proved unsuccessful so far (Garofalo et al. 1991a) no doubt due to the paucity of the inclusions in comparison with neurofibrillary tangles. Nevertheless, the biochemical and molecular characterisation of the inclusions remains an important goal for future research.

Molecular biological techniques provide another approach to understanding the role of ubiquitin in neurodegeneration. Heggie et al. (1989) have examined ubiquitin gene expression in a small group of MND patients and controls. The ubiquitin C gene, which is under the control of a heat shock promotor, was more highly expressed in the spinal cord of MND cases compared to controls. Is ubiquitin over-expressed as a consequence of increased turnover because it is involved in the degradation of abnormal proteins, or is ubiquitin involved in some other response of the damaged neuron? Is the increased expression of ubiquitin cytoprotective? Ubiquitin is expressed by damage or "stressed" cells, as is heat shock protein 72 (HSP 72), one of the major heat shock proteins which are synthesised in response to cell damage of various kinds and which are cytoprotective. Surviving neurons in MND do not, however, express HSP 72 at a high level as judged by immuno-cytochemistry, although HSP 72 immunoreactive structures which represent poly-glucosan bodies (corpora amylacea) are present in normal and ALS spinal cord gray matter, but are more abundant in MND than in controls (Garofalo et al. 1991b). Antibodies to HSP 72 do not label ubiquitin-IR inclusions. This may indicate that the presence of ubiquitin is indeed concerned with degradation of abnormal proteins, rather than being a non-specific expression of a stress response.

Many questions remain to be answered. What protein(s) are associated with the ubiquitin? If ubiquitin is associated with altered neurofilaments, why are the inclusions not identified by antibodies to neurofilaments? What is the relationship between MND and other disorders in which Lewy bodies or similar structures are found? Ultimately it may be possible to identify key molecular events which trigger cytoskeletal degeneration and ubiquitination of cellular proteins – although at present there is no direct evidence that the presence of immunoreactive ubiquitin represents ubiquitin-protein conjugates, or indeed is related to this particular function of ubiquitin. Nevertheless, identification of abnormal proteins in MND motor neurons will enable us to move closer to the biochemical processes associated with cell death, and thus to the pathogenic mechanisms of the disease.

Conclusions

The discovery of characteristic ubiquitin-immunoreactive inclusions in vulnerable neurons in MND has provided a new basis for investigating the molecular consequences and possibly the pathogenic mechanisms of cell damage. New insights into the biology of ubiquitin and the heat shock response may explain why ubiquitin accumulates in degenerating motor neurons, although its presence may yet turn out to be a non-specific response. The presence of filamentous structures within these inclusions probably indicates that ubiquitin is associated with altered cyto-skeletal components of some sort, but MND is strikingly different from other neurodegenerative diseases in that antibodies against cytoskeletal proteins do not identify the skein-like inclusions, although occasionally hyaline bodies or Lewy body-like inclusions are labelled by anti-neurofilament antibodies. At present there is no evidence that abnormalities of neurofilament phosphorylation or processing play a pathogenic role in MND, although this possibility has not been absolutely discounted.

The next few years of research are likely to bring important new information on the clinical, pathological and molecular significance of the findings described in this chapter. The study of molecular pathology has been decisive in locating genetic abnormalities in Alzheimer's disease and the prion disorders, and there is hope that this approach will also be fruitful in MND.

References

Anderton BH, Breinburg D, Downes MJ et al. (1982) Monoclonal antibodies show that neurofibrillary tangles and neurofilaments share antigenic determinants. Nature 298:84–86

Bachmair A, Varshavsky A (1989) The degradation signal in a short-lived protein. Cell 56:1019–1032

Bachmair A, Finley D, Varshavsky A (1986) In vivo half-life of a protein is a function of its amino-terminal residue. Science 234:179–186

Bancher C, Brunner C, Lassmann H et al. (1989a) Accumulation of abnormally phosphorylated tau precedes formation of neurofibrillary tangles in Alzheimer's disease. Brain Res 477:90–99

Bancher C, Lassmann H, Budka H et al. (1989b) An antigenic profile of Lewy bodies: immunocytochemical indication for protein phosphorylation and ubiquitination. J Neuropathol Exp Neurol 48:81–93

Bond U, Schlesinger MJ (1985) Ubiquitin is a heat shock protein in chicken embryo fibroblasts. Mol Cell Biol 5:949–956

Brown P, Goldfarb LG, Gajdusek DC (1991) The new biology of spongiform encephalopathy: infectious amyloidoses with a genetic twist. Lancet 337:1019–1022

Bunina TL (1962) On intracellular inclusions in familial amyotrophic lateral sclerosis. Zhurnal Nevropatologii i Psikhiatrii Imeni S.S. Korsakova 62:1293–1299 (in Russian)

Carpenter S (1968) Proximal axonal enlargement in motor neuron disease. Neurology 18:841–851

Chau V, Tobias JW, Bachmair A et al. (1989) A multiubiquitin chain is confined to specific lysine in a targeted short-lived protein. Science 243:1576–1583

Chin DT, Carlson N, Kuehl L, Rechsteiner M (1986) The degradation of guanidinated lysozyme in reticulocyte lysate. J Biol Chem 261:3883–3890

Chou SM (1979) Pathognomy of intraneuronal inclusions in ALS. In: Tsubaki T, Toyokura Y (eds) Amyotrophic lateral sclerosis. Tokyo University Press, Tokyo and University Park Press, Baltimore, pp 135–177

Ciechanover A, Heller H, Elias S, Haas AL, Hershko A (1980) ATP-dependent conjugation of reticulocyte proteins with the polypeptide required for protein degradation. Proc Natl Acad Sci USA 77:1365–1368

Ciechanover A, Heller H, Katz-Etzion R, Hershko A (1981) Activation of the heat stable polypeptide of the ATP-dependent proteolytic system. Proc Natl Acad Sci USA 78:761–765

Ciechanover A, Finley D, Varshavsky A (1984) Ubiquitin dependence of selective protein degradation demonstrated in the mammalian cell cycle mutant ts85. Cell 37:57–66

Ciechanover A, Wolin SL, Steitz JA, Lodish HF (1985) Transfer RNA is an essential component of the ubiquitin- and ATP-dependent proteolytic system. Proc Natl Acad Sci USA 82:1341–1345

Clark AW, Parhad IM, Griffin JW, Price DL (1984) Neurofilamentous axonal swellings as a normal finding in spinal anteror horn of man and other primates. J Neuropathol Exp Neurol 43:253–262

Cole GM, Timiras PS (1987) Ubiquitin-protein conjugates in Alzheimer's lesions. Neurosci Lett 79:207–212

Collinge J, Harding AE, Owen F et al. (1989) Diagnosis of Gerstmann–Sträussler syndrome in familial dementia with prion protein gene analysis. Lancet ii:15–17

Cork LC, Adams RJ, Griffin JW, Price DL (1982a) Hereditary canine spinal muscular atrophy: a canine model of human motor neuron disease. In: Animal models of inherited metabolic diseases. Alan R Liss, New York, pp 449–458

Cork LC, Griffin JW, Chi Choy MS, Padula CA, Price DL (1982b) Pathology of motor neurons in accelerated hereditary canine spinal muscular atrophy. Lab Investigation 46:89–99

Cork LC, Troncoso JC, Klavano G et al. (1988) Neurofilamentous abnormalities in motor neurons in spontaneously occurring animal disorders. J Neuropathol Exp Neurol 47(4):420–431

Delisle MB, Carpenter S (1984) Neurofibrillary axonal swellings and ALS. J Neurol Sci 63:241–252

Delisle MB, Gorce P, Hirsch E, Hauw JJ, Rascol A, Bouissou H (1987) Motor neuron disease, parkinsonism, and dementia. Report of a case with diffuse Lewy body-like intracytoplasmic inclusions. Acta Neuropathol 75:104–108

Ferber S, Ciechanover A (1986) Transfer RNA is required for conjugation of ubiquitin to selective substrates of the ubiquitin- and ATP-dependent proteolytic system. J Biol Chem 261:3128–3134

Ferber S, Ciechanover A (1987) Role of arginine-tRNA in protein degradation by the ubiquitin pathway. Nature 326:808–811

Fried VA, Smith HT, Hildebrandt E, Weiner K (1987) Ubiquitin has intrinsic proteolytic activity: implications for cellular regulation. Proc Natl Acad Sci USA 84:3685–3689

Galloway PG, Grundke-Iqbal I, Iqbal K, Perry G (1988) Lewy bodies contain epitopes both shared and distinct from Alzheimer neurofibrillary tangles. J Neuropathol Exp Neurol 47:654–663

Ganoth D, Leshinsky E, Eytan E, Hershko A (1988) A multicomponent system that degrades proteins conjugated to ubiquitin. J Biol Chem 263:12412–12419

Garofalo O, Kennedy PGE, Swash M et al. (1991a) Ubiquitin and heat shock expression in amyotrophic lateral sclerosis. Neuropathol Appl Neurobiol 17:39–46

Garofalo O, Hajimohammadreza I, Leigh PN (1991b) Development of methods for the identification of abnormal proteins in ALS. In: Rose FC (ed) New evidence in MND/ALS research. Advances in ALS/MND, 2. Smith-Gordon, London, pp 139–141

Gibb WRG, Luthert PJ, Janota I, Lantos PL (1989) Cortical Lewy body dementia: clinical features and classification. J Neurol Neurosurg Psychiatr 52:185–192

Gold B (1987) The pathophysiology of proximal neurofilamentous giant axonal swellings: implications for the pathogenesis of amyotrophic lateral sclerosis. Toxicology 46:125–139

Gonda DK, Bachmair A, Wunning I, Tobias JW, Lane, WS, Varshavsky A (1989) Universality and structure of the N-end rule. J Biol Chem 264:16700–16712

Grundke-Iqbal I, Iqbal K, Tung Y-C, Quinlan M, Wisniewski HM, Binder LI (1986) Abnormal phosphorylation of the microtubule-associated protein tau in Alzheimer cytoskeletal pathology. Proc Natl Acad Sci USA 83:4913–4917

Haas AL, Warms JVB, Hershko A, Rose IA (1982) Ubiquitin-activating enzyme. J Biol Chem 257:2543–2548

Haas AL, Bright PM, Jackson VE (1988) Functional diversity among putative E2 isozymes in the mechanism of ubiquitin-histone ligation. J Biol Chem 263:13268–13275

Haugh MC, Probst A, Ulrich J, Kahn J, Anderton BH (1986) Alzheimer neurofibrillary tangles contain phosphorylated and hidden neurofilament epitopes. J Neurol Neurosurg Psychiatr 49:1213–1220

Heggie P, Burdon T, Lowe J et al. (1989) Ubiquitin gene expression in brain and spinal cord in motor neuron disease. Neurosci Lett 102:343–348

Hershko A (1991) The ubiquitin pathway for protein degradation. Trends Biochem 16:265–268
Hershko A, Ciechanover A, Heller H, Haas AL, Rose IA (1980) Proposed role of ATP in protein breakdown: conjugation of proteins with multiple chains of the polypeptide of ATP-dependent proteolysis. Proc Natl Acad Sci USA 77:1783–1786
Hershko A, Ciechanover A, Rose IA (1981) Identification of the active aminoacid residue of the polypeptide of ATP-dependent protein breakdown. J Biol Chem 256:1525–1528
Hershko A, Eytan E, Ciechanover A, Haas AL (1982) Immunochemical analysis of the turnover of ubiquitin-protein conjugates in intact cells. J Biol Chem 257:13964–13970
Hershko A, Heller H, Elias S, Ciechanover A (1983) Components of ubiquitin-protein ligase system. J Biol Chem 258:8206–8214
Hershko A, Heller H, Eytan E, Kaklij G, Rose IA (1984a) Role of the alpha-amino group of protein in ubiquitin-mediated protein breakdown. Proc Natl Acad Sci USA 81:7021–7025
Hershko A, Leshinsky E, Ganoth D, Heller H (1984b) ATP-dependent degradation of ubiquitin-protein conjugates. Proc Natl Acad Sci USA 81:1619–1623
Hershko A, Heller H, Eytan E, Reiss Y (1986) The protein substrate binding site of the ubiquitin-protein ligase system. J Biol Chem 261:11992–11999
Hirano A, Kurland LT, Sayre GP (1967) Familial amyotrophic lateral sclerosis: a subgroup characterized by posterior and spinocerebellar tract involvement and hyaline inclusions in the anterior horn cells. Arch Neurol 16:232–243
Hirano A, Donnefeld H, Sasaki S, Nakano I (1984a) Fine structural observations on neurofilamentous changes in amyotrophic lateral sclerosis. J Neuropathol Exp Neurol 43:461–470
Hirano A, Nakano I, Kurland LT, Mulder DW, Holley PW, Saccomanno G (1984b) Fine structural study of neurofibrillary changes in a family with amyotrophic lateral sclerosis. J Neuropathol Exp Neurol 43:471–480
Hoffman PN, Griffin JW, Price DL (1984) Control of axonal caliber by neurofilament transport. J Cell Biol 99:705–714
Hough R, Pratt G, Rechsteiner M (1986) Ubiquitin-lysozyme conjugates. J Biol Chem 261: 2400–2408
Hough R, Pratt G, Rechsteiner M (1987) Purification of two high molecular weight proteases from rabbit reticulocyte lysate. J Biol Chem 262:8303–8313
Hsiao KK, Baker HF, Crow TJ et al. (1989) Linkage of a prion protein missense variant to Gerstmann–Sträussler syndrome. Nature 338:342–345
Hughes JT, Jerrome D (1971) Ultrastructure of anterior horn motor neurones in the Hirano–Kurland–Sayre type of combined neurological system degeneration. J Neurol Sci 13:389–399
Jentsch S, Mcgrath JP, Varshavsky A (1987) The yeast DNA repair gene RAD6 encodes a ubiquitin-conjugating enzyme. Nature 329:13964–13970
Kahn J, Anderton BH, Gibb WRG, Lees AJ, Wells FR, Marsden CD (1985) Neuronal filaments in Alzheimer's Pick's, and Parkinson's diseases. N Engl J Med 313:520–521
Kato T, Hirano A, Donnenfeld H (1987) A Golgi study of the large anterior horn cells of the lumbar cords in normal spinal cords and in amyotrophic lateral sclerosis. Acta Neuropathol 75:34–40
Kato T, Katagiri T, Hirano A, Kawanami T, Sasaki H (1989) Lewy body-like hyaline inclusions in sporadic motor neuron disease are ubiquitinated. Acta Neuropathol 77:391–396
Kew JJM, Leigh PN (1992) Dementia with motor neuron disease. In: Rossor M (ed) Unusual dementias, Baillière Tindall, London (Baillière's clinical neurology 1(3))
Kihira T, Yoshida S, Uebayashi Y, Yase Y, Yoshimasu F (1991) Involvement of Onuf's nucleus in ALS. Demonstration of intraneuronal conglomerate inclusions and Bunina bodies. J Neurol Sci 104:119–128
Kondo A, Iwaki T, Tateishi J, Kirimoto K, Morimoto T, Oomura I (1986) Accumulation of neurofilaments in a sporadic case of amyotrophic lateral sclerosis. Jpn J Psychiatr Neurol 40:677–684
Kosik KS, Joachim CL, Selkoe DJ (1986) Microtubule-associated protein tau is a major antigenic component of paired helical filaments in Alzheimer's disease. Proc Natl Acad Sci USA 83:4044–4048
Kusaka H, Imai T, Hashimoto S, Yamamoto T, Maya K, Yamasaki M (1988) Ultrastructural study of chromatolytic neurons in an adult-onset sporadic case of amyotrophic lateral sclerosis. Acta Neuropathol 75:523–528
Kuzahara S, Mori H, Izumiyama N, Yoshimura M, Ihara Y (1988) Lewy bodies are ubiquitinated. A light and electron microscopic immunocytochemical study. Acta Neuropathol 75:345–353
Leigh PN, Anderton BH, Dodson A, Gallo J-M, Swash M, Power DM (1988) Ubiquitin deposits in anterior horn cells in motor neuron disease. Neurosci Lett 93:197–203

Leigh PN, Dodson A, Swash M, Brion J-P, Anderton BH (1989a) Cytoskeletal abnormalities in motor neuron disease: an immunocytochemical study. Brain 112:521–535

Leigh PN, Probst A, Dale GE et al. (1989b) New aspects of the pathology of neurodegenerative disorders as revealed by ubiquitin antibodies. Acta Neuropathol 79:61–72

Leigh PN, Whitwell H, Garofalo O et al. (1991) Ubiquitin-immunoreactive intraneuronal inclusions in amyotrophic lateral sclerosis: morphology, distribution and specificity. Brain 114:775–788

Leung DW, Spencer SA, Cachianes G et al. (1987) Growth hormone receptor and serum-binding protein: purification, cloning and expression. Nature 330:537–543

Love S, Saitoh T, Quijada S, Cole GM, Terry RD (1988) Alz-50, ubiquitin and tau immunoreactivity of neurofibrillary tangles, Pick bodies and Lewy bodies. J Neuropathol Exp Neurol 47:393–405

Lowe J, Mayer RJ (1990) Ubiquitin, cell stress and disease of the nervous system. Neuropathol Appl Neurobiol 16:281–292

Lowe J, Aldridge F, Lennox G et al. (1989) Inclusion bodies in motor cortex and brainstem of patients with motor neuron disease are detected by immunocytochemical localisation of ubiquitin. Neurosci Lett 105:7–13

Lowe J, Blanchard A, Morrell K et al. (1988a) Ubiquitin is a common factor in intermediate filament inclusion bodies of diverse type in man, including those of Parkinson's disease, Pick's disease, and Alzheimer's disease, as well as Rosenthal fibres in cerebellar astrocytomas, cytoplasmic bodies in muscle, and Mallory bodies in alcoholic liver disease. J Pathol 155:9–15

Lowe J, Lennox G, Jefferson D (1988b) A filamentous inclusion body within anterior horn neurons in motorneurone disease defined by immunocytochemical localisation of ubiquitin. Neurosci Lett 94:203–210

Manetto V, Sternberger NH, Perry G, Sternberger LA, Gambetti P (1988) Phosphorylation of neurofilaments is altered in amyotrophic lateral sclerosis. J Neuropathol Exp Neurol 47:642–653

Manetto V, Abdul-Karim FW, Perry G, Tahaton M, Autilio Cambetti L, Gambetti P (1989) Selective presence of ubiquitin in intracellular inclusions. Am J Pathol 134:505–513

Matsumoto S, Hirano A, Goto S (1990) Ubiquitin-immunoreactive filamentous inclusion in anterior horn cells of Guamanian and non-Guamanian amyotrophic lateral sclerosis. Acta Neuropathol 80:233–238

Mayer AN, Wilkinson KD (1989) Detection, resolution and nomenclature of multiple ubiquitin carboxyl-terminal esterases from bovine calf thymus. Biochemistry 28:166–172

Migheli A, Autilio-Gambetti L, Gambetti P, Mocellini C, Vigliani MC, Schiffer D (1990) Ubiquitinated filamentous inclusions in spinal cord of patients with amytrophic lateral sclerosis. Neurosci Lett 114:5–10

Miller CCJ, Brion J-P, Calvert R et al. (1986) Alzheimer's paired helical filaments share epitopes with neurofilament side arms. EMBO J 5:269–276

Mizusawa H, Matsumoto S, Yen S-H, Hirano A, Rojas-Corona RR, Donnenfeld H (1989) Focal accumulation of phosphorylated neurofilaments within anterior horn cell in familial amyotrophic lateral sclerosis. Acta Neuropathol 79:37–43

Mizusawa H, Nakamura H, Wakayama I, Yen S-HC, Hirano A (1991) Skein-like inclusions in the anterior horn cells in motor neuron disease. J Neurol Sci 105:14–21

Mori H, Kondo J, Ihara Y (1986) Ubiquitin is a component of paired helical filaments in Alzheimer's disease. Science 235:1641–1646

Munoz DG, Green C, Perl D, Selkoe DJ (1988) Accumulation of phosphorylated neurofilaments in anterior horn motoneurons of ALS patients. J Neuropathol Exp Neurol 47:9–18

Murayama S, Ookawa Y, Mori H et al. (1989) Immunocytochemical and ultrastructural study of Lewy body-like hyaline inclusions in familial amyotrophic lateral sclerosis. Acta Neuropathol 78:143–152

Murayama S, Bouldin TW, Suzuki K (1991) Immunocytochemical and ultrastructural studies of Werdnig–Hoffman disease. Acta Neuropathol 81:408–417

Murayama S, Mori H, Ihara Y, Bouldin TW, Suzuki K, Tomonaga M (1990a) Immunocytochemical and ultrastructural studies of lower motor neurons in ALS. Ann Neurol 27:137–148

Murayama S, Mori H, Ihara Y, Tomonaga M (1990b) Immunocytochemical and ultrastructural studies of Pick's disease. Ann Neurol 27:394–405

Nakazato Y, Yamazaki H, Hirato J, Ishida Y, Yamaguchi H (1990) Oligodendroglial microtubular tangles in olivopontocerebellar atrophy. J Neuropathol Exp Neurol 49:521–530

Neary D, Snowden JS, Mann DMA, Northern B, Goulding PJ, MacDermott N (1990) Frontal lobe dementia and motor neuron disease. J Neurol Neurosurg Psychiatry 53:23–32

Nickel BE, Davie JR (1989) Structure of polyubiquitinated histone H2A. Biochemistry 28:964–968

Okamato K, Hirai S, Yamazaki T, Sun X, Nakazato Y (1991a) New ubiquitin-positive intraneuronal inclusions in the extra-motor cortices in patients with amyotrophic lateral sclerosis. Neurosci Lett 129:233–236

Okamoto K, Hirai S, Ishiguro K, Kawarabayashi T, Takatama M (1991b) Light and electron microscopic and immunohistochemical observations of the Onuf's nucleus of amyotrophic lateral sclerosis. Acta Neuropathol 81:610–614

Owen F, Poulter M, Lofthouse R et al. (1989) Insertion in prion protein gene in familial Creutzfeldt–Jakob disease families. Lancet i:51–52

Papp MI, Kahn JE, Lantos PL (1989) Glial cytoplasmic inclusions in the CNS of patients with multiple system atrophy (striatonigral degeneration, olivopontocerebellar atrophy and Shy–Drager syndrome). J Neurol Sci 94:79–100

Perry G, Friedman R, Shaw G, Chau V (1987) Ubiquitin is detected in neurofibrillary tangles and senile plaque neurites of Alzheimer disease brains. Proc Natl Acad Sci USA 84:3033–3036

Pickart CM, Rose IA (1985) Functional heterogeneity of ubiquitin-carrier proteins. J Biol Chem 260:1573–1581

Pickart CM and Vella AT (1988) Levels of active ubiquitin carrier proteins decline during erythroid maturation. J Biol Chem 263:12028–12035

Poltorak M, Freed WJ (1989) Immunoreactive phosphorylated epitopes on neurofilaments in neuronal perikarya may be obscured by tissue preprocessing. Brain Research 480:349–354

Rechsteiner M (1987) Ubiquitin-mediated pathways for intracellular proteolysis. Ann Rev Cell Biol 3:1–30

Reiss Y, Kaim D, Hershko A (1988) Specificity of binding of NH_2-terminal residue of proteins to ubiquitin-protein ligase. J Biol Chem 263:2693–2698

Rivett AJ, Skilton HE, Rowe AJ, Eperon IC, Sweeney ST (1991) Components of the multicatalytic proteinase complex. Biomed Biochem Acta 4(6):4450–4487

Rivett AJ (1989) The multi-catalytic proteinase of mammalian cells. Arch Biochem Biophys 268:1–8

Robinson PA, Anderton BH (1988) Neurofilament probes – a review of neurofilament distribution and morphology. Rev Neurosci 2:1–40

Rosenfeld J, Dorman ME, Griffin JW, Sternberger LA, Sternberger NH, Price DL (1987) Distribution of neurofilament antigens after axonal injury. J Neuropathol Exp Neurol 46:269–282

Sasaki S, Kamei H, Yamane K, Murayama S (1988) Swelling of neuronal processes in motor neuron disease. Neurology 38:1114–1118

Sasaki S, Yamane K, Sakuma H, Murayama S (1989) Sporadic motor neuron disease with Lewy body-like hyaline inclusions. Acta Neuropathol 78:555–560

Schiffer D, Autilio-Gambetti L, Chio A et al. (1991) Ubiquitin in motor neuron disease: a study at the light and electron microscope level. J Neuropathol Exp Neurol 50:463–473

Schlaepfer WW (1987) Neurofilaments: structure, metabolism, and implication in disease. J Neuropathol Exp Neurol 46:117–129

Schmidt ML, Carden MJ, Lee VM-Y, Trojanowski JQ (1987) Phosphate-dependent and independent epitopes in the axonal swellings of patients with motor neuron disease and controls. Lab Invest 56:282–294

Schochet JS, Hardmann JM, Ladewig PP, Earle KM (1969) Intraneuronal conglomerates in sporadic motor neuron disease. Arch Neurol 20:548–553

Siegelman M, Bond MW, Gallatin WM et al. (1986) Cell surface molecule associated with lymphocyte homing is a ubiquitinated branched-chain glycoprotein. Science 231:823–829

Sobue G, Hashizume Y, Yasuda T (1990) Phosphorylated high molecular weight neurofilament protein in lower motor neurons in ALS and other neurodegenerative diseases involving ventral horn cells. Acta Neuropathol 79:402–408

Sternberger LA, Sternberger NH (1983) Monoclonal antibodies distinguish phosphorylated and non-phosphorylated forms of neurofilaments in situ. Proc Natl Acad Sci USA 80:6126–6130

Thorne AW, Sautier P, Briand G, Crane-Robinson C (1987) The structure of ubiquitinated histone H2B. EMBO J 6:1005–1010

Toyoshima I, Yamamoto A, Masamune O, Satake (1989) Phosphorylation of neurofilament proteins and localisation of axonal swellings in motor neuron disease. J Neurol Sci 89:269–277

Wightman G, Anderson VER, Martin J et al. (1992) Hippocampal and neocortical ubiquitin-immuno-reactive inclusions in amyotrophic lateral sclerosis with dementia. Neurosci Lett 139:269–274

Wilkinson KD, Keunmyoung L, Deshpande S, Duerksen-Hughes P, Boss JM, Pohl J (1989) The neuron-specific protein PGP 9.5 is a ubiquitin carboxyl-terminal hydrolase. Science 246:670–673

Wohlfart G (1959) Degenerative and regenerative axonal changes in the ventral horns, brain stem and cerebral cortex in amyotrophic lateral sclerosis. Acta Univ Lund (New Series 2) 56:1–13

Wood JG, Mirra SS, Pollock NJ, Binder NJ, Binder LI (1986) Neurofibrillary tangles of Alzheimer's disease share antigenic determinants with the axonal microtubule-associated protein tau. Proc Natl Acad Sci USA 83:4040–4043

Yarden Y, Escobedo JA, Kuang WJ et al. (1986) Structure of the receptor for platelet-derived growth factor helps define a family of closely related growth factor receptors. Nature 323:226–232

Zhaung Z, McCauley R (1989) Ubiquitin is involved in the in vitro insertion of monoamine oxidase B into mitochondrial outer membranes. J Biol Chem 264:14594–14596

8 Neurochemistry of Motor Neuron Disease

S. Malessa and P. N. Leigh

Introduction

The studies of Oleh Hornykiewicz which identified nigrostriatal dopamine deficiency as the major neurochemical abnormality in Parkinson's disease led directly to the introduction of effective symptomatic treatment for that condition. Replacement therapy as applied to Parkinson's disease seems unlikely to be helpful in MND, but it is still possible that replacement therapy might be of temporary benefit, particularly if combined with treatment designed to slow or halt the progress of the condition. The emphasis, therefore, of recent neurochemical studies has been to identify neurochemical abnormalities which might provide clues to pathogenic mechanisms. Of particular interest has been the possibility that the balance between excitatory and inhibitory amino acids might be altered in MND (and other neurodegenerative disorders), resulting in excitotoxicity mediated via glutamate and other endogenous or exogenous excitotoxins.

Many of the neurochemical changes which have been identified in MND can be predicted from the pathology of the disease, and are largely a consequence of neuronal loss. In this chapter we review the literature on neurochemical changes in MND, assess areas of controversy and discuss some of the technical and methodological difficulties which may underlie them. Further discussion of neurochemical changes in relation to pathogenesis and treatment will be found in Chapter 10.

Table 8.1 summarises the important neurotransmitters and neuropeptides present in the spinal cord, indicating substances derived from descending pathways and those present in intrinsic ventral horn motor neurons and interneurons. The anatomy and pharmacology of motor systems in the spinal cord is discussed in Chapters 13 and 15.

While it is recognised that plasma or CSF concentrations of neuroactive substances provide an imperfect reflection of their concentrations in the CNS, and of their functional activities, plasma and CSF samples are relatively easy to acquire, and accordingly there is much information on the plasma and CSF levels of classical neurotransmitters, neuropeptides and amino acids. Unfortunately, many of these studies have yielded conflicting results.

Excitatory Amino Acids in Plasma and CSF in MND

Patten et al. (1978) found increases in fasting plasma total aromatic amino acids, total basic amino acids, and in tyrosine and ornithine, but not in glutamate or aspartate, in 12 MND patients who were compared to 12 neurologically abnormal

Table 8.1. Neurotransmitters and peptides in ventral horn of normal spinal cord

a Neurotransmitter:	ACh	NE	DA	5-HT	GLU	ASP	GLY	GABA
Intrinsic cells	+[1]	–	–	–	+[1]	+[2]	+[3]	+[4]
Descending pathways	+[5]	+[6]	+[7]	+[8]	+[9]	+[9]		

b Peptide:	SP	TRH	ENK	SOM	OXY	CGRP	END	CCK	NPY
cell bodies	–	–	–	–	–	+	+	–	–
fibres	+[11]	+[11]	+[11]	+[11]	+[12]	+[11]	+[10]	+[12]	+[11]

Neurotransmitters and peptides present (+) in the ventral horn of normal spinal cord. Neuropeptide studies either used immunocytochemistry or in situ hybridisation in human autopsy tissue. Lack of peptide-staining in cell bodies might be attributed to post-mortem factors in the absence of colchicine pre-treatment. Immunoreactive fibres may be derived from supraspinal levels as suggested for SP, TRH (Johansson et al. 1981), oxytocin, vasopressin and enkephalin or from dorsal root ganglia, or both.
Abbreviations: ACh, acetylcholine; ASP, aspartate; CCK, cholecystokinin; CGRP, calcitonin gene-related peptide; DA, dopamine; END, endothelin; ENK, enkephalin; GABA, gamma-aminobutyric acid; GLU, glutamate; GLY, glycine; 5-HT, 5-hydroxytryptamine; NE, noradrenaline; NPY, neuropeptide Y; OXY, oxytocin; SOM, somatostatin; SP, substance P; TRH, thyrotropin-releasing hormone.
References: 1, Houser et al. 1983; 2, Ottersen et al. 1984; 3, Davidoff et al. 1967; 4, Barber et al. 1982; 5, Spann and Grofova 1989; 6, Moore and Card 1984; 7, Bjöklund and Skagerberg 1979; 8, Johansson et al. 1981; 9, Young et al. 1983; 10, Giaid et al. 1989; 11, Gibson et al. 1988; 12, Schoenen et al. 1985a.

controls. CSF amino acid concentrations were altered, with raised levels of total basic amino acids and essential amino acids, and elevated lysine and leucine levels. Urinary amino acid excretion was not significantly different in MND patients and controls.

Patten and colleagues (1982) did, however, find a significant inverse correlation between serum aspartate levels and "activity" of the disease, the latter assessed by dividing the severity index (a simple measure of functional disability) by the duration of the disease in years. Levels of branched chain amino acids (BCAA) were positively correlated with disease duration, probably because these amino acids are metabolised in muscle, and advanced disease is associated with decreased muscle mass. It is doubtful whether such correlations, based on a small sample of patients and controls, are meaningful bearing in mind the variability of subsequent studies of plasma and CSF amino acids. Patten et al. (1978) noted a trend towards higher CSF aspartate levels in MND than in controls, and suggested that "excessive neuronal activity in ALS, including fibrillations and fasciculations, might be due in part to metabolic imbalance caused by an increase of this excitatory amino acid transmitter".

It is now widely accepted that glutamate and aspartate function are excitatory neurotransmitters (Lucas and Newhouse 1957; Olney et al. 1971; see Chapter 10. Although Patten et al. (1978) found no change in plasma or CSF glutamate levels in MND, Plaitakis et al. (1984), pursuing the idea that excitotoxicity might contribute to neuronal death in neurodegenerative disorders, found that fasting plasma glutamate (but not aspartate) levels were raised by nearly 100% in MND patients compared with healthy controls and patients with a variety of neurological disorders. Furthermore, plasma glutamate and aspartate levels were abnormally elevated after an oral glutamate load (Plaitakis and Caroscio 1987). Perry et al.

(1990) also found that plasma glutamate levels were elevated by about 30% in patients with MND compared with controls, but they attributed this to the greater age of their MND subjects, since plasma glutamate levels were positively correlated with increasing age.

Elevated plasma glutamate levels may, of course, have no relevance to CNS neurotransmitter glutamate, but CSF amino acid levels are thought to bear a closer relationship to biochemical processes in the CNS. In the case of glutamate, however, interpretation of altered CSF and tissue levels is confounded by the complexity of CNS glutamate metabolism, as discussed below and in Chapter 13.

CSF glutamate and aspartate levels have been variously determined as unchanged (Meier and Schott 1988; Perry et al. 1990), or significantly increased (Rothstein et al. 1990) in MND (Tables 8.2 and 8.3). In the study by Rothstein et al. (1990), CSF glutamate was elevated by 190%, and aspartate 100% in MND. Others have detected only trace amounts of aspartate in serum and CSF (Perry et al. 1990), whereas Rothstein et al. (1990) found levels of 4–8 μmol/l in CSF (Table 8.2). Furthermore, CSF levels of glutamate in the two studies differed 15-fold in the control groups, and 40-fold in the MND groups.

These differences are probably explained by technical factors such as the speed of centrifugation, sulphosalicylic acid deproteinisation, and the method of amino acid analysis. Automated HPLC probably provides the most satisfactory method of amino acid analysis, although CSF glutamate and aspartate levels are 10-fold lower when measured by this technique compared with ion exchange chromatography (Rothstein et al. 1991).

Other possible sources of error lie in the selection of patients. Older individuals have significantly higher levels of free aspartate, glycine, GABA, valine, isoleucine, leucine, phenylalanine, and 3-methyl histidine, and significantly lower levels of free phosphoethanolamine, serine, GABA, homocarnosine, and conjugated GABA and β-alanine than individuals below the age of 40 (Ferraro and Hare 1985). CSF amino acid levels also differed significantly in men and women. Plasma glutamate concentrations tend to be higher in older subjects, and have been attributed to elevated plasma glutamate concentrations in MND patients to differences in age between MND and control groups (Perry et al. 1990). However, in the study of Rothstein et al. (1990) in which there were marked increases in CSF concentrations of aspartate and glutamate, there were no significant differences in age and sex between the MND and control groups.

In summary, it is evident that subjects should be fasted (preferably overnight), that blood and CSF should immediately be cooled to 0–4 °C, centrifuged at around 20 000*g* for at least 10 minutes, and then either analysed immediately or frozen at −70°C. It is not clear whether samples should be deproteinised immediately on collection, or precisely what concentration of sulphosalicylic acid should be used for this step, but for automated HPLC analysis, it is acceptable to omit deproteinisation, since this step may spuriously elevate glutamate levels (Lundqvist et al. 1989). At present, the data on plasma and CSF glutamate and aspartate can neither be used to support the excitotoxin hypothesis nor to refute it, and it is arguable whether such data will ever tell us a great deal about synaptic events within the CNS. It is thus important to discuss studies on excitatory and other amino acids in human post-mortem tissue, and to survey other aspects of neurochemistry relevant to these problems.

Table 8.2. Summary of major studies on serum and CSF amino acids (AAs) in MND

Authors	MND patients			Controls			Methodology	Changes in plasma: MDN compared with controls	Changes in CSF, MND compared with controls* (see Table 8.3)	Concentrations of some specific AAs in MND CSF in μmol/1 ± SEM or SD (where given)						Comment
	n	Mean age (range)	M:F	*n*	Mean age (range)	M:F				GLU	ASP	GLY	THRE	VAL	LEU	
1. Patten et al. (1982)	12	50 (28–68)	6:6	12*	48 (30–75)	6:6	Subjects fasted for 8–10 h; CSF placed on ice for automated AA analysis on same day.	↑ total aromatic AA ↑ total basic AA ↑ tyrosine ↑ ornithine	↑ total basic AA ↑ total essential AA ↑ lysine ↑ leucine	3.8 ±2.2	1.1 ±1.4*	5.7 ±3*	31 ±4*	17 ±5*	13 ±4*	
				*Heterogeneous control group of neurologically abnormal patients						*Not stated whether SEM or SD						
2. De Belleroche et al. (1984)	11	55 (40–74)	7:4	9*	53 (35–71)	3:6	Not state whether fasting sample. Sample acidified with sulphosalicylic acid, filtered, frozen and stored at −20 °C before automated AA analysis on a Beckman analyser.	ND	↑ threonine ↑ glycine ↑ alanine ↑ alanine ↑ valine ↑ isoleucine ↑ phenylalanine ↑ non-polar R group AA	–	–	13.7 ± 1.6 (SEM)	40.6 ± 2 (SEM)	22.3 ± 1.8 (SEM)	17 ± 1.7 (SEM)	No significant change in glutamate, although glutamate and glutamine levels were combined. Positive correlations were identified between levels of glycine and alanine and activity of disease.
				*Patients with lumbar pain undergoing myelography for diagnosis of prolapsed intervertebral disc												
3. Plaitakis & Caroscio (1987)	22	56 (31–70)	NS	81*	NS	NS	Blood sample collected into EDTA after overnight fast, centrifuged and deproteinised, plasma stored at −80 °C. analysed with Technicon TSM AA analyser *or* by HPLC. Oral loading with monosodium glutamate (MSG) after overnight fast in 18 MND patients and normal and neurologically abnormal controls.	↑ glutamate (+100%) MND patient showed impaired clearance of glutamate following oral load, with elevated serum aspartate levels, compared with controls	ND							No changes in other plasma AAs in MND compared with controls. Leucocyte glutamate dehydrogenase (LGDH) unchanged, despite impaired glutamate tolerance after oral load of MSG.
				*Comparing 45 'normal' controls, 5 patients with Friedreich's ataxia, 20 with spinocerebellar degeneration, and 11 with neuromuscular disorders												

Table 8.2. Continued.

Authors	MND patients			Controls			Methodology	Changes in plasma: MDN compared with controls	Changes in CSF, MND compared with controls* (see Table 8.3)	Concentrations of some specific AAs in MND CSF in μmol/1 ± SEM or SD (where given)						Comment
	n	Mean age (range)	M:F	*n*	Mean age (range)	M:F				GLU	ASP	GLY	THRE	VAL	LEU	
4. Meier and Schott (1988)	5	50 (45–70)	4:1	17	48 (29–72)	10:7	Typical MND/ALS patients with UMN & LMN signs. CSF obtained at "routine" lumbar puncture, centrifuged, supenatant stored overnight at 4 °C, and an aliquot frozen at −70 °C until analysed by micro thin layer chromatography.	ND	No significant changes	27.2	5.6	19.6	49.4	18.5	11.1	Difficult to interpret in view of very small number of MND samples.
5. Perry et al. (1990)	28 (plasma) 17 (CSF)	60 (32–82) 57 (32–75)	20:8 11:6	48 (plasma) (healthy controls) 80 (CSF) (various neurological disorders)	45 (30–79) 50 (NS)	NS NS	Typical MND/ALS with UMN and LMN signs, with strict diagnostic and exclusion criteria. NB: MND subjects significantly older than controls. Serum and CSF obtained after overnight fast. Serum and CSF deproteinised with sulphosalicylic acid. AA analysis on a Technicon AA analyser.	↑ glutamate ↑ glutamine ↓ threonine ↑ cystine ↓ GABA ↑ methylhistidine ↓ phenylalanine ↓ methionine ↓ histidine	↑ glutamine ↑ alanine ↑ valine ↑ isoleucine ↑ leucine ↑ tryptophan ↑ ethanolamine ↑ lysine ↑ arginine	0.2 ± 0.1 (SD)	TR	5.2 ± 2.5 (SD)	32.8 ± 8.4 (SD)	23.6 ± 7.0 (SD)	15.8 ± 3.9 (SD)	↑ plasma glutamate attributed to greater age of MND group, since glutamate levels rise with increasing age. No evidence of raised glutamate, aspartate, or glycine levels in MND CSF. A few MND patients with longer duration of disease (4+ years) showed increases in serum glutamate up to 200%). ?Immediate deproteinisation of CSF may reduce detectable AAs. This study does not support the notion that a systemic defect in glutamate metabolism causes MND. Analysis by automated ion exchange resin.

Table 8.2. Continued.

Authors	MND patients			Controls			Methodology	Changes in plasma: MND compared with controls	Changes in CSF, MND compared with controls* (see Table 8.3)	Concentrations of some specific AAs in MND CSF in μmol/1 ± SEM or SD (where given)						Comment
	n	Mean age (range)	M:F	*n*	Mean age (range)	M:F				GLU	ASP	GLY	THR	VAL	LEU	
6. Rothstein et al. (1990)	18	52 (31–75)	NS	28* *comprising 16 patients with diverse neurological disorders, 2 with 1° biliary cirrhosis, and 10 with hepatic encephalopathy.	53 (22–80)	NS	Strict inclusion criteria, all MND/ALS patients having UMN and LMN signs. CSF (?not fasting) placed on ice immediately after LP and either assayed for AAs immediately or stored at −80 °C. Analysis of sulphosalicylic acid-treated samples was performed by automated ion-exchange chromatography on a Beckman 6300 AA analyser.	ND	↑ aspartate (100%) ↑ glutamate (190%) ↑ threonine ↑ serine	8.4 ± 1.4 (SEM)	8.4 ± 1.2 (SEM)	26.6 ± 3.7 (SEM)	44.4 ± 3.5 (SEM)	21.7 ± 1.5 (SEM)	14.9 ± 0.9 (SEM)	Increases were observed in CSF glycine (+38%), alanine (+29%) valine(+19%), lysine (+28%) and arginine (+29%) in MND. CSF concentrations of N-acetyl aspartyl glutamate (NAAG) and N-acetyl aspartate (NAA) were *increased* in MND, but *decreased* concentrations of NAAG and NAA were found in ventral horn of post mortem samples of spinal cord in 8 MND patients compared with 9 controls. Analysis by Automated ion exchange resin.
7. Rothstein et al. (1991)	15	NS	NS	8	NS	NS	Same inclusion criteria as (6)	ND	↑ glutamate (230%) ↑ aspartate (154%)	0.8 ± 0.31	0.91 ± 0.09					Automated HPLC.
8. Perry et al. (1991)	29	60 (32–82)	NS	49	59.6 (32–85)	NS	Strict inclusion criteria; all MND/ALS patients had UMN and LMN signs. Fasting plasma deproteinised within 30 minutes of collection with sulphosalicylic acid or stored at −70 °C until analysed. AA analysis performed on Technicon AA analyser to give concentrations of cystine, cystinylglycine, and total cysteine.	No changes in plasma concentrations of crystine, cystinylglycine, total cysteine, or inorganic sulphate in MND compared with control subjects.								Studies of post-mortem brain and spinal cord samples showed no differences in L-cysteine concentrations in MND patients versus controls.

Table 8.3. Comparison of CSF amino acid levels in MND patients and controls

Author	F/NF	TAU	ASP	THR	SER	GLU	GLY	ALA	VAL	CYS	ILE	LEU	LYS	ARG
Patten et al. (1978)	F	–	–	–	–	–	–	–	–	ND/ TR	–	↑ (30%)	↑ (20%)	–
De Belleroche et al. (1990)	NF	–	NE	↑ (25%)	–	–*	↑ (58%)	↑ (38%)	↑ (30%)	NE	↑ (60%)	–	–	NE
Meier et al. (1988)	NF	–	–	–	–	–	–	–	–	NE	–	–	–	–
Perry et al. (1990)	F	–	ND/ TR	–	–	–	–	↑ (22%)	↑ (36%)	ND/ TR	↑ (34%)	↑ (29%)	↑ (34%)	↑ (18%)
Rothstein et al. (1990)	NF	NE	↑ (100%)	↑ (35%)	↑ (36%)	↑ (190%)		–	–	–	–	–	↑ (28%)	–

–, no statistically significant change;
↑ (%), statistically significant increase in MND patients (compared with controls (% increase));
NE, not examined;
ND/TR, not detected, or only trace amount;
F/NF, fasting/not fasting;
*, glutamate and glutamine assayed together.

Glutamate, Glutamate Dehydrogenase (GDH) and Excitotoxicity

Robinson (1968) who first measured glutamate and aspartate concentrations in post-mortem tissue found normal levels in two MND patients and reductions in two others. Patten et al. (1982) found no change in the ventral horn, although local changes may have been masked by the fact that they pooled cervical and lumbar grey matter. Perry et al. (1987) and Plaitakis et al. (1988) identified a widespread deficiency of brain and spinal cord glutamate in MND. Our own observations support the notion of a glutamate deficiency which extends beyond the usual areas of pathological damage, since we found reduced levels in cervical and lumbar spinal cord, and in white and grey matter (Malessa et al. 1991). As regards aspartate, significant reductions have been noted in the mediodorsal thalamus (Perry et al. 1987) and the spinal cord (Plaitakis et al. 1988; Malessa et al. 1991).

Interest in GDH in relation to MND originally arose from observations of Plaitakis et al. (1982, 1984) and Duvoisin et al. (1983) which identified a partial deficiency of fibroblast and leucocyte GDH in some patients with olivopontocerebellar atrophy (OPCA). GDH is a key enzyme in glutamate metabolism, catalysing the reaction:

$$\alpha\text{-ketoglutarate} + NH_4^+ + NADH \rightarrow \text{L-glutamate} + H_2O + NAD^+$$

The enzyme is thought to act mainly in the direction of glutamate synthesis, and thus to have an important role in detoxification of ammonia (Chee et al. 1979).

Changes in GDH levels may be relatively non-specific since slightly lower leucocyte, fibroblast or muscle GDH activities have been noted in a wide variety of neurodegenerative disorders (Sheu et al. 1985; Aubby et al. 1988). Nevertheless, OPCA with GDH deficiency was associated with high fasting plasma levels of glutamate, and impaired glutamate clearance after an oral glutamate load (Plaitakis et al. 1982). Because OPCA can be accompanied by motor system degeneration (Plaitakis et al. 1984), Plaitakis and Caroscio (1987) measured plasma glutamate levels fasting and after an oral glutamate load, and observed that in both situations plasma glutamate levels were significantly raised in MND patients compared with healthy and neurologically abnormal controls. However, they did not find altered leucocyte GDH activity in their MND subjects (Plaitakis & Caroscio 1987). Subsequently Hugon et al. (1989) found that about 60% of MND patients had decreased leucocyte GDH activity (i.e. a deficiency of more than two standard deviations from the control mean). However, the remaining 40% of MND subjects had normal leucocyte GDH levels.

Because measurements of brain GDH levels might be expected to be more informative than leucocyte GDH activity, we studied GDH activity in the spinal cord of MND patients and controls (Malessa et al. 1988, 1991). In homogenates of cervical cord, GDH activity was increased by more than 50% of control levels in the lateral and ventral white matter, and in the dorsal horn (where it was increased more than 38%), whereas GDH activity was normal in the anterior horn and in the dorsal white matter corresponding to the dorsal columns (Malessa et al. 1988, 1991). The same pattern was seen in the lumbar enlargement where, however, the changes were less prominent.

Most brain and spinal cord GDH is present in astrocytes (Kaneko et al. 1987). As yet there is no evidence that the enzyme is selectively associated with glutamatergic pathways, and it may have little importance in the formation of neurotransmitter glutamate (Filla et al. 1986; Yudkoff et al. 1991). The obvious interpretation of our findings on GDH levels in the spinal cord was that increased levels of GDH were due to reactive gliosis. Histochemical studies have suggested that GDH is increased in the lateral corticospinal tracts in MND, and in other areas associated with reactive gliosis (Osterberg and Wattenberg 1962). Nevertheless, we were surprised that GDH was not increased in the ventral horn, where gliosis is often marked, and that GDH activity was elevated in the dorsal horns. We cannot exclude the possibility that even these changes reflect gliosis, but they do not correlate well with the distribution of astrocytosis in MND.

If there is a widespread deficiency of glutamate, as suggested by the evidence reviewed in the previous section, it is most likely to be related to the metabolic functions of glutamate. Plaitakis and Caroscio (1987) have suggested that there may be a generalised defect in the transport system linked to glutamate oxidation, or in the oxidative process itself. A widespread decrease in cerebral uptake of (^{18}F)2-fluoro-2-deoxy-D-glucose measured by positron emission tomography (PET) has been reported in patients with MND/ALS, but not in patients with only lower motor neuron involvement (Dalakas et al. 1987). This could be taken as indirect support of this hypothesis but more recent PET studies have not confirmed the presence of widespread abnormalities in MND patients (Kew et al. 1993).

Impaired glutamate tolerance in MND is unlikely to be due to defective transamination by aspartate aminotransaminase (AAT), since aspartate levels increase alongside those of glutamate, and leucocyte AAT activity is normal in MND (Plaitakis and Caroscio 1987; Hugon et al. 1989). Orally administered glutamate is metabolised mainly in the liver, and in the heart, kidney and muscles and does not cross the blood–brain barrier. Decreased peripheral GDH activity could be associated with impaired glutamate tolerance but it is not easy to see how this would cause any change in CNS glutamate, and the evidence available suggests that CNS (specifically, spinal cord) GDH levels are increased in MND. This change could conceivably be accompanied by excess glutamate formation in the CNS.

The question arises as to how this might lead to increases in synaptic glutamate. Clearly, post-mortem studies do not allow distinction between the various glutamate "pools". The large neuronal compartment, which contains 85%–98% of the total glutamate pool in brain, is composed mainly of the metabolic pool but includes the fraction available for release as well as the precursor glutamate for GABA, while the remainder (about 20%) is localised to astrocytes (Fonnum 1984; Plaitakis 1990). In the large compartment, glutamate can be synthesised from glucose, one of its many precursor substances. Since none of the enzymes involved in glutamate synthesis is known to be specific for the transmitter pool, this fraction probably accounts for only 20%–30% of the total endogenous glutamate content.

The changes in CSF and tissue glutamate levels might be due to an altered distribution of glutamate between the extracellular and the intracellular space consequent upon a decrease in the active re-uptake of synaptically released glutamate (Plaitakis 1990). This might lead to accumulation of glutamate at the synaptic cleft, to reduced tissue levels, and possibly to increased CSF levels. It is worth mentioning also that acetylcholine can inhibit the depolarisation-evoked release of glutamate via muscarinic receptors (Marchi et al. 1989). Given the cholinergic deficit and the loss of muscarinic receptors in MND dorsal grey matter, it is possible that loss of cholinergic inhibition might lead to excessive release of glutamate.

There is now more evidence that glutamate transport may be abnormal in MND (Rothstein et al. 1992). High-affinity sodium-dependent glutamate transport into synaptosomes was measured in several regions of the brain and in the spinal cord of 13 MND subjects, and in carefully matched controls without neurological disease. High-affinity glutamate uptake was decreased in the motor and somatosensory cortex and spinal cord, but not in the visual cortex of the MND subjects. The maximal velocity (V_{max}) of glutamate transport was decreased in MND motor cortex and spinal cord, but not in the striatum or hippocampus. The affinity of glutamate for the transporter (K_t) was not altered in MND tissue, and there were no changes in V_{max} in Alzheimer's disease or Huntington's disease brain, even in those regions (such as the striatum in Huntington's disease) where there is severe cell loss. High-affinity uptake of GABA and phenylalanine was not altered in MND, suggesting that the defect in glutamate transport is not a non-specific generalised loss of amino acid transport.

These abnormalities are likely to be secondary to the loss of glutamatergic synapses from the brain and spinal cord in MND. Only the demonstration of altered function of glutamate carriers pre-dating disease onset can fully resolve this issue, although secondary changes in glutamate transport could contribute to neuronal damage even if the causal lesion is elsewhere. The cloning of glutamate receptor subtypes (Nakanishi 1992; Sommer and Seeburg 1992), the localisation of the GluR5 subunit gene to chromosome 21 close to the putative familial MND locus (Eubanks et al. 1993), and the cloning of glia and neuronal glutamate carriers (Amara 1992; Pines et al. 1992; Kanai and Hediger 1992) provide new approaches and tools for examining the basis of the excitotoxic hypothesis in MND.

Finally, increased GDH activity could be detrimental to neuronal survival in MND by increasing synaptic (neurotransmitter) glutamate. GDH activity can be stimulated by BCAA such as leucine and valine, which are excellent precursors to glutamate nitrogen, and this effect can influence synaptic glutamate levels (Yudkoff et al. 1991) although the function of the GDH pathway which is present in synaptosomes at relatively high specific activity is uncertain. In astrocytes, the GDH reaction may be an important route for glutamate disposal (Yu et al. 1982). Thus although BCAA may stimulate the activity of GDH and lead to increased amounts of glutamate (probably initially in glia) it is not certain whether this glutamate would be available for neurotransmission (Yudkoff et al. 1991). Alternatively, stimulation of GDH activity might increase the rate of glutamate oxidation, and thus protect against glutamate excitotoxicity (Plaitakis 1990).

N-Acetyl-Aspartyl-Glutamate (NAAG) and MND

Other findings relevant to the excitotoxin hypothesis stem from recent studies on CSF and tissue levels of the acidic dipeptide N-acetyl-aspartyl-glutamate (NAAG) and its metabolite N-acetyl-aspartate (NAA). NAAG has been localised by immunocytochemistry to a number of neuronal systems, including motor neurons of the rat spinal cord and brain stem. NAAG can be synthesised from glutamate and aspartate, and has been colocalised with glutamate in some areas of rat brain (Anderson et al. 1986). It is present in highest concentration in the spinal cord (Koller and Coyle 1984), and fulfils the major criteria for a neurotransmitter or neuromodulatory substance in the CNS (Coyle et al. 1986; Zollinger et al. 1988). NAAG has an intrinsic neuroexcitatory action on spinal neurons, mediated by NMDA (N-methyl-D-aspartate) receptors (Westbrook et al. 1986). NAAG is converted to NAA and glutamate by the enzyme N-acetyl-alpha-linked acidic dipeptidase (NAALA-Dase).

Preliminary reports suggested that the concentration of NAAG is decreased in cervical spinal cord in MND (Constantakakis and Plaitakis 1988). More recently, Rothstein et al. (1990), in parallel with their studies on CSF amino acid concentrations in MND, measured CSF levels of NAAG and NAA, and spinal cord levels of NAAG, NAA, and NAALADase. CSF levels of NAAG and NAA were significantly increased in MND patients compared to the control group, which included patients with a wide variety of systemic and neurological disorders. In contrast, the concentrations of NAAG and NAA in MND spinal cord were decreased, although NAALADase activity was unchanged.

The functional significance of these findings is uncertain. Increased CSF levels of NAAG and NAA could be due to increased synthesis and release from the nerve terminals of descending pathways, with defective uptake due to loss of ventral horn neurons, if such an uptake mechanism exists. Decreased tissue levels of NAAG and NAA most likely reflect loss of spinal cord motor neurons, although Rothstein et al. (1990) found no correlation between changes in tissue NAAG and spinal ChAT levels. If the degenerative process in MND were associated with decreased uptake of NAAG (and perhaps of amino acids such as glutamate and aspartate) by neurons or glia one might predict an increased extracellular pool of these substances, with corresponding increases in CSF concentrations, but with either normal or decreased tissue levels. Such excitotoxic compounds could contribute to neuronal damage, as suggested by Plaitakis (1990) and others.

Plasma Cysteine and Metabolism of Sulphur-Containing Compounds

Heathfield and colleagues (1990) recently reported that plasma cysteine levels were raised, and that plasma inorganic sulphate levels were lower, in MND subjects than in controls. In addition, they identified a defect in sulphur oxidation in MND, using S-carboxymethyl-L-cysteine as a metabolic probe (Steventon et al. 1988). The altered ratio between cysteine and inorganic sulphate was also found in patients with Parkinson's disease and Alzheimer's disease, and defects in sulphur oxidation were likewise present in Parkinson's disease (Steventon et al. 1988). These observations led Heathfield et al. (1990) to propose that defective oxidation of toxic sulphur-containing compounds might lead to neuronal degeneration, either through failure to detoxify exogenous or endogenous sulphur-containing compounds, or because L-cysteine might itself act as a neurotoxin (Olney et al. 1990), as discussed in more detail in Chapter 10. In contrast, Perry et al. (1991) found no differences in fasting plasma levels of cystine (the product of cysteine oxidation), cystinyl glycine, total cysteine, or inorganic sulphate, in MND subjects compared to controls, although they used a different method to measure cysteine levels. In addition they measured cysteine and taurine levels in nine brain regions and in the cervical spinal cord in MND and control tissue samples, and found neither an increase in cysteine levels, nor a decrease in taurine levels. The latter would be predicted if MND were associated with a defect in sulphur oxidation attributable to altered activity of the enzyme cysteine dioxygenase, since taurine is formed in brain from cysteine, via cysteine sulphinate, hypotaurine and taurine (Yamaguchi and Hosokawa 1987). In fact, taurine levels are increased in MND brain and spinal cord (Perry et al. 1987, 1991; Malessa et al. 1991). Changes in cysteine, inorganic sulphate and sulphur oxidation are discussed in more detail in Chapter 10.

Glycine and GABA

Glycine is an inhibitory neurotransmitter in the spinal cord and brain stem. Glycine receptors in the rat, labelled by (^{3}H)strychnine, show the greatest density in grey matter of spinal cord; their number decreases progressively in regions more rostral in the neuraxis (Zarbin et al. 1981). Double-labelling experiments in the rat have shown glycinergic receptors on large, cholinergic neurons in the ventral horn (Geyer et al. 1987). In the spinal cord, glycine is thought to be released by propriospinal fibres and segmental interneurons including Renshaw cells. Glycine, rather than GABA, plays a major role in post-synaptic inhibition (Fagg and Foster 1983). Strychnine-sensitive glycine receptors are present on anterior horn motor neurons, and their activation produces increased chloride conductance and hyperpolarisation. As discussed in Chapter 10, glycine also interacts with glutamate at a strychnine *insensitive* allosteric site on the N-methyl-D-aspartate (NMDA) receptor. De Belleroche et al. (1984) first reported that CSF glycine concentrations were elevated in MND, and drew attention to a report of non-ketotic hyperglycinaemia in three brothers who presented with a syndrome of spastic paraparesis with lower limb wasting (Bank and Morrow 1972). De Belleroche and colleagues (1990) subsequently studied 19 patients with MND, including 11 with typical upper and lower motor neuron signs (ALS) and 8 with lower motor neuron signs only (PMA). A control group consisted of patients with a variety of muscular disorders, including late-onset spinal muscular atrophy, and post-poliomyelitis syndrome. Although the MND patients showed slightly elevated fasting plasma levels of isoleucine and leucine, there were no differences between MND and control groups in the levels of other plasma amino acids. After an oral glycine load, however, plasma glycine levels remained abnormally high at 4 hours in the MND-ALS subjects but not in the PMA group. This effect was not entirely specific, since one patient with late-onset spinal muscular atrophy, and one patient with the post-poliomyelitis syndrome, also showed significantly impaired glycine clearance at 4 hours. No differences in the serum glycine levels were evident at 1.5 hours in the MND patients who showed delayed clearance at 4 hours. CSF glycine levels were raised between 2- and 6-fold over known baseline values at 1.5 and 2.5 hours respectively, suggesting that the oral glycine load led to prompt changes in glycine concentrations within the CNS.

De Belleroche and colleagues (1990) proposed that glycine might potentiate the action of other endogenous or exogenous excitotoxins on the NMDA receptor, and therefore might contribute to excitotoxic neuronal damage. However, it is doubtful whether brain or spinal cord glycine concentrations resulting from normal dietary fluctuations of glycine intake or even from an oral glycine load would achieve CNS levels sufficient to modify glutamate neurotransmission at the NMDA receptor. The analogy with familial non-ketotic hyperglycinaemia with a motor system disorder may be misleading, since the patients described by Bank and Morrow had elevated fasting plasma glycine levels in addition to impaired glycine clearance following an oral load. In addition, these individuals (unlike MND patients) showed normal fasting serine levels, and there was no elevation of plasma serine at 1.5 or 4 hours after the glycine load, suggesting a defect in glycine–serine interconversion. Clinically, the patients with hyperglycinaemia did not have a rapidly progressive MND-like syndrome. Thus the significance of altered glycine tolerance in MND is unclear at present, and further studies are needed to assess the importance of these observations.

Comparisons of glycine levels in post-mortem tissue are unreliable unless death-to-freezing intervals and post-mortem conditions of cases and controls are carefully matched. When compared, no change in glycine levels was found in MND brains and spinal cord (Robinson 1968; Boehme et al. 1976; Yoshino et al. 1979; Perry et al. 1981; Patten et al. 1982; Plaitakis et al. 1988; Malessa et al. 1991) whereas one would predict elevated tissue glycine levels if there were a deficiency in glycine to serine interconversion in MND. Glycine binding to the strychnine-sensitive receptor is reduced in the ventral horn of the spinal cord in MND (Hayashi et al. 1981). It is assumed that (^{3}H)strychnine-binding sites are, at least in part, localised to motor neurons and that the observed changes are related to the loss of these cells in MND. Using glycine as a ligand, receptor binding is also decreased (Whitehouse et al. 1983; Gillberg & Aquilonius 1985).

An imbalance between excitatory and inhibitory neurotransmitters could contribute to neuronal damage. GABA, which is known to mediate presynaptic inhibition of primary afferent fibres, might be particularly important. Moreover, GABA could be involved in post-synaptic forms of motor neuron inhibition. GABA levels in autopsied human brain can be assessed with reasonable confidence, since in rodent brain GABA content remains stable for many hours after an initial rapid increase within the first 2 hours after death. Studies on GABA levels in MND are inconclusive. Normal values have been reported in the motor cortex (Yoshino et al. 1979), frontal and cerebellar cortex, while GABA was reduced in lumbar spinal cord (Plaitakis et al. 1988). In many of the 13 brain regions analysed by Perry et al. (1987), GABA levels were low, and significant decreases were noted in caudate nucleus, cerebellar cortex, pons and inferior olivary nucleus. We found no major changes in the spinal cord (Malessa et al. 1991). The discrepancies may partly be explained by the wide range of GABA concentrations measured in control samples.

Support for abnormalities of GABA neurotransmission in MND derives from studies of parvalbumin immunoreactive cells which represent a subset of GABAergic interneurons in the cerebral cortex. Counts of parvalbumin positive neurons in the primary motor cortex in MND were decreased compared to controls (Nihei et al. 1992). If this observation is correct, it contrasts with the situation in Huntington's disease, in which this population of neurons is selectively spared (Beal 1992). GABAergic neurons may be more sensitive to toxins acting at AMPA (amino-3-hydroxy-5-methyl-4-isoxazole propionate) and kainate (KA) receptors than an NMDA receptor (Beal 1992). AMPA/KA receptors may be more important than NMDA receptors in mediating excitotoxic motor neuron damage (Kuncl et al. 1992), and the non-NMDA receptor antagonists CNQX (6–cyano-7-nitro quinoxaline-2,3-diare) or NBQX (2,3-dihydroxy-6-nitro-7-sulfamoxybenzo(F)quinoxaline) block the toxic effects of glutamate upon motor neurons in tissue culture, and protect cultured spinal cord neurons against the toxic effects of CSF from MND patients (Kuncl et al. 1992; Couratier et al. 1993). Different aspects of excitotoxicity are discussed also in Chapter 10, on neurotoxicity, and in relation to the pharmacology of motor neurons in Chapter 15.

Although taurine cannot yet be classed as a classical inhibitory transmitter, it is known to act as a stabiliser of membrane excitability and to exert a depressant effect on neuronal firing (Oja and Kontro 1983). Taurine interacts with several neurotransmitters including catecholamines, acetylcholine and glutamate. Little is known about the origin and distribution of taurine in normal human spinal cord, but it is probably formed from cysteine via the action of cysteine dioxygenase, as discussed above. The highest concentrations of taurine are found in the anterior horn and exceed the mean white matter values by approximately 30%. In autopsied human brain, taurine content remains unchanged even with prolonged storage (Perry et al. 1981). In MND tissue, taurine levels have been described as approximately normal (Robinson 1968; Plaitakis et al. 1988), or as significantly increased in cortical (Yoshino et al. 1979), several subcortical regions, and in whole sections of cervical spinal cord (Perry et al. 1987). We found an increase in cervical, but not in lumbar spinal cord taurine, notably in lateral and ventral white matter comprising affected fibre tracts (Malessa et al. 1991).

Since the neurobiological functions of taurine are as yet poorly understood, the implications of this change are not known (Yoshino et al. 1979). The increase in taurine points to disturbances of the folate cycle. As has been pointed out, a defect in sulphur oxidation due to an abnormality of sulphur dioxygenase should result in decreased rather than increased CNS taurine levels.

In summary, studies on the glycine content of the brain and spinal cord suggest that the tissue levels of this inhibitory transmitter are not significantly altered in MND brain and spinal cord, while glycine receptor densities are diminished in the ventral horn, probably reflecting loss of anterior horn cells. GABA levels and the number of benzodiazepine-binding sites tend to be lower than control values, while taurine is probably increased in cortex and cervical spinal cord. The possible interaction between glycine and the NMDA receptor remains speculative, and it is premature to conclude that an abnormality of glycine metabolism is of pathogenic significance in MND.

Spinal Cholinergic Systems

The traditional marker of cholinergic cells is acetylcholine esterase (AChE), the enzyme responsible for acetylcholine catabolism. Since any non-cholinergic cells are capable of hydrolysing acetylcholine (Eckenstein and Sofroniew 1983) this marker lacks specificity. In contrast, the transmitter-synthesising enzyme, choline-acetyltransferase (ChAT) is expressed exclusively by cholinergic cells (Rossier 1977), and is thus considered a definitive marker of this neuronal population. Moreover, ChAT-activity has been shown to remain stable in autopsy tissue (Spokes and Koch 1978). Tritiated hemicholinium can also be used as a pre-synaptic marker of cholinergic cells, since it binds to the sodium-dependent, high-affinity choline-uptake site (Vickroy et al. 1985).

The first post-mortem studies on cholinergic systems in MND relied on the histochemical visualisation of AChE and showed decreased staining for the enzyme in the neuropil of the anterior horn and the nucleus of the hypoglossal nerve, but normal AChE content in the surviving motor neurons (Robinson 1966; Friede 1969) perhaps reflecting loss of dendrites which may precede degeneration of perikarya. This would be in keeping with the notion that dendritic atrophy is an early feature of neuronal degeneration in MND, as suggested by Karpati et al. (1988). A quantitative study on AChE activity (Nagata et al. 1982) confirmed a decrease in this enzyme and in the spinal cord ventral grey matter showed that the dorsal grey matter is similarly affected (Table 8.4), although the reason for the latter is unclear.

Choline acetyltransferase (ChAT) in MND spinal cord is reduced in the ventral and also in the dorsal spinal cord grey matter in MND (Gillberg et al. 1982; Nagata et al. 1982; Kanazawa 1977) (Table 8.4). Depletion of large anterior horn motor neurons almost certainly accounts for most of the observed decrease in ChAT activity, since ChAT has been localised to large and small motor neurons, and to synaptic terminals in cell bodies in the rat (Houser et al. 1983) and cat (Kimura et al. 1981). ChAT is also present in synaptic terminals and probably in interneurons in the rat and cat (Borges and Iversen 1986; Kanazawa et al. 1979), and significant ChAT activity has been detected neurochemically in the dorsal horn of the human spinal cord (Scatton et al. 1984).

Loss of cholinergic neurons would be expected to lead to changes in cholinergic receptors in the spinal cord and, as summarised in Table 8.4, the density of muscarinic cholinergic binding site labelled by (^{3}H)quinuclidinylbenzilate (QNB) and (^{3}H)-N-methylscopolamine is decreased in MND (Whitehouse et al. 1983; Gillberg and Aquilonius 1985; Manaker et al. 1988). These studies reveal that the greatest reductions in muscarinic binding sites are found in lamina IX of the spinal cord, and indicate a strong correlation between reduction in muscarinic receptor

Table 8.4. Changes in markers for cholinergic neurotransmission in the spinal cord in MND

	Cervical		Lumbar		Reference
	Ventral horn	Dorsal horn	Ventral horn	Dorsal horn	
AChE	−49%	−31%	nd	nd	Nagata et al. 1982
ChAT	−60%	nd	−75%	nd	Nagata et al. 1982
	−57–85%	nd	−76–89%	−80%	Gillberg et al. 1982
Muscarinic receptors (change in B_{max})	−51%	nd	−53%	−31%	Nagata et al. 1982
	−66%	−44%	nd	nd	Whitehouse et al. 1983
	−59%	−40%	−65%	−29%	Manaker et al. 1988

AChE, acetylcholinesterase; ChAT, choline acetyltransferase.

binding and the degree of motor neuron loss in the anterior horns. However, as can be seen from Table 8.4, decreases in cholinergic markers are also found in the dorsal horn of the spinal cord, and the pathological correlate of this is uncertain. Most of the muscarinic binding sites in the spinal cord are non-M_1 binding sites, and it is these which are decreased in MND spinal cord (Whitehouse et al. 1983). There is no evidence that nicotonic receptors are altered in MND (Gillberg and Aquilonius 1985). One would, however, expect to find a decrease in (^{3}H)-hemicholinium-3 ((^{3}H)HC-3) binding in MND spinal cord, since this provides a pre-synaptic marker for cholinergic cells, but Manaker et al. (1988) found no significant change.

In summary, the studies quoted above demonstrate a loss of cholinergic markers in MND spinal cord, most probably secondary to the degeneration of motor neurons. Moreover, these studies point to a spinal cholinergic deficit in non-motor areas.

Monoamines

The monoaminergic innervation of the spinal cord is considered to be derived from supraspinal sources (see Chapter 13). Projections containing noradrenaline arise from the medullary cell groups A1 and A2 and the pontine locus coeruleus complex (A5/A6/A7) respectively (Nygren and Olson 1977). Dopaminergic descending projections originate in the hypothalamic A11 cell group; they terminate throughout the dorsal and ventral horn (Bjöklund and Skagerberg 1979).

The medullary and pontine raphe nuclei (B1–B3) are the principal source of serotonergic projections to the spinal cord (Holstege and Kuypers 1982); in some of these neurons, substance P and/or thyrotropin-releasing hormone (TRH) are co-localised with 5-HT. Serotonergic terminals are seen in close apposition to motor neurons and to the intermediolateral cell column as well as in intermediate grey matter (Appel et al. 1987; Bowker et al. 1983). The coerulospinal and raphe-spinal pathways are thought to mediate an overall facilitation on motor neurons, as evidenced by enhanced excitability of these cells after iontophoretic application of 5-HT and noradrenaline (White and Neumann 1985; Bédard et al. 1987).

The concentration of 5-HT has been reported to be normal in cervical spinal cord homogenates of 6 MND patients, although levels of 5-hydroxyindole acetic acid (5-HIAA), the main metabolite of serotonin, were decreased in the same tissue samples (Ohsugi et al. 1987). The density of 5-HT fibres detected by immunocytochemistry is normal in MND spinal cord (Schoenen et al. 1985b), and in line with these observations, we found no major changes, although there was a slight increase in 5-HT content in the cervical spinal cord (lateral white matter and lateral part of the ventral horn), whereas levels of 5-HIAA were somewhat low in MND. The molar ratio of 5-HIAA/5-HT was decreased in all areas investigated (cervical, thoracic and lumbar segments), and notably in the ventral grey matter of cervical spinal cord. In human tissue, post-mortem decreases in 5-HIAA concentrations have been found to exceed slightly the changes in 5-HT (Lackovic et al. 1988). Since in our study death-to-freezing intervals of control and MND samples were matched, we interpreted the lowered 5-HIAA/5-HT ratios as evidence of decreased release of 5-HT in MND spinal cord, possibly secondary to depletion of target cells.

Relatively little is known about changes in 5-HT receptors in MND, although a striking increase (up to 140%) in 5-HT_{IA} receptor density assessed by binding of (^{3}H)-8-hydroxy-N,N-dipropyl-2-aminotetraline ((^{3}H)-8-OHDPAT) binding has been detected in laminae IX of the spinal cord at cervical, thoracic, lumbar and sacral levels (Manaker et al. 1988). These authors also noted a lower affinity of the 5-HT_{IA} binding site.

In view of the fact that 5-HT and TRH are co-localised in raphe neurons projecting to the spinal cord (Johansson et al. 1981) it is interesting that TRH receptors in lamina IX are reduced by nearly 90%. These observations may imply that 5-HT_{IA} and TRH receptors have different localisation, despite the co-localisation of the neurotransmitter molecules. A likely explanation is that post-synaptic TRH receptors of anterior horn cell motor neurons are lost as these cells die, but that 5-HT_{IA} receptors are located elsewhere, and that the observed increase reflects denervation supersensitivity in receptors which are not located on cells or processes undergoing degeneration. Changes in TRH receptors will be discussed in more detail below.

No major changes have so far been recorded in the concentration of dopamine or its major metabolites in MND spinal cord (Ohsugi et al. 1987), but dopamine concentrations in spinal cord are close to the limits of detection using conventional high pressure liquid chromatography (HPLC) with electrochemical detection, so it is premature to conclude that dopamine concentrations are indeed normal. In contrast, we have observed that noradrenaline levels may be increased in lumbar and thoracic cord, although one previous study (Ohsugi et al. 1987) did not identify any change in noradrenaline or 3,4-dihydroxyphenylglycol (3-MHPG) in MND spinal cord tissue. It is interesting that increased sympathetic activity has been detected in MND, and the CSF may contain elevated noradrenaline levels (Chida et al. 1989).

Dopaminergic and β-adrenergic receptors have been investigated in MND spinal cord tissue by means of radioactive-labelled spiroperidol and alprenolol and pindolol binding (Hayashi et al. 1981; Manaker et al. 1988). No differences in receptor densities between controls and MND patients were apparent.

In summary, serotonin content seems to be normal in ALS spinal cord although turnover may be reduced. Noradrenaline concentration may be increased in MND spinal cord, an observation which is consistent with physiological changes and CSF studies, but which needs to be confirmed in a larger series of patients. No definitive data are available on dopamine concentrations in MND.

Neuropeptides

Neuropeptides are of importance both with regard to membrane action (i.e. as putative neurotransmitters or modulators) or as possible trophic factors involved in maintenance of nerve and muscle integrity. Studies in MND may help to decide whether a decrease in certain peptides may precede cell loss, or whether cell groups that are selectively spared share a certain pattern of peptide content or peptidergic innervation. These questions can be approached by studying those nuclei that are selectively (but not totally) preserved in MND, such as Onuf's nucleus located in the ventral horn of sacral segments S2–S3 (Gibson et al. 1988) and the oculomotor nuclei III, IV and VI.

Most peptides are concentrated in the dorsal horn, and undergo little or no change in MND (Table 8.5), and will not be referred to in the text. We will comment on three selected peptides: thyrotropin-releasing hormone (TRH), substance P (SP), calcitonin gene-related peptide (CGRP) and endothelin. Another neuropeptide of potential importance in MND is N-acetyl-aspartyl-glutamate (NAAG), which has been discussed in relation to glutamate in the preceding section. Experimental work suggests a trophic role for TRH, SP, CGRP and possibly endothelin, and some data is available on changes in MND tissue.

Table 8.5. Changes in neuropeptides in MND spinal cord

Peptides	Cervical			Lumbar			Source
	Ventral horn	Whole spinal cord	Dorsal horn	Ventral horn	Whole spinal cord	Dorsal horn	
CGRP			nc			nc	Gibson et al. 1988
ENK			nc			nc	Gibson et al. 1988
	nd		nd	nc		nc	Schoenen et al. 1985a,b
GAL	nc		nc	nc		nc	Gibson et al. 1988
SOM			nc			nc	Gibson et al. 1988
SP	−58%		−67%	nc		nc	Gillberg et al. 1982
	nd		nd			nc	Schoenen et al. 1985a,b
	nc		nc	nc		nc	Dietl et al. 1989
TRH	−50%		nc	nd		nd	Mitsuma et al. 1984
		−62%			−63%		Jackson et al. 1986
	−75%		nc	nd		nd	Banda et al. 1986
	nc		nc	nc		nc	Court et al. 1989
			nc	nc		nc	Gibson et al. 1988

Immunolabelling of peptides is indicated as reduced (−) or no change (nc); quantitative results are listed as % loss compared to control mean values. CGRP, calcitonin gene-related peptide; ENK, enkephalin; GAL, galanin; SOM, somatostatin; SP, substance P; TRH, thyrotropin-releasing hormone.

Thyrotropin-Releasing Hormone (TRH)

Spinal TRH is thought to originate in medullary raphe nuclei where it may be co-localised with serotonin and/or substance P (Johansson et al. 1981). TRH-positive fibres are concentrated in the intermediolateral cell column and in lamina IX (Gibson et al. 1988). Unlike most other peptides, TRH shows a higher concentration in the ventral than in the dorsal horn (Bennett et al. 1986), while TRH receptor density decreases slightly in autopsy tissue analysed between 5 and 25 hours post-mortem (Bennett et al. 1986). Iontophoretic application of TRH increases motor neuron excitability (White 1985).

A trophic role for TRH is suggested by the following studies:

1. TRH enhances choline acetyltransferase and creatine-kinase activities in cultured ventral, but not in dorsal, horn neurons (Schmidt-Achert et al. 1984).
2. Intrathecal administration of a TRH analogue in rats led to increased CAT activity in the ventral, but not the dorsal horn (Fone et al. 1988).
3. Treatment with TRH following sciatic nerve dissection in rats increased the number of surviving ventral horn cells (Banda and Means 1989).
4. Addition of TRH to cultures of anterior horn cells promotes neurite outgrowth (Iwasaki et al. 1989).

TRH in MND

In autopsy tissue analysed between 5 and 25 hours post-mortem, concentrations of TRH were found to decrease slightly (Bennett et al. 1986). When measured in fresh MND tissue, TRH concentrations were found to be decreased by 50% or more (Table 8.5). Jackson and colleagues (1986) described a concomitant loss in protein and concluded that "TRH neurons are not preferentially affected, but TRH and tissue protein are lost together as the disease progresses". On the other hand, given that TRH partly co-localises with serotonin, which shows no decrease, those neurons containing only TRH seem to disappear selectively, or earlier, than those comprising both compounds. The discrepancy between changes in TRH and serotonin might indicate differential subcellular storage compartments, and possibly turnover rates, of TRH and 5-HT.

Manaker and colleagues (1985) reported a decrease of TRH receptors in lamina II and IX and later confirmed their findings in a larger series of patients (Manaker et al. 1988). The finding in the anterior horn suggests a post-synaptic localisation of TRH-binding sites that are lost consequent to the depletion of motor neurons. The decrease in TRH receptors in the dorsal horn is not easily explained; it is possible that they are localised to afferent fibres of the corticospinal tract which terminates throughout the spinal grey matter.

Substance P

In the spinal cord, substance P (SP) is present in three, possibly four different systems: (a) primary sensory fibres, particularly nociceptive afferents; (b) inter-neurons or propriospinal neurons; (c) descending afferents, some of which may contain TRH and 5-HT. The highest fibre density has been demonstrated immuno-cytochemically in lamina I, in the intermediolateral cells column and in the sacral parasympathetic nuclei (Schoenen et al. 1985b), while the anterior horn seems to be sparsely innervated by SP-positive fibres. SP is involved in transmission of nociceptive stimuli and is thought to influence somatic motor reflexes (Krivoy et al. 1980). A close morphological relationship between motor neurons and SP-positive fibres and terminals has been demonstrated in monkey and human spinal cord (De Lanerolle and Lamotte 1982). Evidence for a trophic role of SP is suggested by several studies:

1. SP stimulates the neurite outgrowth in embryonic chick dorsal root ganglia and embryonic rat ventral horn neurons (Iwasaki et al. 1989).
2. An SP antagonist (D-Pro^{2},D-Try^{7,9}) substance P was shown to have neurotoxic properties (Hökfelt et al. 1981).
3. Intrathecal application of an SP antagonist preferentially affects ventral horn cells of rat spinal cord (Gordh et al. 1986).

Substance P in MND

Initial studies showed a decrease in substance P in ventral (Schoenen et al. 1985a,b) and dorsal (Gillberg et al. 1982) horn of MND spinal cord. Rarification of SP-immunoreactive fibres in the ventral horn has been confirmed on a total of 32 MND cases examined (Gibson et al. 1988). Another recent study reported on normal SP-immunoreactivity in MND spinal cord (Dietl et al. 1989). Differences in the antiserum used, and in techniques and tissues, may account for these discrepancies. Schoenen and colleagues (1985a), who described a loss of SP-positive fibres in the ventral horn even in segments where motor neuron degeneration was not prominent, conclude that a deficit in SP might precede loss of anterior horn cells. A marked decrease in binding of (^{125}I)substance P in lamina IX probably reflects loss of motor neurons (Dietl et al. 1989).

Calcitonin Gene-Related Peptide (CGRP)

CGRP is a 37-amino acid neuropeptide that results from alternative processing of the calcitonin gene primary transcript in neural tissue (Rosenfeld et al. 1983). In the CNS, CGRP-immunoreactivity has been shown in sensory, autonomic and motor systems (Rosenfeld et al. 1983). CGRP co-localises to substance P in the human peripheral nervous system (Franco-Cereceda et al. 1987), to glutamate decarboxylase in Purkinje cells of rat cerebellum (Kawaki et al. 1987) and to choline acetyltransferase in rat hypoglossal, facial and ambiguus nuclei (Takami et al. 1985). In the ventral horn of human spinal cord, CGRP-immunoreactivity was found to be localised to large anterior horn cells tentatively identified as motor neurons (Gibson et al. 1984; but see Harmann et al. 1988).

CGRP has been proposed to function as a motor neuron-derived trophic factor, since it enhances the synthesis of junctional acetylcholine receptors in cultures of embryonic chick myotubes (New and Mudge 1986; Fontaine et al. 1986).

In MND, none of the remaining motor neurons in lamina IX displayed CGRP-immunoreactivity (Gibson et al. 1988). Studies of CGRP gene expression are in progress using in situ hybridisation, and preliminary indications are that CGRP mRNA may be decreased in motor neurons in MND. This raises the possibility that a defect in synthesis or transport of CGRP may contribute to the attrition of skeletal muscles, since the formation of AChE receptors at the neuromuscular junction seems to depend on such anterograde factors rather than activity (New and Mudge 1986).

Neurotrophic Factors and Motoneuronal Death

Neurotrophic molecules regulate the survival of neurons during normal embryonic development, since neurons compete for limited amounts of target-derived neurotrophic factors. By analogy, Appel (1981) has proposed that reduced availability of neurotrophic factors may lead to degeneration of motor neurons in MND. Among the many neurotrophic factors known so far, only two seem to have a specific effect on motor neurons. McManaman et al. have purified a putative neurotrophic factor from extracts of rat skeletal muscle termed CDF choline acetyltransferase (CAT) development factor, which enhances the expression of CAT in embryonic rat spinal cord cultures (McManaman et al. 1988). CDF supports the survival of motor neurons in the developing chick embryo during the period of naturally occurring cell death.

Motor neuron degeneration in the facial nucleus can be prevented by application of purified CNTF (ciliary neurotrophic factor; Sendtner et al. 1990). CNTF is a cytosolic molecule that is expressed by Schwann cells and astrocytes but not by target tissues. The time course of CNTF expression during development and its tissue distribution suggest that it acts as a "lesion factor" in postnatal animals. More detailed discussion of the possible role of neurotrophic factors in the pathogenesis and treatment of MND in provided in Chapter 12.

Summary and Conclusions

In summary, recent studies have shown a glutamatergic deficit in MND that is not restricted to the motor system. The mechanisms leading to this change are entirely speculative at present, and include neuronal loss, reduced glutamate synthesis (from glucose), and altered distribution of glutamate with accumulation in the extracellular space, due either to decreased re-uptake or to loss of cholinergic inhibition of glutamate release. Neither the "metabolic" nor the "excitotoxic" hypotheses adequately explain selective vulnerability in MND, although this might arise from interactions between glutamate and other amino acids, including glycine.

The changes observed in CNS and CSF glutamate and in glutamate uptake may be secondary phenomena resulting from neuronal degeneration, and it is difficult to see how further understanding of dynamic aspects of excitatory amino neurotransmission can be gained from studies of post-mortem tissue or CSF. Perhaps positron emission tomographic (PET) studies may provide the way forward, although ligands for excitatory amino acids and their receptors are not yet available.

A number of other neurochemical abnormalities have been detected in MND, none of them providing convincing insights into pathogenic mechanisms. Studies on neurotrophic factors may provide such information, and lead to new therapeutic approaches, as discussed in Chapter 18.

References

Amara SG (1992) A tale of two families. Nature 360:420–474

Anderson KJ, Managhan DT, Cangro CB, Namboodiri MAA, Neale JH, Cotman CW (1986) Localization of N-acetylaspartylglutamate-like immunoreactivity in selected areas of the rat brain. Neurosci Lett 72:14–20

Aoki C, Milner TA, Sheu KRF et al. (1987) Regional distribution of astrocytes with intense immunoreactivity for glutamate dehydrogenase in rat brain: implications for neuron–glia interactions in glutamate transmission. J Neurosci 7:2214–2231

Appel SH (1981) A unifying hypothesis for the cause of amyotrophic lateral sclerosis, parkinsonism, and Alzheimer's disease. Ann Neurol 10:499–505

Appel NM, Wessendorf MW, Elde RP (1987) Thyrotropin-releasing hormone in spinal cord: coexistence with serotonin and with substance P in fibres and terminals apposing identified preganglionic sympathetic neurons. Brain Res 415:137–143

Arawaka Y, Sendtner M, Thoenen H (1990) Survival effect of ciliary neurotrophic factor (CNTF) on chick embryonic motoneurons in culture: comparison with other neurotrophic factors and cytokines. J Neurosci 10:3507–3515

Aubby D, Saggu HK, Jenner P, Quinn NP, Harding AE, Marsden CD (1988) Leucocyte glutamate dehydrogenase activity in patients with degenerative neurological disorders. J Neurol Neurosurg Psychiatry 51:839–902

Banda RW, Means ED (1989) Effect of thyrotropin-releasing hormone on neural loss secondary to axotomy. Neurology 39 (Suppl 1):401

Banda RW, Kubek JM, Means ED (1986) Decreased content of thyrotropin releasing hormone (TRH) in cervical ventral horns of patients with amyotrophic lateral sclerosis (ALS). Neurology 36 (Suppl):139

Bank WJ, Morrow G (1972) A familial spinal cord disorder with hyperglycinemia. Arch Neurol 27:136–144

Barber RP, Vaughn JE, Roberts E (1982) The cytoarchitecture of GABAergic neurons in rat spinal cord. Brain Res 238:305–328

Beal MF (1992) Does impairment of energy metabolism result in excitotoxic neuronal death in neurodegenerative illnesses? Ann Neurol 31:119–130

Bédard P, Tremblay LE, Barbeau H et al. (1987) Action of 5-hydroxytryptamine, substance P, thyrotropin-releasing hormone and clonidine on motoneuron excitability. Can J Neurol Sci 14:506–509

Bennett GW, Nathan PA, Wong KK, Marsden CA (1986) Regional distribution of immunoreactive thyrotropin-releasing hormone and substance P, and indolamines in human spinal cord. J Neurochem 46:1718–1724

Bjöklund A, Skagerberg G (1979) Evidence for a major spinal cord projection from the diencephalic A11 dopamine cell group in the rat using transmitter-specific fluorescent retrograde tracing. Brain Res 177:170–175

Boehme DH, Marks N, Fordice MW (1976) Glycine levels in degenerated human spinal cord. J Neurol Sci 27:347–352

Borges LF, Iversen SD (1986) Topography of choline acetyltransferase immunoreactive neurons and fibres in the rat spinal cord. Brain Res 362:140–148

Bowker RM, Westlung KN, Sullivan MC, Wilber JF, Coulter JD (1983) Descending serotonergic, peptidergic and cholinergic pathways from the raphe nuclei: a multiple transmitter complex. Brain Res 288:33–48

Chee PY, Dahl JL, Fahien LA (1979) The purification and properties of rat brain glutamate dehydrogenase. J Neurochem 33:53–60

Chida K, Sakamaki S, Takasu T (1989) Alteration in autonomic function and cardiovascular regulation in amyotrophic lateral sclerosis. J Neurol 236:127–130

Constantakakis E, Plaitakis A (1988) N-acetylaspartate and N-acetylaspartylglutamate are altered in the spinal cord in amyotrophic lateral sclerosis. Ann Neurol 24:478

Couratier P, Hugon J, Sindou P, Vallat J-M, Dumas M (1993) Cell culture evidence for neuronal degeneration in amyotrophic lateral sclerosis being linked to glutamate AMPA/kainate receptors. Lancet 341:265–268

Court JA, McDermott JR, Gibson AM et al. (1989) Raised thyrotropin-releasing hormone, proglutamylamino peptidase, and proline endopeptidase are present in the spinal cord of wobbler mice but not in human motor neuron disease. J Neurochem 49:1084–1090

Coyle JT, Blakely R, Zaczek R et al. (1986) Acidic peptides in brain: do they act at putative glutamatergic synapses? Adv Exp Biol 203:357–384

Dalakas MC, Hatazawa J, Brooks RA, Di Chiro G (1987) Lowered cerebral glucose utilisation in amyotrophic lateral sclerosis. Ann Neurol 22:580–586

Davidoff RA, Graham LT, Shank RP, Werman R, Aprison MH (1967) Changes in amino acid concentrations associated with loss of spinal interneurons. J Neurochem 14:1025–1031

De Belleroche J, Recordati A, Rose FC (1984) Elevated levels of amino acids in the CSF of motor neuron disease patients. Neurochem Pathol 2:1–6

De Belleroche J, Lane RJM, Bandopadhyay R, Rose FC (1990) Abnormalities in amino acid metabolism in amyotrophic lateral sclerosis. In: Rose FC, Forbes HN (eds) Amyotrophic lateral sclerosis. New advances in toxicology and epidemiology. Smith-Gordon, London, pp 261–264

De Lanerolle N, Lamotte CC (1982) The morphological relationships between substance P immunoreactive processes and ventral horn neurons in the human and monkey spinal cord. J Comp Neurol 207:305–313

Dietl MM, Sanchez M, Probst A, Palacios JM (1989) Substance P receptors in the human spinal cord: decrease in amyotrophic lateral sclerosis. Brain Res 483:39–49

Duvoisin RC, Chokroverty S, Lepore F et al. (1983) Glutamate dehydrogenase deficiency in patients with olivopontocerebellar atrophy. Neurology 33:1322–1326

Eckenstein F, Sofroniew MV (1983) Identification of central cholinergic neurons containing both choline acetyltransferase and acetylcholinesterase and of central neurons containing only acetylcholinesterase. J Neurosci 3:2286–2291

Eubanks JH, Puranam RS, Kleckner NW, Bettler B, Heineann SF, McNamara JO (1993) The gene encoding the glutamate receptor subunit GluR5 is located on human chromosome 21q21.1–22.1 in the vicinity of the gene for familial amyotrophic lateral sclerosis. Proc Natl Acad Sci USA 90:178–182

Fagg GE, Foster AC (1983) Amino acid neurotransmitters and their pathways in the mammalian central nervous system. Neuroscience 9:701–719

Ferraro TN, Hare TA (1985) Free and conjugated amino acids in human CSF: influence of age and sex. Brain Research 338:53–60

Filla A, De Michele G, Brescia Morra V et al. (1986) Glutamate dehydrogenase in human brain: regional distribution and properties. J Neurochem 46:422–424

Finocciaro G, Taroni F, DiDonato S (1986) Glutamate dehydrogenase in olivopontocerebellar atrophies: leucocytes, fibroblasts and muscle mitochondria. Neurology 36:550–553

Fone KCP, Dix P, Tomlinson DR, Bennett GW, Marsden CA (1988) Spinal effects of chronic intrathecal administration of the thyrotropin-releasing hormone analogue (CG3509) in rats. Brain Res 455:157–161

Fonnum F (1984) Glutamate: a neurotransmitter in mammalian brain. J Neurochem 42:1–11

Fontaine B, Klarsfeld A, Hökfelt T, Changeux J-P (1986) Calcitonin gene-related peptide, a peptide present in spinal cord motor neurons, increases the number of acetylcholine receptors in primary cultures of chick embryo myotubes. Neurosci Lett 71:59–65

Franco-Cereceda A, Henke H, Lundberg JM, Petermann JB, Hökfelt T, Fischer JA (1987) Calcitonin gene-related peptide (CGRP) in capsaicin-sensitive substance P-immunoreactive sensory neurons in animal and man: distribution and release by capsaicin. Peptides 8:399–410

Friede RL (1969) Enzyme histochemical observations in amyotrophic lateral sclerosis. In: Norris FM, Kurland NY (eds) Motor neuron disease. Contemporary neurology symposia vol II. Grune and Stratton, New York, pp 218–234

Geyer SW, Gudden W, Betz H, Gnahn H, Weindl A (1987) Co-localization of choline acetyltransferase and postsynaptic glycine receptors in motor neurons of rat spinal cord demonstrated by immunocytochemistry. Neurosci Lett 82:11–15

Giaid A, Gibson SJ, Ibrahim NBN et al. (1989) Endothelin 1, an endothelium-derived peptide, is expressed in neurons of the human spinal cord and dorsal root ganglia. Proc Natl Acad Sci USA 86:7634–7638

Gibson SJ, Polak JM, Bloom SR et al. (1984) Calcitonin gene-related peptide immunoreactivity in the spinal cord of man and of eight other species. J Neurosci 4:3101–3111

Gibson SJ, Polak JM, Katagiri T et al. (1988) A comparison of the distribution of eight peptides in spinal cord from normal controls and cases of motor neuron disease with special reference to Onuf's nucleus. Brain Res 474:255–278

Gillberg PG, Aquilonius SM (1985) Cholinergic, opioid and glycine receptor binding sites localized in human spinal cord by in vitro autoradiography: changes in amyotrophic lateral sclerosis. Acta Neurol Scand 72:299–306

Gillberg PG, Aquilonius SM, Eckern SA, Lundqvist G, Winblad B (1982) Choline acetyltransferase and substance P-like immunoreactivity in the human spinal cord: changes in amyotrophic lateral sclerosis. Brain Res 250:394–397

Gordh JT, Post C, Olsson Y (1986) Evaluation of the toxicity of subarachnoid clonidine, guanfacine and a substance P-antagonist on rat spinal cord and nerve roots: light and electron microscopic observations after chronic intrathecal administration. Anesth Analg 65: 1301–1311

Harmann PA, Chung K, Briner RP, Westlund KN, Carlton SM (1988) Calcitonin gene-related peptide (CGRP) in the human spinal cord: a light and electron microscopic analysis. J Comp Neurol 269:371–380

Hayashi H, Suga M, Satake M, Tsubaki T (1981) Reduced glycine receptor in the spinal cord in amyotrophic lateral sclerosis. Ann Neurol 9:292–294

Heathfield MT, Fearn S, Steventon GB, Waring RH, Williams AC, Sturman SG (1990) Plasma cysteine and sulfate levels in patients with motor neuron, Parkinson's and Alzheimer's disease. Neurosci Lett 110:216–220

Hökfelt T, Vincent S, Hellsten I, Rosell S, Goldstein M, Cuello L (1981) Immunohistochemical evidence for a "neurotoxic" action of (D-Pro^2,D-Try^{7,9}) substance P, an analogue with substance P activity. Acta Physiol Scand 113:571–573

Holstege JC, Kuypers HGJM (1982) The anatomy of brainstem pathways to the spinal cord in cat. A labelled amino acid tracing study. In: Kuypers HGJM, Martin GF (eds) Descending pathways to the spinal cord. Progress in brain research. vol 57, Biomedical Press, Amsterdam, pp 145–175

Houser CR, Crawford GD, Barber RP, Salvaterra PM, Vaughn JE (1983) Organisation and morphological characteristics of cholinergic neurons: an immunocytochemical study with a monoclonal antibody to choline acetyltransferase. Brain Res 266:97–119

Hugon J, Tabaraud F, Rigaud M, Vallat JM, Dumas M (1989) Glutamate dehydrogenase and aspartate aminotransferase in leukocytes of patients with motor neuron disease. Neurology 39:956–958

Iwasaki Y, Kinoshita M, Ikeda K, Takamiya K, Shiojima (1989) Trophic effect of various neuropeptides on the cultured ventral spinal cord of rat embryo. Neurosci Lett 101:316–320

Jackson IMB, Adelman LS, Munsat TL, Forte S, Lechan RM (1986) Amyotrophic lateral sclerosis: thyrotropin-releasing hormone and histidyl proline diketopoperazine in the spinal cord and cerebrospinal fluid. Neurology 36:1218–1223

Johansson O, Hökfelt T, Jeffcoate SL (1981) Immunohistochemical support of three putative transmitters in one neuron: coexistence of 5-hydroxytryptamine, substance P- and thyrotropin releasing hormone-like immunoreactivity in medullary neurons projecting to the spinal cord. Neuroscience 6:1857–1881

Kanai Y, Hediger MA (1992) Primary structure and functional characterisation of a high affinity glutamate transporter. Nature 360:467–471

Kanazawa I (1977) Neurotransmitters in motor neuron disease. Jpn J Clin Med 35:4025–4029

Kanazawa I, Sutoo D, Oshima I, Saito S (1979) Effect of transection on choline acetyltransferase, thyrotropin releasing hormone and substance P in the rat cervical spinal cord. Neurosci Lett 13:325–330

Kaneko T, Akiyama H, Mizuno N (1987) Immunohistochemical demonstration of glutamate dehydrogenase in astrocytes. Neurosci Lett 77:171–175

Karpati G, Carpenter S, Durham H (1988) A hypothesis for the pathogenesis of amyotrophic lateral sclerosis. Rev Neurol 144:672–675

Kawaki Y, Emson PC, Hillyard CJ et al. (1987) Immunocytochemical evidence for the coexistence of calitonin gene-related peptide and glutamate decarboxylase-like immunoreactivities in the Purkinje cells of the rat cerebellum. Brain Res 409:371–373

Kew JJM, Leigh PN, Playford ED et al. (1993) Cortical function in amytrophic lateral sclerosis. Brain 116:655–680

Kimura H, McGeer PL, Peng JH, McGeer EG (1981) The cholinergic system studied by choline acetyltransferase immunocytochemistry in the cat J Comp Neurol 200:151–201

Koller KJ, Coyle JT (1984) Ontogenesis of N-acetyl-aspartate and N-acetyl-aspartyl-glutamate in rat brain. Dev Brain Res 15:37–140

Krivoy WA, Couch JR, Steward JM, Zimmerman E (1980) Modulation of cat monosynaptic reflexes by substance P. Brain Res 202:356–372

Kuncl RW, Jin L, Rothstein JD (1992) Chronic glutamate toxicity in motor neurons from organotypic spinal cord cultures. Society for Neuroscience Abstracts, vol 18, p 756

Lackovic Z, Jakupcevic M, Bunarevic I, Damjanov M, Relja M, Kostovic I (1988) Serotonin and norepinephrine in the spinal cord of man. Brain Res 443:199–203

Lucas DR, Newhouse JP (1957) The toxic effect of sodium L-glutamate on the inner layers of the retina. Arch Ophthalmol 58:193–201

Lundqvist C, Blomstrand C, Hamberger A, Wikkelso C (1989) Liquid chromatographic separation of CSF amino acids after precolumn fluorescence derivatisation. Acta Neurol Scand 79:273–279

Malessa S, Leigh PN, Hornykiewicz O (1988) Glutamate dehydrogenase in amyotrophic lateral sclerosis. Lancet ii:681–682

Malessa S, Leigh PN, Bertel O, Sluga E, Hornykiewicz O (1991) Amyotrophic lateral sclerosis: glutamate dehydrogenase and transmitter amino acids in the spinal cord. J Neurol Neurosurg Psychiatr 54:984–988

Manaker S, Shulman LH, Winokur A, Rainbow TC (1985) Autoradiographic localization of thyrotropin-releasing hormone receptors in amyotrophic lateral sclerosis. Neurology 35:1650–1653

Manaker S, Calne SB, Winokur A (1988) Alterations in receptors for thyrotropin-releasing hormone, serotonin, and acetylcholine in amyotrophic lateral sclerosis. Neurology 38:1464–1474

Marchi M, Bocchieri P, Garbarino L, Raiteri M (1989) Muscarinic inhibition of endogenous glutamate release from rat hippocampus synaptosomes. Neurosci Lett 96:229–234

McManaman JL, Crawford FG, Stewart SS, Appel SH (1988) Purification of a skeletal muscle polypeptide which stimulates choline acetyltransferase activity in cultured spinal cord neurons. J Biol Chem 263:5890–5897

Meier DH, Schott K-J (1988) Free amino acid pattern of cerebrospinal fluid in amyotrophic lateral sclerosis. Acta Neurol Scand 77:50–53

Misra CH, Olney JW (1975) Cysteine oxidase in brain. Brain Research 97:117–126

Mitsuma T, Nogimori T, Adachi K, Mukoyama M, Ando K (1984) Concentrations of immunoreactive thyrotropin-releasing hormone in spinal cord of patients with amyotrophic lateral sclerosis. Am J Med Sci 287:34–36

Moore RY, Card JP (1984) Noradrenaline-containing neuron systems. In: Bjöklund A, Hökfelt T (eds) Handbook of chemical neuroanatomy, vol 2, part 1. Excerpta Medica, Amsterdam, pp 131–132

Nagata Y, Okuya M, Watanabe R, Honda M (1982) Regional distribution of cholinergic neurons in human spinal cord transections in patients with and without motor neuron disease. Brain Res 244:223–229

Nakanishi S (1992) Molecular diversity of glutamate receptors and implications for brain function. Science 258:597–603

New HV, Mudge AW (1986) Calciton gene-related peptide regulates muscle acetylcholine receptor synthesis. Nature 323:809–811

Nihei K, McKee AC, Kowall NW (1992) GABAergic local circuit in neurons degenerate in the motor cortex of amyotrophic lateral sclerosis patients. Society for Neuroscience Abstracts, vol 18, p 1992

Nygren L-G, Olson L (1977) A new major projection from locus coeruleus: the main source of noradrenergic nerve terminals in the ventral and dorsal columns of the spinal cord. Brain Res 132:85–93

Ohsugi K, Adachi K, Mukoyama M, Ando K (1987) Lack of change in indoleamine metabolism in spinal cord of patients with amyotrophic lateral sclerosis. Neurosci Lett 79:351–354

Oja SS, Kontro P (1983) Taurine. In: Lajtha A (ed) Handbook of neurochemistry, vol 3, 2nd ed. Plenum Press, New York, pp 501–533

Olney JW, Ho OL, Rhee V (1971) Cytotoxic effects of acidic and sulphur containing amino acids in the infant mouse central nervous system. Exp Brain Res 14:61–76

Olney JW, Zorumski C, Price MT, Labruyère J (1990) L-cysteine, a bicarbonate-sensitive endogenous excitotoxin. Science 248:596–599

Osterberg OP, Wattenberg LW (1962) Oxidative histochemistry of reactive astrocytes. Arch Neurol 7:211–218

Ottersen OP, Storm-Mathisen J (1984) Neurons containing or accumulating transmitter amino acids. In: Bjöklund A, et al. (eds) Handbook of chemical neuroanatomy, vol 3, part III. Excerpta Medica, Amsterdam, pp 219–222

Patten BM, Harati Y, Acosta L (1978) Free amino acid levels in amyotrophic lateral sclerosis. Ann Neurol 3:305–309

Patten BM, Kurlander HM, Evans B (1982) Free amino acid concentrations in spinal tissue from patients dying of motor neuron disease. Acta Neurol Scandinav 66:594–599

Peppard RF, Guttman M, Martin WRW, Clark C, Eisen A, Calne DB (1989) Dopaminergic deficit demonstrated in caucasian amyotrophic lateral sclerosis using positron emission tomography. Neurology 39 (suppl 1):400 (abstract)

Perry TL, Hansen S, Gandham SS (1981) Postmortem changes of amino compounds in human and rat brain. J Neurochem 36:406–412

Perry TL, Hansen S, Jones K (1987) Brain glutamate deficiency in amyotrophic lateral sclerosis. Neurology 37:1845–1848

Perry TL, Krieger C, Hansen S, Eisen A (1990) Amyotrophic lateral sclerosis: amino acid levels and cerebrospinal fluid. Ann Neurol 28:12–17

Perry TL, Krieger C, Hansen S, Tabatabaei A (1991) Amyotrophic lateral sclerosis: fasting plasma levels of cysteine and inorganic sulfate are normal as are brain contents of cysteine. Ann Neurol 41:487–490

Pines G, Danbolt NC, Bjøras M, Zhang Y, Bendahan A, Eide L, Koepsell H, Storm-Mathisen J, Seeberg E, Kanner BI (1992) Cloning and expression of a rat brain L-glutamate transporter. Nature 360:464–467

Plaitakis A (1990) Glutamate dysfunction and selective motor neuron degeneration in ALS: an hypothesis. Ann Neurol 28:3–8

Plaitakis A, Caroscio JT (1987) Abnormal glutamate metabolism in amyotrophic lateral sclerosis. Ann Neurol 22:575–579

Plaitakis A, Berl S, Yahr MD (1982) Abnormal glutamate metabolism in an adult-onset degenerative neurological disorder. Science 216:193–196

Plaitakis A, Constantakakis E, Smith J (1984) The neuroexcitotoxic amino acids glutamate and aspartate are altered in the spinal cord and brain in amyotrophic lateral sclerosis. Ann Neurol 24:446–449

Plaitakis A, Nicklaus WJ, Desnick RJ (1988) Glutamate dehydrogenase deficiency in three patients with spinocerebellar syndromes. Ann Neurol 7:297–303

Robinson N (1966) A histochemical study on motor neuron disease. Acta Neuropathol 7:101–110

Robinson N (1968) Chemical changes in the spinal cord in Friedreich's ataxia and motor neuron disease. J Neurol Neurosurg Psychiatr 31:330–333

Rosenfeld MG, Mermod J-J, Amara SG et al. (1983) Production of a novel neuropeptide encoded by the calcitonin gene via tissue-specific RNA processing. Nature 304:129–135

Rossier J (1977) Choline acetyltransferase: a review with special reference to its cellular and subcellular localisation. Int Rev Neurobiol 20:284–337

Rothstein JD, Tsai G, Clawson L et al. (1990) Excitatory amino acid metabolism in amyotrophic lateral sclerosis. Abstr Soc Neurosci 15 Ann Neurol 28:18–25

Rothstein JD, Kunel R, Chaudhry V et al. (1991) Excitatory amino acids in ALS: an update. Ann Neurol 30:224–225

Rothstein JD, Martin LJ, Kuncl RW (1992) Decreased glutamate transport by the brain and spinal cord in amyotrophic lateral sclerosis. N Engl J Med 326:1464–1468

Saper CB, Wainer BH, German DC (1987) Axonal and transneuronal transport in the transmission of neurological disease: potential role in system degenerations, including Alzheimer's disease. Neurosci 23:389–398

Scatton B, Dubois A, Javoy-Agid F, Camus A (1984) Autoradiographic localisation of muscarinic cholinergic receptors at various segmental levels of the human spinal cord. Neurosci Lett 49:239–245

Schmidt-Achert K, Askanas V, Engel WK (1984) Thyrotropin-releasing hormone enhances choline acetyltransferase and creatine kinase in cultured spinal ventral horn neurons. J Neurochem 43:586–589

Schoenen J, Lotstra F, Vierendeels G, Reznik M, Vanderhaeghen J-J (1985a) Substance P, enkephalins, somatostatin, cholecystokinin, oxytocin and vasopressin in human spinal cord. Neurology 35:881–890

Schoenen J, Reznik M, Delwaide PJ, Vanderhaeghen J-J (1985b) Etude imunocytochimique de la distribution spinale de substance P, des enképhalines, de cholécystokinine et de sérotonine dans la sclérose latérale amyotrophique. CR Soc Biol 179:528–534

Sendtner M, Kreutzberg GW, Thoenen H (1990) Ciliary neurotrophic factor prevents the degeneration of motor neurons after axotomy. Nature 345:440–441

Sheu KFR, Cedarbaum JM, Harding BS, Clarke DD, Blass JP (1985) Measurements of pyruvate and glutamate dehydrogenase as well as fumarase in platelet lysates from patients with Parkinson's and related disorders. VIII International Symposium on Parkinson's Disease p 38

Sommer B, Seeburg PH (1992) Glutamate receptor channels: novel properties and new clones. TiPS 13:291–296

Spann BM, Grofova I (1989) Origin of ascending and spinal pathways from the nucleus tegmenti pedunculopontinus in the rat. J Comp Neurol 283:13–27

Spencer PS, Nunn PB, Hugon J et al. (1987) Guam amyotrophic lateral sclerosis-parkinsonism-dementia linked to a plant excitant neurotoxin. Science 237:517–522

Spokes EGS, Koch DJ (1978) Post-mortem stability of dopamine, glutamate decarboxylase and choline acetyltransferase in the mouse brain under conditions simulating the handling of human autopsy material. J Neurochem 31:381–387

Steventon G, Williams AC, Waring PH, Pall HS, Adams D (1988) Xenobiotic metabolism in motor neuron disease. Lancet ii:644–647

Takami K, Kawai Y, Shiosaka S et al. (1985) Immunohistochemical evidence for the coexistence of calcitonin gene-related peptide and choline acetyltransferase-like immunoreactivity in the neurons of the rat hypoglossal, facial and ambiguus nuclei. Brain Res 328:368–389

Vickroy TW, Roeske WR, Gehlert DR, Wamsley JK, Yamamura HI (1985) Quantitative light microscopic autoradiography of (^{3}H)hemicholinium-3 binding sites in the rat central nervous system: a novel biochemical marker for mapping the distribution of cholinergic nerve terminals. Brain Res 329:368–373

Westbrook GL, Mayer ML, Namoodiri MAA, Neale JH (1986) High concentrations of N-acetylaspartyl-glutamate (NAAG) selectively activate NMDA receptors on mouse spinal cord neurons in cell culture. J Neurosci 6(11):3385–3392

White SR (1985) Serotonin and co-localized peptides: effects on spinal motor neuron excitability. Peptides 6 (Suppl 2):123–127

White SR, Neumann RS (1985) Facilitation of spinal motor neurone excitability by 5-hydroxytryptamine and noradrenaline. Brain Res 335:63–70

Whitehouse PJ, Wamsley JK, Zarbin MA, Price DL, Tourtelotte WW, Kuhar MJ (1983) Amyotrophic lateral sclerosis: alterations in neurotransmitter receptors. Ann Neurol 14:8–16

Yamaguchi K, Hosokawa Y (1987) Cysteine dioxygenase. Methods Enzymol 143:395–403

Yoshino Y, Koike H, Akai K (1979) Free amino acids in motor cortex of amyotrophic lateral sclerosis. Experientia 35:219–220

Young A, Penney JB, Dauth GW, Bromberg MB, Gilman S (1983) Glutamate or aspartate as a possible neurotransmitter of cerebral corticofugal fibres in the monkey. Neurology 33:1513–1516

Yu AC, Schonsbue A, Hertz L (1982) Metabolic fate of ^{14}C-labelled glutamate in astrocytes in primary cultures. J Neurochem 39:954–960

Yudkoff M, Nissim I, Nelson D, Lin Z-P, Erecinska M (1991) Glutamate dehydrogenase reaction as a source of glutamic acid in synaptosomes. J Neurochem 57:153–160

Zarbin MA, Wamsley JK, Kuhar MJ (1981) Glycine receptor: light microscopic autoradiographic localization with (^{3}H)strychnine. J Neurosci 1:532–547

Zollinger M, Amsler U, Do KQ, Streit P, Cuénod M (1988) Release of N-acetylaspartylglutamate on depolarisation of rat brain slices. J Neurochem 51:1919–1923

9 Calcium and Aluminium in the Chamorro Diet: Unlikely Causes of Alzheimer-Type Neurofibrillary Degeneration on Guam

J. C. Steele and D. B. Williams
with the help of L. F. Heitz, W. J. Zolan, H. R. Wood, R. H. Randall, R. A. Stephenson, C. M. Parker

The 15 small islands of the Marianas comprise an archipelago that stretches northwards from Guam to Japan. For more than a century, the indigenous population of the Mariana Islands has been susceptible to a progressive fatal paralysis, beginning in adult life. This disorder is a neurodegenerative disease that the Marian people call Lytico-Bodig and which Kurland and colleagues termed the amyotrophical lateral sclerosis/parkinsonism-dementia complex of Guam (ALS/PDC) (Elizan et al. 1966). Initially, aggregations of ALS/PDC in certain Chamorro families suggested that the disease might be inherited (Kurland and Mulder 1954). However, a steady decline in the annual incidence of ALS/PDC since the early 1960s, and its occurrence among non-Chamorro immigrants to Guam indicate that the cause is environmental, and associated with a long latency from the initial exposure to the causative agent or agents and development of the disease (Reed et al. 1975; Garruto et al. 1985; Plato et al. 1986). There are two other foci of ALS/PDC, one in the Kii Peninsula of Japan (Shiraki and Yase 1975) and the other in western New Guinea (Gajdusek 1963). Because these three locales are geographically, ethnically and culturally distinct, investigators have sought shared factors which could explain the presumed common aetiology and pathogenesis of the disease in these different regions of the world.

In 1963 Kimura, Yase and colleagues suspected an environmental cause for ALS/PDC in the Kii Peninsula after identifying low calcium concentrations in river water and drinking water from villages where the condition was endemic (Kimura et al. 1963). To test their hypothesis, Yase visited Guam with Iwata and Sasajima in 1976 and studied the concentrations of 10 elements in Guam's water and soils. They had anticipated that calcium and magnesium concentrations might be high because of extensive limestone (coral) deposits there, but on finding low calcium concentrations in water samples from southern Guam, they concluded that "the environmental factor of ALS on Guam (i.e. mineral deficiency) bears a resemblance to that of the South Kii Peninsula in Japan" (Iwata et al. 1978).

Additional support for this hypothesis came from the finding that the water and soils of the western New Guinea villages also had low calcium concentrations, and that this was the only consequent mineral abnormality found in the environment in all three areas (Garruto et al. 1984). On the Kii peninsula and in western New Guinea magnesium concentrations were low in drinking water, but on Guam magnesium concentrations in drinking water were as high or higher than those from many other parts of the world. Soil aluminium concentration tended to be high in all three areas, although equally high values were also present in soils from the northern Mariana Islands (Saipan, Alamagan and Pagan) and the Philippines (Ilocos Norte and Ilocos Sud), where ALS/PDC is not reported (Garruto et al. 1984).

Geology, Villages and Water Sources of Guam and Rota

Guam is the largest island between Hawaii and the Philippines and between Japan and New Guinea (Fig. 9.1). Its land mass is 225 square miles in area and it is shaped like a peanut (Fig. 9.2). There are 18 villages on Guam located in northern, central and southern parts of the island. The principal villages of Guam and Rota, and the prevalence of ALS in 1953, are shown in Figure 9.3.

Guam's northern half is a broad limestone plateau which slopes southwestward to the island's narrow centre. Because limestone is a porous rock, no surface drainage system of rivers or streams has developed on Guam. Rainwater falling on the plateau quickly percolates downward to the level of impermeable volcanic rock, and forms a fresh water lens system where limestone dips below sea-level. By contrast, the southern half of the island consists of a volcanic upland. A mountain chain, 300–400 m high, runs parallel with the west coast and is partly overlain by thick limestone deposits. The western side of the southern mountain chain consists of a

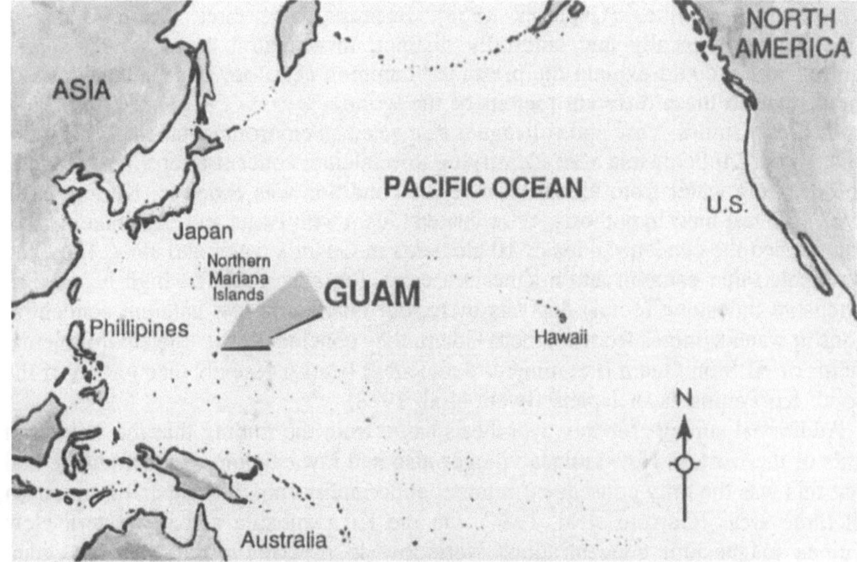

Fig. 9.1. Location of Guam in the western Pacific Ocean.

Fig. 9.2. The island of Guam from the air.

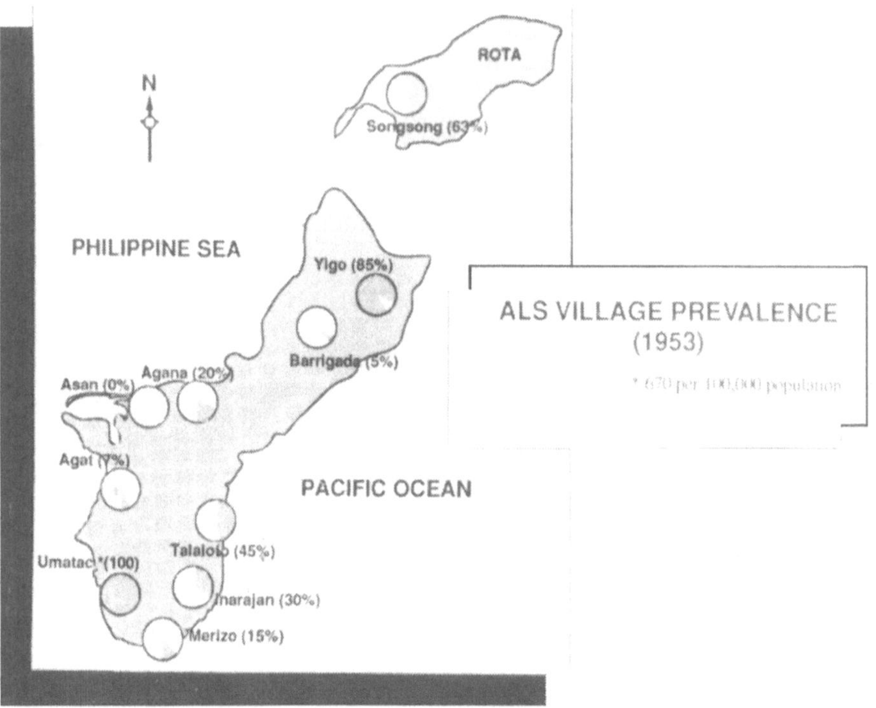

Fig. 9.3. Prevalence of amyotrophical lateral sclerosis in the principal villages of Guam and Rota.

steep scarp with numerous short, high-gradient streams draining to the coast. Numerous springs and seeps occur at sites where volcanic rock contacts the capping limestone at the surface. Potable water on Guam is thus derived from rivers and springs in the south and from deep wells in the north. Though some of Guam's 18 villages, such as Umatac, still depend on piped water from local sources, most are now supplied by an island-wide distribution system originating from wells in the north (Ward et al. 1965; Stephanson 1979).

Piga Spring is a remote water source that issues from a fault in the Mount Bolanos scarp, 1.6 km behind Umatac village (Fig. 9.4) at an elevation of 100 m, where a small lens of limestone forms an outcrop. Piga Spring's outwelling is like other springs in southern Guam except that its texture is slippery because the water is soft. In the early twentieth century, the U.S. Navy built a water system for Umatac village, taking water by pipe system from the La Sa Fua river in an adjacent valley. However, because that river sometimes dried up, and the Piga Spring does not, the spring was chosen as Umatac's main water source after 1953, and 72 700 l per day were distributed to village households. Before 1953, the Piga Spring flowed into a minor tributary of the Umatac River. It was deep in the jungle, said to be guarded by an unfriendly spirit, and avoided by villagers.

Rota is 45 miles north-east of Guam and its 1500 Chamorro residents suffer the same high rates of ALS/PDC as in Umatac. With an area of 85:2 km^2, Rota is the fourth largest of the Mariana Islands. Its volcanic core is entirely covered by limestone terraces which rise to 490 m. Since the early 1930s all households in the island's principal village of Songsong have received piped water from the Matanhanum water cave, a reservoir formed of rainwater which has percolated through the island's central limestone plateau. Before the Japanese developed this distribution system, the residents of Rota village depended on water from wells dug into a water lens close to the shore. Only one such well, at Muchon, on the north side of the island is still used. Rota has two rivers, but both are distant from Songsong village and they are not used as drinking water sources by villagers (Stephanson and Moore 1980).

Fig. 9.4. Umatac village, showing the limestone outcrop above the village, and Umatac Bay (see Fig. 9.5).

We have analysed the mineral content of the drinking water in these villages in order to consider whether disturbances of mineral intake are important in the pathogenesis of ALS/PDC. During 1984 and 1985, water samples were obtained at monthly intervals from four rivers in southern Guam. Two of the rivers, La Sa Fua and Umatac are in the Umatac district, and two are in the adjacent district of Merizo (Zolan and Ellis-Neill 1986). In 1986 the pattern of water distribution to households in the village of Umatac and nearby areas was studied. In 1987 water samples were obtained from the Piga spring and analysed for calcium content; the concentration of five other elements also measured. In 1987 and 1988 samples were collected from representative domestic water sources on Rota. All three water samples were analysed at the Water and Energy Research Institute at the University of Guam, using a standard titrametric method (American Public Health Association).

During 1985, families in the Umatac district with high rates of ALS/PDC were identified from NINCDS records. With the help of the village Mayor, five family farms in three valleys of the Umatac district were selected for intensive study (Fig. 9.5). These studies consisted of detailed dietary histories and testing of both soils and subsistence foods for mineral content. Comparable samples from Palau and Jamaica, two distinct and remote islands, in which ALS/PDC is unknown, were used as controls (Crapper McLachlan et al. 1989; Steele et al. 1990).

The calcium levels in diverse well, river and spring sources on Guam reported by previous researchers and ourselves are shown in Table 9.1. On Guam more than 10 mg/l of calcium was present in 90% of all the water sources tested by Iwata et al. (1978) and Garruto et al (1984). All of 9 wells had calcium values in excess of 65 mg/l. Zolan and Ellis-Neill found that calcium levels in river water varied seasonally (1986). Of 14 rivers tested 11 had at least one calcium level greater than 21.5 mg/l. Of the 8 springs tested 5 had values in excess of 60 mg/l. Only the Piga spring had calcium concentrations lower than 10 mg/l.

Table 9.2 shows water quality for Piga Spring, and the Umatac households it has supplied since 1953. Calcium and magnesium are extremely low but the spring's water contains relatively high concentrations of sodium (72–126 mg/l), moderate amounts of silicon (20–24 ppm), and traces of metals (Iwata et al. 1978; Garruto et al. 1984). In Rota all tested water supplies had high calcium levels characteristic of borderline hard water (Table 9.3).

Detailed review of the diets of the five selected Umatac families with a high incidence of ALS/PDC revealed that they had all had access to, and had eaten many calcium-rich foods (specifically including milk, fish and taro) and their diets were comparable with those enjoyed on other parts of Guam, and on other Micronesian islands on which ALS/PDC does not occur (Steele et al. 1990).

Root crops and vegetables from the southern region of Guam had calcium concentrations comparable with the high levels that were present in similar foods grown on Palau and Jamaica. These diverse subsistence foods provide ample calcium and do not contain excess aluminium. For example, 500 grams (wet weight) of taro tuber core, a main starch staple of the traditional Chamorro diet, provides 500 calories and 700 mg of calcium but only 14 mg of aluminium (Crapper McLachlan et al. 1989).

Compared with soils of Palau and Jamaica, the soil in the valleys of southern Guam averaged a 42-fold higher yield of elutable aluminium. However, the concentrations of aluminium in vegetables grown in these soils were essentially the same as on both the other islands, and were not excessive (Crapper McLachlan et al. 1989).

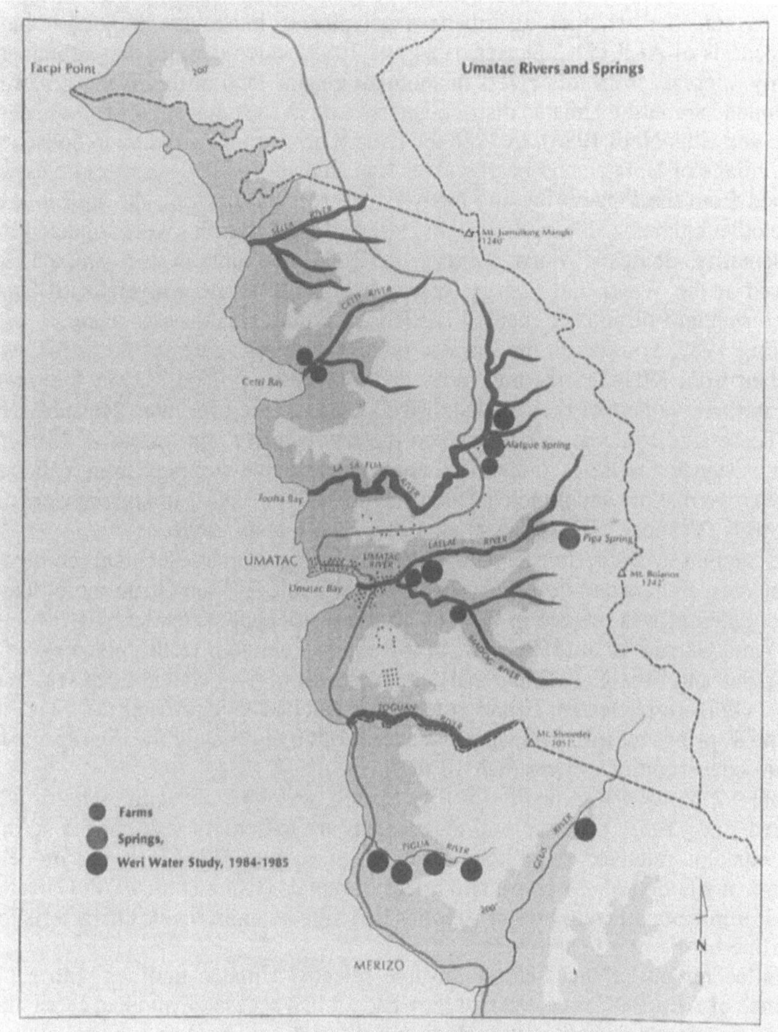

Fig. 9.5. Rivers and springs from which the water supply of Umatac, Guam is derived.

Implications of These Findings

The village of Umatac has been a focus of ALS/PDC epidemiological research since the early 1950s because of the consistently high rate of disease there. Umatac is an ancient coastal village colonised by the Spanish soon after Magellan's discovery of Guam in 1521. Folk tales tell of paralysis among families of the village in the nineteenth century, and death certificates record that ALS was common there after 1898 when the island was ceded to the United States (Kurland and Mulder 1954; Mulder et al. 1954). Umatac is surrounded on three sides by mountains and on the

Table 9.1. Calcium levels (mg/l) in diverse raw water sources on Guam

Village	Water source	Iwata[a] (1978)	Garruto[a] (1984)	Zolan[a] (6/1984–6/1985)
Northern Guam				
Dededo	Dededo Well	72.9	82.0	
	Finegayan Well	80.0	79.0	
Yigo	Agafa Gumas Well	76.0	80.0	
	Yigo Well	84.0	88.0	
	Marbo Cave	68.0		
	Mataguac Spring	24.0	22.0	
	Chunge-1 Spring	24.0		
	Chunge-2 Spring	28.8		
	Santa Rosa Spring	10.0		
Central Guam				
Agana	A-1 Well	128.0	123.0	
	A-14 Well	140.0		
	A-15 Well	112.0		
	Agana Spring	115.2	87.0	
Asan	Asan Spring	88.0	87.0	
Mangilao	Mangilao Well	65.6	72.0	
	Fenna Reservoir-1	24.8		
	Fenna Reservoir-2	25.6		
	Fenna Beach	23.2		
	Fonte River	21.6		
Southern Guam				
Agat	Upper Taleyfac River	36.0		
	Upper Namo River	69.0		
San Rita	Santa Rita Spring	68.0	67.0	
Umatac	Lower La Sa Fua River	15.0		
	Upper La Sa Fua River			20–55
	Laelae River		47.0	
	Lower Laelae River		44.0	
	Lower Laelae River		23.0	
	Madog River		26.0	
	Umatac River	49.6		40–69
	Alatgue Spring	80.0	83.0	
	Piga Spring-1	3.2	1.2	
	Piga Spring-2	4.0	1.3	
Merizo	Pigua River			20–58
	Geus River	34.4	44.0	18–51
	Siligan Spring	61.6	60.0	
Inarajan	Inarajan River	7.2		
	Fintasa River		2.1	
	Malojojo Well	104.8		
Talofofo	Talofofo River	104.4		
	Upper Talofofo River		52.0	
	Upper Ugum		7.0	
	Ylig River	36.0	42.0	

[a]Ca was determined by the EDTA titration method

fourth by the sea. The village remains small, self-sufficient and traditional in outlook. Its population is less than 1000 and the majority are Chamorro. In nearby coastal valleys the villagers farm and raise subsistence crops.

Table 9.2. Calcium and magnesium contents of water from Piga Spring and households of Umatac village (mg/l)

Investigator	Date	Sample site		Calcium	Magnesium
		Spring	Household		
Iwata et al.	1976	+		4.0	1.2
Garruto et al.	1984	+		1.3	0.17
			+	1.4	0.29
Zolan/Steele	1986		+	2.2	
Heitz/Wood	1987	+		0.28	0.03

Table 9.3. Calcium contents of water sources on Rota

Source	Date of Collection		
	Jan 1987	May 1987	Feb 1988
Matanhanum Water Cave		36.3	
Songsong Village	36.4[a]	35.9[a]	43.8[a]
Tatacho Village	37.3[a]		41.7[a]
Airport			39.6
River 1	27.5	29	
River 2		31.2	
Muchon Well			108.3

[a]Household water sample piped from the Matanhanum water cave

In 1956 the U.S. National Institutes of Health established a research station on Guam to monitor and investigate ALS/PDC. A longitudinal study confirmed that the disease was remarkably common among the five extended families of Umatac, and from 1945 to 1982 more than 10% of all patients in the NINCDS registry were from this village (Fig. 9.6). A house-to-house survey in 1987 found that the disease was still prevalent, but that ALS was less common, and that the clinical syndromes of parkinsonism and dementia predominated. These syndromes are most common among the older residents (Steele et al. 1989). When Yase and Garruto investigated ALS/PDC on Guam, they too studied the village and environs of Umatac. Their finding of unusual mineral concentrations in Piga spring water in the Umatac region suggested to them that those perturbations might be causative factors in the disease (Garruto and Yase 1986). This concept was supported by their subsequent experimental studies in nonhuman primates, which incriminated a low calcium–high aluminium diet as a causative factor in neuronal degeneration. Yase observed argyrophilic bodies, chromatolysis, and axonal swelling of anterior horn cells in macaques (*Macaca fuscata*) after 7 months on a low calcium and low magnesium diet when aluminium lactate was added to drinking water. Electron microscopy showed 10 nm straight neurofilaments. These neuropathological changes were generally similar to those found in animals with aluminium-induced neurofibrillary accumulations (Yase 1988). Garruto found chromatolysis, intraneuronal inclusions, axonal spheroids, loss of axons and myelin, and well-formed neurofibrillary tangles resembling those of ALS/PDC in cynomolgus monkeys (*Macaca fascicularis*) after 4 years of a specially formulated low-calcium diet (0.32% calcium) (Garruto et al. 1989).

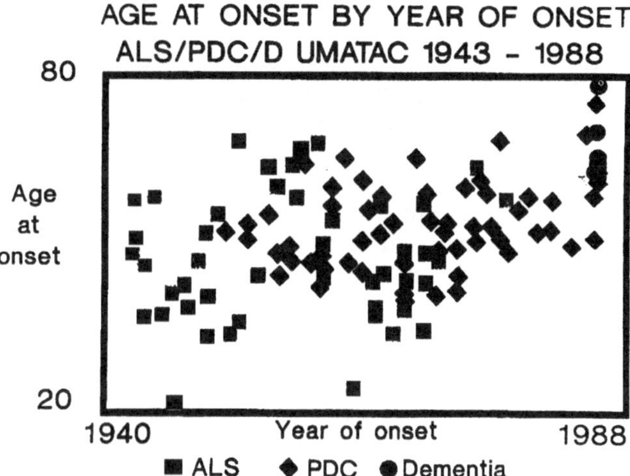

Fig. 9.6. Scattergram showing relation of age at onset of ALS/PDC to year of onset of the disease in Umatac, Guam between 1943 and 1988.

Aluminium by itself can be neurotoxic, and it is implicated in neurofibrillary degeneration and Alzheimer's disease (editorial 1989). Recently Martyn et al. found that high concentrations of aluminium in drinking water appeared to be related to an increased risk of Alzheimer's disease in England and Wales (Martyn et al. 1989).

However, despite these experimental and epidemiological findings, it is difficult to reconcile the proposed mineral hypothesis with the unremarkable mineral concentrations found in foods and drinking water of Umatac, other Guam villages, and the island of Rota, where ALS/PDC also occurs. Thus calcium concentrations in most water sources on Guam and Rota do not appear to be unusually low, and aluminium concentrations in drinking water are not increased. Of 56 potable water sources tested the mean aluminium concentration was only 19 μm/l, 10 times less than the average aluminium content of 17 samples of municipal drinking water in the city of Toronto, Canada (Crapper McLachlan et al. 1989).

Springs and water caves on Guam and Rota occur at junctures of porous limestone and impermeable volcanic rock. Piga spring, the only water source with unusually low calcium concentration appears to be unique because it comes from reservoirs within Mount Bolanos. Geologists postulate that surface drainage and groundwater on the ridge above moves downward along internal fractures (large void spaces) developed by geological faulting. Water contact with the volcanic pyroclastic rocks that form the walls of the fractures adds silicon and trace amounts of iron, lead and zinc to the water. However, because there is little limestone in this area, and the pyroclastic rocks are non-calcareous, the calcium content of the water remains extremely low. At the spring itself, although the water exits are close to a limestone outcrop, the contact with limestone there is too brief to dissolve calcium in significant quantities.

If the single water source with a low calcium concentration is insufficient to explain the more widespread occurrence of ALS/PDC, could the diet of susceptible Guamanians be implicated in some way? Our data from a hyperendemic region of Guam does not suggest that the current or recent diet of our subjects was deficient in calcium or that it provided an excess of elemental aluminium. Indeed, the diet appeared to be nutritionally equivalent to diets from other Micronesian islands in which there is no evidence of endemic degenerative neurological disease (Steele 1984).

During 1986, Parker (a research nutritionist) and Quinata-Guzman conducted an investigation of the traditional diet of Chamorro people before and during World War II in three southern Guam villages which have different rates of ALS/PDC: Umatac (high ALS/PDC), Merizo (high ALS/low PDC) and Asan/Maina (low ALS/PDC). They were unable to identify any differences in dietary practice or food types between villages, or between families, with and without ALS/PDC. The average calcium intake exceeded the minimal recommendation (Food and Agriculture Organization) of 450 mg/day in all families, even during World War II when many foods were scarce (Parker 1988; Steele & Quinata-Guzman 1990).

After World War II the Chamorro diet altered to include a much higher proportion of imported foods. But the new, Western-style foods which Chamorros chose were less nutritious than those of the traditional diet and by 1983 Kretsch et al. (Kretsch and Todd 1985; Todd and Kretsch 1985) found that for a significant proportion of the Guamanian population the contemporary diet was deficient in calcium (mean dietary intake 321 mg/day compared with the 1980 United States Recommended Dietary Allowance of 800 mg/day), vitamin A, and ascorbic acid. The prevalence rates of ALS/PDC were highest just after World War II when dependence on imported foods, with their apparently inadequate amounts of calcium, was at a minimum, and dependence on traditional food sources was greatest (Garruto et al. 1985). If calcium deficiency is the culpable factor of ALS/PDC, among Chamorros who are perhaps susceptible and metabolically predisposed, then the rates of disease would be expected to have risen as the diet altered and calcium deficiency increased, rather than steadily declining as they have. Finally, among individuals suffering from ALS or PDC, there is no laboratory evidence for disturbed calcium or vitamin D metabolism (Yanigihara et al. 1983).

For almost 50 years physicians have been challenged by the enigma of ALS on Guam and for 35 years NIH scientists have sought its cause, now strongly presumed to be environmental. Some scientists attribute the disease to the Chamorro diet; to calcium deficiency, to aluminium excess or to neurotoxins in cycad seeds (Duncan et al. 1990; Kurland 1972). Our studies of the past 6 years do not incriminate any of these factors, and we conclude that the cause of this Alzheimer-related, aluminium-associated disease is still unknown.

Our search for the cause of ALS/PDC continues. We agree with Kurland that "clarification of the etiology of ALS and parkinsonism-dementia on Guam will have widespread repercussions on the aetiology of parkinsonism, the presenile dementias and ALS elsewhere" (Kurland 1972). If we can discover the cause of this disappearing disease on these distant tropical islands, we are likely to find a key to understanding other related sporadic neurodegenerative diseases, and aging of the nervous system, in all other parts of the world.

Conclusions

Alzheimer-type neurofibrillary degeneration of nerve cells is remarkably common among Chamorro indigenes of Guam, and many also suffer from amyotrophic lateral sclerosis or the parkinsonism-dementia complex (ALS/PDC). There is evidence that a low-calcium, high-aluminium diet may cause similar neurofibrillary degeneration in experimental animals. As the sources of drinking water in Guam and the two other Pacific ALS/PDC foci are all reported to be low in calcium, while the soils of these regions have high aluminium concentrations, previous investigators proposed that ALS/PDC could be the consequence of mineral dysmetabolism.

The results of several investigations fail to support that hypothesis. Of all the water sources tested on the island of Guam and the neighbouring island of Rota only one, the Piga spring, has a low calcium concentration. This spring has only recently become a water source for the village of Umatac, which is known to have had a high frequency of ALS/PDC for many decades. In addition, soil aluminium concentration varies markedly within one hyperendemic region of Guam (Umatac), the traditional subsistence foods of Guam are not deficient in calcium, and neither they nor potable water sources contain excessive concentrations of aluminium. Calcium metabolism is not known to be abnormal in affected individuals. This evidence does not suggest that quantitative abnormalities of dietary calcium or aluminum are likely causes of the neurofibrillary degeneration in ALS/PDC of the Mariana Islands.

Acknowledgements

The authors gratefully acknowledge the Guam Lytico-Bodig Association and NIH Grant AG08802, "The epidemiology of dementia in Micronesia", which supported these studies.

References

American Public Health Association – American Water Works Association – Water Pollution Control Federation. Standard methods for the examination of water and waste water, 15th ed, pp 185–187

Crapper McLachlan DR, McLachlan CD, Krishnan B, Krishnan SS, Dalton AJ, Steele JC (1989) Aluminum and calcium in soil and food from Guam, Palau and Jamaica: implications for amyotrophic lateral sclerosis and parkinsonism-dementia syndromes of Guam. Environ Geochem Health 11:45–53

Duncan MW, Steele JC, Kopin IJ, Markey SP (1990) 2-Amino-3-(methylamino)-propanoic acid (BMAA) in cycad flour: An unlikely cause of amyotrophic lateral sclerosis and parkinsonism-dementia on Guam. Neurology 40:767–772

Editorial (1989) Aluminium and Alzheimer's disease. Lancet i:82–83

Elizan TS, Hirano A, Abrams BM, Need RL, VanNuis C, Kurland LT (1966) Amyotrophic lateral sclerosis and parkinsonism-dementia complex of Guam. Neurological Re-evaluation. Arch Neurol 14:356–368

Gajdusek DC (1963) Motor neuron disease in natives of New Guinea. N Engl J Med 268:474–476

Garruto R, Yase Y (1986) Neurodegenerative disorders of the western Pacific: the search for mechanisms of pathogenesis. Trends Neurosci 9:368–374

Garruto RM, Yanagihara R, Gajdusek DC, Arion DM (1984) Concentrations of heavy metals and essential minerals in garden soils and drinking water in the western Pacific. In: Chen KM, Yase Y (eds) Amyotrophic lateral sclerosis in Asia and Oceania. National Taiwan University, Taipei, pp 265–330

Garruto RM, Yanagihara R, Gajdusek DC (1985) Disappearance of high-incidence amyotrophic lateral sclerosis and parkinsonism-dementia on Guam. Neurology 35:193–198

Garruto RM, Shankar S, Yanagihara R, Salazar A, Amyx H, Gajdusek C (1989) Low-calcium, high-aluminium diet-induced motor neuron pathology in cynomolgus monkeys. Acta Neuropathol 78:210–219

Iwata S, Sasajima K, Yase Y, Chen KM (1978) Report of investigation of the environmental factors related to occurrence of amyotrophic lateral sclerosis in Guam island. Overseas field research in 1976–77 supported by the Japanese Ministry of Education (Unpublished), pp 1–10. (Available at WERI Library, University of Guam).

Kimura K, Yase Y, Higasahi Y et al. (1963) Epidemiological and geomedical studies on ALS. Dis Nerv Syst 24:155–159

Kretsch MJ, Todd KS (1985) Nutrient intake of Guamanian women. Fed Proc 44:756 (abstract)

Kurland LT (1972) An appraisal of the neurotoxicity of cycad and the etiology of amyotrophic lateral sclerosis of Guam. Fed Proc 31:1540–1542

Kurland LT, Mulder DW (1954) Epidemiological investigations of amyotrophic lateral sclerosis. I. Preliminary report on geographic distribution, with special reference to the Mariana Islands, including clinical and pathological observations. Neurology (Minneap) 4:355–378, 5:438–448

Martyn C, Osmond C, Edwardson J, Barker D, Harris E, Lacey R (1989) Geographical relation between Alzheimer's disease and aluminium in drinking water. Lancet i:59–62

Mulder D, Kurland L, Iriarte L (1954) Neurologic diseases on the island of Guam. U.S. Armed Forces Med J 5:1724–1739

Parker C. (1988) Dietary history in three villages of Guam prewar and wartime (1925–1950), with special reference to the role of calcium intake in motor neuron disease. M.Sc. thesis, Food Sciences and Human Nutrition, University of Hawaii

Plato CP, Garruto RM, Fox PM, Gajdusek CD (1986) Amyotrophic lateral sclerosis and parkinsonism-dementia on Guam: A 25-year prospective case–control study. Am J Epidemiol 124:643–656

Reed DM, Torres JM, Brody JA (1975) Amyotrophic lateral sclerosis and parkinsonism-dementia on Guam, 1945–1972. II. Familial and genetic studies. Am J Epidemiol 101:302–310

Shiraki H, Yase Y (1975) Amyotrophic lateral sclerosis in Japan. North-Holland, Amsterdam, pp 353–419 (Handbook of clinical neurology vol 22, 16)

Steele JC (1984) Micronesia: Health status and neurological diseases In: Chen KM, Yase Y (eds) Amyotrophic lateral sclerosis in Asia and Oceania. National Taiwan University, Taipei, pp 173–182

Steele JC, Quinata-Guzman T (1990) The Chamorro diet: an unlikely cause of neurofibrillary degeneration on Guam. In: Rose,Norris (eds) ALS: new advances in toxicology and epidemiology. Smith-Gordon, UK

Steele J, Lavine L, Workman A et al. (1989) Current prevalence of amyotrophic lateral sclerosis and parkinsonism-dementia complex on Guam. Abstract XXIV CCNS Meeting, Ottawa Ontario 73

Steele JC, Guzman TQ, Driver MG et al. (1990) Nutritional factors in amyotrophic lateral sclerosis on Guam: Observations from Umatac. In: Hudson AJ (ed) Amyotrophic lateral sclerosis: Concepts in pathogenesis and etiology. University of Toronto Press, Toronto, pp 195–225

Stephanson R (1979) Freshwater use customs on Guam. WERI, University of Guam. 8:1–150

Stephanson R, Moore D (1980) Freshwater use customs on Rota. WERI, University of Guam 17:1–145

Todd KS, Kretsch MJ (1985) Food intake patterns of Guamanian women. Fed Proc 44:756 (abstract)

Ward P, Hoffard S, Davis D (1965) Hydrology of Guam. Geological survey professional paper 403–H, US Government Printing Office, Washington, DC, pp 1–28

Yanigihara R, Garruto RM, Gajdusek DC et al. (1983) Calcium and vitamin D metabolism in Guamanian chamorros with amyotrophic lateral sclerosis and parkinsonism-dementia. Ann Neurol 15:42–48

Yase Y (1988) Metal studies of ALS. Further development. In: Tsubaki, Yase (eds) Amyotrophic lateral sclerosis: recent advances in research and treatment. Excerpta Medica, Amsterdam, pp 59–65 (International congress series 769)

Zolan W, Ellis-Neill L (1986) Concentration of aluminium, manganese, iron and calcium in four southern Guam rivers. Water and Energy Research Institute (WERI), University of Guam. 64:1–67

10 Toxicology of Motor Systems

P. B. Nunn

Introduction

There is no shortage of toxic factors that have been proposed as aetiological agents in motor neuron disease. Of exogenous origin lead, mercury (elemental and organic) selenium and manganese are most cited (Tandan and Bradley 1985). The incidence of the amyotrophic lateral sclerosis (motor neuron disease)-parkinsonism-dementia complex (ALS-PD) of Guam has been related to nutritional deficiencies of calcium and magnesium resulting in the accumulation of manganese and aluminium (Garruto and Yase 1986). Specific amino acids have been implicated in neurolathyrism (Rao et al. 1964), in Guamanian ALS-PD (Spencer et al. 1987) and more generally in neuronal death in various parts of the central nervous system that are not explicitly implicated in motor neuron disease (Olney 1978; Barinaga 1990; Perl et al. 1990). Epidemiological studies have linked exposure to organic solvents with an increased incidence of motor neuron disease (Hawkes et al. 1989), although it is likely that such substances also cause more generalised central nervous system dysfunctions (see Johnson 1987).

Glutamate acts as both an exogenous and as an endogenous neurotoxin (Olney 1971; Olney et al. 1972; Plaitakis et al. 1988); sulphur-containing amino acids (Olney et al. 1971) and quinolinate (Stone et al. 1987) have been implicated as experimental neurotoxins, with evidence that the latter may play a role in Huntington's chorea. Abnormalities of glutamate (Plaitakis and Caroscio 1987), glycine tolerance (Lane et al. 1993) and elevated cysteine and reduced sulphate concentrations in plasma (Heathfield et al. 1990) have been detected while screening patients with sporadic motor neuron disease.

It has been necessary to make a selection from this wealth of apparently unconnected observations and emphasis has been placed on those agents for which there is good evidence implicating specific mechanisms in the mode of action. Thus, glutamate and the sulphur amino acids have been selected; three excitatory amino acids that are implicated in neurolathyrism (which affects the populations of parts of East Africa and the Indian sub-continent), Guamanian ALS-PD and the recent outbreak of shellfish poisoning in Canada are discussed.

Mechanistic Basis of Neuronotoxicity

General Considerations

There is considerable evidence that motor neurons that do not make sufficient cellular interactions during development do not survive. A current view is that neuronal survival is dependent upon the availability of neurotrophic factors, perhaps released by muscle, following successful neuromuscular connections (Oppenheim 1987). The progressive loss of neurons that occurs within the ageing central nervous system in man (Calne et al. 1986) could occur as a result of the gradual loss of neurotrophic compounds or be due to other, presently unknown, factors. In neurodegenerative diseases the focus of attention is upon an increased rate of neuronal death rather than upon neuronal death itself (Agid and Blin 1987). So far as the part that neurotoxins might play in this process is concerned, the concept of an agent whose presence diminishes neuronal survival over a period of decades is extremely difficult to approach experimentally.

Extrapolating an evaluation of the chronic toxicity of a compound from animal experiments to human populations exposes uncertainties concerning variables such as dosage, interaction with other environmental or endogenous factors, life span of the animal selected and differential species susceptibility. Much current work on neurodegenerative diseases has adopted a different experimental strategy, which is to seek mechanistic detail by acute experimentation in simplified systems and to extrapolate the mechanisms thus discovered to the chronic picture seen in man.

The work discussed in this chapter is founded upon recent progress in four main areas of research; the detailed functions of central nervous system excitatory amino acid receptors; the homeostatic regulation of intraneuronal Ca^{2+} concentrations; the transport of amino acids and the regulation of amino acid metabolism in man.

Neuroexcitatory Amino Acid Receptors in the Central Nervous System

It is advantageous at this stage to present a brief explanation of excitatory amino acid receptors and of cellular Ca^{2+} homeostasis in order to refer freely to these topics with respect to the experimental work that will be discussed subsequently.

An excellent review covers the development and current status of research on excitatory amino acid receptors (Monaghan et al. 1989). The following account is restricted to those features that are relevant to an understanding of the subsequent discussion of neurotoxicity. Five major receptors are currently recognised; receptors (a)–(c) control ion channels that also permit the passage of water.

(a) NMDA (N-methyl-D-aspartate) receptors. These receptors regulate an ion channel that is permeable to Na^+ and Ca^{2+}. The receptor is gated by Mg^{2+} in a voltage-dependent manner and modulated by Zn^{2+}. It is antagonised by MK-801 (non-competitively) and competitively by D-2-amino-5-phosphonovalerate (AP5) and by D-2-amino-7-phosphonoheptanoate (AP7).

(b) AMPA (α-amino-3-hydroxy-5-methyl-isoxazole-4-propionate) receptors. These receptors regulate an ion channel that is permeable to Na^+. The activity of this receptor was formerly named for quisqualate (Watkins and Evans 1981), but AMPA is the more specific agonist (Monaghan et al. 1989).

(c) Kainate receptors. These receptors have properties similar to those of AMPA receptors. Some doubt exists as to the separate existence of the two receptors (see discussion by Young and Fagg 1990). Nonetheless, the anatomical distribution of

binding sites for the two compounds is effective evidence of the functional distinction of kainate and AMPA receptors.

(d) AP4 (L-2-amino-4-phosphonobutyrate) receptors. The antagonist effects of AP4 to evoked potentials in a variety of neurophysiological preparations distinguishes this receptor from other excitatory amino acid receptors.

(e) ACPD (metabotrophic) receptors. The receptors activated by 1-amino-cyclopentenyl-1,3-dicarboxylate (ACPD) stimulate the metabolism of inositol phosphates and, as a result, the mobilisation of Ca^{2+} from intracellular stores.

The Excitotoxic Hypothesis of Neuronal Death

Olney and colleagues established a strong correlation between the ability of acidic amino acids to excite neurons and to cause a characteristic neuropathology both in vivo and in vitro. As a result Olney (Olney et al. 1971; Olney 1978) proposed the "excitotoxic hypothesis of neuronal death", which holds that a depolarisation mechanism underlies the neurotoxic properties of glutamate and other acidic amino acids; thus the term "excitotoxin" was coined. Depolarisation was envisaged to result in the activation of ATP-dependent homeostatic mechanism, which would deplete the energy reserves of neurons in a vain attempt to restore ionic balance, with cell death ensuing when these reserves became exhausted.

When chick embryo retinas were treated with neurotoxic amino acids Na^+ and Cl^- entered neurons and were accompanied osmotically by water, which gave rise to characteristic intraneuronal vacuoles (Olney et al. 1986). While this simple mechanism sufficed for short-term (i.e. 30 min) experiments, the neurotoxicity seen following overnight exposure was clearly dependent upon the presence of Ca^{2+} in the extracellular medium (Choi 1985, 1987).

Since the NMDA receptor is permeable to Na^+, Ca^{2+} and water its activation by glutamate and other compounds whose activities are mediated by this receptor would explain both the early and late phases of glutamate-like neurotoxicity.

In contrast non-NMDA receptors regulate ion channels that allow the entry of Na^+, Cl^- and water only. However, depolarisation of the neuronal membrane following activation of AMPA and kainate receptors also activates voltage-dependent Ca^{2+} channels, allowing Ca^{2+} entry into the cell by indirect means. The cellular response to increased Ca^{2+} entry is a rise in intracellular Ca^{2+} and the stimulation of phospholipase C to produce diacylglycerol and inositol triphosphate (Eberhard and Holz 1988; Berridge and Irvine 1989). An increase in the concentration of intracellular Ca^{2+} would also occur via the activity of the metabotrophic receptor (Sladeczek et al. 1988).

Thus the effects of single acidic amino acids on neurons can be very complex, especially so since the specificity of receptor agonists may be concentration-dependent (Bridges et al. 1989). A valuable analysis of the compound effects of glutamate on neurons via NMDA, AMPA, kainate and metabotrophic receptors is available (Mayer and Miller 1990). Other excitatory amino acids would be expected to elicit similar, though less comprehensive, responses depending upon the concentration of the agonist and its specificity.

Glutamate as an Endogenous and Exogenous Factor in Neurological Disease

General Considerations

Glutamate finds itself considered both as a potential endogenous and exogenous neurotoxin because of its high concentration in the central nervous system, its high content in plant food proteins (FAO 1970) and its use (as the monosodium salt) for enhancing flavour in manufactured foods and as a condiment.

The concentration of glutamate in brain tissue after homogenisation is about 10 mM, but there is good evidence that much of the glutamate in the central nervous system is located in synaptic vesicles (Burger et al. 1989) and histological evidence confirms this observation (see Headley and Grillner 1990).

Neurotoxicity of Exogenous Glutamate

In work that laid the foundations for the excitotoxic hypothesis, Olney's group showed that monosodium glutamate caused extensive hypothalamic damage in infant mice (Olney 1971) and in infant rhesus monkeys (Olney et al. 1972). The brain damage that results from systemically administered glutamate is selective for those brain regions, known as the circumventricular organs, that lie outside the blood–brain barrier. Neonatal animals are more sensitive to the effects of glutamate than are adults. The lesions induced by glutamate involve the rapid swelling of neuronal cell bodies and dendrites followed by degenerative changes in intracellular organelles and the clumping of nuclear chromatin. Axons are not affected in glutamate neurotoxicity.

As a result of these and other experiments the addition of monosodium glutamate to the food of infants has been a contentious issue (Olney 1969). Does the use of the compound as a flavour enhancer produce excitotoxic damage in the young? There is strong evidence to support this argument experimentally, but the 20-year-old debate is still unresolved (Barinaga 1990). An allied subject concerns the consumption of Chinese food containing considerable amounts of monosodium glutamate, which has given rise to the colourfully named "Chinese Restaurant Syndrome" in which a variety of excitatory features are displayed in adults (Schaumburg et al. 1969). However, a double-blind trial in which 3 g of monosodium glutamate (or a placebo) was fed to volunteers showed no difference in symptomatology (Morselli and Garatini 1970).

The very high local concentrations of glutamate that result from its confinement in presynaptic vesicles means that its adventitious release could have profound effects on neighbouring tissue. The release of endogenous glutamate has been cited as a pathological feature in a number of neurological disorders, in which brain damage occurs as a result of anoxia-ischaemia, hypoglycaemia and seizures (Rothman and Olney 1987; Choi and Rothman 1990).

The relevance of these observations concerning the neurotoxicity of exogenous and endogenous glutamate to motor neuron disease is unclear, but there is certainly evidence that some abnormalities of glutamate metabolism are involved in neurodegeneration, including motor neuron disease itself.

Glutamate Metabolism and Neurological Disease

Olivopontocerebellar atrophy is a good example of a neurological syndrome in which abnormal glutamate metabolism appears to play an aetiological role. This adult-onset degenerative disease is associated with the demise of neurons in the brainstem, cerebellum, spinal cord and substantia nigra. Patients suffering from this disorder have an abnormal tolerance to ingested glutamate, whether administered in protein or as the free amino acid. After ingestion plasma glutamate and aspartate concentrations are elevated, but other amino acids are not affected (Plaitakis et al. 1982).

The results are compatible with a deficiency of hepatic (and probably peripheral) glutamate dehydrogenase; a deficiency of the enzyme, which catalyses the following reaction, was shown in leucocyte homogenates.

$$\text{Glutamate} + H_2O + NAD(P)^+ = \text{2-oxoglutarate} + NH_4^+ + NAD(P)H + H^+$$

The extent to which elevated concentrations of glutamate (and of aspartate) might result in nerve cell death depends upon the ease with which these amino acids could enter the central nervous system. Were increased plasma concentrations of glutamate and aspartate to cause increased entry of these amino acids following feeding the condition would represent an attractive model by which repetitive neurotoxic insults could be inflicted upon the central nervous system several times each day.

A reduced activity of glutamate dehydrogenase in the central nervous system is more difficult to explain. Here the enzyme is involved both in the synthesis and the deamination of glutamate, depending upon the cell type (see Plaitakis 1990). Certainly reduced deamination of glutamate could result in increased amounts of the amino acid at synapses (Plaitakis et al. 1982), which could lead to neurodegeneration by an excitotoxic mechanism.

Glutamate Metabolism in Motor Neuron Disease

There is considerable evidence that abnormalities of glutamate metabolism occur in sporadic motor neuron disease. The evidence for glutamate toxicity in MND has been discussed in more detail in Chapter 8. Thus, increases in the fasting plasma glutamate concentration and an abnormal glutamate tolerance have been reported. The concentration of glutamate (presumably the neurotransmitter pool) is reduced in several parts of the central nervous system (Perry et al. 1987; Plaitakis et al. 1988; see also Plaitakis 1990). However, increases in the concentration of glutamate (190%) and of aspartate (100%) in cerebrospinal fluid (Rothstein et al. 1990) have proved a controversial matter (Perry et al. 1990) that is not easily resolved (Young 1990) (see Chapter 8). Increases in the concentration of glutamate and aspartate of plasma and cerebrospinal fluid are of potential aetiological interest as these compounds could cause excitotoxic injury to motor neurons.

There are more generalised changes to be seen in motor neuron disease that are not so easily explained by this mechanism. For example, Rothstein et al. (1990) report reductions in the concentrations of N-acetylaspartylglutamate (NAAG; 60%) and of N-acetylaspartate (NAA; 40%) in ventral horns of patients with the sporadic disease compared with control values. Additionally, the many studies of the concentrations of amino acids in the body fluids of sufferers from the sporadic disease have reported quite variable results (reviewed by Perry et al. 1990) and, without doubting the analytical procedures involved, it may be that there is considerable metabolic heterogeneity in the populations studied, despite a consistent clinical picture.

Abnormalities of glutamate and of glycine tolerance (Lane et al. 1993) seen in patients with motor neuron disease appear to reflect a deficiency of the normal homeostatic regulation of the concentrations of these compounds in plasma. This function, which would normally regulate plasma amino acid concentrations within close limits and return them to that level after feeding, is mainly a hepatic activity. For example, the glycine oxidase system that would be responsible for removing excess glycine is located in liver (Yoshida and Kikuchi 1973).

A major function of hepatic glutamate dehydrogenase in liver is to deaminate glutamate formed by transamination of 2-oxoglutarate with amino acids entering the liver from peripheral tissues (McGilvery and Goldstein 1983). The 2-oxoglutarate thus formed is recycled, whereas the NH_4^+ is converted to carbamoyl phosphate for incorporation into the urea cycle. Similarly, the elevation of plasma cysteine that is discussed below has also been attributed to a deficiency of cysteine dioxygenase (Heathfield et al. 1990), which appears to be specifically located in liver (Griffiths 1987). Whereas the previously mentioned enzymes are located in mitochondria, cysteine dioxygenase is a cytosolic enzyme.

From these observations it appears that the changes in the concentrations of amino acids that occur in plasma (and perhaps in cerebrospinal fluid) in patients with sporadic motor neuron disease might be secondary to a loss of hepatic homeostatic function. In this respect it may be relevant that morphological abnormalities have been found in about half of a group of patients with sporadic motor neuron disease (Masui et al. 1985).

Other Endogenous Factors: Sulphur-Containing Amino Acids

General information on the mammalian metabolism of sulphur amino acids may be found in two excellent reviews (Cooper 1983; Griffiths 1987). It must, however, be emphasised that several problematic areas of this subject remain.

Amino Acids Containing Oxidised Sulphur Atoms

It has been known for some time that amino acids containing oxidised sulphur atoms are powerful neuroexcitants in the central nervous system neurons (see classification by Krnjevic 1965). Homocysteate and cysteate were among the most active neuroexcitants then known and to these were added subsequently cysteine sulphinate (Olney et al. 1971) and cysteine-S-sulphonate (Biscoe et al. 1975). All these compounds are excitotoxins in the mouse pup or by injection directly into mouse brain (Olney et al. 1975) and, with the exception of cysteine-S-sulphonate, all the compounds are normal constituents of the central nervous system, though cysteate is not a mammalian product (Griffiths 1987). There is some evidence that homocysteate may play a role as a neurotransmitter in some parts of the central nervous system (Do et al. 1986).

Modern pharmacological analysis has revealed that all four compounds referred to above act at both NMDA and non-NMDA receptors (Foster and Fagg 1984). Thus, for the majority of these compounds the available evidence is that their neurotoxicity is mediated *via* excitatory amino acid receptors whose relationship to mechanisms of neurodegeneration have been described above.

The only compound in this group that has been positively associated with a human neurological disease state is cysteine-S-sulphonate, which is found in body fluids in patients with a genetic disorder associated with a deficiency of sulphite oxidase (Johnson and Wadman 1989). This enzyme catalyses the oxidation of sulphite (formed by desulphuration of cysteine) to sulphate. As a result, there is an increase in the tissue concentration (probably mainly in liver) of sulphite, which reacts non-enzymically with intracellular cysteine to form cysteine-S-sulphonate. In one reported case (Shih et al. 1977) this compound was detected in blood and urine, but not in cerebrospinal fluid. On the other hand, sulphite was detected in all three body fluids. Thiosulphate, formed from the reaction of sulphate and sulphite, was present in urine in greater amount than in controls, and the plasma cysteine concentration was lower than control values.

In an interesting discussion the authors attempt to relate in turn the possible reasons for the neurological defects that were observed and which, in another case where neuropathology was performed (Rosenblum 1968), were clearly related to a diffuse loss of neurons, and other pathological features, in many parts of the brain. Thus, increased sulphite concentrations in cerebrospinal fluid might be associated with the destruction of thiamine, which would impair brain glucose oxidation; deficiencies in plasma cystine (a precursor of proteins and of glutathione) and of sulphate (as a precursor of sulphate esters in myelin and in mucopolysaccharides) would generate further metabolic maladies.

It is clear from these data that, despite the effectiveness of cysteine-S-sulphonate as an excitotoxin (Olney et al. 1975), the absence of the compound from cerebrospinal fluid appears to mitigate against it being involved directly in the pathophysiology of this condition. Thus in this single case it seems unlikely that an excitotoxic mechanism accounts for the neurodegeneration that occurs in this disorder. In the light of recent work it is unfortunate that no determination of plasma sulphate was made in this study. From the accompanying data it would be anticipated that this value would have been lower than that in control plasma.

Cysteine and Cystine

Cysteine is an effective neurotoxin when administered orally in high doses (3 mg/g) to 8–10-day-old mice (Olney and Ho 1970). In short-term (2–3 h) experiments the pattern of the neuropathology induced by this compound strikingly resembles that seen following the administration of glutamate. The cellular changes also appear identical. When lower doses of cysteine are given orally or subcutaneously a more generalised necrosis is seen after 24 h that spreads into the cerebral cortex, hippocampus, amygdala and thalamus (Olney et al. 1972). This type of neuropathology is not encountered with glutamate or other excitotoxins. A similar, but not identical, double pattern of neurotoxicity was observed after 4-day-old rats were injected subcutaneously with a single dose of cysteine (Lund Karlson et al. 1981).

Cysteine is not itself neuroexcitatory to cat spinal neurons (see Olney 1978) and the possibility that some aspects of its neurotoxicity are mediated *via* its metabolites (e.g. cysteine sulphinate and cysteate; see Griffiths 1987) must be considered (Olney et al. 1972). Since both cysteine sulphinate and cysteate are themselves excitotoxins that act at NMDA and non-NMDA receptors this mechanism is consonant with the observed glutamate-like neuropathology. However, oxidation of cysteine to cystine, which is the predominant extracellular form, is also likely to occur. The time interval (6–48 h) from administration of the amino acid to perfusion-fixing the animals would have been sufficient to permit a significant amount of metabolism to occur.

The more diffuse neuropathology resulting from the administration of cysteine is not easily explained by an excitatory amino acid receptor-mediated mechanism. It has been suggested (Lund Karlson et al. 1981) that, following administration of the amino acid, an analogue was formed that inhibited glutamate transport and that the toxic action of the extracellular glutamate that resulted was responsible for the observed neurotoxicity. A variation of this interesting idea is discussed below.

The neurotoxicity of cysteine was reinvestigated recently (Olney et al. 1990). Rat pups were administered the compound subcutaneously and some were protected by prior administration of MK-801. This compound (1 g/kg) fully protected from cysteine neurotoxicity the rats so treated. It is not clear from these experiments whether the antagonist was active against cysteine or a metabolic product because of the length of time that elapsed (6 h) from injecting the compound to perfusion-fixing the brain. Use of the chick embryo retinal preparation permitted a more complete pharmacological analysis of the mechanism of action of cysteine, using a time-scale (30 min) that reduced the possibility of metabolism. These data suggest that cysteine activates NMDA receptors at lower concentrations (2 mM) and non-NMDA receptors at higher concentrations. The very great surprise was that the neurotoxicity of cysteine was bicarbonate-dependent in this preparation. Moreover, electrophysiological recording in cultured rat hippocampal neurons revealed that the inward current was increased by about three times in the presence of bicarbonate (20 mM) and this effect was blocked completely by AP5.

These experiments demonstrate that cysteine does act as a neuroexcitatory amino acid, principally at NMDA receptors. By analogy with the neurotoxic amino acid from *Cycas circinalis* (Nunn and O'Brien 1989), cysteine reacts with bicarbonate to form a carbamate, which is readily detectable by ^{1}H NMR spectroscopy (Nunn and O'Brien, unpublished data). This adduct lacks the charge configuration that is typical of neuroexcitatory amino acids (Watkins and Olverman 1987). It is not clear from these data whether carbamate formation alone offers an explanation of the neurotoxicity of cysteine.

Cystine and Glutamate Transport

Concentration on excitatory amino acid neuropharmacology has obscured to some extent significant recent progress in cystine transport. It is well established that cystine (the oxidised form of cysteine) is transported in the proximal tubule of the kidney by a transport system that is shared with arginine, lysine and ornithine (Segal et al. 1977). However, in some cell types (hepatocytes, hepatoma cell line HTC and human fibroblasts) good evidence has been obtained of a system for cystine transport that is shared with aspartate and glutamate (Makowske and Christensen, 1982). Cystine is transported effectively as a monoaminodicarboxylate by this system, i.e. both carboxyl groups are ionised but only one of the two amino groups carries a positive charge. A characteristic of this transport system is that (a) it allows competition between cystine and glutamate for entry into the cell; and (b) it exchanges extracellular cystine for intracellular glutamate (Murphy et al. 1990; see also Anon 1989).

This transport system has been promulgated recently as a mechanism by which glutamate might accelerate neuronal death (Anon 1990). Extracellular glutamate inhibits the uptake of extracellular cystine which, following intracellular reduction, would be used as a source of intracellular cysteine for glutathione synthesis. In addition to the cell types already mentioned this system is present in primary cultures of rat cortical neurons (Murphy et al. 1990). These cells degenerate when exposed for 24–72 h to culture media containing reduced concentrations of cystine, to increased concentrations of glutamate or to a number of compounds that are well known for their excitotoxic properties. These compounds i.e. quisqualate, homocysteate, β-ODAP and ibotenate, inhibited the glutamate/cystine transport system. The neurotoxic properties of glutamate, quisqualate and homocysteate were inversely proportional to the cystine concentration of the culture medium, which suggested that the inhibitors competed with cystine for transport. Prolonged exposure of the cells to glutamate or homocysteate depleted neuronal glutathione.

These results are intriguing and, as the majority of brain glutathione is located in glia (Slivka et al. 1987), an attack by inhibitors on the transport system that is linked directly to the neuronal supply of glutathione would be expected to have devastating effects upon these cells. This expection is fully realised by the results obtained in this interesting work.

Cysteine, Sulphate and Motor Neuron Disease

A fascinating proposal has been made (Heafield et al. 1990) concerning the occurrence of deranged cysteine metabolism, together with a deficiency of plasma sulphate, in motor neuron disease (although the evidence on which this notion was based has been challenged (Perry et al. 1991; see Chapter 7). The latter feature was also apparent in patients with Parkinson's disease and in Alzheimer's disease. The metabolic relationship between cysteine and sulphate is established by cysteine dioxygenase, followed by transamination of the product (cysteine sulphinate) to yield ultimately pyruvate and sulphate.

Cysteine + O_2 = cysteine sulphinate + H^+
Cysteine sulphinate + 2-oxoglutarate = β-sulphinylpyruvate + glutamate
β-Sulphinylpuruvate + H_2O = pyruvate + SO_3^{2-}
SO_3^- + 2e = SO_4^{2-}

Cooper (1983) has discussed the relative contributions of this pathway for the catabolism of cysteine and that which occurs via 3-mercaptopyruvate in mitochondria.

If a deficiency of cysteine dioxygenase is causally associated with motor neuron disease it would be expected that, with cysteine metabolism inhibited, metabolites such as the excitotoxic homocysteate (derived from methionine) might accumulate in body tissues (see Griffiths 1987). Alternatively, the low blood sulphate seen in motor neuron disease, but also in Parkinson's disease and Alzheimer's disease (Heathfield et al. 1990), might be the more metabolically important parameter.

There is some evidence that low concentrations of plasma sulphate are associated with neurological disease in man. For example in Africa the consumption of inadequately washed cassava, which contains a cyanogenic glycoside that releases CN after ingestion, reduces the concentration of sulphate in body fluids of people who consume this crop by forming thiocyanate (Cliff et al. 1985). Such populations are especially sensitive to depletion of sulphate because of their poor sulphur intake resulting from reliance on a foodstuff that contains very low levels of sulphur amino acids (FAO 1970). Dietary methionine and cysteine are the major sources of metabolic sulphate.

Another condition where perturbed sulphate metabolism may be present is in patients with sulphite oxidase deficiency. The reaction of sulphite (produced from cysteine sulphinate as described above) with tissue sulphate would form thiosulphate which, like cyanide, would be expected to reduce sulphate availability.

These examples suggest that in neurological diseases of an idiopathic nature and where metabolic disturbances prevail a reduction in plasma sulphate is a common feature. Whether there is a causal relationship between reduced sulphate availability and accelerated nerve cell death demands further investigation.

Exogenous Factors

General Considerations

The idea that motor neuron disease may result from exposure to a number of exogenous factors derived from dietary intake or following environmental exposure, pervades much of the literature on the aetiology of motor neuron disease (for example, see Tandan and Bradley 1985). And yet the specific instances that can be cited where there is convincing evidence of the involvement of toxic agents in neurological diseases are few.

The neurotoxic effect of a group of non-protein amino acids contained in foodstuffs has provided evidence that such compounds represent potential environmental risks (Meldrum and Garthwaite 1990). However, present evidence is that the populations at risk are localised and no evidence is available that these compounds, or others of similar structure, are aetiological agents of sporadic motor neuron disease (but see Barinaga 1990). It is relevant to this discussion that, in the original assignment of subclasses of glutamate receptors in the central nervous system (Watkins and Evans 1981), two of the three compounds that most specifically activated these receptors (kainate and quisqualate) were obtained from plants.

The justification for studying the effects of neurotoxins of the types described is that, if the mechanisms by which these substances act were known, it should lead to uncovering the lesions that characterise sporadic motor neuron disease. The mechanisms of action of food-borne amino acids are likely to be diverse. They are considered in the following sections under the names of their plant vectors. By chance, the three compounds discussed appear to act at three different excitatory amino acid receptor sites.

Lathyrus sativus Seed

It is well established that consumption of *Lathyrus sativus* seed (chickling pea; grass pea) for a period of weeks or months causes a chronic neurological disease in man (Sarma and Padmanaban 1969; Hugon et al. 1988). This disorder presents clinically as an upper motor neuron dysfunction. The component of the seed that is thought to be the principal causative agent is β-N-oxalyl-L-α,β-diaminopropionic acid (β-ODAP; synonym β-oxalylamino-L-alanine, BOAA, Rao et al. 1964; Murti et al. 1964). The relative importance of this and other potential neurotoxins that are contained in the seed has been reviewed (Nunn 1989).

The mode of action of β-ODAP is likely to be complex as the compound affects a number of different systems in experimental brain preparations. The significance of these diverse effects is difficult to evaluate at present. However, the best established mechanism by which neurotoxicity of β-ODAP could be expressed is via excitotoxicity (Olney et al. 1976) and this area is explored in this section.

Shortly after its discovery β-ODAP was shown to be a powerful excitant to cat spinal motor neurons and Betz cells (Watkins et al. 1966). The neuroexcitatory effects of this compound are mediated in the frog hemisected spinal cord preparation via non-NMDA receptors (Pearson and Nunn 1981) and this response was also demonstrated in cultured foetal mouse spinal cord nerons (MacDonald and Morris 1984). In the foetal mouse spinal cord–dorsal root ganglia preparation β-ODAP rapidly causes typical excitotoxic damage to neurons (Nunn et al. 1987). The natural D-isomer is not effective in this respect, which is in accordance with its non-excitatory nature in the frog spinal cord preparation (S. Pearson and P.B. Nunn, unpublished data). Binding studies using rat brain preparations confirmed these data and provided additional evidence that the compound binds to AMPA receptors at low concentrations and to both AMPA and kainate receptors at higher concentrations (Bridges et al. 1988; Ross et al. 1989). These pharmacological characteristics of β-ODAP are also demonstrable in the chick embryo retina preparation (Zeevalk and Nicklas 1989).

Direct injection of the compound into the cerebrospinal fluid of monkeys (Mani et al. 1971) and rats (Chase et al. 1985) causes extensive neuropathological damage. However, it is disappointing that, despite considerable effort, the macaque has not proved to be a suitable model in which to induce chronic irreversible neurological dysfunction by feeding *L. sativus* and β-ODAP. This model showed signs of reversible changes that may represent early stages of neurolathyrism (Spencer et al. 1986; Hugon et al. 1988).

This evidence suggests that, if β-ODAP expresses its neurotoxicity by a excitotoxic mechanism involving non-NMDA receptors, the entry of Na^+ and Cl^- and water into neurons, accompanied by an influx of extracellular Ca^{2+} following the activation of voltage-dependent Ca^{2+} channels, would provide a satisfying mechanism by which the compound's neurotoxicity would be expressed. An additional factor may be that, if kainate receptors are located pre-synaptically, activation of these receptors by higher concentrations of β-ODAP might occur. This reaction would be expected to release glutamate from pre-synaptic stores. It may be significant in this context that β-ODAP stimulates the release of glutamate by rat brain synaptosomes (Gannon & Terrian 1989). The apparent specificity of β-ODAP towards human upper motor neurons remains an enigma.

Cycas circinalis Seed

Whiting (1963) suggested that the custom of the native Chamorro people of Guam using flour derived from *Cycas circinalis* seed for culinary purposes might be related to the high incidence of ALS-PD seen on the island. Following a purposeful search for neurotoxic amino acids in the seed Vega and Bell (1967) isolated L-α-amino-β-methylaminopropionic acid (MeDAP; synonym β-methylamino-L-alanine, BMAA). This amino acid proved to be neurotoxic to day-old chicks, rats and mice (Bell et al. 1967), an observation that stimulated speculation as to its aetiological role in Guamanian ALS-PD.

Attempts to induce chronic neurotoxicity in rats with MeDAP were not successful (Polsky et al. 1972), but when macaques were fed the compound for several weeks they developed a motor dysfunction and some parkinsonian features. Histological damage to upper (mainly) and to lower motor neurons was apparent in treated animals (Spencer et al. 1987). Interestingly the acute convulsant effects of MeDAP in rats are associated with cerebellar lesions (Seawright et al. 1990). No cerebellar lesions were observed in macaques (P.S. Spencer, personal communication).

The compound was stereospecifically neuronotoxic to cultured foetal mouse spinal cord, with only the natural L-form being effective (Nunn et al. 1987), and the neuronal vacuolation that occurred was a typical excitotoxic response. This effect was antagonised by MK-801, suggesting that it was mediated by NMDA receptors (Spencer et al. 1987). The vacuolation was accompanied by the appearance of dark, shrunken, cells and both responses were concentration-dependent and antagonised by AP7 (Ross et al. 1987). The mediation of NMDA receptors in the neurotoxicity of MeDAP was confirmed in the embryonic chick retina preparation (Zeevalk and Nicklas 1989).

In cultured foetal mouse cortical neurons a somewhat different pathopharmacology was evident (Weiss and Choi 1988). Although the vacuolation that occurred at relatively high concentrations of the amino acid (up to 3 mM) was sensitive to NMDA receptor antagonists, a subpopulation of neurons containing the enzyme NADPH diaphorase were preferentially damaged at lower concentrations (100 μM). This effect was more effectively antagonised by kynurenate (an antagonist at non-NMDA receptors) than by AP5 suggesting that, in this preparation, MeDAP causes specific cell damage by activating non-NMDA receptors by a mechanism that is distinct from that responsible for neuronal vacuolation.

Although MeDAP activates excitatory amino acid receptors in these experiments it lacks the monoaminodicarboxylate structure that characterises agonists to these receptors (Watkins et al. 1990). The discovery that the activity of MeDAP was dependent upon the presence of bicarbonate (in excess of 6 mM) offered an explanation for this observation (Weiss and Choi 1988). Neuronotoxicity induced by NMDA, β-ODAP and quisqualate was not bicarbonate-dependent under the same conditions.

Two explanations of bicarbonate-dependence have been proposed. The α-amino group of L-MeDAP reacts with bicarbonate to yield a α-carbamate that is a structural analogue of NMDA (Nunn and O'Brien 1989). This structure results from an apparent inversion of configuration about the α-carbon atom following carbamate formation. Additionally, a β-methylcarbamate is formed that is an analogue of glutamate (Myers and Nelson 1990). The relative importance of these two adducts as neurotoxins has not been established.

It may be possible for MeDAP to act at NMDA receptors in an indirect fashion. These receptors are deactivated by binding Zn^{2+} and the ion blocks the neurotoxicity of NMDA in a concentration-dependent manner (Peters et al. 1987). MeDAP is a highly effective chelator of Zn^{2+} (Nunn et al. 1989) and, if it were to remove Zn^{2+} from NMDA receptors, the sensitivity of the receptors to endogenous excitatory amino acids would be expected to increase.

At this juncture it is impossible to evaluate the extent to which MeDAP may be involved in the aetiology of Guamanian ALS-PD and current debate is concerned with whether sufficient of the compound exists in *Cycas circinalis* flour to justify its serious consideration as a food-borne neurotoxin (Duncan et al. 1990). However, the importance of MeDAP also lies in the mechanisms by which it precipitates neuronal death and these remain to be elucidated. The possibility that compounds of similar structure to MeDAP that also are activated by carbamate formation has prompted a new search for additional neurotoxins of this type. The observations of Weiss et al. (1989) that the neurotoxicity of other dibasic amino acids from plants, i.e. α,β-diaminopropionic acid and α,β-diaminobutyric acid are also bicarbonate-dependent, are pertinent to this discussion. So too is the report that cysteine neurotoxicity is mediated by bicarbonate (Olney et al. 1990).

Nitschia pungens

In 1987 an outbreak occurred in Canada of an acute illness in which unusual neurological signs and symptoms occurred in over 100 people following the ingestion of cultivated mussels from Prince Edward Island, Vancouver (Perl et al. 1990). The agent responsible was rapidly identified as domoic acid (Wright et al. 1989), a powerful neuroexcitatory amino acid (Biscoe et al. 1975) that had first been isolated from *Chondria* spp (see Takemoto 1978). The compound binds powerfully to kainate receptors (Young and Fagg 1990). The alga has been used as an antihelminth in Japan; this activity of the plant resided in its content of domoic acid (Takemoto 1978).

The neurological defects observed in this outbreak were varied (Teitelbaum et al. 1990) and included loss of short-term memory, headache, seizures, spastic hemiparesis, coma and death in some cases. A pure motor or sensory motor deficit was a feature in some patients, but it was not associated with loss of motor neurons either in the brain stem or spinal cord in those affected by the toxin who subsequently succumbed. This result was surprising as spinal motor neurons contain receptors (probably kainate receptors) that are stimulated by domoate (Biscoe et al. 1975). However, cells in the hippocampus showed significant degradation and this effect was reproducible in rats. Such animals also suffer a longlasting anterograde amnesia for spatial information (Sutherland et al. 1990).

Thus the main features of this accidental poisoning appear to mimic pathological features that are seen in laboratory animals that have been acutely poisoned with domoate. The extent to which this episode would correspond with the effects of lower amounts of the compound being administered for a prolonged time period (which would mimic more closely the situation experienced with *L. sativus*) cannot be evaluated. Preparations used as an antihelminth contain about 20 mg of domoate (see Perl et al. (1990) for discussion) whereas estimates of the amounts consumed by those affected by consuming contaminated muscles ranged from 60 to 290 mg.

Non-NMDA Receptors on Glia

Although the neurotoxicity of excitatory amino acids has been interpreted here in terms of their actions at receptor sites on neuronal bodies and dendrites recent evidence has revealed the presence of quisqualate (AMPA) and kainate receptors, but not NMDA receptors, on Type-2 astrocytes (Usowicz et al. 1989). If these receptors are also present on glia throughout the central nervous system, it seems likely that a re-evaluation of the neurotoxicity generated by stimulation of non-NMDA receptors will be necessary.

Summary

Strong evidence has been presented that a number of endogenous and exogenous compounds kill nerve cells in a variety of in vitro and in vivo preparations following specific receptor-mediated activity. Because of experimental difficulties most work has concentrated upon those systems that are sensitive to acute neurotoxins. Indeed, the "excitotoxic theory of cell death" is based upon relatively short-term effects of excitatory amino acids, which cause nerve cell death by overwhelming the homeostatic mechanisms that obtain after normal excitatory responses in neurons. That such a mechanism can be a cause of human neurodegeneration was seen in a recent instance of acute neurotoxicity, resulting from the incorporation of domoic acid into the human food chain, in which the effects seen in man were, to a large extent, predictable on the basis of acute experiments in experimental animals and in tissues derived from them. On this basis the "excitotoxic theory" (including the involvement of Ca^{2+}) offers a satisfying mechanism of action of the acute effects of domoic acid.

A more complex task confronts the application of these concepts to chronic nerve cell death. Here application of the excitotoxic mechanism becomes difficult to sustain purely in terms of ionotrophic reactions. However, some compounds may act by receptor-mediated reactions that regulate nerve cell metabolism via metabotrophic receptors. Similarly, it is now clear that the opportunity exists for neurotoxins to act at sites other than receptors on nerve cell bodies and dendrites; the modulation of non-NMDA receptors on astrocytes by exogenous or endogenous compounds might allow the complex interactions between neurons and glia to be modified. To date this aspect seems not to have been investigated for any toxin that would be expected to be active at these receptor subtypes. Inhibition of neuronal cysteine transport, which would deplete neuronal glutathione and result in neurons being subjected to oxidative stress, is caused by some amino acids that are well known for their neuroexcitatory behaviour; a deficiency of sulphate, which perhaps may be required for some specific neuronal process, appears to be both spontaneous and toxic in a number of chronic neurological diseases.

Although there is evidence of the involvement of excitatory amino acids in chronic neurological diseases, including motor neuron disease, it is still unclear as to which of the mechanisms discussed plays a primary role in the demise of central nervous system neurons. It seems probable that the "excitotoxic" explanation of cell death is a simplification that is applicable in a pure form only to some acute neurotoxins and it is now time to evaluate the relative importance of this mechanism with respect to other mechanisms as the prime aetiological feature of the nerve cell death that occurs in chronic neurological disease in man.

References

Agid Y, Blin J (1987) Nerve cell death in degenerative diseases of the central nervous system: clinical aspects. In: Bock G, O'Conner M (eds) Selective neuronal death. Ciba Foundation Symposium 126. Wiley, New York, pp 3–29

Anon (1989) Amino acid transport and glutathione metabolism. Nutr Rev 47:26–28

Anon (1990) Antagonism to neuronal cysteine uptake: a factor in neuronal damage by glutamate. Nutr Rev 48:440–442

Barinaga M (1990) Amino acids: how much excitement is too much? Science 247:20–22

Bell EA, Vega A, Nunn PB (1967) A neurotoxic amino acid in seed of *Cycas circinalis*. In: Whiting MG (ed) Toxicity of Cycads: Implications for neurodegenerative diseases and cancer. Fifth Conference Third World Medical Research Foundation, New York, 1988, pp XI-1–XI-4

Berridge MJ, Irvine RF (1989) Inositol phosphates and cell signalling. Nature 341:197–205

Biscoe TJ, Evans RH, Headley PM, Martin M, Watkins JC (1975) Domoic and quisqualic acids are potent amino acid excitants of frog and rat spinal neurones. Nature 247:166–167

Bridges RJ, Kadri MM, Monaghan DT, Nunn PB, Watkins JC, Cotman CW (1988) Inhibition of ^{3}H-AMPA binding by the excitotoxin β-N-oxalyl-L-α,β-diaminopropionic acid. Eur J Pharmacol 145:357–359

Bridges RJ, Stevens DR, Kahle JS, Nunn PB, Kadri MM, Cotman CW (1989) Structure–function studies of N-oxalyl-diaminopropionic acids and excitatory amino acid receptors: evidence that β-ODAP is a selective non-NMDA agonist. J Neurosci 9:2073–2079

Burger PM, Mehl E, Cameron PL et al. (1989) Synaptic vesicles immunoisolated from rat cerebral cortex contain high levels of glutamate. Neuron 3:715–720

Calne DB, Eisen A, McGeer E, Spencer P (1986) Alzheimer's disease, Parkinson's disease, and motoneuron disease: abiotropic interaction between ageing and environment? Lancet ii:1067–1070

Chase RA, Pearson S, Nunn PB, Lantos PL (1985) Comparative toxicities of α- and β-N-oxalyl-L-diaminopropionic acids to rat spinal cord. Neurosci Lett 55:89–94

Choi DW (1985) Glutamate neurotoxicity in cortical cell culture is calcium dependent. Neurosci Lett 58:293–297

Choi DW (1987) Ionic dependence of glutamate neurotoxicity. J Neurosci 7:369–379

Choi DW, Rothman SM (1990) The role of glutamate neurotoxicity in hypoxic-ischemic neuronal death. Ann Rev Neurosci 13 171–82

Cliff J, Lundqvist P, Martensson J, Rosling H, Sorbo B (1985) Association of high cyanide and low sulphur intake in cassava-induced spastic paraparesis. Lancet ii:1211–1213

Cooper AJL (1983) Biochemistry of sulphur-containing amino acids. Ann Rev Biochem 52:187–222

Do KO, Herrling PL, Streit P, Turski WA, Cuenod M (1986) In vitro release and electrophysiological effects of in situ homocysteic acid, an endogenous N-methyl-D-aspartic acid agonist, in the mammalian striatum. J Neurosci 6:2236–2244

Duncan MW, Steele JC, Kopin IJ, Sanford PM (1990) 2-amino-3-(methylamino)-propanoic acid (BMAA) in cycad flour: an unlikely cause of amyotrophic lateral sclerosis and parkinsonism-dementia of Guam. Neurology 40:767–772

Eberhard DA, Holz RW 1988 Intracellular Ca^{2+} activates phospholipase C. Trends Neurosci 11:517–520

FAO (1970) Amino-acid content of foods and biological data on proteins. Food and Agricultural Organization of the United Nations, Rome

Foster AC, Fagg GE (1984) Acidic amino acid-binding sites in mammalian neuronal membranes: their characteristics and relationships to synaptic receptors. Brain Res Rev 7:103–164

Gannon RL, Terrian DM (1989) BOAA selectively enhances L-glutamate release from guinea pig hippocampal mossy fiber synaptosomes. Neurosci Lett 107:289–294

Garruto RM, Yase Y (1986) Neurodegenerative disorders of the Western Pacific: the search for mechanisms of pathogenesis. Trends Neurosci 9:368–374

Griffiths OW (1987) Mammalian sulfur amino acid metabolism: an overview. In: Jakoby WB, Griffiths OW (eds) Methods in enzymology, vol 143. Academic Press, New York, pp 366–376

Hawkes CH, Cavanagh JB, Fox AJ (1989) Motoneuron disease: a disorder secondary to solvent exposure? Lancet i:73–76

Headley PM, Grillner S (1990) Excitatory amino acids and synaptic transmission: the evidence for a physiological function. Trends Pharmacol Sci 11:205–211

Heathfield MT, Fearn S, Steventon GB, Waring RH, Williams AC, Sturman SG (1990) Plasma cysteine and sulphate levels in patients with motor neurone, Parkinson's and Alzheimer's disease. Neurosci Lett 110:216–220

Hugon J, Ludolph A, Roy DN, Schaumburg HH, Spencer PS (1988) Studies on the etiology and pathogenesis of motor neurone diseases. II. Clinical and electrophysiological features of pyramidal dysfunction in macaques fed *Lathyrus sativus* and IDPN. Neurology 38:435–442

Johnson BL (1987) Prevention of neurotoxic illness in working populations. Wiley, Chichester

Johnson JL, Wadman SK (1989) Molybdenum cofactor deficiency. In: Scriver CR, Beaudet AL, Sly WS, Valle D (eds) The metabolic basis of inherited disease. McGraw-Hill, New York

Krnjevic K 1965 Actions of drugs on single neurones in the cerebral cortex. Br Med Bull 21:10–14

Lane RJ, Bandopadhyay R, de Belleroche J (1993) Abnormal glycine metabolism in motor neurone disease: studies on plasma and cerebrospinal fluid. J R Soc Med 86(9):501–505

Lund Karlson R, Grofova I, Malthe-Sorenssen D, Fonnun F (1981) Morphological changes in rat brain induced by L-cysteine injection in newborn animals. Brain Res 208:167–180

MacDonald JF, Morris ME (1984) Lathyrus excitotoxin: mechanism of neuronal excitation by L-2-oxalylamino-3-amino- and L-3-oxalylamino-2-aminopropionic acid. Exp Brain Res 57:158–166

Makowske M, Christensen HN (1982) Contrasts in transport systems for anionic amino acids in hepatocytes and a hepatoma cell line HTC. J Biol Chem 257:5663–5670

Mani KS, Sriramachari S, Rao SLN, Sarma PS (1971) Experimental neurolathyrism in monkeys. Ind J Med Res 59:880–885

Masui Y, Mozai T, Kakehi K (1985) Functional and morphometric study of the liver in motor neurons disease. J Neurol 232:15–19

Mayer, ML, Miller RJ (1990) Excitatory amino acid receptors, second messengers and regulation of intracellular Ca^{2+} in mammalian neurones. Trends Pharmacol Sci 11:254–260

McGilvery RW, Goldstein G (1983) Biochemistry: a functional approach, 3rd edn. Saunders, Philadelpia, pp 578–589

Meldrum B, Garthwaite J (1990) Excitatory amino acid neurotoxicity and neurodegenerative disease. Trends Pharmacol Sci 11:379–387

Monaghan DT, Bridges RJ, Cotman CW (1989) The excitatory amino acid receptors: their classes, pharmacology and distinct properties in the function of the central nervous system. Ann Rev Pharmacol Toxicol 29:365–402

Morselli PL, Garatini S (1970) Monosodium glutamate and the Chinese restaurant syndrome. Nature 227:611–612

Murphy TH, Schnaar RL, Coyle JT (1990) Immature cortical neurones are uniquely sensitive to glutamate toxicity by inhibition of cysteine uptake. FASEB J 4:1624–1633

Murti VVS, Seshadri TR, Venkitsubramanian TA (1964) Neurotoxic compounds of the seeds of *Lathyrus sativus*. Phytochemistry 3:73–78

Myers TG, Nelson SD (1990) Neuroactive carbamate adducts of β-N-methylamino-L-alanine and ethylenediamine. J Biol Chem 265:10193–10195

Nunn PB (1989) *Lathyrus sativus* toxins: identification and possible mechanisms. In: Spencer PS (ed) The grass pea: threat and promise. Third World Medical Research Foundation, New York, pp 89–96

Nunn PB, O'Brien P (1989) The interaction of L-methylaminoalanine with bicarbonate: ^{1}H NMR study. FEBS Lett 251:31–35

Nunn PB, O'Brien P, Pettit LD, Pyburn SI (1989) Complexes of zinc, copper and nickel with the non-protein amino acid α-amino-β-methylaminopropionic acid: a naturally occurring neurotroxin. J Inorg Biochem 37:175–183

Nunn PB, Seelig M, Zagoren JC, Spencer PS (1987) Stereospecific acute neuronotoxicity of "uncommon" plant amino acids linked to human motor system diseases. Brain Res 410:375–379

Olney JW (1969) Monosodium glutamate. Science 165:1028–1029

Olney JW (1971) Glutamate-induced neuronal necrosis in the infant mouse hypothalamus. An electron microscopic study. J Neuropathol Exp Neurol 30:75–90

Olney JW (1978) Neurotoxicity of excitatory amino acids. In: McGeer EG, Olney JW, McGeer PL (eds), Kainic acid as a tool in neurobiology. Raven Press, New York, pp 95–121

Olney JW, Ho O-L (1970) Brain damage in infant mice following oral intake of glutamate, aspartate or cysteine. Nature (Lond) 227:609–610

Olney JW, Ho OL, Rhee V (1971) Cytotoxic effects of acidic and sulphur-containing amino acids in the infant mouse central nervous system. Exp Brain Res 14:61–76

Olney JW, Ho LH, Rhee V, Schainker B (1972) Cysteine-induced brain damage in infant and fetal rodents. Brain Res 45:309–313

Olney JW, Misra CH, de Gubareff T (1975) Cysteine-S-sulphate: brain damaging metabolite in sulphite oxidase deficiency. J Neuropathol Exp Neurol 34:167

Olney JW, Misra CH, Rhee V (1976) Brain and retinal damage from the lathyrus excitotoxin β-N-oxalyl-L-α,β-diaminopropionic acid (ODAP). Nature 264:659–661

Olney JW, Price MT, Samson L, Labruyere J (1986) The role of specific ions in glutamate neurotoxicity. Neurosci Lett 65:65–71

Olney JW, Sharpe LG, Feigin RD (1972) Glutamate-induced brain damage in infant primates. J Neuropathol Exp Neurol 31:464–488

Olney JW, Zorumski C, Price MT, Labruyere J (1990) L-cysteine, a bicarbonate-sensitive endogenous excitotoxin. Science 248:596–599

Oppenheim RW (1987) Muscle activity and motor neuron death in the spinal cord. In: Bock G, O'Connor M (eds) Selective neuronal death. Ciba Foundation Symposium 126. Wiley, Chichester, pp 96–112

Pearson S, Nunn PB (1981) The neurolathyrogen, β-N-oxalyl-L-α,β-diaminopropionic acid, is a potent agonist at "glutamate preferring" receptors in frog spinal cord. Brain Res 206:178–182

Perl TM, Bédard L, Kosatsky T, Hockin JC, Todd ECD, Remis RS (1990) An outbreak of toxic encephalopathy caused by eating mussels contaminated with domoic acid. N Engl J Med 322:1775–1780

Perry TL, Hansen S, Jones K (1987) Brain glutamate deficiency in amyotrophic lateral sclerosis. Neurology 37:1845–1848

Perry TL, Krieger C, Hansen S, Eisen A (1990) Amyotrophic lateral sclerosis: amino acid levels in plasma and cerebrospinal fluid. Ann Neurol 28:12–17

Perry TL, Krieger C, Hansen S, Tabatabaei A (1991) Amyotrophic lateral sclerosis: fasting plasma cysteine and inorganic sulfate are normal as are brain contents of cysteine. Ann Neurol 41:487–490

Peters SJ, Koh J, Choi DW (1987) Zinc selectively blocks the action of N-methyl-D-aspartate on cortical neurones. Science 236:589–593

Plaitakis A (1990) Glutamate dysfunction and selective motor neuron degeneration in amyotrophic lateral sclerosis: a hypothesis. Ann Neurol 28:3–8

Plaitakis A, Berl S, Yahr MD (1982) Abnormal glutamate metabolism in an adult-onset degenerative neurological disorder. Science 219:193–196

Plaitakis A, Caroscio JT (1987) Abnormal glutamate metabolism in amyotrophic lateral sclerosis. Ann Neurol 22:575–579

Plaitakis A, Constantakakis E, Smith J (1988) The neuroexcitotoxic amino acids glutamate and aspartate are altered in the spinal cord and brain in amyotrophic lateral sclerosis. Ann Neurol 24:446–449

Polsky FI, Nunn PB, Bell EA (1972) Distribution and toxicity of α-amino-β-methylaminopropionic acid. Fed Proc 31:1473–1475

Rao SLN, Adiga PR, Sarma PS (1964) Isolation and characterization of β-N-oxalyl-L-α,β-diaminopropionic acid: a neurotoxin from the seeds of *Lathyrus sativus*. Biochemistry 3:432–436

Rosenblum WI (1968) Neuropathologic changes in a case of sulphite oxidase deficiency. Neurology 18:1187–1196

Ross SM, Seelig M, Spencer PS (1987) Specific antagonism of excitotoxic action of "uncommon" amino acids assayed in organotypic mouse cortical cultures. Brain Res 425:120–127

Ross SM, Roy DN, Spencer PS (1989) β-N-oxalylamino-L-alanine action on glutamate receptors. J Neurochem 53:710–715

Rothman SM, Olney JW (1987) Excitotoxicity and the NMDA receptor. Trends Neurosci 10:299–302

Rothstein JD, Tsai G, Kuncl RW (1990) Abnormal excitatory amino acid metabolism in amyotrophic lateral sclerosis. Ann Neurol 28:18–25

Sarma PS, Padmanaban G (1969) Lathyrogens. In: Leiner IE (ed) Toxic constituents of plant foodstuffs. Academic Press, New York

Schaumburg HH, Byck R, Gerstl R, Mashman JH (1969) Monosodium glutamate: its pharmacology and role in the Chinese restaurant syndrome. Science 163:826–828

Seawright AA, Brown AW, Nolan CC, Cavanagh JB (1990) Selective degeneration of cerebellar cortical neurones caused by the cycad neurotoxin, L-β-methylaminoalanine (L-BMAA), in rats. Neuropathol Appl Neurobiol 16:153–169

Segal S, McNamara PD, Pepe LM (1977) Transport interaction of cystine and dibasic amino acids in renal brush border vesicles. Science 197:169–171

Shih VE, Abroms IF, Johnson JL et al. (1977) Sulphite oxidase deficiency: biochemical and clinical investigations of a hereditary metabolic disorder in sulphur metabolism. N Engl J Med 297:1022–1028

Sladeczek F, Recasesn M, Bockaert J (1988) A new mechanism for glutamate receptor action: phosphoinositide hydrolysis. Trends Neurosci 11:545–549

Slivka A, Mytilineou C, Cohen G (1987) Histochemical evaluation of glutathione in brain. Brain Res 409:275–284

Spencer PS, Roy DN, Ludolph AC, Hugon J, Dwivedi MP (1986) Lathyrism: evidence for the role of the neuroexcitatory amino acid BOAA. Lancet ii:1066–1067

Spencer PS, Nunn PB, Hugon J et al. (1987) Linkage of Guam amyotrophic lateral sclerosis-parkinsonism-dementia to a plant excitant neurotoxin. Science 237:517–522

Stone TW, Connick JH, Winn P, Hastings MH, English M (1987) Endogenous excitotoxic agents. In: Bock G, O'Connor M (eds) Selective neuronal death. Ciba Foundation Symposium 126. Wiley, Chichester, pp 204–220

Sutherland RJ, Hoesing JM, Whishaw IQ (1990) Domoic acid, an environmental toxin, produces hippocampal damage and severe memory impairment. Neurosci Lett 120:221–223

Takemoto T (1978) Isolation and structural identification of naturally occurring excitatory amino acids. In: McGeer MG et al. (eds) Kainic acid as a tool in neurobiology. Raven Press, New York, pp 1–15

Tandan R, Bradley WG (1985) Amyotrophic lateral sclerosis. Part 2. Etiopathogenesis. Ann Neurol 18:419–431

Teitelbaum JS, Zatorre RJ, Carpenter S, Gendron D, Evans AC, Gjedde A, Cashman NR (1990) Neurological sequelae of domoic acid intoxication due to the ingestion of contaminated mussels. N Engl J Med 322:1781–1787

Usowicz MM, Gallo V, Cull-Candy SG (1989) Multiple conductance channels in type-2 cerebellar astrocytes activated by excitatory amino acids. Nature 339:380–383

Vega A, Bell EA (1967) α-amino-β-methylaminopropionic acid, a new amino acid from seeds of *Cycas circinalis*. Phytochemistry 6:759–762

Watkins JC, Curtis DR, Biscoe TJ (1966) Central effects of β-N-oxalyl-α,β-diaminopropionic acid and other *Lathyrus* factors. Nature 211:637

Watkins JC, Evans RH (1981) Excitatory amino acid transmitters. Ann Rev Pharmacol Toxicol 21:165–204

Watkins JC, Krogsgaard-Larsen P, Honore T (1990) Structure-activity relationships in the development of excitatory amino acid receptor agonists and competitive antagonists. Trends Pharmacol Sci 11:25–33

Watkins JC, Olverman HJ (1987) Agonists and antagonists for excitatory amino acid receptors. Trends Neurosci 10:265–272

Weiss JH, Choi DW (1988) β-N-methylamino-L-alanine neurotoxicity: requirement for bicarbonate as a cofactor. Science 241:973–975

Weiss JH, Christine CW, Choi DW (1989) Bicarbonate dependence of glutamate receptor activation by β-N-methylamino-L-alanine: channel recording and study with related compounds. Neuron 3:321–326

Whiting MG (1963) Toxicity of Cycads. Econ Botany 17:271–302

Wright JLC, Boyd RK, De Freitas ASW et al. (1989) Identification of domoic acid, a neuroexcitatory amino acid, in toxic mussels from eastern Prince Edward Island. Can J Chem 67:481–490

Yoshida T, Kikuchi G (1973) Major pathways of serine and glycine catabolism in various organs of the rat and cock. J Biochem 73:1013–1022

Young AB (1990) What's the excitement about excitatory amino acids in amyotrophic lateral sclerosis? Ann Neurol 28:9–10

Young AB, Fagg GE (1990) Excitatory amino acids in the brain: membrane binding and receptor autoradiographic approaches. Trends Pharmacol Sci 11:126–133

Zeevalk GD, Nicklas WJ (1989) Acute excitotoxicity in chick retina caused by the unusual amino acids BOAA and BMAA: effects of MK-801 and kynurenate. Neurosci Lett 102:284–290

11 Theories of Causation

S.H. Appel, J.I. Engelhardt, R.G. Smith and E. Stefani

Introduction

Amyotrophic lateral sclerosis (ALS) or motor neuron disease is a devastating human illness of unknown aetiology. Many different disorders have been associated with certain facets of the clinical syndrome of ALS, and thus the concept of different ALS syndromes with diverse aetiologies has been proposed (Rowland 1982). It is true that certain intoxications such as lead, viral infections such as enterovirus, endocrine dysfunction such as hyperparathyroidism, genetic disturbances such as hexosaminidase deficiency, as well as other disturbances, may simulate ALS. However, these cases together comprise only a small percentage of the total number of patients presenting with sporadic amyotrophic lateral sclerosis, and do not appear to offer insights into the aetiology of the sporadic disease. Similarly, although prior trauma has been more frequently encountered in ALS patients than in controls (Kurtzke and Beebe 1980; Gawel et al. 1983), there are no data to explain how such trauma participates in either the aetiology of the disease or the pathogenesis of motoneuron destruction.

Axonal transport has been shown to be altered in ALS (Breuer et al. 1987), but given the many different biochemical abnormalities which can impair axoplasmic transport, it is more likely that impaired axonal flow is a secondary consequence rather than a primary cause of motoneuron damage and destruction.

In the discussion which follows, we present the evidence both for and against various aetiological theories of sporadic ALS. We have excluded a discussion of inherited disorders since 90%–95% of patients with classical ALS lack a family history and are considered to be sporadic cases.

Toxins

Since the turn of the century, attention has been drawn to the possibility of a relationship between ALS and metal intoxication, particularly lead (Wilson 1907). Early reports suggested that ALS patients may have an increased exposure to heavy metals (Currier and Haerer 1968; Campbell et al. 1970), and two case–control studies found that exposure to heavy metals was a significant risk factor (Felmus et al. 1976; Roelofs-Iverson et al. 1984). However, subsequent larger studies indicated that lead was not a significant risk factor (Kondo and Tsubaki 1981; Kurtzke and Beebe 1980). Furthermore other groups have not been able to find an abnormal lead concentration in any fluid, or in erythrocytes from patients with ALS (Stober et al. 1983). There is no question that lead can cause a reversible, ALS-like disease with upper and lower motoneuron involvement in the absence of elevated blood lead levels (Boothby et al. 1974). The problem is whether lead is a risk factor in sporadic

ALS. The answer at the present time appears to be that it is not. The problem in assessing the reliability of many of these studies is the markedly different techniques used to assess lead levels, and the fact that such levels may vary as much as a hundred-fold (Currier and Haerer 1968; Petkau et al. 1974; Conradi et al. 1976, 1980a, 1982a; Manton and Cook 1979; Barry and Mossman 1979; Pierce-Ruhland and Patten 1980). It is also possible that increased lead concentrations in ALS tissue, if true, still might be of secondary rather than of primary relevance (Mandybur and Cooper 1979).

Another metal, namely mercury, can produce ALS-like syndrome (Kantarjian 1961; Barber 1978; Adams et al. 1983), but once again there is no evidence to suggest that mercury is either a risk factor in sporadic ALS or is involved in the pathogenesis of this disorder.

The high soil manganese levels in areas endemic for Western Pacific ALS, together with elevated concentrations of aluminium, and low concentrations of calcium and magnesium, have been considered an important aetiologic factor (Yase 1972). Neutron activation analysis of spinal cords of Western Pacific ALS patients demonstrates an increase in calcium and aluminium, and a significant positive correlation between calcium and manganese levels (Yoshimasu et al. 1980). Furthermore, increased manganese has been identified in spinal cord from patients with ALS (Yase 1972; Miyata et al. 1983). Osteoporosis, fractures, skeletal abnormalities, and ectopic calcifications are all more common in sporadic ALS patients than in controls (Patten and Engel 1982). Secondary hypoparathyroidism may give rise to symptoms and signs clinically similar to motor neuron disease (Patten and Engel 1982). Increased calcium levels have been reported in Western Pacific ALS spinal cords (Gajdusek 1985; Kilness and Hochberg 1977). Despite the many claims of altered metal content, the significance of these abnormalities in ALS patients remains unclear, and therapeutic trials of chelating agents including ethylenedi-amine tetra-acetic acid (EDTA), penicillamine, thioctic acid and dimercaprol are not effective at altering the course of sporadic ALS (Campbell et al. 1970; Currier and Haerer 1968; Conradi et al. 1982a,b; Kurlander and Patten 1979; Campbell 1955). Furthermore, the idea that chronic nutritional deficiency of calcium and magnesium may secondarily lead to the deposition of calcium, aluminium and other elements in neurons is based upon the presumed low concentrations of calcium in drinking water in the Western Pacific. This observation has been challenged by an extensive study of the southern rivers of Guam where high concentrations of calcium characteristic of borderline-hard waters and mean aluminium concentrations similar to those in rivers of southeastern USA have been found (Zolan and Ellis-Neill 1986). Thus, intraneuronal accumulation of aluminium and other metals in ALS may not be a primary manifestation of nutritional deprivation or excess. This topic is reviewed critically by Steele and Williams in Chapter 9.

The toxic theory of ALS has more recently been supported by the investigations of Spencer and his colleagues (1987). These investigators associated the decline in the high incidence of amyotrophic lateral sclerosis among the Chamorro population of the Western Pacific islands of Guam and Rota with the decreased utilisation of the cycad nut as a traditional source of food and medicine following the westernisation of the Chamorro people after World War II. An amino acid, alpha-aminobetamethyl aminoalanine (L-BMAA) has been implicated as a cause of Guamanian ALS. L-BMAA can give rise to cortical motor neuronal dysfunction, parkinsonian features, behavioural anomalies, and chromatolytic and degenerative changes of motor neurons in cerebral cortex and spinal cord following its oral administration to macaque monkeys. Furthermore, L-BMAA can cause pathological changes in the CNS normally associated with excitotoxic amino acids. In tissue culture experiments Spencer et al. (1987), documented that an NMDA receptor antagonist, MK801, reversed the L-BMAA-induced neuronal changes. These data not only provide a potential explanation for Guamanian ALS, but also raise the possibility that endogenous excitatory amino acids such as glutamate may play a role in the pathogenesis of motoneuron destruction in sporadic ALS. The significant question with respect to the Guamanian data is why several decades seem to intervene between the presumed exposure to the cycad nut and the subsequent development of ALS. Furthermore, the Chamorro practice was to extensively wash the cycad nut flour prior to ingestion, and the residual flour would be expected to be substantially free from soluble L-BMAA.

Further suggestive evidence for a role of the excitotoxin glutamate is the finding of significant increase in the plasma glutamate of 22 patients with early stage ALS (Plaitakis and Caroscio 1987). Glutamic acid was also found to be reduced in the brain and the cervical cord of ALS patients (Perry et al. 1987b). Plaitakis et al. (1988a) also studied glutamate levels in the CNS of ALS patients and found the greatest decrease of glutamate in the cervical and lumbar spinal cord. Aspartate levels were also significantly reduced, but only in the spinal cord. A positive correlation was shown between the changes in glutamate and aspartate levels as well as the significant alterations in the glutamate : diglutamine ratio in ALS spinal cords. These investigators acknowledge the fact that the abnormalities may result from neuronal cell loss, but they prefer the hypothesis that a generalised change in glutamate is the primary defect and alterations in glutamate are responsible for the neurodegeneration. At the present time, several therapies such as branched-chain amino acids (Plaitakis et al. 1988b), the use of MK801, and another NMDA receptor antagonist, dextromethorphan, are all being tried in experimental trials to try to halt the inexorable progression of ALS based upon this excitotoxic theory.

Additional circumstantial evidence for the toxic theory of ALS comes from studies of lathyrism, which presents as a pure upper motoneuron syndrome in man. For many years, lathyrism has been linked to the ingestion of the chickling pea (*Lathyrus sativus*), but only recently has the active toxin been identified as another unusual amino acid, beta-N-oxalylamino-L-alanine (L-BOAA) (Spencer et al. 1986), which is a potent agonist of the quisqualate and kainate type of glutamate receptors. The oral administration of the intact chickling pea, an alcoholic extract of the seed, or L-BOAA itself, to macaques could reproduce features of the human disease (Hugon et al. 1988).

Ageing

The concept of "premature ageing" as an explanation of the aetiology of ALS, is based upon certain findings common to both normal ageing and ALS: progressive neuronal loss in ventral horns, a decrease in motoneuron RNA content, and a decrease in motor unit number as measured electrophysiologically (Tomlinson and Irving 1977; Uemura and Hartmann 1978; McComas et al. 1973). The ability of a motoneuron to sprout in response to botulinum toxin is inversely related to age, as is its regenerative capacity after a crush lesion (Pestronk et al. 1980). These age-related changes provide circumstantial evidence for accelerated ageing as a significant aetiological factor. Because of the relationship of ageing and the ability to repair DNA damage (Gensler and Bernstein 1981; Hart et al. 1979), Bradley proposed that abnormalities in specific DNA repair mechanisms in motoneurons might be a final common pathway in ALS (Bradley and Krasin 1982a,b). Studies of DNA repair in cultured fibroblasts (Tandon et al. 1985) and lymphoblastoid cells (Lambert et al. 1986) from ALS patients have shown that repair of methyl methane sulfonate (MMS)-induced alkylation may be deficient. However, cultures of ALS fibroblasts are not overly sensitive to radiation (Robbins et al. 1985) and patients do not have exaggerated responses to other alkylating agents such as cyclophosphamide (Bradley and Krasin 1982b). At present, no current experiments provide convincing support for the ageing theory of ALS.

Trophic Factors

Studies on the development and maintenance of the neuromuscular system have documented the importance of retrograde control of motoneuron structure and function. In the developing embryo vertebrate motor system a two-fold excess of motoneurons is present (Hamburger 1975). These extra motoneurons subsequently die, apparently after failing to compete successfully with other motoneurons for a limited supply of a survival factor, produced by muscle and regulated by muscle innervation and membrane polarisation (Hamburger 1975; Pittman and Oppenheim 1979). Studies from our own laboratory have documented the ability of trophic factors derived from muscle to prevent motoneuron cell death (Oppenheim et al. 1988). Considerable evidence also suggests that motoneuron sprouting is controlled by the same or similar factors (Brown et al. 1981). Furthermore, loss of motoneurons in man after amputation also supports the importance of muscle-derived factors in the continued survival of adult motoneurons (Kawamura and Dyck 1981).

Nerve growth factor does not appear to be responsible for either developmental or maintenance influences in the motor system. Other factors purified in our own laboratory appear to be of more relevance (McManaman et al. 1988). However, at the present time it remains to be demonstrated whether such factors will have similar effects in an adult organism in vivo. Clearly abnormalities in the motor neurotrophic system may be a final common pathway for the development of motor system diseases such as ALS (Appel 1981). However, at the present time there is no evidence to suggest that ALS is due to a specific deficiency in a motor system trophic factor, or that the target of an immunological response is either a trophic factor itself or a trophic factor receptor. Nevertheless, interruption of production, processing, or secretion by muscle, or of interaction with its receptor on presynaptic terminals; uptake by such terminals; retrograde transport; or post-uptake processing of a necessary survival factor might lead to motoneuron dysfunction and death. Thus, even though no current data support the primacy of trophic factors or their potential role in initiating motoneuron destruction, such factors could play a role in reversing or ameliorating an injury initiated by some other causative process.

Viruses

A frequently proposed cause for neurological disease of unknown aetiology is the probability of viral infection, and ALS is no exception. There is ample circumstantial evidence that ALS could be induced by a virus, primarily from the existence of other non-inflammatory chronic viral infections in the CNS, e.g. those giving rise to progressive multifocal leukoencephalopathy, subacute sclerosing panencephalitis, kuru, and Creutzfeldt–Jakob disease. A type C murine RNA virus causes a transmissible lower motor neuron disease without inflammation (Gardner et al. 1973; Officer et al. 1973). The existence of a clear relation between the development of progressive muscular atrophy late in the neuronal history of patients who have had acute paralytic poliomyelitis many years previously has been known for many decades. Although poliomyelitis attacks motoneurons, this "postpolio syndrome" of progressive lower motoneuron weakness does not appear to depend on renewed or persistent infection (Dalakas et al. 1986).

Attempts to document a viral aetiology of ALS have centred on ultrastructural identification of viral inclusions, viral cultures, and in situ hybridisation to define viral specific nucleic acids, or viral specific enzymes and antigens. The ultrastructural analyses of ALS motoneurons have defined a number of inclusions, the most prominent of which are Bunina bodies (Bunina 1962). These are eosinophilic intracytoplasmic inclusions that may be spherical, elongated or ribbon shaped and can be demonstrated in a significant number of sporadic ALS patients (Iwata and Hirano 1979). The origin of these inclusions is unclear, and thus they are discussed in this section on viral aetiologies. They are not pathognomonic for ALS because similar structures have been reported in motoneurons in neurolathyrism, vinca alkaloid intoxication, and Chediak–Higashi disease (review by Iwata and Hirano 1979). Leigh et al. (1988) and Lowe et al. (1988), demonstrated other inclusion bodies closely associated with classical Bunina bodies, which stained positively for ubiquitin. Since ubiquitin is involved in degradation of intracellular proteins by targeting abnormal proteins for proteolysis (Ciechanover et al. 1984), it is likely that these inclusions represent abnormalities in the cytoskeletal apparatus rather than evidence of viral inclusions. Other studies documenting changes in the cytoskeleton in ALS include the demonstration that microtubules are altered in ALS motoneurons (Binet and Meininger 1988). In addition, abnormalities have been noted in the phosphorylation of neurofilaments in ALS (Munoz et al. 1988; Manetto et al. 1988). All of these cytoskeletal changes appear to be secondary to neuronal injury, and resemble comparable changes in disorders such as Alzheimer's disease (Mori et al. 1987; Cole and Timiras 1987; Perry et al. 1987a). They provide no specific evidence for a viral aetiology.

Efforts to culture viruses from patients with ALS have been largely unsuccessful, although adeno-associated virus was cultured from 2 of 11 ALS cases (Kascsak et al. 1982), and the virus closely related to the Russian spring–summer encephalitis virus was isolated from the CSF of a patient with ALS (Muller and Hilgenstock 1975). Attempts have been made to transmit ALS to primates employing human ALS patient tissues. However, even in cases of ALS with dementia, such efforts have been unsuccessful (Salazar et al. 1983). Using the technique of in situ hybridisation, ALS tissue has been examined for the presence of viral nucleic acids without success (Viola et al. 1982; Kohne et al. 1981). Brahic et al. (1985)

documented the presence of picornavirus sequences in one patient with ALS and in one control among 17 patients. Polio virus, herpes virus, and measles virus antigens have been found in jejunal biopsies of ALS patients (Pertschuk et al. 1977), but the significance of this finding is unclear.

Extensive serologic studies have been carried out to identify exposure to specific viral antigens in ALS, but no differences with control populations have been demonstrated (Catalano 1972; Lehrich et al. 1974; Cremer et al. 1976; Gardner et al. 1976; Jokelainen et al. 1977; Kascsak et al. 1978; Kurent et al. 1979). Cell-mediated immune responses have been analysed and the responses to polio virus were increased over controls (Cunningham-Rundles et al. 1977; Kott et al. 1979; Bartfeld et al. 1982a).

The most interesting data supporting a viral aetiology concerns reports of motor system involvement with known viral agents. One of four patients who received injections of human growth hormone to treat dwarfism and had developed Creutzfeldt–Jakob disease also developed both clinical and pathological evidence of motor neuron involvement (Brown et al. 1985). Hoffman et al. (1985) reported the case of a 26-year-old homosexual man with typical ALS and AIDS. The human immunodeficiency virus was isolated from his blood. However, the combination of ALS and AIDS could have been fortuitous, and there may be no specific causal relationship between ALS and the presence of HIV. Alternatively, autoimmunity rather than viral infection could have been the mechanism of motoneuron destruction. HIV infection has been commonly associated with dementia, meningitis, vacuolar myelopathy and distal predominantly sensory polyneuropathy (Petito et al. 1985; Hollander and Stringari 1987; Cornblath and McArthur 1988; Price et al. 1988). Although uncommon, corticospinal tracts as well as motoneurons may be involved (Horoupian et al. 1984), and motor neuropathy without sensory dysfunction has been reported in AIDS patients (Przedborski et al. 1986; Cornblath et al. 1987). Clearly HIV infection may on rare occasions simulate motor neuron disease, but the pertinent, still unanswered question is whether these rare cases represent a significant factor in the development of sporadic ALS.

Human T-lymphotrophic virus-type I (HTLV-I), the causative agent of adult T-cell leukaemia/lymphoma, also appears to be the cause of tropical spastic paraparesis, a chronic myelopathy reported in several different regions in the world. It is the suspected aetiological agent in the chronic myeloneuropathy endemic to Jamaica, Martinique, Colombia, the Seychelles, the Ivory Coast of Africa, and a similar clinical condition reported in Kagoshima, Japan. The chronic progressive myelopathy associated with HTLV-I clearly compromises corticospinal tracts, but anterior horn cells are spared. Serological evidence of HTLV-I infection was found in one patient with ALS, one control subject from Guam, one patient from the United States, and all seven Jamaican patients with polymyositis (Mora et al. 1988). However, the absence of lower motoneuron involvement, the presence of sensory and sphincter abnormalities (Osame et al. 1987; Vernant et al. 1987), and the seronegativity of all except one ALS patient makes it unlikely that this specific retrovirus is a major aetiological factor in sporadic ALS.

Thus, the data do not provide convincing evidence for the viral theory of ALS. Broad spectrum antiviral therapies have, to date, been ineffective in the treatment of ALS. Nevertheless, the absence of positive evidence does not disprove a viral aetiology, and the search for a virus that causes ALS will undoubtedly continue until a cause for the disease is forthcoming.

Metabolic Abnormalities

A wide range of metabolic abnormalities has been reported in patients with ALS, and a wide range of metabolic alterations can be associated with motoneuron dysfunction. For example, hypoglycaemia, noted with insulin-secreting islet cell tumours, can also be associated with weakness and amyotrophy (Barris 1953; Tom and Richardson 1951). Some researchers have observed glucose intolerance in ALS patients (Matthews 1958; Ionasescu and Luca 1964; Steinke and Tyler 1964; Collis and Engel 1968; Mueller and Quick 1970; Gotoh et al. 1972; Koerner 1976; Saffer et al. 1977; Murai et al. 1983), while others have denied this association (Cumings 1962; Quick and Greer, 1967; Colin Brown and Kater 1969; Friedman et al. 1969; Astin et al. 1975; Moxley et al. 1983; Harno et al. 1984). Although many different mechanisms have been proposed to explain the abnormality, the lack of consistent abnormalities of glucose tolerance in ALS patients and the presence of abnormal glucose tolerance in patients with other wasting neuromuscular diseases, raise serious questions about the clinical significance of the original observations. Alterations in pancreatic exocrine function have also been reported based upon the decreased volume and bicarbonate content of secretin-stimulated duodenal contents and the decreased uptake of triglycerides in the face of normal free fatty acid absorption (Quick and Greer 1967). However, these abnormalities could not be confirmed (Colin Brown and Kater 1969; Friedman et al. 1969; Utterback et al. 1970; McEwan-Alvarado et al. 1971).

Abnormalities in folic acid metabolism have been reported in ALS patients (Yoshino 1984). Levels of tetrahydrofolate in ALS motor cortex have been reported to be low, while 5-methyltetrahydrofolate levels are normal (Yoshino et al. 1979). These changes would be consistent with a decrease in the activity of 5-methyltetrahydrofolate homocysteine methyltransferase (MTHM). An abnormality in this enzyme would trap folic acid as 5-methyltetrahydrofolate and limit nucleic acid synthesis, thereby providing an explanation for the decreased levels of neuronal RNA (Davidson et al. 1981; Davidson and Hartmann 1981a,b; Mann and Yates 1974). However, inhibitors of MTHM cause neuropathologic effects similar to subacute combined degeneration, involving sensory neurons without specific compromise of lower motoneurons (Gandy et al. 1973; Kondo et al. 1981; Scott et al. 1981). Thus the decrease in motoneuron RNA and the changes in the base composition of RNA obtained from motoneurons in ALS (Davidson and Hartmann 1981b) cannot be readily explained as a motor system-specific abnormality in folic acid metabolism, and must have some other explanation. Similarly, the increased taurine levels in the CSF of patients with ALS (Yoshino et al. 1979) are probably due to motoneuron loss rather than due to alterations in folic acid metabolism. Numerous other changes in biochemical constituents of the spinal cord of ALS patients may also be secondary to motoneuron loss. These findings include decreased cholinergic, benzodiazepine, and glycine receptors (Hayashi et al. 1981; Nagata et al. 1982; Whitehouse et al. 1983; Gillberg and Aquilonius 1985), and decreased levels of choline acetyltransferase (Gillberg et al. 1982). Of interest is the fact that several neurotransmitter receptors are diminished in the dorsal horn as well as the anterior horn, suggesting involvement of neurons other than upper and lower motoneurons.

An abnormality in androgen metabolism was suggested (Weiner 1980) based upon the fact that ALS occurs more frequently in men than women. Reductions in plasma testosterone and variable increases in oestrogen have been reported in an X-linked recessive form of spinal and bulbar muscular atrophy in which patients may have testicular atrophy as well as gynaecomastia (Imai et al. 1980; Hausmanowa-Petrusewicz et al. 1983). No abnormalities, however, have been reported in androgen metabolism in ALS patients. Motor nuclei controlling extraocular muscles and the motoneurons controlling urinary and anal sphincters are routinely spared in ALS. Weiner (1980) postulated that the eye muscles are spared because their motoneurons do not have androgen receptors while other somatic motoneurons do (Sar and Stumpf 1977). However, urinary and anal sphincters are also routinely spared, yet motoneurons innervating the sphincters have more androgen receptors than other lumbar motoneuron pools (Breedlove and Arnold 1983) and are under androgen control (Arnold and Garski 1984). Thus neither androgen metabolism nor androgen receptor localisation can adequately explain the selective sparing of eye movements and bladder.

The clinical report that thyrotrophin-releasing hormone (TRH) may increase mobility and strength and lessen spasticity in ALS (Engel et al. 1983a) prompted extensive studies of TRH metabolism in ALS. The concentration in the CSF was reported to be reduced by Engel et al. (1983b) and to be unchanged by others (Jackson et al. 1986). This latter group also found no change in spinal cord TRH levels, while others found the concentrations to be decreased (Mitsuma et al. 1984; Mitsuma et al. 1986). Subsequent clinical studies have not been able to confirm the original observation of Engel (Brooke et al. 1986; Mitsumoto 1986; Imoto et al. 1984; Stober et al. 1985) and thus alterations in TRH metabolism are no longer considered pertinent to the aetiology of ALS.

Cyclic nucleotide metabolism has been reported to be abnormal in ALS, with decreased levels of cyclic AMP in the CSF of ALS patients (Brooks et al. 1976). Other studies have documented depressed levels of homovanillic acid, suggesting an abnormality in dopamine metabolism (Mendell et al. 1971). This observation is of interest in light of the recent demonstration by PET scanning of a subclinical dopaminergic deficit in the striatum of patients with ALS (Peppard et al. 1989). Other studies have documented lowered cerebral glucose utilisation in most brain regions, employing fluorodeoxyglucose with PET scanning (Dalakas et al. 1987). Perhaps the most straightforward explanation of the many metabolic abnormalities which have been reported is the possibility that the pathology in ALS are not limited to the motor system, but may involve tissues both outside and within the nervous system other than those commonly associated with classical motor system dysfunction. The recent reports of abnormalities in collagen bundles and the presence of amorphous material in skin biopsies of ALS patients (Ono et al. 1989) and the presence of abnormal liver function tests (Masui et al. 1985; Nakano et al. 1987) would support this point of view. The latter group documented swollen mitochondria and possible inclusion bodies on liver biopsies. Furthermore, within the CNS, involvement of the spinocerebellar pathways, although clinically silent, is a regular feature of sporadic ALS (Averback and Crocker 1982; Swash et al. 1988), thereby suggesting that a full understanding of the aetiology, or at least the pathogenesis of ALS, will require an understanding of the involvement of structures other than upper and lower motoneurons.

Autoimmunity

Recent circumstantial evidence supports the potential role of autoimmune factors in the aetiology of ALS. There is an increased incidence of autoimmune disorder (Appel et al. 1986) and paraproteinaemias (Shy et al. 1986; Latov et al. 1988) in ALS. The underlying thesis in our own study was that one would expect to see an increase in other immune-related disorders if ALS were an immune-related disorder itself, similar to the increased incidence of rheumatoid arthritis in Simpson's original studies of myasthenia gravis patients (Simpson 1969). In fact, of 58 patients studied, past or current history of thyroid disease was present in 19% of these patients, and 23% of an additional 47 patients had elevated titers of anti-thyroid microsomal antibodies.

Shy et al. reported that 5% of 202 patients with motor neuron disease had paraproteinaemia (Shy et al. 1986), whereas only 1% of 100 controls had paraproteinaemia. When paraproteinaemia and monoclonal gammopathy were associated with sensory motor peripheral neuropathy, the monoclonal protein was found to have antibody activity against myelin-associated glycoprotein (MAG), as well as several unusual gangliosides. However, patients with motor neuron disease lack antibodies to MAG. In a patient with lower motor neuron disease and IgM monoclonal gammopathy, the M protein bound to gangliosides GM_1 and GD_{1b} (Freddo et al. 1986). However, in ALS patients with monoclonal gammopathy of the IgG kappa or IgG lambda variety, there is no evidence of specific reactivity with gangliosides or other known antigens. In fact, the overwhelming number of ALS cases associated with monoclonal gammopathy do not have specific reactivity of their Ig with motoneuron constituents. Thus there is no direct evidence in the majority of cases of ALS that the paraprotein causes the motoneuron disorder. Nevertheless, the coexistence of paraproteinaemia and ALS provides circumstantial evidence that an altered immune system may be responsible for both conditions.

Careful analysis of the histocompatibility antigen (HLA) subgroups in ALS has been undertaken in many different laboratories with no clear-cut subtype overrepresented. High incidences of HLA-A3 and HLA-Bw35 have been described in some patient groups (Jokelainen et al. 1977; Kott et al. 1976, 1979; Antel et al. 1979) but not in others (Hoffman et al. 1971; Behan et al. 1976; Pedersen et al. 1977; Seigmaliet et al. 1979). Differences in rate of disease progression have also been related to HLA subtype (Hoffman et al. 1971; Antel et al 1979). Our own studies in over 120 patients have detected no consistent HLA subtypes in sporadic ALS.

No consistent differences in ALS have been documented with respect to cellular immunity. Total T cell count (OKT11+ cells), T4/T8 ratios, spontaneous blastogenesis, human natural killer cells, and mixed lymphocyte cultures are all normal in ALS patients (Appel et al. 1986; Bartfeld et al. 1982b, 1985; Tavolato et al. 1975; Cashman et al. 1985). Reports of mitogen responsiveness have been controversial. Several investigators found a decrease in response to T cell mitogens (Hoffman et al. 1978; Behan 1979; Aspin et al. 1986), while others have found no change (Appel et al. 1986; Bartfeld et al. 1982b; Antel et al. 1982). Studies of the response to skin testing have also been contradictory (Bartfeld et al. 1982b; Kott et al. 1976; Tavolato et al. 1975; Hoffman et al. 1978; Behan et al. 1977), as have studies of the increased presence of T cells bearing the Ia antigen indicating T cell activation (Appel et al. 1986; Barfeld et al. 1985). Macrophage function in ALS is reported to be abnormal in vivo (Urbanek and Jansa 1974), and putative helper T cell numbers are depressed (Westall et al. 1983), although neither of these findings have been confirmed. Thus, the T cell subpopulations *per se* do not provide circumstantial evidence for an autoimmune aetiology of ALS.

Several studies have reported increased cellular immunity to polio-virus (Cunningham-Rundles et al. 1977; Kott et al. 1979; Bartfeld et al. 1982a; Olarte and Shafer 1985; Kott et al. 1976), myelin basic protein (Kott et al. 1976), and various CNS subfractions (Bartfeld et al. 1982a; Aspin et al. 1986), but such findings may be secondary phenomena. Some reports have documented the presence of circulating immune complexes as well as immune complex deposition in the kidneys of ALS patients (Bartfeld et al. 1982a; Oldstone et al. 1976; Tachovsky et al. 1976; Araga et al. 1984). Deposits of IgG and several components of complement have been localised by immunofluorescent techniques to spinal cord and motor cortex of patients with ALS, as well as several other immunologically mediated neurological disorders (Donnenfeld et al. 1984). The deposits were seen mainly in astrocytes, and no staining of neurons was observed.

The search for autoantibodies against neurons in vitro has been extensive. Among the earliest reports was our demonstration that 9 of 15 ALS sera demyelinated cultured neurons (Bornstein and Appel 1965). Although this result was subsequently confirmed (Field and Hughes 1965), the demyelinating factor in serum was not characterised as an immunoglobulin. Wolfgram and Myers (1973) demonstrated that serum from 70% of patients with ALS was toxic to cultured mouse ventral horn explants. Roisen et al. (1982) confirmed a neuron-specific cytotoxic effect on neonatal mouse ventral horn explants. However, other laboratories could not reproduce such toxic effects either in organotypic cultures (Horwich et al. 1974), or in cultures of neuroblastomas (Lehrich & Couture 1978), chick ciliary ganglion cells (Touzeau and Kato 1983) or cultured human spinal cord neurons (Touzeau and Kato 1986). Serum from ALS patients was found to depress frog ventral root depolarisation amplitudes (Schauf et al. 1980), but the lack of further follow-up precludes an assessment of the validity or significance of this finding.

The most recent example of an attempt to define humoral immune mechanisms in ALS has been the studies of Gurney (Gurney 1984; Gurney et al. 1984). His studies reported a presumed novel growth factor produced by denervated muscle whose biological function was the promotion of nerve terminal sprouting. Patients with ALS were thought to have antibodies against this growth factor. Unfortunately the growth factor turned out to be glucose-6 phosphate isomerase (Chaput et al. 1988; Falk et al. 1988); and no convincing data exist that this "factor", which had also been named neuroleukin, has reproducible growth or sprouting effects. Furthermore, other studies (Ingvar-Marden et al. 1986; Hauser et al. 1986) have failed to document antibodies to a 56 kDa protein constituent specific for ALS. Brown et al. (1987) demonstrated antineural antibodies to 52 and 70 kDa spinal cord proteins more frequently in ALS than in control sera. Serum antibodies against neurofilament protein were reported with neurodegenerative disorders such as ALS and parkinsonism-dementia complex (Bahmanyar et al. 1982, 1983). However, such antibodies are also found in normal individuals, suggesting that antibodies against neurofilament proteins probably have no pathogenic significance (Stefansson et al. 1985).

A recent report employing immunoblot techniques demonstrated antibodies in the serum of ALS patients which reacted with several proteins from fetal rat muscle extracts (Ordonez and Sotelo 1989). The fact that activity was noted with several rather than a single protein derived from fetal muscle argues against a simple pathogenic mechanism involving autoantibody production. More likely it represents the presence of increased immune reactivity in ALS patients and provides further circumstantial evidence for autoimmunity without specifying either the pathogenic importance of the humoral immune alterations or the specific target of the immune attack. Similarly antibodies to gangliosides have been reported in ALS patients without monoclonal gammopathies. With different techniques, from 10% of ALS patients to more than 75% of ALS patients have been reported to possess antiganglioside antibodies (Shy et al. 1987; Pestronk et al. 1988, 1989). The fact that such antibodies are also present in many other presumed autoimmune disorders (Endo et al. 1984), and may be experimentally produced in mice following sciatic nerve injury (Schwartz et al. 1982), makes it extremely likely that these autoantibodies are secondary responses, and are not primarily related to the pathogenesis of the disease. Only the direct demonstration of functional changes in motoneurons produced by antiganglioside antibodies would help define a pathogenic role for these or any other antimotoneuron antibodies that have been reported.

One of the important steps in defining the autoimmune basis of myasthenia gravis and the myasthenic syndrome was the development of animal models to help determine the role of immune factors in the pathogenesis of these diseases (Patrick and Lindstrom 1973; Lang et al. 1981). Similarly, in ALS the availability of an animal model could help define whether immune factors contribute to the mechanism of motoneuron cell death. We have recently reported two distinct animal models of motor neuron disease: *experimental autoimmune motor neuron disease* (EAMND) and *experimental autoimmune grey matter disease* (EAGMD). EAMND is a lower motor syndrome induced in guinea pigs by five monthly injections of purified bovine spinal cord motoneurons (Engelhardt et al. 1989). The gradual onset of weakness of the limbs is associated with electromyographic and morphological evidence of denervation, as well as a loss of spinal cord motoneurons. IgG is located at the neuromuscular junction and in the cytoplasm of spinal cord motoneurons. EAGMD is a more acute disorder involving lower and upper motoneurons induced in guinea pigs by two inoculations of spinal cord ventral horn homogenates (Engelhardt et al. 1990). Denervation is also present as evidenced by EMG and morphological criteria. Within the CNS there are scattered perivascular inflammatory foci, and a loss of spinal cord motoneurons and large pyramidal cells in the motor cortex. IgG is localised at the neuromuscular junction, the cytoplasm of spinal motoneurons early in the disease, and on the external membrane of spinal motoneurons and cortical pyramidal neurons later in the disease. The sera from these animal models contain high titres of antibodies against bovine motoneurons. Perhaps most significant is the fact that the serum, and more specifically the IgG fraction, can cause physiological changes when passively transferred to a normal host (Appel et al. unpublished data).

These findings suggest that an autoimmune mechanism may explain the destruction of motoneurons in experimental disease in animals. By analogy autoimmunity may explain the destruction of motoneurons in ALS in humans. The animal model with both upper and lower motoneuron destruction, EAGMD, also exhibited inflammatory foci in the CNS and a breakdown in blood–brain barrier and suggested a means whereby IgG could invade the CNS, namely direct passage from blood to brain. Inflammatory foci have also been described in the spinal cord and motor cortex in human ALS (Troost et al. 1988; Lampson et al. 1988; McGeer et al. 1988), and suggest a similar means by which IgG could pass from blood to brain to react with upper motoneuron epitopes similar to those on lower motoneurons. Our current studies are aimed at determining how close the parallel may be between the animal and the human condition by defining the specific antigenic target of the antibodies in EAMND and EAGMD, and thereafter determining whether human ALS patients have high titre antibodies against the same constitutent.

Even if the presence of an antibody with motoneuron reactivity could be convincingly demonstrated in ALS, many of the metabolic changes already discussed would have to be explained. For example, how would the changes in cells other than upper and lower motoneurons be explained? How would changes in structures outside the CNS be explained? Perhaps only by implicating an antigenic moiety which is widely distributed, but one whose functional activity is more critical for motoneurons than for other cells, could we provide insight into the extensive devastation noted in ALS.

Still the hypothesis of an autoimmune aetiology has numerous stumbling blocks, not least of which is the fact that immunosuppressive therapy at best may slow the course of ALS (Appel et al. 1988), but it does not halt progression for long periods of time. Furthermore immunosuppression in the form of steroids or plasmapheresis has been relatively ineffective (Pieper and Fields 1957; Keleman et al. 1983). This fact leads us to conclude either that autoimmunity is not the relevant aetiology, or that it is relevant but in a unique way. Our own hypothesis is that autoimmunity may be a relevant factor in initiating the disease, but that the subsequent destruction of the motoneuron may be independent of the continuing presence of the antibody which initiated the process. For example, the initial immune attack could alter the threshold to excitotoxic amino acids (Spencer et al. 1987; Plaitakis and Caroscio 1987; Plaitakis et al. 1988a), increase the number of reactive microglia which release destructive cytokines and superoxide radicals (McGeer et al. 1988), activate proteases and phospholipases within motoneurons, impair trophic function, or interrupt intracellular events such as axoplasmic flow. All of these pathogenic events could occur even if the offending antibodies were removed from the circulation or if the relevant antibodies were sequestered within the soma of the motoneuron and thus not influenced by immunosuppressive therapy.

Clearly, all such possibilities are entirely hypothetical. Yet the availability of animal models of motoneuron destruction provides an opportunity to delineate potential biological mechanisms of motoneuron loss relevant to ALS, just as the animal models of myasthenia gravis (Patrick and Lindstrom 1973) and the myasthenic syndrome (Lang et al. 1981) paved the way for an understanding of the relevant mechanisms responsible for neuromuscular dysfunction in these diseases in humans.

Acknowledgements. We are grateful to our ALS patients, to our ALS clinic coordinator, Vicki Appel, and to the Muscular Dystrophy Association for their support of our MDA ALS Research and Clinical Center at Baylor College of Medicine. We are also grateful to the Muscular Dystrophy Association, the M.H. "Jack" Wagner Memorial Fund, the Cullen Foundation, and Cephalon, Inc., for other support.

Addendum

Since this chapter was first submitted for publication in 1989, additional evidence has been published that has removed some of the previously enumerated "stumbling blocks". Recent studies demonstrate the presence of antibodies to L-type, N-type, and P-type calcium channels in patients with ALS, and support a role for autoimmune mechanisms in the pathogenesis of disease. First, immunoglobulins from patients with ALS passively transferred into mice reproducibly produce an increase in miniature endplate potential frequency at neuromuscular junctions (Appel et al. 1991). These data suggested that ALS IgG increased intracellular calcium leading to increased neurotransmitter release. ALS IgG has been found to augment calcium current through neuronal P-type voltage-gated calcium channels in Purkinje cells (Llinas et al. 1993), and through N-type voltage-gated calcium channels (Stefani et al. unpublished results), both of which are in accord with increased calcium entry. ALS IgG and F_{ab} fragments also interact with L-type voltage-gated calcium channels in skeletal muscle (Delbono et al. 1991a,b, 1993), and lipid bilayers (Magnelli et al. 1993). ALS IgG also bind directly to voltage-gated calcium channels in ELISA assays (Smith et al. 1992), and more specifically to the ionophore-containing α_1-subunit (Kimura et al. 1994).

The titer of calcium channel antibodies correlates with the rate of progression of ALS. Further, evidence for lymphocytic infiltrates in spinal cords of patients with amyotrophic lateral sclerosis (Engelhardt et al. 1993) strengthens the argument that autoimmune processes may be important to ALS pathogenesis. Finally, ALS IgG are selectively cytotoxic to a motoneuron hybrid cell line, and this calcium-dependent cytotoxicity can be blocked by antagonists of N-type and P-type calcium channels and removed by preabsorption with either whole purified voltage-gated calcium channel or with the α_1-subunit (Smith et al. 1994).

In the past five years, other hypotheses have also been suggested. In patients with sporadic ALS, motoneuron death has been hypothesized to result from activation of ligand-gated ion channels and excitotoxicity (Plaitakis et al. 1988a, b; Rothstein et al. 1992). While changes in glutamate and glutamate uptake in spinal cords of patients with ALS may be secondary to motoneuron loss (Appel, 1993), it is also possible that ALS IgG-mediated calcium entry through voltage-gated calcium channels presynaptically into upper motoneurons may increase glutamate release onto spinal motoneurons and further activate both excitotoxic ligand-gated ion channel function and motoneuron injury. Further, the observation that familial ALS (which affects 10% of ALS patients) is linked in some kindreds to mutations in the copper/zinc dependent superoxide dismutase suggests a potential role for free radicals in motoneuron injury (Rosen et al. 1993; Deng et al. 1993). In 1994, important questions concerning the causation of ALS include whether over-activation of voltage-gated calcium channels and/or ligand-gated calcium channels induces free radical production, and whether intracellular processes mediated by free radicals produce a final common pathway of motoneuron death in both sporadic and familial ALS.

References

Adams CR, Ziegler DK, Lin JT (1983) Mercury intoxication simulating amyotrophic lateral sclerosis. JAMA 250:642–643

Antel JP, Medof ME, Richman DP, Aranson BGW (1979) Immunological considerations in amyotrophic lateral sclerosis. In: Rose FC (ed) Clinical neuroimmunology. Blackwell Scientific, Oxford, pp 277

Antel JP, Noronha ABC, Oger JJ-F, Arnason BGW (1982) Immunology of amyotrophic lateral sclerosis. In Rowland LP (ed) Human motor neuron diseases. Raven Press, New York, pp 395–492

Appel SH (1981) A unifying hypothesis for the cause of amyotrophic lateral sclerosis, parkinsonism, and Alzheimer disease. Ann Neurol 10:499–505

Appel SH (1993) Excitotoxic neuronal cell death in amyotrophic lateral sclerosis. Trends Neurosci 16:3–5

Appel SH, Stockton-Appel V, Stewart SS, Kerman RH (1986) Amyotrophic lateral sclerosis: Associated clinical disorders and immunologic evaluations. Arch Neurol 43:234–238

Appel SH, Stewart SS, Apel V et al. (1988) A double-blind study of cyclosporine in amyotrophic lateral sclerosis. Arch Neurol 45:381–386

Appel SH, Engelhardt J, Garcia J, Stefani E (1991) Immunoglobulins from animal models of motor neuron disease and human ALS passively transfer physiological abnormalities of the neuromuscular junction. Proc Natl Acad Sci (USA) 88:647–651

Araga S, Irie H, Trakahashi K (1984) Conglutinin microtiter plate ELISA system for detecting circulating immune complexes. J Neuroimmunol 6:161–168

Arnold AP, Gorski RA (1984) Gonadal steroid induction of structural sex differences in the central nervous system. Ann Rev Neurosci 7:413–432

Aspin J, Harrison R, Jehanli A, Lunt G, Campbell M (1986) Stimulation by mitogens and neuronal membranes from patients with motor neuron disease. J Neuroimmunol 11:31–40

Astin KJ, Wilde CE, Davies-Jones GAB (1975) Glucose metabolism and insulin response in the plasma and CSF in motor neuron disease. J Neurol Sci 25:205–210

Averback P, Crocker P (1982) Regular involvement of Clarke's nucleus in sporadic amyotrophic lateral sclerosis. Arch Neurol 39:155–156

Bahmanyar S, Gajdusek DC, Soleto J, Gibbs CJ Jr (1982) Longitudinal spinal cord sections as substratum for anti-neurofilament antibody detection. J Neurol Sci 53:85–90

Bahmanyar S, Moreau-Dubois ML, Brown P, Cathala F, Gajdusek DC (1983) Serum antibodies to neurofilament antigens in patients with neurological and other diseases and in healthy controls. J Neuroimmunol 5:191–196

Barber FE (1978) Inorganic mercury intoxication reminiscent of amyotrophic lateral sclerosis. J Occup Med 20:667–669

Barris RW (1953) Pancreatic adenoma (hyperinsulinism) associated with neuromuscular disorders. Ann Intern Med 38:124–129

Barry PSI, Mossman DB (1979) Lead concentrations in human tissues. Br J Ind Med 27:339

Bartfeld H, Dham C, Donnenfeld H et al. (1982a) Immunological profile of amyotrophic lateral sclerosis patients and their cell-mediated immune responses to viral and CNS antigens. Clin Exp Immunol 48:137–147

Bartfeld H, Pollack MS, Cunningham-Rundles S, Donnenfeld H (1982b) HLA frequencies in amyotrophic lateral sclerosis. Arch Neurol 39:270–271

Bartfeld H, Dham C, Donnenfeld H (1985) Immunoregulatory and activated T cells in amyotrophic lateral sclerosis patients. J Neuroimmunol 9:131–137

Behan PO, Durward WF, Dick H (1976) Histocompatibility antigens associated with motor neuron disease. Lancet ii:803

Behan PO, Behan WM, Bell E, Lannigan C, McQueen A, More IA (1977) Possible persistent virus in motor neuron disease (letter). Lancet ii:1176

Behan PO (1979) Cell-mediated immunity in motor neuron disease and polio-myelitis. In: Rose FC (ed) Clinical neuroimmunology. Blackwell Scientific, Oxford, pp 259–272

Binet S, Meininger V (1988) Modifications of microtubule proteins in ALS nerve precede detectable histologic and ultrastructural changes. Neurology 38:1596–1600

Boothby JA, deJesus PV, Rowland LP (1974) Reversible forms of motor neuron disease. Arch Neurol 31:18–23

Bornstein MB, Appel SH (1965) Tissue culture studies of demyelination. Ann NY Acad Sci 122:280–286

Bradley WG, Krasin F (1982a) A new hypothesis of the etiology of amyotrophic lateral sclerosis: the DNA hypothesis. Arch Neurol 39:677–680

Bradley WG, Krasin F (1982b) DNA hypothesis of amyotrophic lateral sclerosis. Adv Neurol 36:493–502

Brahic M, Smith RA, Gibbs CJ Jr, Garruto RM, Tourtellote WW, Cash E (1985) Detection of picornavirus sequences in nervous tissue of amyotrophic lateral sclerosis and control patients. Ann Neurol 18:337–343

Breedlove SM, Arnold AP (1983) Sex differences in the pattern of steroid accumulation by motoneurons of the rat lumbar spinal cord. J Comp Neurol 215:211–216

Breuer AC, Lynn MP, Atkinson MB et al. (1987) Fast axonal transport in amyotrophic lateral sclerosis: an intra-axonal organelle traffic analysis. Neurology 37:738–748
Brooke MH, Florence JM, Heller SL et al. (1986) Controlled trial of thyrotropin-releasing hormone in amyotrophic lateral sclerosis. Neurology 36:146–151
Brooks BR, Sode J, Engel WK (1976) Cyclic nucleotide metabolism in neuromuscular disease. UCLA Forum Med Sci 19:101–118
Brown MC, Holland RL, Hopkins WG (1981) Motor nerve sprouting. Ann Rev Neurosci 4:17–42
Brown P, Gajdusek DC, Gibbs JC Jr, Asher DM (1985) Potential epidemic of Creutzfeldt–Jakob disease from human growth hormone therapy. N Engl J Med 313:728–731
Brown RH Jr, Johnson D, Ogonowski M, Weiner HL (1987) Antineural antibodies in the serum of the patients with amyotrophic lateral sclerosis. Neurology 37:152–155
Bunina TL (1962) Intracellular inclusions in familial amyotrophic lateral sclerosis. Korsakov J Neuropathol Psychiatry 62:1293–1299
Campbell AMG (1955) Calcium versenate in motor neurone disease. Lancet ii:376–377
Campbell AMG, Williams ER, Baltrop D (1970) Motor neurone disease and exposure to lead. J Neurol Neurosurg Psychiatry 33:877–885
Cashman NR, Gurney ME, Antel JP (1985) Immunology of amyotrophic lateral sclerosis. Springer Sem Immunopathol 8:141–152
Catalano LW (1972) Herpes virus hominis antibody in multiple sclerosis and amyotrophic lateral sclerosis. Neurology 22:473–478
Chaput M, Claes V, Portetelle D et al. (1988) The neurotrophic factor neurokin is 90% homologous with phosphohexose isomerase. Nature 332:454–455
Ciechanover A, Finley D, Varshavsky A (1984) Ubiquitin dependence of selective protein degradation demonstrated in the mammalian cell cycle mutant. Cell 37:57–66
Cole GM, Timiras PS (1987) Ubiquitin-protein conjugates in Alzheimer's lesions. Neurosci Lett 79:207–212
Colin Brown J, Kater RMH (1969) Pancreatic function in patients with amyotrophic lateral sclerosis. Neurology 19:185–189
Collis WJ, Engel WK (1968) Glucose metabolism in five neuromuscular disorders. Neurology 18:915–923
Conradi S, Ronnevi L-O, Vesterberg O (1976) Abnormal tissue distribution of lead in amyotrophic lateral sclerosis. J Neurol Sci 29:259–265
Conradi S, Ronnevi L-O, Vesterberg O (1980a) Abnormal tissue distribution of lead in amyotrophic lateral sclerosis. Re-estimation of lead in the cerebrospinal fluid. J Neurol Sci 48:413–418
Conradi S, Eriksson H, Ronnevi L-O (1980b) Cholinesterase activity of whole blood and plasma in amyotrophic lateral sclerosis. Acta Neurol Scand 62:191–192
Conradi S, Ronnevi L, Norris F (1982a) Motor neuron disease and toxic metals. Adv Neurol 36:201–231
Conradi S, Ronnevi L-O, Nise G, Vesterberg O (1982b) Long-time penicillamine treatment in amyotrophic lateral sclerosis with parallel determination of lead in blood, plasma and urine. Acta Neurol Scand 65:203–211
Cornblath DR, McArthur J (1988) Predominantly sensory neuropathy in patients with AIDS and AIDS-related complex. Neurology 38:794–796
Cornblath DR, McArthur JC, Kennedy RGE, Witte AS, Griffin JW (1987) Inflammatory demyelinating peripheral neuropathies associated with human T-cell lymphotropic virus type III infection. Ann Neurol 21:32–40
Cremer NE, Norris FH, Shinomoto T, Lennette EH (1976) Antibody titers to Coxsackie viruses in amyotrophic lateral sclerosis. N Engl J Med 295:107–108
Cumings JW (1962) Discussion on motor neurone disease. Biochemical aspects. Proc R Soc Med 55:1023–1024
Cunningham-Rundles S, Dupont B, Posner J, Hansen JA, Good RA (1977) Cell-mediated immune response to polio virus antigen in amyotrophic lateral sclerosis. Fed Proc 36:1190A
Currier RD, Haerer AF (1968) Amyotrophic lateral sclerosis and metallic toxins. Arch Environ Health 17:712–719
Dalakas MC, Elder G, Hallett M et al. (1986) A long-term follow-up study of patients with post-poliomyelitis neuromuscular symptoms. N Engl J Med 314:959–963
Dalakas MC, Hatazawa J, Brooks RA, DiChiro G (1987) Lowered cerebral glucose utilization in amyotrophic lateral sclerosis. Ann Neurol 22:580–586
Davidson TJ, Hartmann HA (1981a) RNA content and volume of motor neurons in amyotrophic lateral sclerosis. II. The lumbar intumescence and nucleus dorsalis. J Neuropathol Exp Neurol 40:187–192

Davidson TJ, Hartmann HA (1981b) Base composition of RNA obtained from motor neurons in amyotrophic lateral sclerosis. J Neuropathol Exp Neurol 40:193–198

Davidson T, Hartmann HA, Johnson PC (1981) RNA content and volume of motor neurons in amyotrophic lateral sclerosis. I. The cervical swelling. J Neuropathol Exp Neurol 40:32–36

Delbono O, Carcía J, Appel SH, Stefani E (1991a) IgG from amyotrophic lateral sclerosis affects tubular calcium channels of skeletal muscle. Am J Physiol 260:C1347–C1351

Delbono O, García J, Appel SH, Stefani E (1991b) Calcium current and charge movement of mammalian muscle: action of amyotrophic lateral sclerosis immunoglobulins. J Physiol 444:723–742

Delbono O, Magnelli V, Sawada T, Smith RG, Appel SH, Stefani E (1993) The Fab fragments from amyotrophic lateral sclerosis IgG affect the calcium channels of skeletal muscle. Am J Physiol 264 (Cell Physiol 33):C537–C543

Deng H-X, Hentati A, Tainer JA, Iqbal Z, Cayabyab A, Hung W-Y et al. (1993) Amyotrophic lateral sclerosis and structural defects in Cu, Zn superoxide dismutase. Science 261:1047–1051

Donnenfeld H, Kascsak RJ, Bartfeld H (1984) Deposits of IgG and C_3 in the spinal cord and motor cortex of ALS patients. J Neuroimmunol 6:51–57

Endo T, Scott DD, Stewart SS, Kundu SK, Marcus DM (1984) Antibodies to glycosphingolipids in patients with multiple sclerosis and SLE. J Immunol 132:1793–1797

Engel WK, Siddique T, Nicoloff JT (1983a) Effect on weakness and spasticity in amyotrophic lateral sclerosis of thyrotropin-releasing hormone. Lancet ii:73–75

Engel WK, Siddique T, Nicoloff JT, Wilbur JF (1983b) TRH levels are reduced in CSF of amyotrophic lateral sclerosis (ALS) and other spastic patients and rise with intravenous treatment. Neurology 53(Suppl 2): 176

Engelhardt JI, Appel SH, Killian JM (1989) Experimental autoimmune motoneuron disease. Ann Neurol 26:368–376

Engelhardt JI, Appel SH, Killian JM (1990) Experimental autoimmune gray matter disease. J Neuroimmunol.

Engelhardt JI, Taiji J, Appel SH (1993) Lymphocytic infiltrates in the spinal cord in amyotrophic lateral sclerosis. Arch Neurol 50:30–36

Falk P, Walker JIN, Redwill AAM, Morgan MJ (1988) Mouse glucose-6-phosphate isomerase and neuroleukin have identical 3' sequences. Nature 332:445–457

Felmus MT, Patten BM, Swanke L (1976) Antecedent events in amyotrophic lateral sclerosis. Neurology 26:167–172

Field EJ, Hughes D (1965) Toxicity of motor neurone disease serum for myelin in tissue culture. BMJ ii:1399–1401

Freddo L, Yu RK, Latov N et al. (1986) Gangliosides GM1 and GD1b are antigens for IgM M-protein in a patient with motor neuron disease. Neurology 36:454–458

Friedman HM, Tzagournis M, Ruppert RD (1969) Pancreatic exocrine and endocrine function in amyotrophic lateral sclerosis. Neurology 19:283

Gajdusek DC (1985) Hypothesis: interference with axonal transport of neurofilament as a common pathogenetic mechanism in certain diseases of the central nervous system. N Engl J Med 312:714–719

Gandy G, Jacobson W, Sidman R (1973) Inhibition of transmethylation reaction in the central nervous system – an experimental model for subacute combined degeneration of the cord. J Physiol 233:1–3

Gardner MD, Henderson BE, Officer JE et al. (1973) A spontaneous lower motor neuron disease apparently caused by indigenous type-C RNA virus in wild mice JNCI 51:1243–1254

Gardner MA, Rasheed S, Klement V et al. (1976) Lower motor neuron disease in wild mice caused by indigenous type C virus and search for a similar etiology in human amyotrophic lateral sclerosis. In: Andrews JM, Johnson RT, Brazier MAP (eds) Amyotrophic lateral sclerosis. Recent research trends. Academic Press, New York, pp 217–234

Gawel M, Zaowalla Z, Rose FC (1983) Antecedent events in motor neuron disease. J Neurol Neurosurg Psychiatry 46:1041–1043

Gensler HL, Bernstein H (1981) DNA damage as the primary cause of aging. Q Rev Biol 56:279–303

Gillberg P-G, Aquilonius S-M, Eckernsas S-A, Lundqvist G, Winblad B (1982) Choline acetyltransferase and substance P-like immunoreactivity in the human spinal cord: changes in amyotrophic lateral sclerosis. Brain Res 250:394–397

Gillberg P-G, Aquilonius S-M (1985) Cholinergic opioid and glycine receptor binding sites localized in human spinal cord by in vitro autoradiography. Acta Neurol Scand 72:299–306

Gotoh F, Kitamara A, Koto A, Kataoka K, Atsuki H (1972) Abnormal insulin secretion in amyotrophic lateral sclerosis. J Neurol Sci 16:201–207

Gurney ME (1984) Suppression of sprouting at the neuromuscular junction by immune sera. Nature 307:546–548

Gurney ME, Belton AC, Cashman N, Antel JP (1984) Inhibition of terminal axonal sprouting by serum from patients with amyotrophic lateral sclerosis. N Engl J Med 311:933–939

Hamburger V (1975) Cell death in the development of the lateral motor column of the chick embryo. J Comp Neurol 160:535–546

Harno K, Rissanen A, Palo J (1984) Glucose tolerance in amyotrophic lateral sclerosis. Acta Neurol Scand 70:451–455

Hart RW, D'Ambrosio SM, Ng KJ, Modak SP (1979) Longevity, stability, and DNA repair. Mech Aging Dev 9:203–223

Hauser SL, Cazenave PA, Lyon-Caen O et al. (1986) Immunoblot analysis of circulating antibodies against muscle proteins in amyotrophic lateral sclerosis and other neurological diseases. Neurology 36:1614–1618

Hausmanowa-Petrusewicz I, Borkowska J, Janczewski Z (1983) X-linked adult form of spinal muscular atrophy. J Neurol 229:175–188

Hayashi H, Suga M, Satake M, Tsubaki T (1981) Reduced glycine receptor in the spinal cord in amyotrophic lateral sclerosis. Ann Neurol 9:292–294

Hoffman PM, Festoff BW, Giron LT, Jr, Hallenbeck LC, Garruto RM, Ruscetti FW (1985) Isolation of LAV/HTLV-III from a patient with amyotrophic lateral sclerosis. N Engl J Med 313:324

Hoffman PM, Robbins DS, Gibbs CJ, Gajdusek DC, Garruto RM, Terasaki OI (1971) Histocompatibility antigens in amyotrophic lateral sclerosis and parkinsonism-dementia on Guam. Lancet ii:717

Hoffman PM, Robbins DS, Nolte MT, Gibbs CJ Jr, Gajdusek DC (1978) Cellular immunity in Guamanians with amyotrophic lateral sclerosis and parkinsonism-dementia. N Engl J Med 299:680–685

Hollander H, Stringari S (1987) Human immunodeficiency virus associated meningitis, clinical course and correlation. Am J Med 83:813–185

Horoupian DS, Pick P, Spigland I, Smith P, Portenoy R, Katzman R, Cho S (1984) Acquired immune deficiency syndrome and multiple tract degeneration in a homosexual man. Ann Neurol 15:502–505

Horwich MS, Engel WK, Chauvin PB (1974) Amyotrophic lateral sclerosis sera applied to cultured motor neurons. Arch Neurol 30:332–333

Hugon J, Ludolph A, Roy DN, Schaumburg HH, Spencer PS (1988) Studies on the etiology and pathogenesis of motor neuron diseases. II. Clinical and electrophysiological features of pyramidal dysfunction in macaques fed *Lathyrus sativus* and IPPN. Neurology 38:435–442

Imai H, Beppu H, Uono M, Narahayashi H (1980) Endocrinological investigation in patients with progressive proximal spinal and bulbar muscular atrophy of late onset (Kennedy–Alter-Sung type). Clin Neurol 20:704–712

Imoto K, Saida K, Iwamura K, Saida T, Nishitani H (1984) Amyotrophic lateral sclerosis: a double-blind crossover trial of thyrotropin-releasing hormone. J Neurol Neurosurg Psychiatry 47:1332–1334

Ingvar-Marden M, Regli F, Steck AJ (1986) Search for antibodies to skeletal muscular proteins in amyotrophic lateral sclerosis. Arch Neurol Scand 74:218–223

Ionasescu V, Luca N (1964) Studies on carbohydrate metabolism in amyotrophic lateral sclerosis and hereditary proximal spinal muscular atrophy. Acta Neurol Scand 40:47–57

Iwata M, Hirano A (1979) Current problems in the pathology of amyotrophic lateral sclerosis. Prog Neuropathol 4:277–298

Jackson I, Adelman LS, Munsat TL, Forte S, Lechan RM (1986) Amyotrophic lateral sclerosis: thyrotropin-releasing hormone and histidyl proline diketopiperazine in the spinal cord and cerebrospinal fluid. Neurology 287:34–36

Jokelainen M, Tiilikainen A, Lapinleimu K (1977) Polio antibodies and HLA antigens in amyotrophic lateral sclerosis. Tissue Antigens 10:259–266

Kantarjian AD (1961) A syndrome clinically resembling amyotrophic lateral sclerosis following chronic mercurialism. Neurology 11:639–644

Kascsak RJ, Shope RE, Donnenfeld H, Bartfeld H (1978) Antibody response to arboviruses. Absence of increased response in amyotrophic lateral sclerosis and multiple sclerosis. Arch Neurol 35:440–442

Kascsak RJ, Carp RI, Vilcek JT, Donenfield H, Bartfeld H (1982) Virological studies in amyotrophic lateral sclerosis. Muscle Nerve 5:93–101

Kawamura Y, Dyck PJ (1981) Permanent axotomy by amputation results in loss of motor neurons in man. J Neuropathol Exp Neurol 40:658–666

Keleman J, Hedlund W, Orlin JB, Berkman EM, Munsat TL (1983) Plasmapheresis with immunosuppression in amyotrophic lateral sclerosis. Arch Neurol 40:752–753

Kilness AW, Hochberg FH (1977) Amyotrophic lateral sclerosis in a high selenium environment. JAMA 237:2843–2844

Kimura F, Smith RG, Nyormoi O, Schneider T, Nastainczyk W, Hofmann F, Stefani E, Appel SH. (1994) Amyotrophic lateral sclerosis patient antibodies label Ca^{2+} channel α_1 subunit. Ann Neurol 35:164–171

Koerner DR (1976) Abnormal carbohydrate metabolism in amyotrophic lateral sclerosis and parkinsonism-dementia on Guam. Diabetes 25:1055–1065

Kohne DE, Gibbs CJ, White L, Tracy SM, Meinke W, Smith RA (1981) Virus detection by nucleic acid hybridization: examination of normal and ALS tissues for the presence of poliovirus. J Gen Virol 56:223–233

Kondo K, Tsubaki T (1981) Case–control studies of motor neuron disease. Arch Neurol 38:220–226

Kondo H, Osborne ML, Kolhouse JF (1981) Nitrous oxide has multiple deleterious effects on cobalamin metabolism and causes decreases in activities of both mammalian cobalamin-dependent enzymes in rats. J Clin Invest 12:1270–1283

Kott E, Livni E, Zamir R, Kuritzky A (1976) Amyotrophic lateral sclerosis: cell-mediated immunity to poliovirus and myelin basic protein in patients with high frequency of HLA-BW35. Neurology 26:376–377

Kott E, Livni E, Zamir R, Kuritzky A (1979) Cell-mediated immunity to polio and HLA antigens in amyotrophic lateral sclerosis. Neurology 29:1040–1044

Kurent JE, Brooks BR, Madden DL, Sever JL, Engel WK (1979) CSF viral antibodies. Evaluation in amyotrophic lateral sclerosis and late-onset postpoliomyelitis progressive muscular atrophy. Arch Neurol 36:269–273

Kurlander HM, Patten BM (1979) Metals in spinal cord tissue of patients dying of motor neuron disease. Ann Neurol 6:21–24

Kurtzke JF, Beebe GW (1980) Epidemiology of amyotrophic lateral sclerosis. I. A case–control comparison based on ALS deaths. Neurology 30:453–462

Lambert WC, Ororodudu AO, Lambert MW (1986) Hypersensitivity of ALS lymphoblastoid cells in culture to the mutagen methyl methane sulfonate. Neurology 36(Suppl 1):136

Lampson LA, Kushner PD, Sobel RA (1988) Strong expression class II major histocompatibility complex (MHC) antigens in the absence of detectable T-cell infiltration in amyotrophic lateral sclerosis. J Neuropath Exp Neurol 47:353

Lang B, Newson-Davis, Wray D, Vincent A, Murphy N (1981) Autoimmune etiology for myasthenia (Eaton–Lambert) syndrome. Lancet 11:224–226

Latov N, Hays AP, Donofrio PD et al. (1988) Monoclonal IgM with unique specificity to gangliosides GM1 and GD1b and to lacto-N-tetraose associated with human motor neuron disease. Neurology 38:763–768

Lehrich JR, Couture J (1978) Amyotrophic lateral sclerosis sera are not cytotoxic to neuroblastoma cells in tissue culture. Ann Neurol 4:384

Lehrich JR, Oger J, Arnason BGW (1974) Neutralizing antibodies to poliovirus and mumps virus in amyotrophic lateral sclerosis. J Neurol Sci 23:537–540

Leigh PN, Anderton BH, Dobson A, Gallo J-M, Swash M, Power DM (1988) Ubiquitin deposits in anterior horn cells in motoneuron disease. J Neurosci 93:197–202

Llinas R, Sugimori M, Cherksey BD, Smith RG, Delbono O, Stefani E, Appel SH (1993) IgG from amyotrophic lateral sclerosis patients increases current through P-type calcium channels in mammalian cerebellar purkinje cells and in isolated channel protein in lipid bilayer. Proc Natl Acad Sci USA 90:11743–11747

Lowe J, Lermoy G, Jefferson D et al. (1988) A filamentous inclusion body within anterior horn neurons in motor neurone disease defined by immunocytochemical localization of ubiquitin. Neurosci Lett 94:203–210

Magnelli V, Sawada T, Delbono O, Smith RG, Appel SH, Stefani E (1993) Amyotrophic lateral sclerosis immunoglobulins action on single skeletal muscle Ca^{2+} channels. J Physiol 461:103–118

Mandybur TI, Cooper GP (1979) Increased spinal cord lead content in amyotrophic lateral sclerosis – possibly a secondary phenomenon. Med Hypoth 5:1313–1315

Manetto V, Sternberger NH, Perry G, Sternberger LA, Gambetti P (1988) Phosphorylation of neurofilaments is altered in amyotrophic lateral sclerosis. J Neuropathol Exp Neurol 47:642–653

Mann DMA, Yates PO (1974) Motor neuron disease: the nature of the pathogenic mechanisms. J Neurol Neurosurg Psychiatry 37:1036–1046

Manton WI, Cook JD (1979) Lead content of cerebrospinal fluid and other tissue in amyotrophic lateral sclerosis (ALS). Neurology 29:611–612

Masui Y, Mozai T, Kakeh K (1985) Functional and morphometric study of the liver in motor neuron disease. J Neurol 232:15–19

Matthews WB (1958) Metabolic disease of the nervous system: clinical aspects. Proc R Soc Med 51:859–863

McComas AJ, Upton ARM, Sica REP (1973) Motor neuron disease and aging. Lancet ii:1477–1480

McEwan-Alvarado G, Hightower N, Carney LR, Barrier CW (1971) Exocrine pancreas function in patients with amyotrophic lateral sclerosis. Dig Dis 16:107–110

McGeer PL, Itagatei S, McGeer E (1988) Expression of the histocompatibility glycoprotein HLA-DR in neurological disease. Acta Neuropathol 76:550–557

McManaman JL, Crawford FG, Stewart AA, Appel SH (1988) Purification of a skeletal muscle polypeptide which stimulates choline acetyltransferase activity in cultured spinal cord neurons. J Biol Chem 263:5890–5897

Mendell JR, Chase TN, Engel WK (1971) Amyotrophic lateral sclerosis. A study of central monoamine metabolism and therapeutic trial of levodopa. Arch Neurol 25:320–325

Mitsuma T, Nogimori T, Adachi K, Mukoyama M, Ando K (1984) Concentrations of immunoreactive TRH releasing hormone in spinal cord of patients with amyotrophic lateral sclerosis. Am J Med Sci 287:34–36

Mitsuma T, Adachi K, Mukoyama M, Ando K (1986) Concentrations of thyrotropin-releasing hormone in the brain of patients with amyotrophic lateral sclerosis. J Neurol Sci 76:277–281

Mitsumoto H, Salgado ED, Negroski D et al. (1986) Amyotrophic lateral sclerosis: effects of acute intravenous and chronic subcutaneous administration of thyrotropin-releasing hormone in controlled trials. Neurology 36:152–159

Miyata S, Nakamura S, Nagata H, Kameyama M (1983) Increased manganese level in spinal cords of amyotrophic lateral sclerosis determined by radiochemical neutron activation analysis. J Neurol Sci 61:283–293

Mora CA, Garruto RM, Brown P et al. (1988) Seroprevalence of antibodies to HTLV-I in patients with chronic neurological disorders other than tropical spastic paraparesis. Ann Neurol 23(Suppl):S192–S195

Mori H, Kondo J, Ihara Y (1987) Ubiquitin is a component of paired helical filaments in Alzheimer's disease. Science 235:1641–1646

Moxley RT, Griggs RC, Forbes GB, Goldblatt D (1983) Influence of muscle wasting on oral glucose tolerance testing. Clin Sci 64:601–609

Mueller PS, Quick DT (1970) Studies of glucose, insulin and lipid metabolism in amyotrophic lateral sclerosis and other neuromuscular disorders. J Lab Clin Med 76:190–201

Muller WK, Hilgenstock F (1975) An uncommon case of amyotrophic lateral sclerosis with isolation of a virus from the CSF. J Neurol 211:11–23

Munoz DG, Greene C, Perl DP, Selkoe DJ (1988) Accumulation of phosphorylated neurofilaments in anterior horn motoneurons of amyotrophic lateral sclerosis patients. J Neuropathol Exp Neurol 47:9–18

Murai A, Miyahara T, Tanaka T (1983) Abnormalities of lipoprotein and carbohydrate metabolism in degenerative diseases of the nervous system – motor neuron disease and spinocerebellar degeneration. Tohoku J Exp Med 139:365–376

Nagata Y, Okuya M, Watanabe R, Honda M (1982) Regional distribution of cholinergic neurons in human spinal cord transections in the patients with and without motor neuron disease. Brain Res 244:223–230

Nakano Y, Hirayama K, Terao K (1987) Hepatic ultrastructural changes and liver dysfunction in amyotrophic lateral sclerosis. Arch Neurol 44:103–106

Officer JE, Tecson N, Estes JD, Fontanilla E, Rangey RW, Gardner MB (1973) Isolation of a neurotropic type-C virus. Science 181:945–945

Olarte MR, Shafer SQ (1985) Levamisole is ineffective in the treatment of amyotrophic lateral sclerosis. Neurology 35:1063–1066

Oldstone MBA, Wilson CB, Perrin LH, Norris FH (1976) Evidence for immune-complex formation in patients with amyotrophic lateral sclerosis. Lancet ii:169–172

Ono S, Mannen T, Toyokura Y (1989) Differential diagnosis between amyotrophic lateral sclerosis and spinal muscular atrophy by skin involvement. J Neurol Sci 91:301–310

Oppenheim RW, Haverkamp LF, Prevette D, McManaman JL, Appel SH (1988) Reduction of naturally occurring motoneuron death in the chick embryo in vivo by a target-derived neurotrophic factor. Science 240:919–922

Ordonez G, Sotelo J (1989) Antibodies against fetal muscle proteins in serum from patients with amyotrophic lateral sclerosis. Neurology 39:683–686

Osame M, Matsumoto M, Usuku K et al. (1987) Chronic progressive myelopathy associated with elevated antibodies to human T-lymphotrophic virus Type I and adult T-cell leukemia-like cells. Ann Neurol 21:117–122

Patrick J, Lindstrom V (1973) Autoimmune response to acetylcholine receptor. Science 180:871–872

Patten BM, Engel WK (1982) Phosphate and parathyroid disorders associated with the clinical syndrome of amyotrophic lateral sclerosis. Adv Neurol 36:181–200

Pedersen L, Platz P, Sersild C, Thomsen M (1977) HLA (SD and LD) in patients with amyotrophic lateral sclerosis (ALS). J Neurol Sci 31:313

Peppard RJ, Guttman M, Martin WRW, Eisen A, Calne DB (1989) Dopaminergic deficits demonstrated in Caucasian amyotrophic lateral sclerosis using positron emission tomography. Neurology 39(Suppl 1):400

Perry G, Friedman R, Shaw G, Chau V (1987a) Ubiquitin is detected in neurofibrillary tangles and senile plaque neurites of Alzheimer's disease brains. Proc Natl Acad Sci USA 84:3033–3036

Perry TL, Hansen S, Jones K (1987b) Brain glutamate deficiency in amyotrophic lateral sclerosis. Neurology 37:1845–1848

Pertschuk LP, Cook AW, Gupta JK et al. (1977) Jejunal immunopathology in amyotrophic lateral sclerosis and multiple sclerosis. Identification of viral antigens by immunofluorescence. Lancet i:1119–1123

Pestronk A, Drachman DB, Griffin JW (1980) Effects of aging on nerve sprouting and regeneration. Exp Neurol 70:65–82

Pestronk A, Adams RN, Clawson L et al. (1988) Serum antibodies to GM1 ganglioside in amyotrophic lateral sclerosis. Neurology 38:1457–1461

Pestronk A, Adams RN, Cornblath D et al. (1989) Patterns of serum antibodies to GM1 and GD1a ganglioside in ALS. Ann Neurol 25:98–102

Petito CK, Navia BA, Cho EJ, Jordan BD, George DC, Price RW (1985) Vascular myelopathy pathologically resembling subacute combined degeneration in patients with the acquired immunodeficiency syndrome. N Engl J Med 312:874–879

Petkau A, Sawatzky A, Hillier CR, Hoogstraten J (1974) Lead content of neuromuscular tissue in amyotrophic lateral sclerosis: case report and other considerations. Br J Ind Med 31:275–287

Pieper SJL, Fields WS (1957) Failure of ALS to respond to intrathecal steroid and vitamin B_{12}. Arch Neurol 19:522–526

Pierce-Ruhland R, Patten BM (1980) Muscle metals in motor neuron disease. Ann Neurol 8:193–195

Pittman RH, Oppenheim RW (1979) Cell death of motoneurons in the chick embryo spinal cord. J Comp Neurol 187:425–446

Plaitakis A, Caroscio JT (1987) Abnormal glutamate metabolism in amyotrophic lateral sclerosis. Ann Neurol 22:575–579

Plaitakis A, Constantakakis E, Smith J (1988a) The neuroexcitotoxic amino acids glutamate and aspartate are altered in the spinal cord and brain in amyotrophic lateral sclerosis. Ann Neurol 24:446–449

Plaitakis A, Smith J, Mandelei J, Yahr MD (1988b) Pilot trial of branched-chain amino acids in amyotrophic lateral sclerosis. Lancet i:1015–1018

Price RW, Brew R, Sidtis J, Rosenblum M, Scheck AC, Cleary P (1988) The brain in AIDS CNS HIV-I infections and AIDS dementia complex. Science 239:586–592

Przedborski S, Lilsnard C, Hildebrand J (1986) HTLV III and vacuolar myelopathy. N Engl J Med 315:63

Quick DT, Greer M (1967) Pancreatic dysfunction in patients with amyotrophic lateral sclerosis. Neurology 17:112–116

Robbins JH, Otsuka F, Tarone RE, Polinski RJ, Brumback RA, Nee LE (1985) Parkinson's disease and Alzheimer's disease: hypersensitivity to X-rays in cultured cell lines. J Neurol Neurosurg Psychiatry 48:916–923

Roelofs-Iverson RA, Muldur DW, Elveback LR, Kurland LT, Molguard CA (1984) ALS and heavy metals: a pilot case–control study. Neurology 34:393–395

Roisen FJ, Bartfeld H, Donnenfeld H, Baxter J (1982) Neuron specific in vitro cytotoxicity of sera from patients with amyotrophic lateral sclerosis. Muscle Nerve 5:48–53

Rosen DR, Siddique T, Patterson D et al. (1993) Mutations in Cu/Zn superoxide dismutase gene are associated with familial amyotrophic lateral sclerosis. Nature 362:59–62

Rothstein JD, Martin LJ, Kuncl RW (1992) Decreased glutamate transport by the brain and spinal cord in amyotrophic lateral sclerosis. N Engl J Med 326:1464–1468

Rowland LP (1982) Diverse forms of motor neuron diseases. Adv Neurol 36:1–13

Saffer D, Morley J, Bill PLA (1977) Carbohydrate metabolism in motor neurone disease. J Neurol Neurosurg Psychiatry 40:533–537

Salazar AM, Masters CL, Gajdusek C, Gibbs CJ (1983) Syndromes of amyotrophic lateral sclerosis and dementia: relation to transmissible Creutzfeldt–Jakob disease. Ann Neurol 14:17–26

Sar M, Stumpf WE (1977) Androgen concentration in motor neurons of cranial nerves and spinal cord. Science 187:77–80

Schauf CL, Antel JP, Arnason BGW, Davis FA, Rooney MW (1980) Neuroelectric blocking activity and plasmapheresis in amyotrophic lateral sclerosis. Neurology 30:1011–1013

Schwartz M, Sela BA, Esher N (1982) Antibodies to gangliosides and myelin autoantigens are produced in mice following sciatic nerve injury. J Neurochem 38:1192–1195

Scott JM, Dinn JJ, Wilson P, Wier DG (1981) Pathogenesis of subacute combined degeneration: A result of methyl group deficiency. Lancet ii:334–339

Seigmaliet J, Cadilhac J, Lapinski H (1979) HLA and amyotrophic lateral sclerosis. Sem Hop Paris 55:1239

Shy ME, Rowland LP, Smith T et al. (1986) Motor neuron disease and plasma cell dyscrasia. Neurology 36:1429–1436

Shy ME, Evans VA, Lublin FD et al. (1987) Anti-GM1 antibodies in motor neuron disease patients without plasma cell dyscrasia. Ann Neurol 22:167

Simpson JA (1969) Myasthenia gravis: a new hypothesis. Scott Med J 5:419–436

Smith RG, Hamilton S, Hofmann F et al. (1992) Serum antibodies to skeletal muscle-derived L-type calcium channels in patients with amyotrophic lateral sclerosis. N Engl J Med 327:1721–1728

Smith RG, Alexianu ME, Crawford G, Nyormoi O, Stefani E, Appel SH (1994) The cytotoxicity of immunoglobulins from amyotrophic lateral sclerosis patients on a hybrid motoneuron cell line. Proc Natl Acad Sci USA 91:3393–3397

Spencer PS, Roy DN, Ludolph A, Hugon J, Dwivedi MP, Schaumberg HH (1986) Lathyrism: Evidence for role of the neuroexcitatory amino acid BOAA. Lancet i:1066–1067

Spencer PS, Nunn PB, Hugon J (1987) Guam amyotrophic lateral sclerosis-parkinsonism-dementia linked to a plant excitant neurotoxin. Science 237:465–564

Stefansson K, Marton LS, Dieperink ME, Malnar GK, Schlaepfer WW, Helgason CM (1985) Circulating autoantibodies to the 200 kDa protein of neurofilaments in the serum of healthy individuals. Science 228:1117–1119

Steinke J, Tyler HR (1964) The association of amyotrophic lateral sclerosis (motor neuron disease) and carbohydrate intolerance, a clinical study. Metabolism 13:1376–1381

Stober T, Stelte W, Kunze K (1983) Lead concentrations in blood, plasma erythrocytes, and cerebrospinal fluid in amyotrophic lateral sclerosis. J Neurol Sci 61:21–26

Stober T, Schimrigk K, Dietzsch S, Theilen T (1985) Intrathecal thyrotropin-releasing hormone therapy of amyotrophic lateral sclerosis. J Neurol 232:13–14

Swash M, Scholtz CL, Vowles G, Ingram DA (1988) Selective and asymmetric vulnerability of corticospinal and spinocerebellar tracts in motor neuron disease. J Neurol Neurosurg Psychiatry 51:785–789

Tachovsky TG, Lisak RP, Koprovski AN, Theofilotoulos AM, Dixon FJ (1976) Circulating immune complexes in multiple sclerosis and other neurologic diseases. Lancet ii:977–999

Tandon R, Robison SH, Munzer JS, Bradley WG (1985) Deficient DNA repair in amyotrophic lateral sclerosis cells. Neurology 35(Suppl 1):73

Tavolato BF, Licandro AC, Saia A (1975) Motor neurone disease: an immunological study. Eur Neurol 13:433–440

Tom MI, Richardson JC (1951) Hypoglycemia from islet cell tumor of pancreas with amyotrophy and cerebrospinal nerve cell changes. J Neuropathol Exp Neurol 10:57–66

Tomlinson BE, Irving D (1977) The number of limb motor neurons in the human lumbosacral cord throughout life. J Neurol Sci 34:213–219

Touzeau G, Kato AC (1983) Effects of amyotrophic lateral sclerosis sera on cultured cholinergic neurons. Neurology 33:317–322

Touzeau G, Kato AC (1986) ALS serum has no effect on three enzymatic activities in human spinal cord neurons. Neurology 36:573–576

Troost D, Vanden Oord JJ, deJong JMBV (1988) Analysis of the inflammatory infiltrate in amyotrophic lateral sclerosis. J Neuropath Appl Neurobiol 14:255–256

Uemura E, Hartmann HA (1978) Age-related changes in RNA content and the volume of the human hypoglossal neuron. Brain Res Bull 3:207–211

Urbanek K, Jansa P (1974) Amyotrophic lateral sclerosis. Abnormal cellular inflammatory response. Arch Neurol 30:186–187

Utterback RA, Cummins AJ, Cape CA, Goldenberg J (1970) Pancreatic function in amyotrophic lateral sclerosis. J Neurol Neurosurg Psychiatry 33:544–547

Vernant JC, Maurs L, Gessain A et al. (1987) Endemic tropical spastic paraparesis associated with human T-lymphotropic virus type I. A clinical and seroepidemiological study of 25 cases. Ann Neurol 21:123–130

Viola MV, Lazarus, Antel J, Roos R (1982) Nucleic acid probes in the study of amyotrophic lateral sclerosis. Adv Neurol 36:317–329

Weiner LP (1980) Possible role of androgen receptors in amyotrophic lateral sclerosis. A hypothesis. Arch Neurol 37:129–131

Westall FC, Rubin R, Nieder J, Jablecki C (1983) Low percentage T-micron cells in amyotrophic lateral sclerosis. Immunol Lett 7:139–140

Whitehouse PJ, Wamsley JK, Zarbin MA, Tourrellottee WW, Kuhar MJ (1983) Amyotrophic lateral sclerosis: alterations in neurotransmitter receptors. Ann Neurol 14:8–16

Wilson SAK (1907) The amyotrophy of chronic lead poisoning: Amyotrophic lateral sclerosis of toxic origin. Rev Neuronal Psychiatry 5:441–445

Wolfgram F, Myers L (1973) Amyotrophic lateral sclerosis: effect of serum on anterior horn cells in culture. Science 179:579–580

Yase Y (1972) The pathogenesis of amyotrophic lateral sclerosis. Lancet ii:292–296

Yoshimasu F, Yasui M, Yase Y et al. (1980) Studies on amyotrophic lateral sclerosis by neutron activation analysis. II. Comparative study of analytical results on Guam PD, Japanese ALS, and Alzheimer's disease cases. Fol Psychiatry Neurol 34:75–82

Yoshino Y (1984) Possible involvement of folate cycle in the pathogenesis of amyotrophic lateral sclerosis. Neurochem Res 9:387–391

Yoshino Y, Koike H, Akai K (1979) Free amino acids in motor cortex of amyotrophic lateral sclerosis. Experientia 35:219–220

Zolan WJ, Ellis-Neill L (1986) Concentration of aluminium, manganese, iron and calcium in four southern Guam rivers. University of Guam, Agana, Technical report 64:68

12 Neurotrophic Factors and Neurodegeneration

J.E. Martin

Introduction

Neurotrophic factors (NTFs) are molecules which act to promote the differentiation of neurons and to maintain their phenotype. The discovery by Hamburger and Levi-Montalcini in 1949 that the number of neurons in dorsal root ganglia was related to the size of the target field suggested that there exists a mechanism of neuronal support by which factors produced by target organs are able to support neurons projecting to them. The isolation of nerve growth factor (NGF) proved that such a factor could indeed promote the growth of a specific type of neuron (Levi-Montalcini and Hamburger 1953), and the importance of NGF in neuronal development was indicated in experiments showing that specific antibodies against NGF could interfere with the normal development of dorsal root ganglion neurons (Johnson et al. 1980). The neurotrophic hypothesis (Davies 1991; Gage et al. 1991) that has developed from these and other pioneering studies can be summarised as follows:

1. Neurons are supported and regulated by their respective neurotrophic factors.
2. Proper maintenance of these neurons depends on an adequate supply and utilisation of these neurotrophic factors.
3. Interference with neurotrophic factors support or utilisation will result in defective development, performance or degeneration of target neurons.

A more specific application of the neurotrophic theory applies during the normal loss of neurons during development (Oppenheim 1991). This proposes that survival of neurons during this phase is actively determined by neurotrophic factors produced in limiting amounts by target tissues (Barde 1989).

Neurotrophic factors may act as mitogens, or as factors promoting differentiation, cell death, and the extent or direction of neurite outgrowth (neurotropic agents). The full range of such activities is not required, however, for a substance to be classified as a neurotrophic factor. Indeed, different neurotrophic factors have diverse effects in different tissues at different stages of development, in vivo and in vitro. The fibroblast growth factors (FGFs) for example have a striking mitogenic effect on fibroblasts and endothelial cells, but also act as neurotrophic factors, promoting division, differentiation and survival of particular groups of neurons (Wagner 1991). Understanding of the mechanisms of action of neurotrophic factors is incomplete, and the complexity of their actions at various sites makes generalisation about the properties of these agents difficult, particularly since their actions on a specific group of neurons in vitro may differ from those in vivo.

NGF is the best characterised and perhaps the most specific neurotrophic factor. Other factors include brain-derived neurotrophic factor (BDNF), neurotrophins -3,

-4 and -5 (NT3, NT4, NT5), ciliary neurotrophic factor (CNTF), the fibroblast growth factor (FGFs), platelet-derived growth factor (PDGF), insulin-like growth factor (IGF) and epidermal growth factor (EGF) (Hofer and Barde 1988; Maisonpierre et al. 1990b; Barde 1989; Berkemeier et al. 1991; Wagner 1991; Richardson 1991; Barres and Raff 1993).

The Neurotrophins

The neurotrophins are related members of a gene family. They are target-derived agents which are transported to the neural cell bodies innervating that target tissue. The neurotrophins are closely related structurally, with about 50% homology of their amino acid sequences. This structural similarity, leading to cross-reactivity of antibodies raised against neurotrophic factors, accounts for the unreliability of some early reports of the actions, tissue levels and distribution of certain members of the family (Whittemore and Seiger 1987). All members have conserved domains which determine their structure and binding to a common low-affinity receptor (Rodriguez-Tebar et al. 1990), and variable regions, probably external loops, which determine their specificity and binding to specific high-affinity receptors. The members of the neurotrophin family are NGF, BDNF, NT3, NT4 and NT5.

Nerve Growth Factor

Levi-Montalcini and Hamburger first described a factor produced by mouse sarcoma cells that elicited extensive growth of chick embryo neurons (Levi-Montalcini and Hamburger 1953). This substance was found to be present in snake venoms and in large quantities in the mouse submaxillary gland and was termed nerve growth factor (NGF) (Levi-Montalcini and Angeletti 1968; Bocchini and Angeletti 1969).

NGF is a protein dimer composed of two identical polypeptide chains each 118 amino acids long. Each chain contains three antiparallel pairs of β-strands, which are linked by loops, and three central disulphide bonds derived from the six cysteine residues which are a constant feature of all neurotrophins (McDonald et al. 1991; Lo 1992). The dimer is held together by hydrophobic bonds. A region on the surface of the dimer rich in positively charged residues may represent the low affinity receptor binding site, where the receptor itself has a cysteine-rich negatively charged site responsible for binding NGF (Welcher et al. 1991). The amino-terminus structure of NGF appears to be flexible. Two serine residues (the first two amino acid residues) are essential for high efficacy action of NGF, whereas the subsequent seven amino acid residues can be deleted with little effect on activity. The functional specificity of the related neurotrophins appears to be conferred by variation in the external loop regions (Ibanez et al. 1991a,b). Thus, highly conserved areas may be responsible for binding to the low affinity receptor p75 (Rodriguez-Tebar et al. 1990), whilst high affinity receptor binding (*trk* oncogene products) may well be mediated by the specific external loop regions (Park 1991; Lo 1992). High affinity binding of NGF, however, requires expression of both low affinity and high affinity receptors (Hempstead et al. 1991).

NGF is important in the development and maintenance of sympathetic and neural crest-derived sensory neurons, both in vivo and in vitro (Levi-Montalcini and Angeletti 1968; Gunderson and Barrett 1980; Johnson et al. 1980). Withdrawal of NGF by administration of antibodies or autoimmunisation to NGF prevents neural crest-derived sensory neuronal and sympathetic neuronal development in vivo (Levi-Montalcini and Booker 1960; Johnson et al. 1980). Placode-derived sensory neurons, however, are not NGF-dependent (Pearson , J. et al. 1983). NGF may also act as neurotrophic agent for cholinergic neurons of the basal forebrain and striatum (Gnahn et al. 1983; Mobley et al. 1985; Dreyfus 1989). These neurons are immunoreactive for the NGF receptor during development and in the adult (Taniuchi et al. 1986), and can retrogradely transport exogenous NGF injected into the hippocampus (Schwab et al. 1979). NGF promotes sprouting of rat cholinergic neurons in vivo, enhances neuronal cholinergic function and promotes cholinergic neuronal graft survival (Gage et al. 1991).

The exposure of animals to NGF antibodies in utero results in loss of up to 80% of mammalian dorsal root ganglion (DRG) neurons (Johnson et al. 1980, 1982, 1983; Ruit et al. 1992). A minority of DRG neurons survive, however, and these neurons are of larger diameter than the susceptible population (Johnson et al. 1980; Ruit et al. 1992). The sensory neurons that are NGF-dependent project to laminae I and II of the dorsal horn, and are thus concerned with nociception and thermoreception. The NGF-independent DRG neurons project mainly to laminae III and IV (mechanoreceptors) and the ventral horn (muscle afferents) (Ruit et al. 1992). Many of the smaller DRG neurons are also known to express the neuropeptide substance P (Verge et al. 1989), and DRG neurons containing substance P are known to be especially vulnerable to the effects of NGF deprivation (Ross et al. 1981; Lindsay and Harmar 1989). Thus, different subsets of sensory neurons may display differing sensitivity to the effects of NGF deprivation. In contrast, when NGF is used to rescue DRG cells from naturally occurring cell death or from death following axotomy, its effects are less selective (Ruit et al. 1992). Thus, the selectivity of action of NGF on groups of sensitive neurons may be dose-dependent, with physiological levels providing greatest discrimination.

NGF is thought to act in the guidance of developing and regenerating axons (neurotropism). NGF mRNA has been detected in many epithelia and in mesenchyme, but during development NGF does not appear in these target areas until after sensory axons first arrive (Davies et al. 1987; Bothwell 1991). Similarly, in the mouse heart and salivary gland, which have a rich sympathetic innervation, expression of NGF mRNA does not occur prior to the ingrowth of sympathetic axons (Korsching and Thoenen 1988). Thus, there appears to be no target-derived trophic effect of NGF at early stages of development, but it may act to provide trophic support to neurons and their connections after the establishment of innervation (Davies et al. 1987).

Schwann cells in peripheral nerve express low levels of NGF, and upregulate this expression in response to axotomy (Heumann et al. 1987). This response may be mediated by interleukin-1 released by activated microglia at the site of injury (Lindholm et al. 1987). A useful experimental model of the effects of NGF on damaged neurons has been the transection of the fimbria/fornix pathway. This leads to degeneration and death of many of the septal and diagonal band neurons which project to the ipsilateral hippocampus. An explanation of this cell death is that axotomy interrupts the supply of NGF to these neurons from post-synaptic neurons or glial cells in the hippocampus (Gage et al. 1986). These cells can be rescued from degeneration and death by administration of purified NGF by infusion into the lateral ventricles (Kromer 1987; Gage et al. 1986).

NGF is not thought to have a neurotrophic effect on motor neurons. Administration of NGF does not prevent motor neuron death during development (Oppenheim et al. 1992) nor does it increase the size or synthesis of neurotransmitters by motor neurons (Yan et al. 1988). NGF does not rescue motor neurons from cell death following axotomy (Miyata et al. 1986; Yan et al. 1988). NGF receptors are, however, expressed on motor neurons for a limited period during development (Raivich et al. 1985, 1987; Yan and Johnson 1988), and NGF is retrogradely transported from motor neuron nerve terminals towards the cell body only at this period during rat development (Yan et al. 1988). The significance of this NGF receptor expression and NGF uptake by motor neurons is not yet clear.

The biochemical effects of NGF have been much studied using the rat phaeochromocytoma cell line, PC12 (Tsao et al. 1990). This is a clonal cell line which is not dependent on NGF for survival and which has not been exposed to NGF. When NGF is added to the medium of PC12 cells, there is a change of phenotype from chromaffin-type to neuronal-type, with extension of neurites (Greene and Tischler, 1976; Greene and Rein, 1977; Greene and Tischler, 1982). It is thus possible to study the molecular events following NGF-induced neuronal differentiation in the PC12 cell line, and this has proved fruitful in the study of signal transduction mechanisms from receptor binding to phenotypic change.

NGF has a variety of metabolic effects on cells in vivo and in vitro (Gage et al. 1991; Halegoua et al. 1991). Primary biochemical events within the cytoplasm link receptor activation to activation signals to the nucleus and the subsequent modulation of gene activity (Roberts 1992). A secondary event within cells stimulated by growth factors is the expression of primary response genes or immediate early genes (Halegoua et al. 1991). The signal transduction process is discussed in more detail below.

Sympathetic and sensory neurons in tissue culture die when deprived of NGF, therefore NGF may function during normal development and adult life as the active suppresser of an intrinsic cell death programme (Martin et al. 1988). This is not proved, however, and NGF may act to prevent cell death by a variety of mechanisms (Edwards et al. 1991).

Brain-Derived Neurotrophic Factor

Brain-derived neurotrophic factor (BDNF) was first identified in mammalian brain by Barde and colleagues (1982). Administration of BDNF can reduce naturally occurring cell death in certain populations of neurons in the developing embryo (Hofer and Barde 1988). Leibrock and colleagues (1989) showed that, although BDNF mRNA is also present in relatively low levels in the mammalian brain, expression is highest in the hippocampus, particularly in the CA3 region (Ernfors et al. 1990b). BDNF mRNA has also been detected in heart, lung and skeletal muscle (Maisonpierre et al. 1990a). BDNF-responsive neurons are all thought be present in, or to project to, the central nervous system (Thoenen 1991)

BDNF has a high (~50%) sequence homology to NGF. As a member of the neurotrophin family it has the characteristic three disulphide bridges stabilising the tertiary structure of the molecule. Despite similarities of structure, in its actions BDNF shows a different pattern of neuronal specificity to NGF (Barde 1989). It supports the survival of embryonic retinal ganglion cells, mesencephalic dopaminergic neurons and spinal motor neurons which are not supported by NGF. Cholinergic neurons of the basal forebrain respond to both NGF and BDNF. BDNF also supports placode- and neural crest-derived sensory neurons which are not responsive to NGF (Davies, 1987). The expression of BDNF is regulated by neuronal activity and glutamate receptor activation (Thoenen 1991; Lindefors et al. 1992). BDNF may rescue motor neurons from axotomy-induced or naturally occurring cell death (Yan et al. 1992; Oppenheim et al. 1992; Sendtner et al. 1992a). BDNF may also protect dopaminergic neurons against neurotoxic damage (Spina et al. 1992). BDNF binds to the low affinity NGF receptor, as do all the neurotrophins (Rodriguez-Tebar et al. 1990), and to the high affinity NGF receptor p145 *trkB*.

Neurotrophins -3, -4 and -5

Neurotrophin-3 (NT3) was isolated by cloning techniques (Ernfors et al. 1990a; Hohn et al. 1990; Maisonpierre et al. 1990b). NT3 was also termed hippocampal-derived neurotrophic factor (Ernfors et al. 1990a,b). The gene encoding NT3 is located on chromosome 12. The highest expression of NT3 mRNA is in neurons of the adult hippocampus, in particular in the pyramidal neurons of the CA1 and CA2 pyramidal cell fields (Hohn et al. 1990; Ernfors et al. 1990b). In the developing embryo, NT3 expression is most prominent in those areas of the CNS where proliferation, differentiation and migration of neurons are occurring (Maisonpierre et al. 1990a). NT3 expression is highest shortly after birth, in contrast with expression of BDNF, which peaks 2–3 weeks after birth (Maisonpierre et al. 1990a). In the adult rat, the only other brain areas which express NT3 mRNA are the taenia tecta and the induseum griseum, both developmentally derived from the hippocampal formation (Ernfors et al. 1990b). In the periphery, NT3 mRNA is present in several tissues, including ovarian secondary follicular epithelium, but with highest levels in the kidney and spleen (Ernfors et al. 1990b; Hohn et al. 1990; Maisonpierre et al. 1990a).

NT3, NT4 and possibly NT5 bind to the low affinity NGF receptor p75, whereas NT3 binds to the high affinity receptor p145 *trkC* and NT4 to p145 *trkB* (Ip et al. 1992a; Meakin and Shooter 1992).

NT4 has been identified in *Xenopus*, viper, rat and in humans, and is encoded by a gene on chromosome 19 (Ip et al. 1992a). NT4 has a 55% homology with NGF, and appears to be expressed most strongly in *Xenopus* ovary (Hallbook et al. 1991). NT4 promotes neurite outgrowth from explanted dorsal root ganglia and, to a lesser extent, the nodose ganglia, but not sympathetic ganglia (Hallbook et al. 1991). Further details of the mode of action, the tissue distribution and the significance of NT4 and NT5 expression will no doubt soon be forthcoming.

Ciliary Neurotrophic Factor and Related Cytokines

CNTF is a small (20–24 Da) polypeptide which was purified from chick ocular tissue and from rat and rabbit sciatic nerve (Adler et al. 1979; Barbin et al. 1984;

Manthorpe et al. 1986). It was identified as a target-derived molecule that supported the survival of avian parasympathetic and ciliary neurons in culture (Adler et al. 1979). CNTF is now known to support retinal, hippocampal, sympathetic, sensory and spinal motor neurons in culture (Barbin et al. 1984; Arakawa et al. 1990). In particular, it can inhibit growth and promote differentiation of sympathetic neurons, and can induce cholinergic differentiation of mature sympathetic neurons (Saadat et al. 1989). CNTF will also lead to upregulation of the expression of the low affinity NGF receptor in a variety of neurons both in vitro and in vivo (Magal et al. 1991; Hagg et al. 1992). The action of CNTF on particular neuronal groups may be enhanced with the addition of basic fibroblast growth factor (bFGF) (Arakawa et al. 1990). bFGF also potentiates the induction of axonal and end-plate sprouting from motor neurons by CNTF (Gurney et al. 1992). CNTF can also promote the differentiation of glial progenitor cells into type II astrocytes (Davis and Yancopoulos 1993). The relationship between data from in vitro studies and the role of CNTF in the support of these neuronal groups in vivo is not yet clear (Arakawa et al. 1990).

There are several reasons why CNTF is unlikely to function as a widely acting target-derived neurotrophic factor, similar to NGF, BDNF and other members of the neurotrophin family (Thoenen 1991). Firstly, CNTF is a cytosolic rather than a secretory protein (Stockli et al. 1989). Secondly, the site of CNTF production does not correspond with the projection fields of CNTF-responsive neurons. Finally, the timing of CNTF expression during development does not coincide with the period of naturally occurring cell death. CNTF and its mRNA have been demonstrated in rat sciatic nerve, spinal cord and optic nerve, principally in Schwann cells and white matter glial cells (Dobrea et al. 1992). The neurotrophic effects of CNTF, however, are not mediated only by non-neuronal cells (Unsicker et al. 1992).

CNTF may act as a lesion factor, being released from cells during injury to promote local regeneration and preventing cell death after axotomy. At birth levels of CNTF and CNTF mRNA are extremely low, but they rise rapidly to adult levels within one week after birth (Stockli et al. 1989). Transection of the facial nerve of adult rats results in chromatolysis of facial nucleus motor neurons and reactive gliosis, but no neuronal degeneration (Tetzlaff et al. 1988). An identical lesion in newborn animals results in degeneration of most facial nucleus motor neurons (Sendtner et al. 1990), and the extent of degeneration produced by nerve transection decreases rapidly after birth (Snider and Thanedar 1989), following the rise in CNTF levels (Stockli et al. 1989; Arakawa et al. 1990). CNTF has been shown to prevent degeneration of facial motor neurons after axotomy (Sendtner et al. 1990) and it can also rescue medial septal neurons from cell death following fimbria-fornix transection (Hagg et al. 1992). In addition, CNTF has been shown to prevent the degeneration of motor neurons in the *pmn* mouse mutant which develops a progressive motor neuronopathy with hindlimb paralysis within the first 3 weeks of life (Schmalbruch et al. 1991; Sendtner et al. 1992b).

CNTF is thought to belong to the leukaemia inhibitory factor (LIF)/interleukin 6 (IL-6)/oncostatin M cytokine group of molecules, with diverse actions on a wide variety of tissues (Thoenen 1991; Patterson 1992; Davis and Yancopoulos 1993). CNTF and LIF share signalling pathways which are mediated, at least in part by the IL-6 receptor component gp130 (Ip et al. 1992b). Other subunits of the CNTF receptor (R) are LIFRβ and CNTRFa (Davis and Yancopoulos 1993). Soluble CNTRFa is found in the CSF and is upregulated during peripheral nerve injury, and may be related to the presence of activated macrophages (Davis and Yancopoulos 1993).

Platelet-Derived Growth Factor

PDGF is one of the major serum factors that are mitogenic for fibroblasts. It was first identified in human platelets from which PDGF (and other factors) are released during platelet activation to act as mitogen for mesenchymal cells during wound repair (Richardson 1991). It consists of covalently linked homo- or heterodimers of two related subunits, the A and B chains. It binds and crosslinks two surface receptors with differing ligand specificities: the alpha receptor can bind A or B chains, whilst the beta can bind only B chains. Thus the various PDGF isoforms may act as agonists or competitive antagonists depending on the subunit composition of the receptor (Heldin and Westermark 1990, Richardson 1991).

PDGF is present in neurons during development and in adulthood (Yeh et al. 1991; Sasahara et al. 1991) but the role of this protein in these sites is not yet clear (Richardson, 1991). It may act as a mitogen and chemoattractant for glial cells, particularly the O2A progenitor cells of the oligodendrocyte lineage. O2A progenitor cells express PDGFaR (the alpha isotype of the PDGF receptor) and are stimulated to divide by all three isoforms of PDGF (Richardson et al. 1990). Axons may therefore attract oligodendrocytes using PDGF as a signal (Barres and Raff 1993). PDGF has been demonstrated in axon terminals (Sasahara et al. 1991), and PDGF may therefore play an additional role in nerve-target interaction (Richardson 1991).

The Fibroblast Growth Factors

The fibroblast growth factors (FGFs) are a family of at least seven closely related peptide growth factors. They were first isolated from bovine pituitary (Gospodarowicz 1974) and then purified from bovine brain (Gospodarowicz et al. 1978). The FGF family of proteins all bind strongly to heparin affinity columns (Shing et al. 1984). On the basis of their heparin affinity during salt elution from binding columns there are two main classes of FGF, the acidic (class I) FGFs (aFGF) and the basic (class II) FGFs (bFGF) (Wagner 1991). aFGF and bFGF have been shown to bind not only to heparin, but also to related extracellular matrix glycosaminoglycans.

aFGF and bFGF have been studied under a variety of names (Gospodarowicz et al. 1987). The gene coding for aFGF is located in chromosome 5 and that for bFGF on chromosome 4 (Jaye et al. 1986; Mergia et al. 1986). Both genes code for polypeptides of 154 amino acids in length (Abraham et al. 1986; Jaye et al. 1986) with bFGF and aFGF showing 55% structural homology. bFGF has been isolated from a wide range of tissues, whereas aFGF has been detected in brain, bone and retina (Gospodarowicz et al. 1987).

FGF receptors of 125 kDa and 145kDa exist and are tyrosine kinases. Both receptors bind aFGF and bFGF has a higher affinity for the 145 kDa receptor (Neufeld and Gospodarowicz 1986). After receptor binding, there is rapid internalisation of aFGF and bFGF and eventual lysosomal degradation. Activation of the receptor leads to MAP kinase and protein kinase C activation (Sano et al. 1992).

Both aFGF and bFGF stimulate neural differentiation in PC12 cells with effects that are similar to the action of NGF (Schubert et al. 1987). The effect of aFGF on PC12 cells is greatly potentiated by heparin, possibly by extending the biological half-life of aFGF by increasing the resistance of aFGF (and bFGF) to proteolysis (Gospodarowicz and Cheng 1986; Damon et al. 1989). aFGF and bFGF also initiate differentiation in chromaffin cells, as does NGF, but in contrast to NGF they do not promote long term survival of these cells (Wagner 1991). Both aFGF and bFGF promote the survival of chick parasympathetic ciliary ganglion cells, but have no effect on sympathetic neurons (Unsicker et al. 1987). In the central nervous system, aFGF and bFGF promote the survival of neurons from the hippocampus, striatum, septum, thalamus and several areas of the cerebral cortex as well as embryonic chick spinal motor neurons; however only aFGF, but not bFGF, will support the survival of retinal ganglion cells and cells from the subiculum (Walicke 1988a,b). aFGF helps to protect adult sensory neurons from axotomy induced cell death, and bFGF promotes the survival of cholinergic neurons in the septum after axonal transection (Otto et al. 1987; Anderson et al. 1988).

bFGF and aFGF are cytoplasmic molecules and do not appear to be secreted during normal activity and functioning of the nervous system (Abraham et al. 1986; Stockli et al. 1989; Klagsbrun 1989; Arakawa et al. 1990). Thus the significance of in vitro studies of neurotrophic support of cortical, hippocampal and embryonic chick spinal motor neurons by FGFs is not certain (Walicke et al. 1986; Arkawa 1990).

bFGF is known to increase choline acetyltransferase (ChAT) activity in spinal cord cell cultures (McManaman et al. 1989), as does interferon-γ (Erkman et al. 1989). This effect is likely to be mediated by astrocytes, which produce CNTF and bFGF in vitro (Ferrara et al. 1988). aFGF and bFGF are mitogenic for astrocytes and stimulate the stellate phenotype of the mature astrocyte and the increased expression of glial fibrillary acidic protein (Wagner 1991). Not all the neurotrophic actions of the FGFs, however, are thought to be mediated through the actions of FGFs on glial cells (Walicke and Baird 1988, Unsicker et al. 1992).

FGFs were initially described as mitogens for fibroblasts and endothelial cells and therefore have a wide range of activity on a variety of tissues during development (Wagner 1991). FGF-like activity appears to be responsible, with TGFβ, for the induction of mesoderm (Munaim et al. 1988) and a related member of the FGF family, the *int*-2 oncogene product, may play a role in the development of the neuroepithelium (Wilkinson et al. 1988).

Many questions concerning the role of FGFs in the development and maintenance of the nervous system remain unanswered. The unusual ability of FGFs to bind to heparin and related extracellular matrix glycosaminoglycans may provide the basis for a unique mechanism of storing, stabilising or presenting these growth factors (Wagner, 1991).

Insulin-like Growth Factors

Insulin-like growth factors I and II (IGF-I and IGF-II) are synthesised in a wide range of tissues and are present at relatively high concentrations in serum and cerebrospinal fluid (Zapf et al. 1981; Haselbacher and Humbel 1982). The receptors for IGF-I and IGF-II show no structural similarity. The IGF-I receptor is a tetrameric structure with two α subunits forming the ligand-binding domain, and the β subunits acting as tyrosine kinases (Cullen et al. 1991). The IGF-II receptor is a single unit with 15 cysteine rich repeat sequences and a small region homologous to the fibronectin collagen binding domain (Pusztai et al. 1993).

IGF-I, IGF-II and insulin itself all help, to a limited extent, to support the survival of chick embryonic motor neurons in culture (Arakawa et al. 1990). The effect of insulin on motor neurons may be mediated by IGF-I receptors, since high concentrations of insulin are required to produce this effect (Arakawa et al. 1990). IGF-I also enhances the regeneration of sensory fibres in the lesioned rat sciatic nerve (Kanje et al. 1989).

The role of these factors is not yet clear for neural or peripheral tissues. They may act as autocrine or paracrine agents rather than true neurotrophic factors.

Epidermal Growth Factor

The epidermal growth factor (EGF) family of structurally related peptides includes EGF, transforming growth factor-α amphiregulin and at least four other agents (Pusztai et al. 1993). The peptides all contain a cysteine-rich domain. They are produced as transmembrane proteins which may then be cleaved by proteases, and released as active molecules into the extracellular space (Gill et al. 1987).

The biological effects of EGF are primarily mitogenic on peripheral non-neuronal tissues (Pusztai et al. 1993), but in PC12 cells EGF will also lead to cell division (Chao, 1992). This contrasts with the effects of NGF which leads to neuronal differentiation with neurite outgrowth and the arrest of cell growth (Greene and Tischler 1976; Greene and Rein 1977; Greene and Tischler 1982). The effects of EGF and NGF on PC12 cells raise questions about the specificity of the signalling mechanisms of neurotrophic factors since, whilst producing very different phenotypic effects, the early biochemical events in the response evoked both by EGF and NGF appear similar (Chao 1992).

Growth Factor Receptors and Signal Transduction Pathways

Most available information on neurotrophic factor receptors and signal transduction pathways concerns NGF and the related neurotrophins (Johnson et al. 1986; Large et al. 1988; Sutter et al. 1990; Meakin and Shooter 1992). Some details are known of the structure and action of the CNTF receptor (Davis and Yancopoulos 1993), but much less is known about the neuronal receptors and events for the other neurotrophic factors. NGF and BDNF, and low-affinity NGF receptors are found on neurons that do not respond either to NGF or BDNF (Yan and Johnson 1988; Ernfors et al. 1988, 1989) suggesting that other neurotrophins may also act via this receptor (Rodriguez-Tebar et al. 1990; Meakin and Shooter 1992). High-affinity binding is achieved, usually in combination with the low-affinity NGF receptor (Hempstead et al. 1989; Hempstead et al. 1991; Anderson 1992) through binding with *trk* proto-oncogene products. There are at least three members of the *trk* family: p140 *trkA*; p145 *trkB*; and p145 *trkC*. The *trkB* receptor binds BDNF, NT3 and NT4, and *trkC* receptor binds NT3 (Park 1991; Meakin and Shooter 1992). The receptors in this family are tyrosine kinases, as are the FGF receptors.

The CNTF receptor complex also appears to link three subunits, with CNTRFa found in association with gp130 and LIFRβ. CNTF interaction with the receptor complex leads to gp130 and LIFRβ subunit phosphorylation with the activation of intracellular kinases (Ip et al. 1992b; Davis and Yancopoulos 1993).

Primary biochemical events within the cytoplasm link receptor activation with activation signals to the nucleus and the subsequent modulation of gene activity (Roberts 1992). Many of these biochemical events involve phosphorylation (Halegoua and Patrick 1980; Aletta et al. 1988; Aletta et al. 1989; Tsao et al. 1990). These events link binding of the growth factor to the receptor tyrosine kinase with subsequent activation of $p21^{ras}$. $p21^{ras}$ activates Raf-1, which in turn activates MAP (mitogen activated protein) kinase kinase (MAPKK) upstream of MAP kinase. MAP kinase then activates $pp90^{rsk}$ (ribosomal S6 kinase) and the latter can translocate to the nucleus to phosphorylate transcription factors such as c-*jun*, c-*fos* and SRF (Gomez and Cohen 1991; Roberts 1992). Understanding of this cascade of signals must, however, be incomplete since it does not explain the selective actions of individual neurotrophic factors on specific cell types (Chao 1992).

Secondary events within cells stimulated by growth factors include the expression of primary response genes or immediate early genes (Halegoua et al. 1991). Primary response genes activated by neuronal growth factors include those coding for structural proteins such as actin, and genes coding for proteins with a regulatory function, in particular transcription factors such as c-*fos* and c-*jun*. These transcription factors are the same factors activated by growth factors via the MAP kinase cascade (Roberts, 1992).

Neurotrophic Factors and Neurodegeneration

Neurodegenerative disease might result from a failure of neurotrophic factor maintenance (Appel 1981). As little work has yet been done on the normal levels of expression of neurotrophic factors in human tissue it is not possible to rule out neurotrophic factor deprivation as a primary pathogenetic mechanism in neurodegenerative diseases such as motor neuron disease (MND). However, intracellular biochemical events that contribute to cell damage or death when neurons are deprived of neurotrophic factors could be relevant to cell damage produced by mechanisms, irrespective of a growth factor-related aetiology.

NGF and Alzheimer's Disease

Ageing of the central nervous system is associated with atrophy and loss of cholinergic neurons of the basal forebrain and with impaired cognitive function

(Bartus et al. 1982; Coyle et al. 1983; Fischer et al. 1989). In rodents these changes may be ameliorated by the implantation of fetal basal forebrain cells or by NGF infusion, both of which improve cholinergic function (Gage et al. 1983; Fischer et al. 1987). In patients with Alzheimer's disease, cholinergic neurons in the hippocampus and nucleus basalis of Meynert degenerate, and there is a strong correlation between the loss of cholinergic activity and the severity of the dementia (Perry et al. 1978; Wilcock et al. 1982; Coyle et al. 1983). In Alzheimer's disease, however, there is no evidence for a lack of NGF, and the level of NGF mRNA is the same as that in age-matched control brains (Goedert et al. 1986). In the surviving neurons of the nucleus basalis of Meynert, however, there is a three-fold increase in the levels of NGF receptor mRNA, suggesting that these neurons may respond to exogenous NGF in the same manner as the aged rats (Ernfors et al. 1990b). Restoration of cognitive function in aged rats with the infusion of NGF, however, raises the possibility that administering exogenous NGF might ameliorate symptoms in Alzheimer's disease, whilst not necessarily invoking trophic factor deprivation as a cause of the disorder (Gage et al. 1983, 1986).

Motor Neuron Disease and Neurotrophic Factors

The loss of trophic factor support has long been regarded as an attractive hypothesis for the death of motor neurons in diseases of the motor neuron (Appel 1981; Fidzianka et al. 1990). The concept of the loss of a neurotrophic factor supporting a specific neuronal population is a convenient explanation for the selective neuronal degeneration seen in motor neuron disease, and has parallels in the selective neuronal degeneration following loss of NGF trophic support of the sensory and autonomic nervous system in vivo (Anand et al. 1991). However, there have been few studies of the levels of neurotrophic factors in human tissues, but assay techniques are being refined and becoming more specific (Whittemore and Seiger 1987). One preliminary study screening for non-specific trophic factor deficits failed to show any such abnormality in human motor neuron disease, but definitive studies are awaited (Ebendal et al. 1989). Informative studies on the levels of neurotrophic factors in human MND await data on the agents, if any, that show specific neurotrophic activity for motor neurons.

Studies of agents which may provide neurotrophic support to motor neurons have failed to identify specific motor neuronal trophic factors operating during adult life in a similar way to the specific support of sensory neurons by NGF. It has become apparent, however, that several neurotrophic factors may act during development, and also during conditions of cell degeneration, to support the survival of motor neurons. Such agents include skeletal muscle extracts (Eagleson and Bennett 1983), astrocytes (Eagleson et al. 1985), CNTF, BDNF, IGF-I, IGF-II, insulin, aFGF and bFGF (Arakawa et al. 1990)

CNTF and Motor Neuron Disease

CNTF is known to support the survival of chick embryonic spinal motor neurons in culture (Arakawa et al. 1990). It is still unclear what role CNTF may play in the rescue and maintenance of motor neurons during normal development, since (at least in rats) it is not expressed during the period of naturally occurring motor neuron death in development (Oppenheim 1992). CNTF can, however, rescue avian motor neurons from death in vitro and in vivo (Arakawa et al. 1990; Nurcombe et al. 1991; Oppenheim et al. 1991) and prevents the death of facial motor neurons in neonatal rats following axotomy (Sendtner et al. 1990). Degeneration of motor neurons in the pmn/pmn (progressive motor neuronopathy) mouse mutant could be retarded by increasing CNTF levels in the circulation, although there is no evidence implicating CNTF in the pathogenesis of this disorder (Sendtner et al. 1992b). Trials of CNTF in human motor neuron disease are currently in progress.

BDNF, Motor Neurons and Parkinson's Disease

BDNF does not promote the survival of chick spinal motor neurons in culture, even at relatively high concentrations (Arakawa et al. 1990). Despite this, BDNF is retrogradely transported by motor neuron axons and can rescue motor neurons from axotomy-induced or naturally occurring cell death (Yan et al. 1992; Oppenheim et al. 1992; Sendtner et al. 1992a). Thus BDNF may have an indirect action on degenerating neurons, possibly via glial cells or a cofactor (Oppenheim et al. 1992; Sendtner et al. 1992a). BDNF is now another candidate for therapeutic trials in MND. BDNF supports the survival of mesencephalic dopaminergic neurons which are not supported by NGF. The administration of BDNF to cultures of such neurons or to cultures of dopaminergic neuroblastoma cells can also protect such cells from the neurotoxic effects of 6-hydroxydopamine (6-OHDA) and N-methyl-4-phenylpyridinium (MPP^+) (Spina et al. 1992). This cytoprotective action appears to be mediated via an upregulation of glutathione reductase (Spina et al. 1992). These studies raise the possibility of using BDNF as a cytoprotective agent during oxidative stress of dopaminergic neurons.

Acknowledgements
The author is a Wellcome Trust Research Fellow and holds the Gillson Scholarship in Pathology of the Worshipful Society of Apothecaries. Our work is supported by the Motor Neurone Disease Association of the United Kingdom.

References

Abraham JA, Mergia A, Whang JL, et al. (1986) Nucleotide sequence of a bovine clone encoding the angiogenic protein, basic fibroblast growth factor. Science 233:545–548

Alder R, Landa KB, Manthorpe M, Varon S (1979) Cholinergic neurotrophic factors: intraocular distribution of trophic activity for ciliary neurons. Science 204:1434–1436

Aletta JM, Lewis SA, Cowan NJ, Green LA (1988) Nerve growth factor regulates both the phosphorylation and steady-state levels of microtubule-associated protein 1.2 (MAP1.2). J Cell Biol 106:1573–1581

Aletta JM, Shelanski ML, Greene LA (1989) Phosphorylation of the peripherin 58 kDa neuronal intermediate filament protein. Regulation by nerve growth factor and other agents. J Biol Chem 264:4619–4627

Anand P, Rudge P, Mathias CJ et al. (1991) New autonomic and sensory neuropathy with loss of adrenergic sympathetic function and sensory neuropeptides. Lancet 337 (8752):1253–1254

Anderson DJ (1992) The highs and lows of an NGF receptor. Curr Biol 2; 461–463

Anderson KJ, Dam D, Lee S, Cotman CW (1988) Basic fibroblast growth factor prevents death of lesioned cholinergic neurons in vivo. Nature 332:360–361

Appel SH (1981) A unifying hypothesis for the cause of amyotrophic lateral sclerosis, parkinsonism and Alzheimer disease. Ann Neurol 10:499–505

Arakawa Y, Sendtner M, Thoenen H (1990) Survival effect of ciliary neurotrophic factor (CNTF) on chick embryonic motoneurons in culture: comparison with other neurotrophic factors and cytokines. J Neurosci 10:3507–3515

Barbin G, Manthorpe M, Varon S (1984) Purification of the chick eye ciliary neurotrophic factor. J Neurochem 43:1468–1478

Barde Y-A (1989) Trophic factors and neuronal survival. Neuron 2:1525–1534

Barde Y-A, Edgar D, Thoenen H (1982) Purification of a new neurotrophic factor from mammalian brain. EMBO J 1:549–553

Barres BA, Raff MC (1993) Proliferation of oligodendrocyte precursor cells depends on electrical activity in axons. Nature 361:258–260

Bartus RT, Dean RL, Beer B, Lippa AS (1982) The cholinergic hypothesis of geriatric memory dysfunction. Science 217:408–417

Berkemeier LR, Winslow JW, Kaplan DR, Nikolics K, Goeddel DV, Rosenthal A (1991) Neurotrophin-5; a novel neurotrophic factor that activates *trk* and *trkB*. Neuron 7:857–866

Bocchini V, Angeletti PU (1969) The nerve growth factor: purification as a 30 000-molecular weight protein. Proc Natl Acad Sci USA 64:787–794

Bothwell M (1991) Tissue localisation of nerve growth factor and nerve growth factor receptors. Curr Top Microbiol Immunol 165:55–70

Buck CR, Martinez HJ, Chao MV, Black IB (1988) Differential expression of the nerve growth factor receptor gene in multiple brain areas. Dev Brain Res 44:259–268

Chao MV (1992) Growth factor signaling: where is the specificity? Neuron 995–997

Coyle JT, Price DL, DeLong MR (1983) Alzheimer's disease: a disorder of cortical cholinergic innervation. Science 219:1184–1189

Cullen KJ, Yee D, Rosen N (1991) Insulin-like growth factors in human malignancy. Cancer Invest 9:443–454

Damon DH, Lobb RR, D'Amore A, Wagner JA (1989) Heparin potentiates the action of acidic fibroblast growth factor by prolonging its biological half life. J Cell Physiol 138:221–226

Davies AM (1987) Molecular and cellular aspects of patterning sensory neurone connections in the vertebrate nervous system. Development 101; 185–208

Davies AM (1991) Nerve growth factor synthesis and nerve growth factor receptor expression in neural development. Int Rev Cytol 128:109–138

Davies AM, Bandtlow C, Heumann R, Korsching S, Rohrer H, Thoenen H (1987) Timing and site of nerve growth factor synthesis in developing skin in relation to innervation and expression of the receptor. Nature 326:353–358

Davis S, Yancopoulos GD (1993) The molecular biology of the CNTF receptor. Curr Opin Neurobiol 3:20–24

Dobrea GM, Unnerstall JR, Rao MS (1992) The expression of CNTF message and immunoreactivity in the central and peripheral nervous system of the rat. Brain Res Dv Brain Res 66:209–219

Dreyfus CF (1989) Effects of nerve growth factor on cholinergic brain neurons. Trends Pharmacol Sci 10:145–149

Eagleson KL, Bennett MR (1983) Survival of purified motor neurons in vitro: effects of skeletal muscle-conditioned medium. Neurosci Lett 38:187–192

Eagleson KL, Raju TR, Bennett MR (1985) Motoneurone survival is induced by immature astrocytes from developing avian spinal cord. Dev Brain Res 17:95–104

Ebendal T, Askmark H, Aquilonius S-M (1989) Screening for neurotrophic disturbances in amyotrophic lateral sclerosis. Acta Neurol Scand 79:188–193

Edwards SN, Buckmaster AE, Tolkovsky AM (1991) The death programme in cultured sympathetic neurones can be suppressed at the post-translational level by nerve growth factor, cyclic AMP and depolarisation. J Neurochem 57:2140–2143

Erkman L, Wuarin L, Cadelli D, Katao AC (1989) Interferon induces astrocyte maturation causing an increase in cholinergic properties of cultured human spinal cord cells. Dev Biol 132:375–388

Ernfors P, Hallbook F, Ebendal T, Shooter EM, Radeke MJ, Misko TP, Persson H (1988) Developmental and regional expression of β-nerve growth factor receptor mRNA in the chick and the rat. Neuron 1:983–996

Ernfors P, Henschen A, Olson L, Persson H (1989) Expression of nerve growth factor receptor mRNA is developmentally regulated and increased after axotomy in rat spinal cord motoneurons. Neuron 2:1605–1613

Ernfors P, Ibanez CF, Ebendal T, Olson L, Persson H (1990a) Molecular cloning and neurotrophic activities of a protein with structural similarities to β-nerve growth factor: developmental and topographical expression in the brain. Proc Natl Acad Sci USA 87:5454–5458

Ernfors P, Wetmore C, Olson L, Persson H (1990b) Identification of cells in rat brain and peripheral tissues expressing mRNA for members of the nerve growth factor family. Neuron 5:511–526

Ferrara N, Ousley F, Gospodarowicz D (1988) Bovine brain astrocytes express basic fibroblast growth factor, a neurotrophic and angiogenic mitogen. Brain Res 462:223–232

Fidzianka A, Goebel HH, Warlo I (1990) Acute infantile spinal muscular atrophy. Muscle apoptosis as a proposed pathogenetic mechanism. Brain 113:433–445

Fischer W, Wictorin K, Bjorklund A, Williams LR, Varon S, Gage FH (1987) Amelioration of cholinergic neuron atrophy and spatial memory impairment in aged rats by nerve growth factor. Nature 329:65–68

Fischer W, Gage FH, Bjorklund A (1989) Degenerative changes in forebrain cholinergic nuclei correlate with cognitive impairment in aged rats. Eur J Neurosci 1:34–45

Gage FH, Bjorklund A, Stenevi U, Dunnett SB (1983) Functional correlates of compensatory collateral sprouting by aminergic and cholinergic afferents in the hippocampal formation. Brain Res 268:39–47

Gage FH, Wictorin K, Fischer W, Williams LR, Varon S, Bjorklund A (1986) Life and death of cholinergic neurons in the septal and diagonal band region following complete fimbria-fornix transection. Neuroscience 19:241–255

Gage FH, Tuszynski MH, Chen KS, Fagan AM, Higgins GA (1991) Nerve growth factor function in the central nervous system. Curr Top Microbiol Immunol 165:72–93

Gill GN, Bertics PJ, Santon JB (1987) Epidermal growth factor and its receptor. Mol Cell Endocrinol 51:169–186

Gnahn H, Hefti F, Heumann R, Schwab ME, Thoenen H (1983) NGF-mediated increase of choline acetyltransferase (ChAT) in the neonatal rat forebrain: evidence for a physiological role of NGF in the brain? Dev Brain Res 9:45–52

Goedert M, Fine A, Hunt SP, Ullrich A (1986) Nerve growth factor mRNA in peripheral and central rat tissue and in the human central nervous system: lesion effects in the rat brain and levels in Alzheimer's disease. Mol Brain Res 1:85–92

Gomez N, Cohen P (1991) Dissection of the protein kinase cascade by which nerve growth factor activates MAP kinases. Nature 353:170–173

Gospodarowicz D (1974) Localisation of a fibroblast growth factor and its effect alone with hydrocortisone on 3T3 cell growth. Nature 249:123–129

Gospodarowicz D, Bialecki H, Greenberg G (1978) Purification of fibroblast growth factor activity from bovine brain. J Biol Chem 253:3736–3743

Gospodarowicz D, Cheng J (1986) Heparin protects basic and acidic FGF from inactivation. J Cell Physiol 128:475–484

Gospodarowicz D, Neufeld G, Schweigerer L (1987) Fibroblast growth factor: structural and biological properties. J Cell Physiol (Suppl) 5:15–26

Greene LA, Rein G (1977) Synthesis, storage and release of acetylcholine by a noradrenergic phaeochromocytoma cell line. Nature 268:349–351

Greene LA, Tischler AS (1976) Establishment of a noradrenergic clonal line of rat adrenal phaeochromocytoma cells which respond to nerve growth factor. Proc Natl Acad Sci USA 73:2424–2428

Greene LA, Tischler AS (1982) Phaeochromocytoma cultures in neurobiological research. Adv Cell Neurobiol 3:373–414

Gunderson RW, Barrett JN (1980) Characterisation of the turning response of dorsal root neurites toward nerve growth factor. J Cell Biol 87:546–554

Gurney ME, Yamamoto H, Kwon Y (1992) Induction of motor neuron sprouting in vivo by ciliary neurotrophic factor and basic fibroblast growth factor. J Neurosci 12:3241–3247

Hagg T, Quon D, Higaki J, Varon S (1992) Ciliary neurotrophic factor prevents neuronal degeneration and promotes low affinity NGF receptor expression in the adult rat CNS. Neuron 8:145–158

Halegoua S, Patrick J (1980) Nerve growth factor mediates phosphorylation of specific proteins. Cell 22:571–581

Halegoua S, Armstrong RC, Kremer NE (1991) Dissecting the mode of action of a neuronal growth factor. Curr Top Microbiol Immunol 165:119–170

Hallbook F, Ibanez CF, Persson H (1991) Evolutionary studies of the nerve growth factor family reveal a novel member abundantly expressed in *Xenopus* ovary. Neuron:845–858

Hamburger V, Levi-Montalcini R (1949) Proliferation, differentiation and degeneration in the spinal ganglia of the chick embryo under normal and experimental conditions. J Exp Zool 111:457–502

Haselbacher G, Humbel R (1982) Evidence for two species of insulin-like growth factor II (IGFII and "big" IGFII) in human spinal fluid. Endocrinology 11:1822–1824

Heldin C-H, Westermark B (1990) Platelet-derived growth factor: mechanism of action and possible in vitro function. Cell Regulation 1:555–566

Hempstead BL, Martin-Zanca D, Kaplan DR, Parada LF, Chao MV (1991) High affinity NGF binding requires coexpression of the *trk* proto-oncogene and the low-affinity NGF receptor. Nature 350:678–683

Hempstead BL, Patil N, Olson K, Chao M (1988) Molecular analysis of the nerve growth factor receptor. Cold Spring Harbor Symp Quant Biol 53:477–485

Hempstead BL, Schleifer LS, Chao MV (1989) Expression of functional nerve growth factor receptors after gene transfer. Science 243:373–375

Heumann R, Lindholm D, Bandtlow C (1987) Differential regulation of mRNA encoding nerve growth factor and its receptor in rat sciatic nerve during development, degeneration, and regeneration: role of macrophages. Proc Natl Acad Sci USA 84:8735–8739

Hofer MM, Barde Y-A (1988) Brain-derived neurotrophic factor prevents neuronal death in vivo. Nature 331:261–262

Hohn A, Leibrock J, Bailey K, Barde Y-A (1990) Identification and characterisation of a novel member of the nerve growth factor/brain-derived neurotrophic factor family. Nature 344:339–341

Ibanez CF, Ebendal T, Persson H (1991a) Chimeric molecules with multiple neurotrophic activities reveal structural elements determining the specificities of NGF and BDNF. EMBO J 10:2105–2110

Ibanez CF, Hallbook F, Soderstrom S, Ebendal T, Persson H (1991b) Biological and immunological properties of recombinant human, rat and chicken nerve growth factors: a comparative study. J Neurochem 57:1033–1041

Ip NY, Ibanez CF, Nye SH et al. (1992a) Mammalian neurotrophin-4: structure, chromosomal localisation, tissue distribution, and receptor specificity. Proc Natl Acad Sci USA 89:3060–3064

Ip NY, Nye SH, Boulton TG et al. (1992b) CNTF and LIF act on neuronal cells via shared signaling pathways that involve the IL-6 signal transducing receptor component gp130. Cell 69:1121–1132

Jaye M, Howk R, Burgess W et al. (1986) Human endothelial cell growth factor: cloning, nucleotide sequence and chromosome localisation. Science 233:541–545

Johnson D, Lanahan A, Buck CR, Sehgal A, Morgan C, Mercer E, Bothwell M, Chao MV (1986) Expression and structure of the human NGF receptor. Cell 47:545–554

Johnson EM, Gorin PD, Brandeis LD, Pearson J (1980) Dorsal root ganglion neurons are destroyed by exposure in utero to maternal antibody to nerve growth factor. Science 210:916–918

Johnson EM, Gorin PD, Osborne PA, Rydel RE, Pearson J (1982) Effects of autoimmune NGF deprivation in the adult rabbit and offspring. Brain Res 240:131–140

Johnson EM, Osborne PA, Rydel RE, Schmidt RE, Pearson J (1983) Characterization of the effects of autoimmune nerve growth factor deprivation in the developing guinea pig. Neuroscience 8:631–642

Kanje M, Skottner A, Sjoberg J, Lundborg G (1989) Insulin-like growth factor-I (IGFI) stimulates regeneration of the rat sciatic nerve. Brain Res 486:396–398

Klagsbrun M (1989) The fibroblast growth factor: structural and biological properties. Prog Growth Factor Res 1:207–235

Korsching S, Thoenen H (1988) Developmental changes of nerve growth factor levels in sympathetic ganglia and their target organs. Dev Biol 126:40–46

Kromer LF (1987) Nerve growth factor treatment after brain injury prevents neuronal death. Science 235:214–216

Large TH, Weskamp G, Helder JC et al. (1988) Structure and developmental expression of the nerve growth factor receptor in the chicken central nervous system. Neuron 2:1123–1134

Leibrock J, Lottspeich F, Hohn A et al. (1989) Molecular cloning and expression of brain-derived neurotrophic factor. Nature 341:149–152

Levi-Montalcini R, Angeletti PU (1968) The nerve growth factor. Physiol Rev 48:534–569

Levi-Montalcini R, Booker B (1960) Destruction of the sympathetic ganglia in mammals by an antiserum to NGF. Proc Natl Acad Sci USA 42:384–391

Levi-Montalcini R, Hamburger V (1953) A diffusible agent of mouse sarcoma producing hyperplasia of sympathetic ganglia and hyperneurotization of viscera in the chick embryo J Exp Zool 123:233–288

Linderfors N, Ballarin M, Ernfors P, Falkenberg T, Persson H (1992) Stimulation of glutamate receptors increases expression of brain-derived neurotrophic factor mRNA in rat hippocampus. Ann NY Acad Sci 648:296–299

Lindholm D, Heumann R, Meyer M, Thoenen H (1987) Interleukin-1 regulates synthesis of nerve growth factor in non-neuronal cells of rat sciatic nerve. Nature 330:658–660

Lindsay RM, Harmar AJ (1989) Nerve growth factor regulates expression of neuropeptide genes in adult sensory neurones. Nature 337:362–364
Lo DC (1992) NGF takes shape. Curr Biol 2:67–69
Magal E, Burnham P, Varon S (1991) Effect of CNTF on low-affinity NGF receptor expression by cultured neurons from different rat brain regions. J Neurosci Res 30:560–566
Maisonpierre PC, Belluscio L, Friedman B et al. (1990a) NT-3, BDNF and NGF in the developing rat nervous system: parallel as well as reciprocal patterns of expression. Neuron 5:501–509
Maisonpierre PC, Belluscio L, Squinto S et al. (1990b) Neurotrophin-3: a neurotrophic factor related to NGF and BDNF. Science 247:1446–1451
Manthorpe M, Skaper SD, Williams LR, Varon S (1986) Purification of adult rat sciatic nerve ciliary neuronotrophic factor. Brain Res 367:282–286
Martin DP, Schmidt RE, DiStefano PS, Lowry OH, Carter JG, Johnson EM (1988) Inhibitors of protein synthesis and RNA synthesis prevent neuronal death caused by nerve growth factor deprivation. J Cell Biol 106:829–844
McDonald NQ, Lapatto R, Murray-Rust J, Gunning J, Wlodawer A, Blundell TL (1991) A new protein fold revealed by a 2.3Å resolution crystal structure of nerve growth factor. Nature 354:411–414
McManaman J, Crawford F, Clark R, Ricker J, Fuller F (1989) Multiple neurotrophic factors from skeletal muscle: demonstration of effects of basic fibroblast growth factor and comparisons with the 22-kilodalton choline acetyltransferase development factor. J Neurochem 53:1763–1771
Meakin SO, Shooter FM (1992) The nerve growth factor family of receptors. Trends Neurosci 15:323–331
Mergia A, Eddy R, Abraham JA, Fiddes JC, Shows TB (1986) The genes for basic and acidic fibroblast growth factors are on different chromosomes. Biochem Biophys Res Commun 138:644–651
Miyata Y, Kashihara Y, Homma S, Kuno M (1986) Effects of nerve growth factor on the survival and synaptic function of Ia sensory neurons axotomised in neonatal rats. J Neurosci 6:2012–2018
Mobley WC, Rustkowski JT, Tennekoon GI, Buchanan K, Johnston MV (1985) Choline acetyltransferase activity in the striatum of neonatal rats increased by nerve growth factor. Science 229:284–287
Munaim SI, Klagsbrun M, Toole B (1988) Developmental changes in fibroblast growth factor in the chicken embryo limb bud. Proc Natl Acad Sci USA 85:8091–8093
Neufeld G, Gospodarowicz D (1986) Basic and acidic fibroblast growth factors interact with the same cell surface receptors. J Biol Chem 261:5631–5637
Nurcombe V, Wreford NG, Bertram JF (1991) The use of the optical dissector to estimate the total number of neurons in the developing chick lateral motor column: effects of purified growth factors. Anat Record 231:416–424
Oppenheim RW (1991) Cell death during development of the nervous system. Ann Rev Neurosci 14:453–501
Oppenheim RW (1992) High hopes of a trophic factor. Nature 358:451–452
Oppenheim RW, Prevette D, Yin QW, Collins F, MacDonald J (1991) Control of embryonic motoneuron survival in vivo by ciliary neutrotrophic factor. Science. 251:1616–1618
Oppenheim RW, Qin-Wei Y, Prevette D, Yan Q (1992) Brain-derived neurotrophic factor rescues developing avian motoneurons from cell death. Nature 360:755–757
Otto D, Unsicker K, Grothe C (1987) Pharmacological effects of nerve growth factor and fibroblast growth factor applied to the transected sciatic nerve on neuron death in adult rat dorsal root ganglia. Neurosci Lett 83:156–160
Park M (1991) Lonesome receptors find their mates. Curr Biol 1:248–250
Patterson PH (1992) The emerging neuropoietic cytokine family: first CDF/LIF, CNTF and IL-6; next ONC, MGF, GCSF? Curr Opin Neurobiol 2:94–97
Pearson J, Johnson EM, Brandeis L (1983) Effects of antibodies to nerve growth factor on intrauterine development of derivatives of cranial neural crest and placode in the guinea pig. Dev Biol 96:32–36
Pearson RCA, Gatter KC, Powell TPS (1983) Retrograde cell degeneration in the basal nucleus of monkey and man. Brain Res 261:321–326
Perry EK, Tomlinson BE, Blessed G, Bergmann K, Gibson PH, Perry RH (1978) Correlation of cholinergic abnormalities with senile plaques and mental test scores in senile dementia. Br Med J 2:1457–1459
Pusztai L, Lewis CE, Lorenzen J, McGee JO'D (1993) Growth factors: regulation of normal and neoplastic growth. J Pathol 169–201
Raivich G, Zimmermann A, Sutter A (1985) The spatial and temporal expression of βNGF receptor expression in the developing chick embryo. EMBO J 4:637–644
Raivich G, Zimmermann A, Sutter A (1987) Nerve growth factor (NGF) receptor expression in chicken cranial development. J Comp Neurol 256:229–245

Richardson WD (1991) PDGF in neurons. Curr Biol 1:162–164

Richardson WD, Raff M, Noble M (1990) The oligodendrocyte-type-2 astrocyte lineage. Semin Neurosci 2:445–454

Roberts TM (1992) A signal chain of events. Nature 360:534–535

Rodriguez-Tebar A, Dechant G, Barde Y-A (1990) Binding of brain-derived neurotrophic factor to the nerve growth factor receptor. Neuron 4:487–492

Ross M, Lofstrandh S, Gorin PD, Johnson EM, Schwartz JP (1981) Use of an experimental autoimmune model to define nerve growth factor dependency of peripheral and central substance P containing neurons in the rat. J Neurosci 1:1304–1311

Ruit KG, Elliott, Osborne PA, Yan Q, Snider WD (1992) Selective dependence of mammalian dorsal root ganglion neurons on nerve growth factor during embryonic development. Neuron 8:573–587

Saadat S, Sendtner M, Rohrer H (1989) Ciliary neurotrophic factor induces cholinergic differentiation of rat sympathetic neurons in culture. J Cell Biol 108:1807–1816

Sano M, Kitajima S (1992) Activation of microtubule-associated protein kinase in PC12D cells in response to both fibroblast growth factor and epidermal growth factor and concomitant stimulation of the outgrowth of neurities. J Neurochem 58:837–844

Sasahara M, Fries JWU, Raines EW (1991) PDGF B-chain in neurons of the central nervous system, posterior pituitary, and in a transgenic model. Cell 64:217–227

Schmalbruch H, Jensen H-JS, Bjaerg M, Kamieniecka Z, Kurland L (1991) A new mouse mutant with progressive motor neuronopathy. J Neuropathol Exp Neurol 50:192–204

Schubert D, Ling N, Baird A (1987) Multiple influences of a heparin-binding growth factor on neuronal development. J Cell Biol 104:635–643

Schwab ME, Otten U, Agid Y, Thoenen H (1979) Nerve growth factor (NGF) in the rat CNS: absence of specific retrograde axonal transport and tyrosine hydroxylase induction in locus coeruleus and substantia nigra. Brain Res 168:473–483

Sendtner M, Kreutzberg GW, Thoenen H (1990) Ciliary neurotrophic factor prevents the degeneration of motor neurons after axotomy. Nature 345:440–441

Sendtner M, Holtmann B, Kolbeck R, Thoenen H, Barde Y-A (1992a) Brain-derived neurotrophic factor prevents the death of motoneurons in newborn rats after nerve section. Nature 360:757–758

Sendtner M, Schmalbruch H, Stockli KA, Carroll P, Kreutzberg GW, Thoenen H (1992b) Ciliary neurotrophic factor prevents degeneration of motor neurons in mouse mutant progressive motor neuronopathy. Nature 358:502–504

Shing Y, Folkman J, Sullivan R, Butterfield C, Murray J, Klagsbrun M (1984) Heparin affinity: purification of a tumour-derived capillary endothelial cell growth factor. Science 223:1296–1298

Snider WD, Thanedar S (1989) Target dependence of hypoglossal motor neurons during development and in maturity. J Comp Neurol 279:489–498

Spina MB, Squinto SP, Miller J, Lindsay RM, Hyman C (1992) Brain-derived neurotrophic factor protects dopamine neurons against 6-hydroxydopamine and N-methyl-4-phenylpyridinium ion toxicity: involvement of the glutathione system. J Neurochem 59:99–106

Stockli KA, Lottspeich F, Sendtner M et al. (1989) Molecular cloning, expression and regional distribution of rat ciliary neurotrophic factor. Nature 342: 920–923

Sutter A, Riopelle RJ, Harris-Warwick RM, Shooter EM (1990) Nerve growth factor receptors. Characterisation of two distinct classes of binding sites on chick embryo sensory ganglia cells. J Biol Chem 254:5972–5982

Taniuchi M, Schweitzer JB, Johnson EJ (1986) Nerve growth factor receptor molecules in rat brain. Proc Natl Acad Sci USA 83:1950–1954

Tetzlaff W, Graeber MB, Bisby MA, Kreutzberg GW (1988) Increased glial fibrillary acidic protein synthesis in astrocytes during retrograde reaction of the rat facial nucleus. Glia 1:90–95

Thoenen H (1991) The changing scene of neurotrophic factors. Trends Neurosci 14:165–170

Tsao H, Aletta JM, Greene LA (1990) Nerve growth factor and fibroblast growth factor selectively activate a protein kinase that phosphorylates high molecular weight microtubule-associated proteins. J Biol Chem 265:15471–15480

Unsicker K, Reichert-Preibsch H, Schmidt R, Pettman B, Labourdette G, Sensenbrenner G (1987) Astroglial and fibroblast growth factors have neurotrophic functions for cultured peripheral and central nervous system neurons. Proc Natl Acad Sci USA 84:5459–5463

Unsicker K, Reichert-Preibsch H, Wewetzer K (1992) Stimulation of neuron survival by basic FGF and CNTF is a direct effect and not mediated by non-neuronal cells: evidence from single cell cultures. Brain Rev Dev Brain Res 65:285–288

Verge VMK, Richardson PM, Benoit R, Riopelle RJ (1989) Histochemical characterizations of sensory neurons with high affinity receptors for nerve growth factor. J Neurocytol 18:583–591

Wagner JA (1991) The fibroblast growth factors: an emerging family of neural growth factors. Curr Top Microbiol Immunol 165:95–118

Walicke PA (1988a) Basic and acidic fibroblast growth factors have trophic effects on neurons from multiple CNS regions. J Neurosci 8:2618–2627

Walicke PA (1988b) Interactions between basic fibroblast growth factor (FGF) and glycosaminoglycans in promoting neurite outgrowth. Exp Neurol 102:144–148

Walicke PA, Baird A (1988) Neurotrophic effect of basic and acidic fibroblast growth factors are not mediated through glial cells. Dev Brain Res 40:711–719

Walicke P, Cowan WM, Ueno N, Baird A, Gullemin R (1986) Fibroblast growth factor promotes survival of dissociated hippocampal neurons and enhances neurite-extension. Proc Natl Acad Sci USA 83:3012–3016

Welcher AA, Bitler CM, Radeke MJ, Shooter EM (1991) Nerve growth factor binding domain of the nerve growth factor receptor. Proc Natl Acad Sci USA 88:159–163

Whittemore SR, Seiger A (1987) The expression, localisation and functional significance of β-nerve growth factor in the central nervous system. Brain Res Rev 12:439–464

Wilcock GK, Esiri MM, Bowen DM, Smith CCT (1982) Alzheimer's disease. Correlation of cortical choline acetyltransferase activity with the severity of dementia and histological abnormalities. J Neurol Sci 57:407–417

Wilkinson DG, Peters G, Dickson C, McMahon AP (1988) Expression of the FGF-related proto-oncogene *int*-2 during gastrulation and neurulation in the mouse. EMBO J 7:691–694

Yan Q, Elliott J, Snider WD (1992) Brain-derived neurotrophic factor rescues spinal motor neurons from axotomy-induced cell death. Nature 360:753–755

Yan Q, Johnson EM (1988) An immunohistochemical study of the nerve growth factor receptor in developing rats. J Neurosci 8:3481–3489

Yan Q, Snider WD, Pinzone JJ, Johnson EM (1988) Retrograde transport of nerve growth factor (NGF) in motoneurons of developing rats: assessment of potential neurotrophic effects. Neuron 1:335—343

Yeh H-J, Rutt KG, Wang Y-X, Parks WC, Snider WD, Deuel TF (1991) PDGF A-chain is expressed by mammalian neurons during development and in maturity. Cell 64:209–216

Zapf J, Froesch ER, Humbel RE (1981) The insulin-like growth factors (IGF) of human serum: chemical and biological characterization and aspects of their possible physiological role. Curr Top Cell Regul 19:257–309

13 Somatic Motoneurons and Descending Motor Pathways. Limbic and Non-limbic Components

G. Holstege

Introduction

Amyotrophic lateral sclerosis (ALS) or motor neuron disease affects primarily the somatic motoneurons in spinal cord and brainstem, i.e. the motoneurons innervating the striated musculature of head and body. Only one nucleus containing somatic motoneurons has been shown to remain unaffected in ALS. This is the so-called nucleus of Onuf, located in the sacral spinal cord, the motoneurons of which innervate the striated musculature of the pelvic floor, including bladder and anal sphincters (Mannen et al. 1977). The sympathetic motoneurons in the intermediolateral cell column of the thoraco-lumbar spinal cord, giving rise to the sympathetic innervation of head and body, are not affected (Hughes 1982). The same is true for the parasympathetic motoneurons in the sacral spinal cord, innervating the bladder and the distal bowel (Mannen et al. 1977). Therefore, bladder and sphincter functions remain intact until the latest stages of the disease. On the other hand, the dorsal vagal nucleus in the caudal medulla oblongata, containing parasympathetic motoneurons innervating large parts of the body, has been reported to be affected in ALS patients (Hughes 1982).

In ALS not only the somatic motoneurons are affected, but also other regions of the brain. The most prominent of these is the corticobulbospinal tract and the neurons giving rise to these fibers, the layer V neurons in the motor cortex. However, observations of gliosis and degenerative changes have also been found in other parts of the forebrain, such as the thalamus, globus pallidus, field of Forel and hypothalamus (Hughes 1982). Apparently, not only the corticospinal tract, but also other motor systems take part in the disease.

Therefore, a thorough understanding of all components of the motor system is necessary to understand the different aspects and stages of ALS. In this chapter a review will be presented of the somatic motoneuronal cell groups in brainstem and spinal cord, with special emphasis on the nucleus of Onuf. Subsequently, the local and bulbospinal projections to these somatic motoneuronal cell groups (the *first system* in our concept) will be described. Next, the descending pathways belonging to the so-called somatic motor system (the *second system*) will be reviewed and, finally, a summary of the many newly discovered pathways related to the limbic system (the *third system*) will be given. In the Conclusions section this concept of the three motor systems will be discussed. It must be emphasized that most of what we know about the structure and function of the central nervous system is based on animal experiments. Therefore, the following survey of the motor system is based on data derived from work on rat, cat and monkey.

Somatic Motoneurons in Spinal Cord and Brainstem

Somatic Motoneurons in the Spinal Cord

The somatic motoneurons innervate striated muscles of body and limbs. They are located in the ventral part of the ventral horn of the spinal cord, called lamina IX by Rexed (1954). The motoneurons innervating one particular muscle form a group, occupying a circumscribed portion of lamina IX (Figs. 13.1–13.3). Rostrocaudally such a cell group can extend from 1 to 19 segments. For example, the medial gastrocnemius and soleus motor nuclei in the cat are located within the confines of one (L7) spinal segment (Burke et al. 1977), while the longissimus dorsi muscle motoneuronal cell group in the cat extends from C8 to L5 (Holstege et al. 1987; Fig. 13.3). The motoneuronal cell groups can be subdivided into a medial and a lateral motor column. The medial motor column is present throughout the length of the spinal cord and its motoneurons innervate the axial muscles, which include the neck muscles (Fig. 13.1). In the cat the lateral motor column is only present at the levels C5 to the upper half of T1 (cervical enlargement; Fig. 13.2) and from L4 to S1 (lumbosacral enlargement). Motoneurons in the cervical and lumbosacral lateral column innervate the muscles of the fore- and hindlimbs respectively.

The axial musculature, innervated by motoneurons in the medial motor column, consists of epaxial and hypaxial muscles. The epaxial muscles are innervated by branches of the dorsal rami of a spinal nerve and the hypaxial muscles by branches of the ventral rami. In the ventral horn, motoneurons innervating epaxial muscles are always located ventral to the motoneurons innervating hypaxial muscles (Sprague 1948; Smith and Hollyday 1983). The epaxial muscles function as extensors and lateral flexors of the head and vertebral column. They also fix the vertebral column and some of them (the rotators) rotate the vertebral column about its longitudinal axis.

Cervical Cord

Motoneurons in the upper cervical cord innervate the neck muscles. Several reports exist on the location of the neck muscle motoneuronal cell groups in the cat, which are summarized in Fig. 13.1 (Holstege and Cowie 1989). The phrenic nucleus occupies a special position among the cervical somatic motoneuronal cell groups, because its motoneurons innervate the diaphragm. Although the diaphragm is an axial muscle, it plays an essential role in respiration, which function is virtually independent of that of the other axial muscles. In the cat the phrenic nucleus is located in the ventromedial part of the ventral horn at the level of the most caudal portion of C4 and throughout the rostrocaudal extent of C5 and C6 (Duron et al. 1979). In the human it is located from C3 to C5 (Feldman 1986).

At the level of the C5 to T1 spinal segments in the cat the medial motor column is located in the ventral portion of the ventral horn. The epaxial muscle motoneurons, for example those innervating the longissimus dorsi, are located in the medial part of this area, while hypaxial muscle motoneurons such as those innervating the

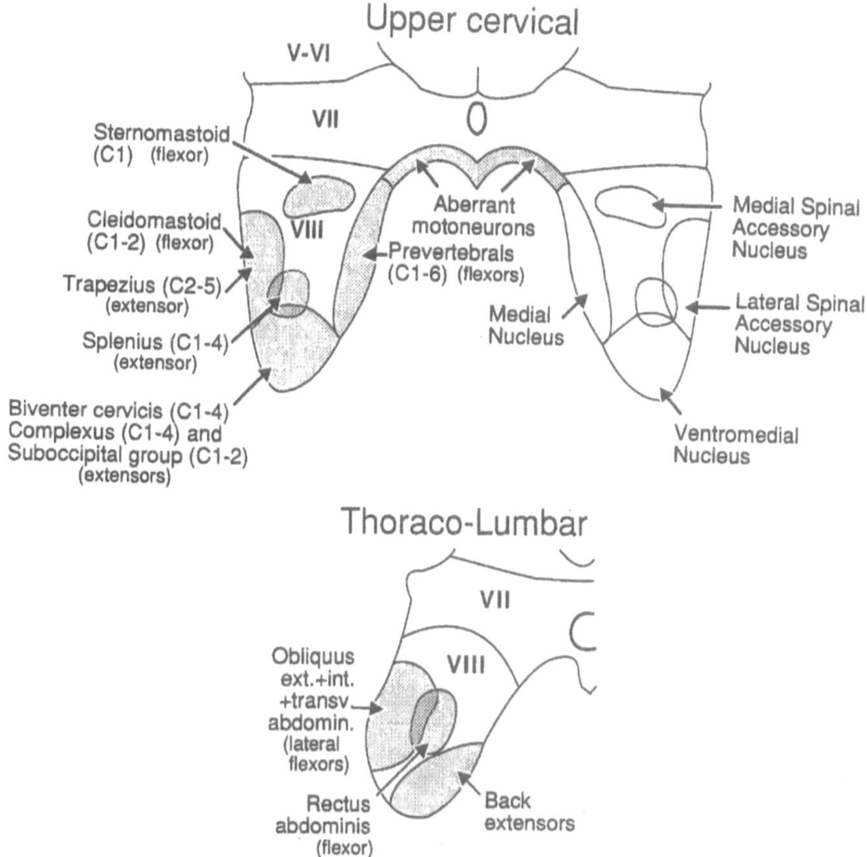

Fig. 13.1. Schematic representation of the combined C1–4 and T10–L2 spinal segments. The motoneuronal cell groups innervating specific neck and axial muscles are shown. The general action of these muscles and the precise cervical cord location of the neck muscle motoneurons are also indicated. It must be emphasized that the cell groups not only contain motoneurons but also interneurons. (From Holstege and Cowie 1989.)

most rostral rectus abdominis, are located just lateral to the longissimus neurons (Holstege et al 1987; Fig. 13.3). Muscles with their origin at the vertebral column (latissimus dorsi) or chest (pectoralis and deltoid muscles), but with insertion on the humerus, produce forelimb movements (Crouch 1969). Therefore they are not considered axial, but limb muscles. Sterling and Kuypers (1967) call them girdle muscles. Their motoneurons take part in the lateral motor column and are located in the ventral part of the ventral horn, lateral to the axial muscle motoneurons, but ventral to the intrinisc limb muscle motoneurons (Sterling and Kuypers 1967; Holstege et al. 1987; Fig. 13.2). Motoneurons innervating muscles intrinsic to the forelimb are located more dorsally in the ventral horn and the motoneurons innervating the most distal (hand) muscles are located most dorsally (Fritz et al. 1986a,b; McCurdy et al. 1987; Fig. 13.2).

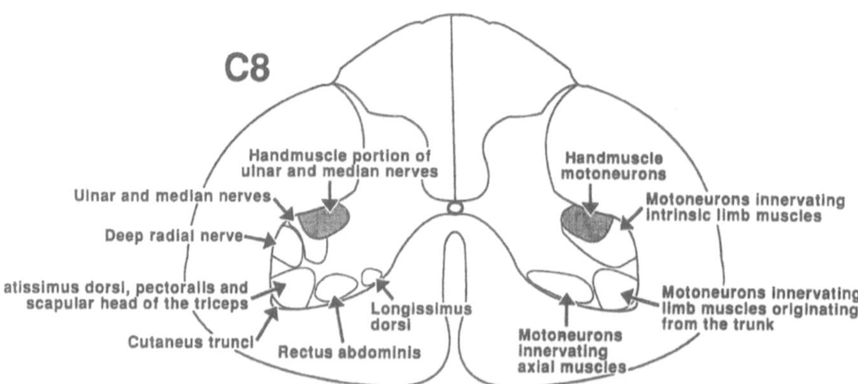

Fig. 13.2. Schematic overview of the location of the motoneuronal cell groups at the C8 level in the cat. The left side of the scheme shows the cell groups, the location of which has been studied using retrograde degeneration or tracing techniques (Sterling and Kuypers 1968; Fritz et al. 1986a,b; Holstege et al. 1987; McCurdy et al. 1987). On the right side of the scheme a more general subdivision into four motoneuronal cell groups has been made.

Thoracic and Lumbar Spinal Cord

At thoracic and upper lumbar levels, in rat and cat all the motoneurons belong to the medial motor column. Many of them innervate the epaxial extensor muscles of the trunk, and are located in greatly overlapping cell columns in the ventromedial portion of the ventral horn, largely segregated from the overlapping cell groups of the motoneurons innervating the hypaxial muscles which lie dorsolateral in the ventral horn (Brink et al. 1979; Smith and Hollyday 1983; Miller 1987; Holstege et al. 1987; Fetcho 1987; Lipski and Martin-Body 1987; Fig. 13.3). The hypaxial muscles include the abdominal (external and internal abdominal oblique, the transversus abdominis and the rectus abdominis) and the internal and external intercostal muscles. The abdominal muscles are involved in postural functions such as flexion and bending of the trunk, but they also play an important role in increasing the intra-abdominal pressure during defecation, vomiting and forced expiration (see Holstege et al. 1987 for review). The intercostal muscles (internal and external) are inserted between adjacent ribs and their contraction decreases the distance between these ribs. The intercostal muscles are important for posture control, but they play a role in respiration also.

The location of the motoneuronal cell groups at the lumbosacral enlargement (L4 to S1 in the cat) is very similar to that of the cervical enlargement. For example, in both enlargements the motoneurons innervating the distal muscles of the limbs are located in the dorsal portions of the ventral horn, while those innervating proximal limb muscles occupy a more ventral position. Furthermore, the motoneurons of the most distal muscles are always located in the caudal part of the enlargement, for example at the level C8–11 for the small hand-muscle motoneuronal cell groups and at the level L7–S1 for the small foot-muscle motoneurons. Trunk muscle motoneurons are always located within the medial column (Brink et al. 1979).

Nucleus of Onuf

The motoneurons of the nucleus of Onuf, as mentioned in the introduction, differ from somatic motoneurons, but resemble autonomic motoneurons, in that they are

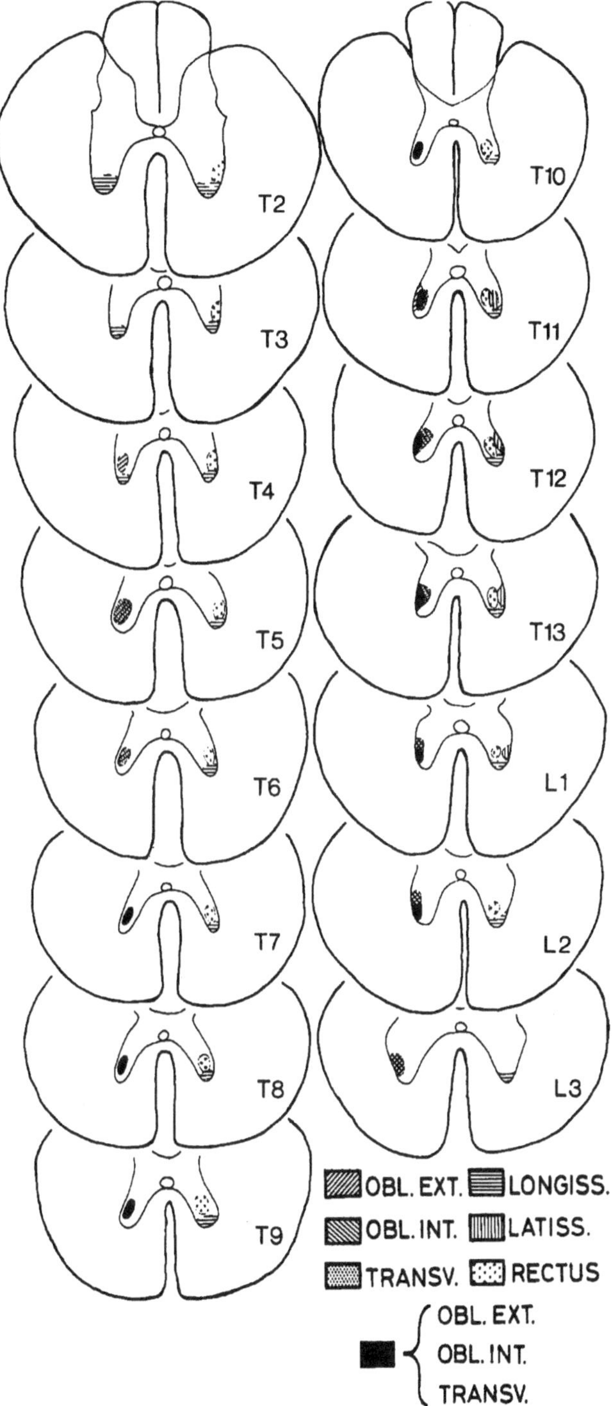

Fig. 13.3. Location of the motoneuronal cell groups innervating the hypaxial abdominal and latissimus dorsi muscles and the epaxial longissimus dorsi muscle. (From Holstege et al. 1987.)

well preserved in the spinal cords of patients who have died from amyotrophic lateral sclerosis (ALS) (Mannen et al 1977, 1982). Because the sacral autonomic (parasympathetic) motoneurons innervating the bladder are also spared in ALS patients, bladder and sphincter functions remain intact until the latest stages of the disease. Since Onuf nucleus motoneurons behave so remarkably differently from all other motoneurons innervating striated musculature, a more detailed survey of this motoneuronal cell group will be presented.

Onufrowicz (1899), who called himself Onuf, described a group X in the ventral horn of the human sacral spinal cord, extending from the caudal S1 to the rostral S3 segments. Romanes (1951) described in the cat a homologous cell group in the caudal half of the first and the rostral half of the second sacral segment and called it group Y. The cell group is now known as nucleus of Onuf (Fig. 13.4). Retrograde HRP tracing studies in the cat (Sato et al. 1978; Mackel 1979; Kuzuhara et al. 1980; Ueyama et al. 1984; Holstege and Tan 1987) demonstrated that Onuf motoneurons, via the pudendal nerve, innervate the striated muscles of the pelvic floor, including the urethral and anal sphincters. Within Onuf's nucleus the dorsomedial motoneurons innervate the anal sphincter, while the ventrolateral motor cells innervate the urethral sphincter (Sato et al. 1978; Kuzuhara et al. 1980; Holstege and Tan 1987; Pullen 1988). The motoneurons in the nucleus of Onuf are characterized by their

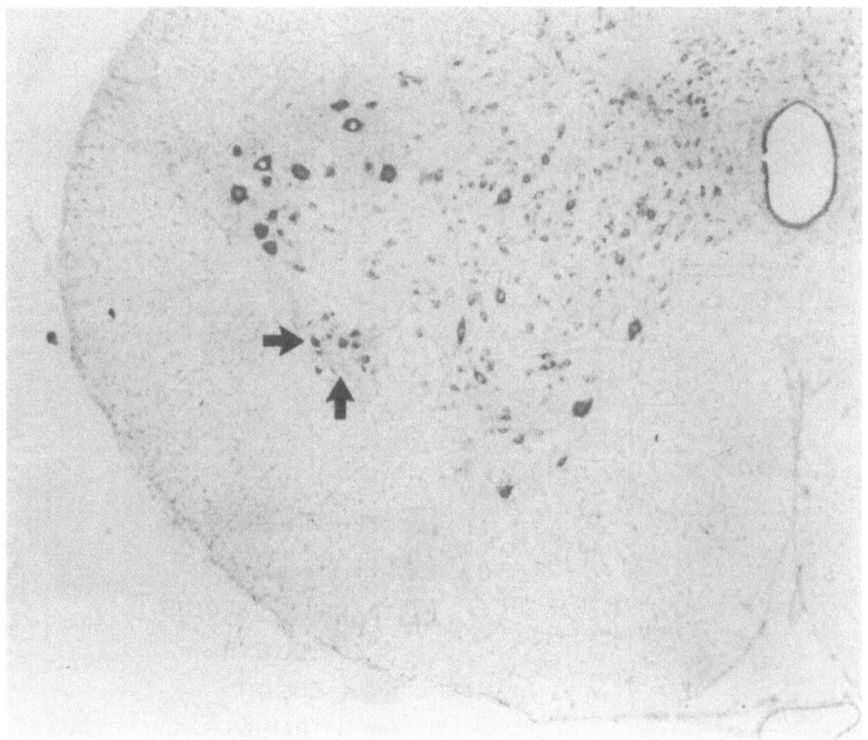

Fig. 13.4. Brightfield photomicrograph of the ventral horn of a section through the left ventral horn of the S1 spinal segment in the cat. The arrows indicate the nucleus of Onuf. (From Holstege and Tan 1987.)

dense packing, their relatively small size (however, see Pullen 1988), and their numerous longitudinal dendrites (Dekker et al. 1973). Although in cat (Sato et al. 1978; Mackel 1979; Kuzuhara et al. 1980; Ueyama et al. 1984; Holstege and Tan 1987), monkey (Roppolo et al 1985) and man (Schrøder 1981) Onuf's nucleus consists of a single motoneuronal pool, in rat it consists of two spatially separate motoneuronal groups, with those innervating the anal sphincter being located at the medial gray border just ventral to lamina X (Schrøder 1980; McKenna and Nadelhaft 1986).

There is evidence that Onuf motoneurons belong to a separate class of motoneurons. On the one hand they are somatic motoneurons, because they innervate striated muscles and are under voluntary control, but on the other hand they are autonomic motoneurons because: (1) cytoarchitectonically they resemble autonomic motoneurons (Rexed 1954; Fig. 13.4); (2) they have an intimate relationship with sacral parasympathetic motoneurons (Holstege and Tan 1987; Nadelhaft et al. 1980; Rexed 1954); (3) they receive direct hypothalamic afferents (Holstege 1987a,b; Holstege and Tan 1987); and (4) they show changes in the same manner as the autonomic motoneurons in Fabry's disease, mannosidosis, Hurler's syndrome, Shy–Drager syndrome and ALS (Sung 1979; Sung et al. 1979; Sung and Mastri 1980). These clinical findings led Tabira et al. (1980) to speculate that Onuf motoneurons and autonomic motoneurons might have common metabolic characteristics.

Somatic Motoneurons in the Brainstem

The motoneurons innervating the muscles of the head, such as the facial, chewing, tongue, pharynx and extraocular muscles are all located in the brainstem. They do not form a continuous rostrocaudal band of motoneurons such as in the spinal cord, but are subdivided into several distinct motoneuronal cell groups. In 53 ALS cases Lawyer and Netsky (1953) found degeneration in the oculomotor nuclei in 4 cases, in the motor trigeminal nucleus in 4 cases, in the nucleus ambiguus in 43 cases, in the dorsal motor nucleus of the vagus in 35 cases and in the hypoglossal nucleus in 50 cases. The following description of the various motor nuclei in the brainstem is based on data obtained in the cat. In the human the location of these motor nuclei is very similar, and is described by Paxinos et al. (1990).

The extraocular muscles are innervated by motoneurons in the oculomotor, trochlear and abducens nuclei, all of which are located dorsomedially in the tegmentum. The oculomotor nucleus is located in the rostral mesencephalon, the trochlear nucleus in the caudal mesencephalon and the abducens nucleus in the pontomedullary transition zone. The oculomotor nucleus contains motoneurons innervating the ipsilateral medial rectus, inferior rectus and inferior oblique muscles and the contralateral superior and levator palpebrae muscles. Trochlear motoneurons innervate the contralateral superior oblique, and abducens motoneurons innervate the ipsilateral lateral rectus muscle (see Evinger, 1988 for review).

The jaw-closing muscles, masseter, temporalis and medial pterygoid muscles, as well as the lateral pterygoid muscle, which is not a jaw-closing muscle, are innervated by motoneurons in the dorsolateral two thirds of the motor trigeminal nucleus. The jaw-opening muscle motoneurons (anterior digastric and mylohyoid) are located in the ventromedial one third of this nucleus (Mizuno et al. 1975; Batini et al. 1976). Motoneurons innervating the tensor tympani, which send their axons via the motor trigeminal nerve, are located just ventral to the motor trigeminal nucleus (Lyon 1975; Mizuno et al. 1982; Keller et al. 1983; Friauf and Baker 1985).

Motoneurons in the facial nucleus innervate the various facial muscles. The lateral and ventrolateral facial subnuclei contain the motoneurons innervating the muscles of the upper and lower mouth respectively. Motoneurons in the dorsomedial facial subnucleus innervate the ear or pinna muscles, and the dorsal portion of the facial nucleus (intermediate facial subnucleus) contains motoneurons innervating the muscles around the eye (Papez 1927; Courville 1966a,b; Kume et al. 1978; Fig. 13.5). Stapedius motoneurons, which send their axons via the facial nerve, are located in cell clusters around the traditional borders of the facial nucleus as well as dorsal to the lateral superior olivary nucleus (Lyon 1978; Shaw and Baker 1983; Joseph et al. 1985).

The somatic motoneurons in the nucleus ambiguus innervate the laryngeal, pharyngeal and soft palate muscles. The nucleus extends for a distance of 5–6 mm caudally from the facial nucleus. Laryngeal motoneurons are located in the caudal two thirds of the nucleus and lie dispersed in the ventrolateral part of the reticular formation. Motoneurons innervating pharynx and soft palate form a compact cell group, the dorsal group of the nucleus ambiguus. It is located 1.5–2.5 m caudal to the facial nucleus. Pharyngeal motoneurons are also located in the more loosely arranged retrofacial part of the nucleus, situated just caudal to the facial nucleus. Furthermore, the retrofacial part of the nucleus ambiguus contains motoneurons innervating the cricothyroid muscles and the upper portion of the oesophagus (Lawn 1966; Yoshida et al. 1981; Holstege et al. 1983; Davis and Nail 1984).

Motoneurons innervating the intrinsic and extrinsic tongue muscles are located in the hypoglossal nucleus, which also contains motoneurons innervating the geniohyoid muscles (Uemura et al 1979; Miyazaki et al. 1981). The extrinsic tongue muscle motoneurons (genioglossus, hyoglossus and styloglossus) are located laterally in the hypoglossal nucleus. The intrinsic muscle motoneurons, which send their axons via the medial branch of the hypoglossal nerve, are located medially and

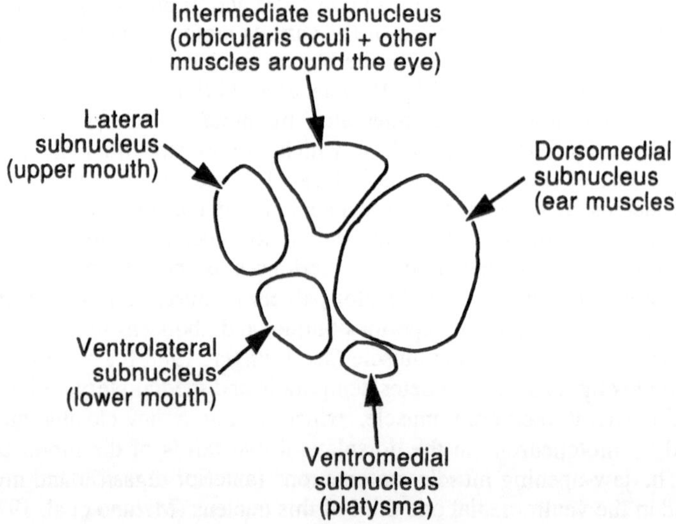

Fig. 13.5. Schematic drawing of a transverse section through the left facial nucleus. The different facial subnuclei and the muscle innervated by the motoneurons in these subnuclei are indicated.

ventrally in the nucleus, while the intrinsic muscle motoneurons, which send their axons via the lateral branch, are located in the dorsal portions of the nucleus (Uemura et al. 1979). This relatively complicated subdivision of the hypoglossal nucleus makes it impossible to subdivide the hypoglossal nucleus into tongue protrusion and tongue retraction regions. Further anatomical and physiological study is necessary to unravel a more precise subdivision within this motoneuronal pool.

Local Projections to Motoneurons

Recurrent Motoneuronal Axon Collateral Projections to Motoneurons

Recurrent collaterals of motoneurons innervating limb muscles terminate directly on local motoneurons innervating the same or synergistic muscles (Cullheim and Kellerth, 1978). Furthermore, motoneuronal axon collaterals project directly on local interneurons (Renshaw cells). Renshaw cells are located in the ventral horn medial to the motor nuclei (Jankowska and Lindström 1971; van Keulen 1979; Fig. 13.6). They have an inhibitory effect on the same or synergistic α and γ motoneuronal cell groups from which they receive their afferents. This phenomenon is known as recurrent inhibition (see Baldissera et al. 1981 for review). Renshaw cells project via propriospinal pathways in the ventral funiculus (Fig. 13.6). Recurrent inhibition is especially strong in motoneuronal cell groups innervating proximal limb muscles, less strong in muscles of more distal parts of the limb (wrist or ankle) and absent in motoneuronal cell groups innervating the most distal limb musculature such as those innervating the phalanges of the forelimb (Hahne et al. 1988) or the small foot muscles of the hindlimb (Cullheim and Kellerth, 1978). Apparently, therefore, recurrent inhibition is primarily concerned with control of the proximal muscles (limb position), rather than of the distal ones (movements of the digits).

Muscle Spindle Afferent Projections to Motoneurons in Spinal Cord and Brainstem

In the spinal cord group Ia muscle spindle afferents have an excitatory effect on motoneurons, innervating the same or synergistic muscle groups (Mendell and Henneman 1971) or via interneurons (Jankowska et al. 1981) onto motoneurons. The group Ia afferent projection system exists in proximal as well as in distal limb muscle control (Ishizuka et al. 1979; Fritz et al. 1978; 1984). Group Ia muscle spindle afferents not only have an excitatory effect on motoneurons, but also on the so-called Ia inhibitory interneurons which in turn have an inhibitory effect on motoneurons innervating muscles, antagonistic to the muscle from which the Ia muscle spindle afferents are derived. The Ia inhibitory interneurons are located in lamina VII of the spinal intermediate zone and project to the antagonist muscle motoneurons, mainly via propriospinal pathways (Jankowska and Lindström 1972; Fig. 13.6). Thus, the Ia afferents of a specific muscle excite the motoneurons of the

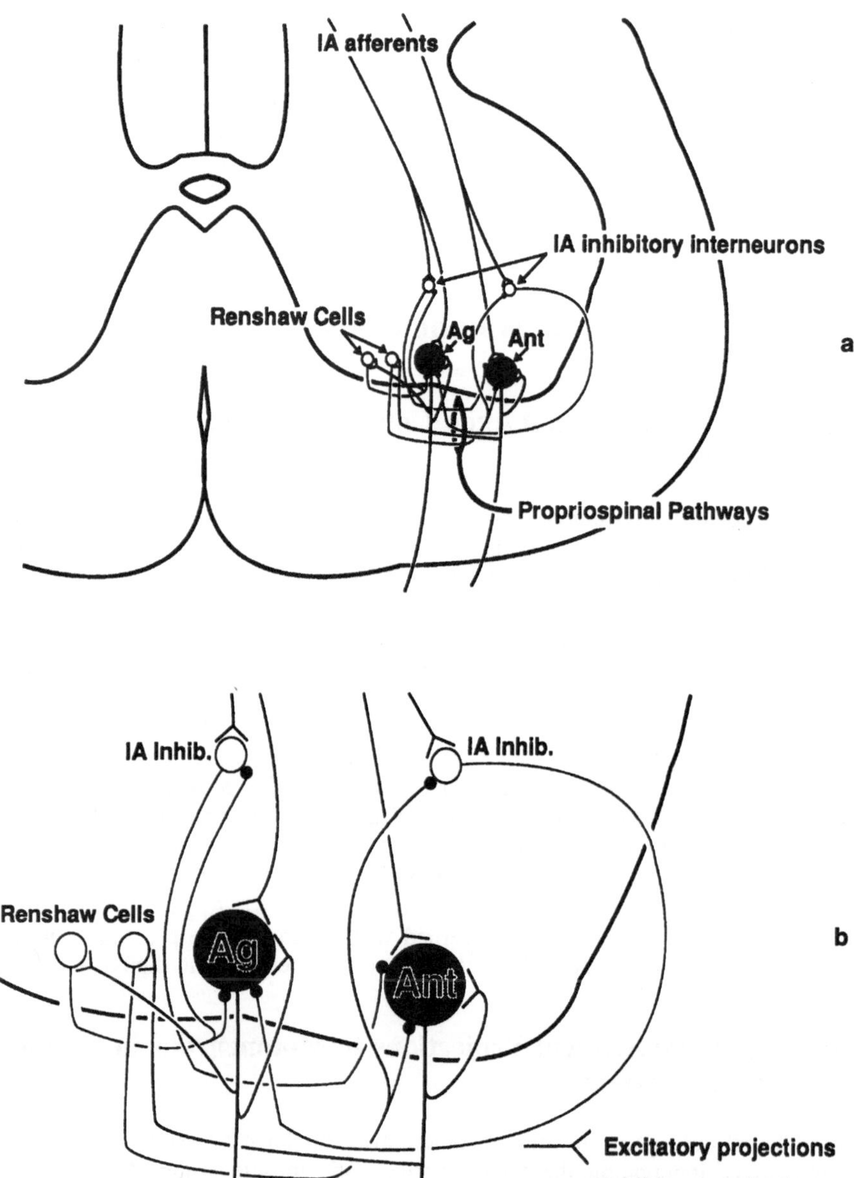

Fig. 13.6. **a** A schematic drawing of the L7 ventral horn showing the recurrent axon collaterals, Renshaw cells, Ia inhibitory interneurons and Ia afferents of two motoneurons innervating an agonist (Ag) and an antagonist (Ant) muscle respectively. Note that many of the neurons project via propriospinal pathways. **b** A magnified view of the different projections is shown. Note that the motoneurons receive inhibitory input from their own axon collaterals and Renshaw cells as well as from the Ia inhibitory interneurons from the antagonist muscle. Excitatory input is derived from Ia afferents. It is known (Cullheim and Kellerth 1978) that recurrent axon collaterals of a proximal muscle motoneuron projects directly onto the somata or dendrites of other motoneurons innervating the same or synergistic muscles. Although indicated as such in the schematic drawing, it is not sure whether a motoneuron projects to its own dendrites or soma. (From Holstege 1991.)

same (homonymous) and synergistic muscles, and, via Ia inhibitory interneurons, inhibit the motoneurons of the antagonistic muscles (see Henneman and Mendell 1981 and Baldissera et al. 1981 for reviews).

The neuronal cell bodies of the muscle spindle afferents are located in spinal sensory ganglia outside the central nervous system. However, the ganglion cells of the muscle spindle afferents of the mouth-closing muscles are located within the central nervous system. They are called mesencephalic trigeminal neurons, and are located at pontine and mesencephalic levels in the border area between periaqueductal gray (PAG) and the dorsally and laterally adjoining tegmentum. Their function is very similar to the spinal ganglion cells, but it is possible, because of their location, that they receive direct afferent projections from the limbic system. For a more detailed description see Holstege (1991).

Propriospinal Pathways

Projections from Inter- and Propriospinal Neurons

With the exception of the Ia afferents, no direct primary afferent projections exist to the motoneurons. For example, stimulation of Ib tendon organ afferents of a specific muscle produces inhibition of the motoneurons of the same and synergistic muscles and excitation of motoneurons of antagonist muscles. These effects are mediated via excitation of interneurons in the intermediate zone, mainly laminae V and VI, which in turn project, via propriospinal pathways, to motoneurons (Czarkowska et al., 1976).

Other primary afferents are derived from the skin and joints, and the group II and III muscle afferents. Their reflex pathways to motoneurons always include interneurons. The last order interneurons, projecting to the motoneurons, enter the funiculus at the same rostro-caudal level as their cell body is located. Within the funiculus they run rostrally and/or caudally, but remain close to the gray matter. They re-enter the spinal gray at the level of their target motoneurons (Jankowska and Roberts, 1972). These parts of the funiculi are called fasciculi proprii or propriospinal pathways. Anatomical studies (Sterling and Kuypers 1968; Rustioni et al. 1971; Molenaar et al. 1974; Molenaar 1978) have indicated that the interneurons, located in different areas of the intermediate zone, project to different motoneuronal cell groups (Fig. 13.7).

According to Baldissera et al. (1981) there is a functional difference between interneurons and propriospinal neurons. Interneurons are intercalated in reflex pathways of limb segments, while propriospinal neurons are located outside the limb segments, but project into them. For example, upper cervical propriospinal neurons relay supraspinal motor information to α-motoneurons in the C5–T1 spinal cord (Illert et al. 1978). These results concur with the anatomical data that neurons in the intermediate zone of C2 project heavily to the C6–T1 motoneuronal cell groups (Holstege 1988b). Alstermark et al. (1981, 1987a,b) demonstrated that upper cervical propriospinal neurons, driven by cortico- and/or rubrospinal fibres, can produce target reaching movements in cats. During this movement the paw is brought in contact with the food. However, direct activation of the C5–T1 inter- and motoneurons from the cortico- and/or rubrospinal tracts can also produce target reaching movements (Alstermark et al. 1987a,b). Such direct activation is essential for food-taking

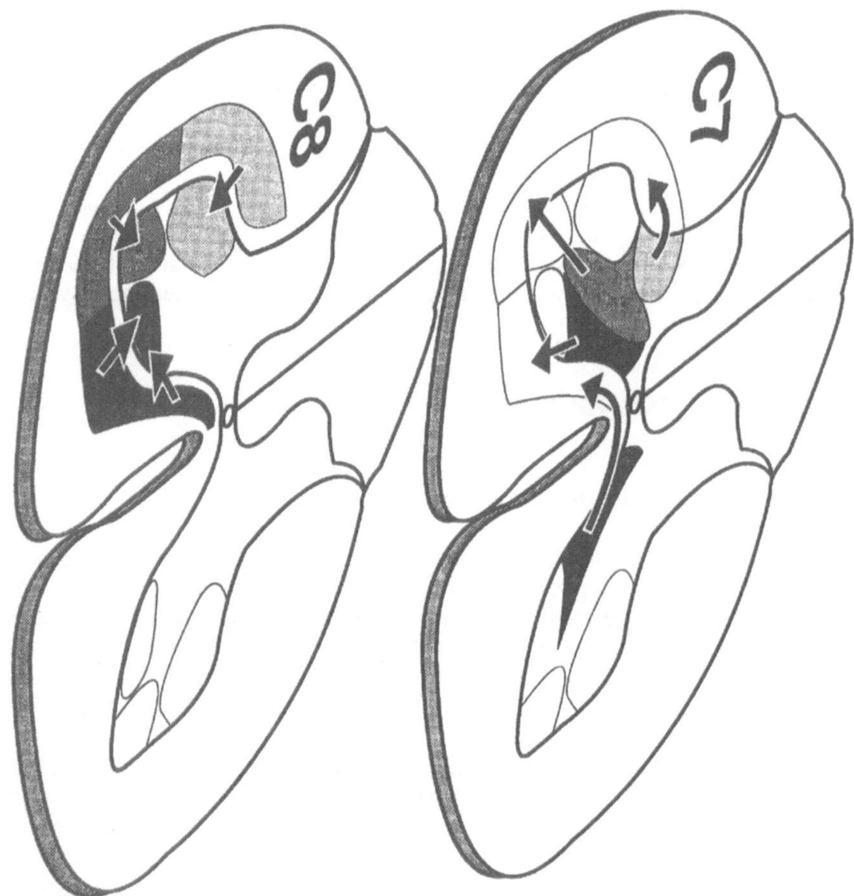

Fig. 13.7. Projection patterns of interneurons in the intermediate gray matter onto anterior horn motor neurons.

movements in cats, consisting of toe grasping and paw supination. Thus the upper cervical propriospinal neurons, when properly stimulated, can produce target-reaching movements, but not the more precise food-taking movements.

Propriospinal neurons as rhythm generators

During the scratch reflex (one limb) or locomotion (all four limbs) the limbs perform rhythmic movements, which are independent of the afferent signals from that limb. The main characteristics of the rhythmic movements of a limb are determined by its so-called spinal generator. During the scratch reflex only one generator is active; during locomotion, all four of them. The spinal generators consist of interneurons, which lie mainly in the lateral part of laminae V, VI and VII over the whole length of the cervical or lumbosacral enlargement. Renshaw cells and Ia inhibitory interneurons are not responsible for the basic pattern of rhythmic changes (see Gelfand et al. 1988 for review). Grillner (1981) hypothesized that the spinal generator of a limb consists of several rhythm generators, each controlling one joint. The regulation of the rhythm generators is performed by means of tonic commands coming from higher brain centers. In all likelihood the diffuse descending systems, originating in the ventromedial medulla oblongata and projecting to all parts of the intermediate zone and motoneuronal cell groups, play an important role in this regulation (see pp 269 and 270).

Long Propriospinal Projections

As pointed out earlier, the column of motoneurons innervating axial muscles extends from the caudal medulla oblongata (neck muscles) to the sacral cord (lower back muscles). Since they are often simultaneously active during certain proximal body movements, long propriospinal projections are necessary to coordinate such axial movements. Giovanelli Barilari and Kuypers (1969) and Molenaar and Kuypers (1978) have shown that there exist direct reciprocal connections between the cervical and lumbosacral spinal cord. The great majority of the neurons giving rise to such long propriospinal projections are located in the medial part of the intermediate zone (lamina VIII and adjoining VII).

Propriobulbar Pathways

The organization of the interneuronal projections to the trigeminal (V), facial (VII), ambiguus (X) and hypoglossal (XII) motor nuclei in the brainstem is not fundamentally different from the interneuronal and propriospinal projections in the spinal cord. This is not the case for the projections to the extraocular motor nuclei of the oculomotor, trochlear and abducens nerves and the ear muscle motoneurons in the facial nucleus, which form part of specific, mainly medially located premotor systems controlling eye, head and ear movements.

Going rostrally from the level of C1, the spinal intermediate zone (laminae V to VIII) is called reticular formation (subnuclei reticulares dorsalis and ventralis of Meessen and Olszewski 1949). It contains interneurons projecting to the motoneurons in the upper cervical cord (Holstege 1988b) and to the V, VII, X and XII motor nuclei (Holstege and Kuypers 1977 and Holstege et al. 1977). Rostral to the level of the obex the reticular formation can be subdivided into a medial and a lateral tegmental field (Fig. 13.8). The lateral tegmental field extends rostrally into the parabrachial nuclei and the nucleus Kölliker-Fuse, and can be considered as the rostral extension of the spinal intermediate zone (Holstege et al. 1977). For example, the projections from the red nucleus and motor cortex in cat and monkey to the bulbar lateral tegmental field are continuous with the projections to the intermediate zone of the spinal cord (see p 287). The lateral tegmental field adjoins the hypoglossal nucleus ventrolaterally, and the facial nucleus dorsomedially, and it surrounds the nucleus ambiguus and the motor trigeminal nucleus. In general, interneurons located medially in the lateral tegmental field project bilaterally to the V, VII and XII motor nuclei, while neurons located laterally project ipsilaterally (Holstege et al. 1977). For an overview of all the different propriobulbar pathways

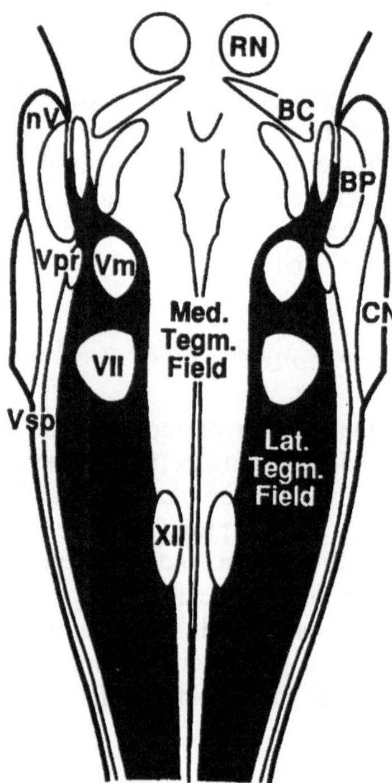

Fig. 13.8. Schematic drawing of the subdivision of the bulbar reticular formation into a medial and lateral tegmental field. The lateral tegmental field can be considered as the rostral extension of the spinal intermediate zone, containing interneurons for the motoneurons in brainstem and spinal cord. The medial tegmental field gives rise to descending pathways involved in postural and orienting movements and in level setting of all neurons in the spinal cord. (From Holstege 1991.)

to the various motor nuclei in the brainstem, see Holstege (1991). An exception will be made for the propriobulbar pathways to the orbicularis oculi, since these pathways play an important role in the neuronal organization of the blink reflex.

The Neuronal Organization of the Blink Reflex

The orbicularis oculi muscle does not contain muscle spindles and uses the overlying skin for its proprioception. The skin overlying the orbicularis oculi muscle and the cornea is innervated by the ophthalmic branch of the trigeminal nerve, the proximal fibers of which terminate in the ventral part of the spinal trigeminal nucleus (Panneton and Burton 1981). Neurons in the ventral part of the spinal trigeminal nucleus project to the blink motoneurons which in the human innervate the orbicularis oculi and, in cat, the orbicularis oculi and the retractor bulbi motoneuronal cell groups (Takeuchi et al. 1979; Panneton and Martin 1983; Holstege et al. 1986a,b). The retractor bulbi nucleus in the cat is a loosely arranged motoneuronal cell group, just dorsal to the superior olivary complex, which innervates the retractor bulbi muscle. The latter is an extraocular muscle divided into four slips, which attach themselves onto the eyeball behind and beside the inferior and superior recti muscles. Retractor bulbi muscles are present in most vertebrates, but not in humans (Bolk et al. 1938). The functional role of the retractor bulbi muscles is purely that of eye protection.

The disynaptic connections between trigeminal nerve afferents on the one hand and orbicularis oculi and retractor bulbi motoneurons on the other probably represent the R1 component of the blink reflex. The blink reflex in the cat consists of two EMG components (R1 and R2) (Lindqvist and Martensson 1970) and has latencies of 9–12 msec (R1) and 15–25 msec (R2). R1 is ipsilateral in all vertebrates; R2 is bilateral in humans (Kugelberg 1952), but ipsilateral in cats (Hiraoka and Shimamura, 1977). Holstege et al. (1986a,b) demonstrated a strong and specific ipsilateral projection to the blink motoneuronal cell groups (orbicularis oculi and retractor bulbi motoneuronal cell groups) from the ventrolateral pontine tegmental field, which they called the pontine blink premotor area (Fig. 13.9). It must be emphasized that this region, which forms part of the lateral tegmental field, lies outside the spinal trigeminal nucleus. This indicates that this projection cannot play a role in the disynaptic R1 component of the blink reflex.

Holstege et al. (1986a,b) also demonstrated specific projections from an area in the medial tegmentum at levels of the hypoglossal nucleus to the blink motoneuronal cell groups, which they called the medullary blink premotor area (Fig. 13.9). This region is not part of the lateral tegmental field, but belongs to the dorsal part of the medullary medial tegmentum, which plays an important role in eye and neck muscle motor control (see p 285). Holstege et al. (1986b) also observed projections from the medullary blink premotor area to the pontine blink premotor area (Fig. 13.9). The projections from the medullary blink premotor area were mainly bilateral, but some ipsilateral projections were also observed (Holstege et al. 1988). Like the pontine blink premotor area, the medullary blink premotor area is not located in the spinal trigeminal nucleus and thus cannot be involved in the disynaptic organization of the R1 blink reflex component.

Both the pontine and medullary blink premotor areas are probably involved in the R2 blink reflex component because (1) the R2 reflex component is not disynaptic, but multisynaptic (Kugelberg 1952; Lindqvist and Martensson 1970; Hiraoka and Shimamura, 1977; Ongerboer de Visser and Kuypers, 1978), and the response consists of several spikes (Berthier and Moore, 1983; Kugelberg, 1952); (2) the R2 blink reflex component, according to Shahani and Young (1972), is responsible for actual closure of the eyelids. For such a motor performance, strong projections to the blink motoneurons are necessary. Holstege et al. (1986a,b) found such connections only from the pontine and medullary blink premotor areas; and (3) the medullary blink premotor area projects specifically to the pontine blink premotor area, indicating that both areas are involved in the same neuronal organization. For a description of the afferent projections to the pontine blink premotor area (from red nucleus, pretectum and medullary blink premotor area) and the medullary blink premotor area (from the superior colliculus and pontine medial tegmentum), which may play an important role in the R2 reflex, see Holstege et al. (1986b, 1988) and Fig. 13.9.

Bulbospinal Interneurons Projecting to Motoneurons

In the section on propriobulbar pathways (p 277) it is stated that the bulbar lateral tegmental field can be considered as the rostral continuation of the spinal intermediate zone. However, the bulbar lateral tegmental field not only contains

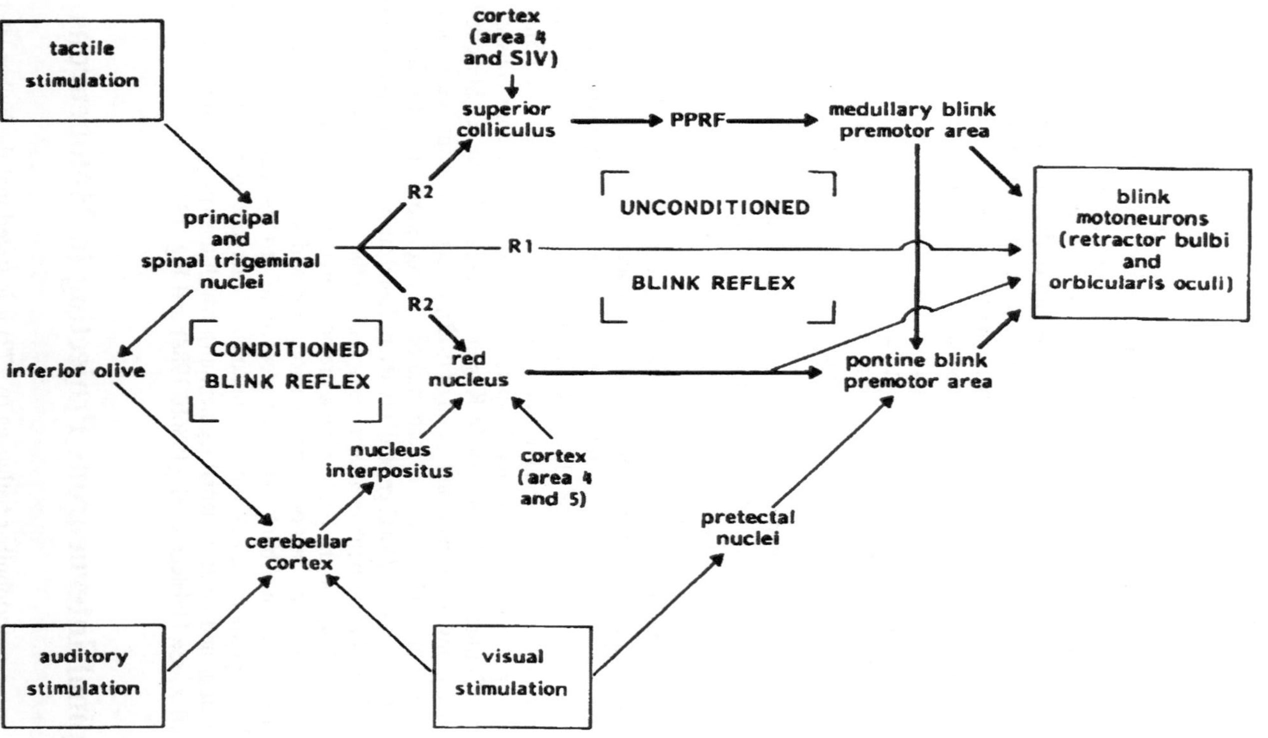

Fig. 13.9. Schematic representation of the pathways possibly involved in the anatomical framework of the R1 and R2 blink reflex components. (From Holstege et al. 1986b.)

interneurons for the motoneuronal cell groups V, VII, X and XII in the brainstem, but also for certain cell groups in the spinal cord, especially those involved in respiration, abdominal pressure and micturition.

Pathways Controlling Respiration and Abdominal Pressure

Chemoreceptors in the carotid body and the pulmonary stretch receptors form the most important peripheral afferents for the respiratory system. From the carotid body, which senses arterial blood gases and pH, fibers terminate in the dorsomedial subnuclei of the solitary tract (Berger 1980). These fibers have the cell bodies in the petrosal ganglion and pass via the glossopharyngeal and carotid sinus nerve. The pulmonary stretch receptors are located in the smooth muscle of the trachea, main bronchi and intrapulmonary airways. Peripheral afferent fibers innervating all these receptors arise from cell bodies in the nodose ganglion and project to the nuclei of the solitary tract (Donoghue et al. 1982). Pre-motor interneurons tend to be located close to the incoming afferent fibers. The same is true for the pre-motor interneurons of the respiratory motor output system. They are located in the caudal medulla, where the vagal nerve enters the brainstem, and not in the spinal cord. Therefore, the medullary projections to the respiratory motoneurons should not be considered as a specific supraspinal control system, but as a propriobulbospinal system.

Physiological studies have demonstrated that the brainstem neurons can be subdivided into inspiratory and expiratory neurons, although in the dorsolateral pons some inspiratory–expiratory phase-spanning neurons exist (see Feldman 1986 for review). From the inspiratory neurons 50–90% project to the spinal cord, while almost all expiratory neurons project to the cord. The inspiratory neurons that project to the spinal cord send excitatory fibers to the phrenic nucleus, while the expiratory neurons send excitatory fibers to the abdominal muscle motor nuclei. The expiratory fibers in the Bötzinger complex send inhibitory fibers to the phrenic nucleus. The importance of these pathways is exemplified by the finding that a transection at the spinomedullary junction completely abolishes respiratory movements of diaphragm, rib cage and abdominal muscles (St. John et al. 1981). Fig. 13.10 illustrates schematically the various pathways controlling respiration and abdominal pressure. For more extensive reviews see Feldman (1986) and Holstege, (1991).

Pathways Involved in Micturition Control

The brainstem, via its long descending pathways to the sacral cord, is vital for coordinating muscle activity of bladder and bladder-sphincter during normal micturition. The importance of the brainstem in micturition control is best shown by patients with spinal cord injuries above the sacral level. They have great difficulty in emptying the bladder because of uncoordinated actions of the bladder and sphincter (detrusor-sphincter dyssynergia). Such disorders never occur in patients with neurologic lesions rostral to the pons, which indicates that the co-ordinatory neurons are located in the pontine tegmentum (Blaivas, 1982). Barrington showed as early as 1925 that these neurons are probably located in the dorsolateral part of the pontine tegmentum, because bilateral lesions in this area in the cat produced inability to empty the bladder. Later studies by Nathan and Smith (1958) supported

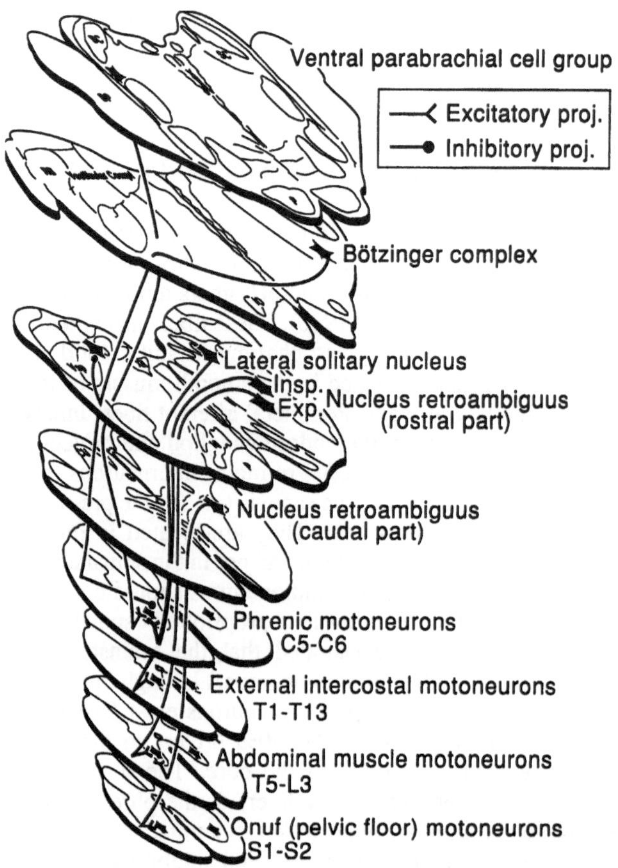

Fig. 13.10. Schematic overview of the pathways controlling respiration and abdominal pressure. Note that from the descending pathways originating in the medulla, only the contralateral ones are indicated, although there exist to a limited extent some ipsilateral pathways. (From Holstege 1991.)

this finding, which led to the concept that micturition can be considered as a spinobulbospinal reflex.

Recent anatomic studies in the rat (Loewy et al. 1979), opossum (Martin et al. 1979b); cat (Holstege et al. 1979, 1986c) and monkey (Westlund and Coulter 1980) have shown that neurons in the dorsolateral pontine tegmentum, medial to the locus coeruleus, project directly and specifically to the sacral intermediolateral cell group (parasympathetic motoneurons) as well as to the sacral intermediomedial cell group, but not to the nucleus of Onuf. The nucleus of Onuf receives specific projections from neurons in more lateral parts of the dorsolateral pontine tegmental field (Holstege et al. 1979, 1986c). The dorsolateral pontine tegmentum does not project to the sacral parasympathetic motoneurons. In order to differentiate between the two different areas in the dorsolateral pons, Holstege et al. (1986c) called them the M- (medial) and L- (lateral) regions. The M-region probably corresponds with Barrington's (1925) area. Neither the M- nor the L-region projects to the lumbar intermediolateral (sympathetic) cell groups.

Electric stimulation in the M-region produces an immediate and sharp decrease in the urethral pressure and pelvic floor EMG, followed after about two seconds by a steep rise in the intravesical pressure (Holstege et al. 1986c), mimicking complete micturition (Fig. 13.11). The decrease in the urethral pressure cannot be caused by a direct M-region projection to the nucleus of Onuf, because such a projection does not exist (Holstege et al. 1979,1986c). A study of Griffiths et al. (1990) suggests that the M- and L-regions may have reciprocal inhibitory connections. Stimulation in the L-region results in strong excitation of the pelvic floor musculature and an increase in the urethral pressure (Holstege et al. 1986c; Fig. 13.12). Bilateral lesions in the M-region result in a long period of urinary retention, during which detrusor activity is depressed and the bladder capacity increases. Bilateral lesions in the L-region give rise to inability to store urine. The urethral pressure decreases and due to absence of the inhibitory influence of the L-region on the M-region detrusor activity increases. The result is that the urine is expelled prematurely because of a combination of increased detrusor activity and decreased urethral pressure. At times separate from the episodes of detrusor activity the urethral pressure is not depressed below normal values (Griffiths et al., 1990). These observations suggest that during the filling phase the L-region has a continuous excitatory effect on the nucleus of Onuf, which inhibits urethral relaxation coupled with detrusor contraction. When micturition takes place, the M-region excites, via a direct pathway, the sacral parasympathetic motoneurons, but at the same time the M-region inhibits the L-region, which disinhibits sphincter relaxation so that micturition can take place.

Although patients with neurological lesions in the brain rostral to the pons never experience detrusor-sphincter dyssynergia, they suffer from lack of control of the initiation of micturition. This raises the question of what determines the beginning of the micturition act. Obviously, precise information about the degree of bladder filling is conveyed to supraspinal levels, but specific sacral projections to the pontine micturition center have not been demonstrated. This suggests that other structures, rostral to the pontine micturition center, determine the initiation of micturition. Such structures would be expected to project specifically to the M-region of the pontine micturition center. Many clinical studies indicate that cortical (the medial frontal gyrus and anterior cingulate lobe) as well as subcortical structures (septum, preoptic region of the hypothalamus, amygdala and periaqueductal gray (PAG)) are involved in control of the beginning of micturition. In the cat, only two structures have been demonstrated to project specifically to the M-region, the preoptic area (Holstege 1987b) and the PAG (Blok and Holstege unpublished observations). Stimulation in the preoptic area produces micturition-like contractions (Grossman and Wang, 1956), but it is not known whether it determines the beginning of micturition. It is possible that regions other than the preoptic area and PAG also project to the M-region. Furthermore, the fact that the pelvic floor, including the intrinsic external urethral sphincter, is under voluntary control, suggests that direct cortical projections to the nucleus of Onuf may exist. However, such projections have not been demonstrated convincingly. Fig. 13.13 gives a schematic overview of the spinal and supraspinal structures involved in micturition control and their role in the neuronal framework of micturition. For a more detailed review, see Holstege (1990).

Descending Pathways of Somatic Motor Control Systems

As pointed out earlier (p 260), the somatic motoneurons in the cervical and lumbosacral enlargements of the spinal cord can be subdivided into lateral and medial columns. At upper cervical, thoracic and upper lumbar levels all motoneur-

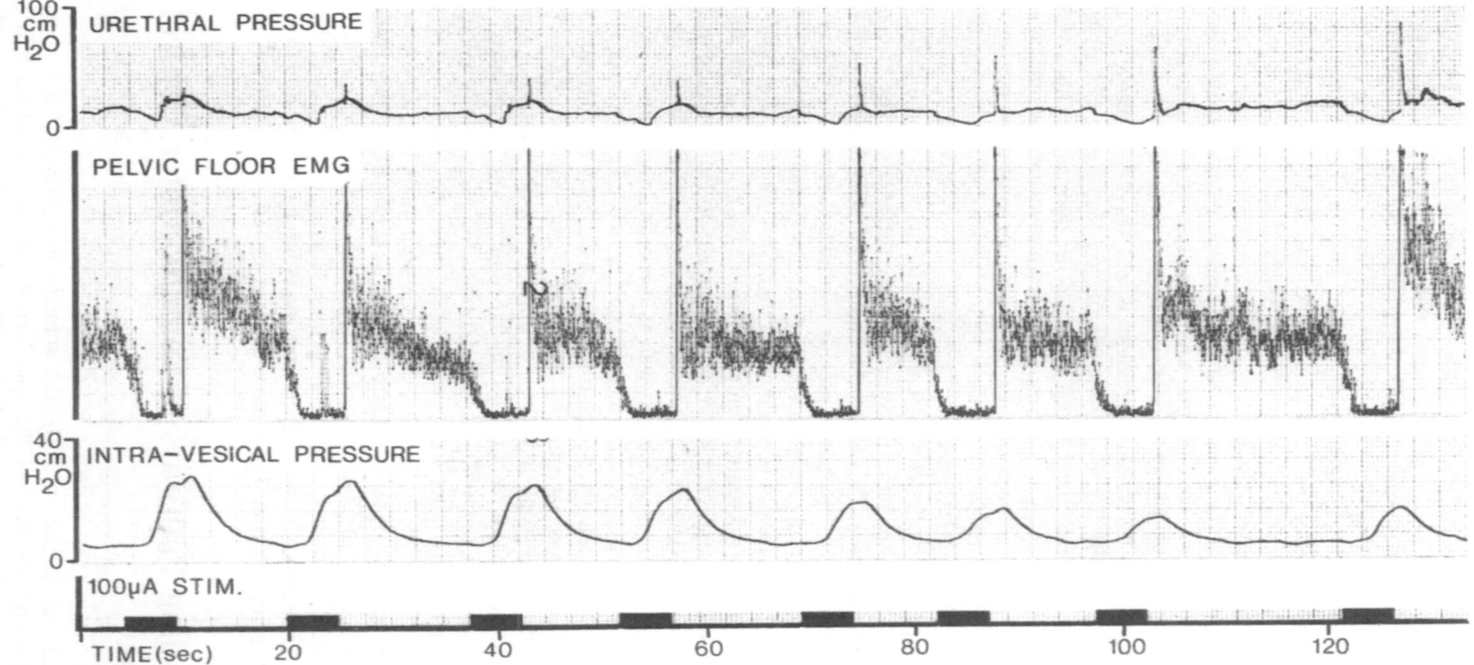

Fig. 13.11. Recordings of urethral pressure, pelvic floor EMG, intravesical pressure and stimulus timing during M-region stimulation in the cat. Note the immediate fall in urethral pressure and pelvic floor EMG after the beginning of the stimulus and the steep rise in the intravesical pressure about 2 seconds after the beginning of the stimulus. This pattern mimics complete micturition. (From Holstege et al. 1986c.)

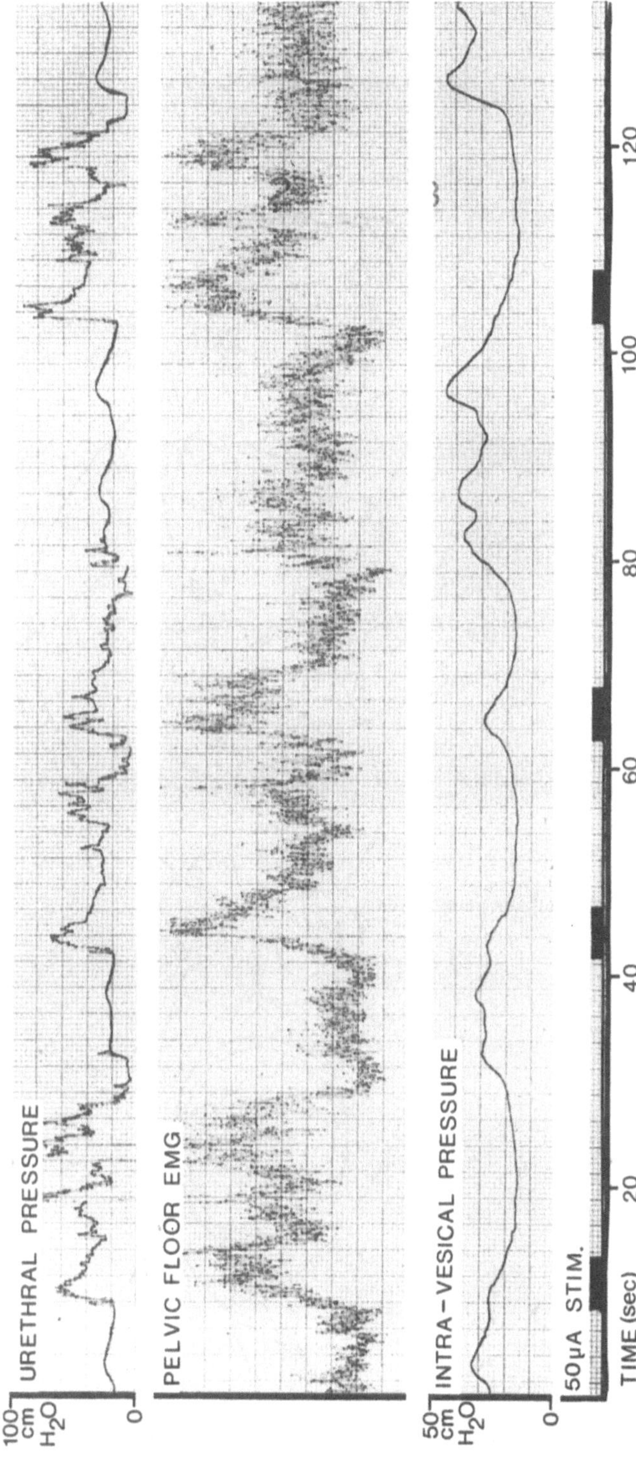

Fig. 13.12. Recordings of urethral pressure, pelvic floor EMG, intravesical pressure, and stimulus timing during L-region stimulation in the cat. At the beginning of each period of stimulation there is an immediate increase in the urethral pressure and the pelvic floor EMG. Note that the spontaneous detrusor contractions tend to be inhibited by the stimulation. (From Holstege et al. 1986c.)

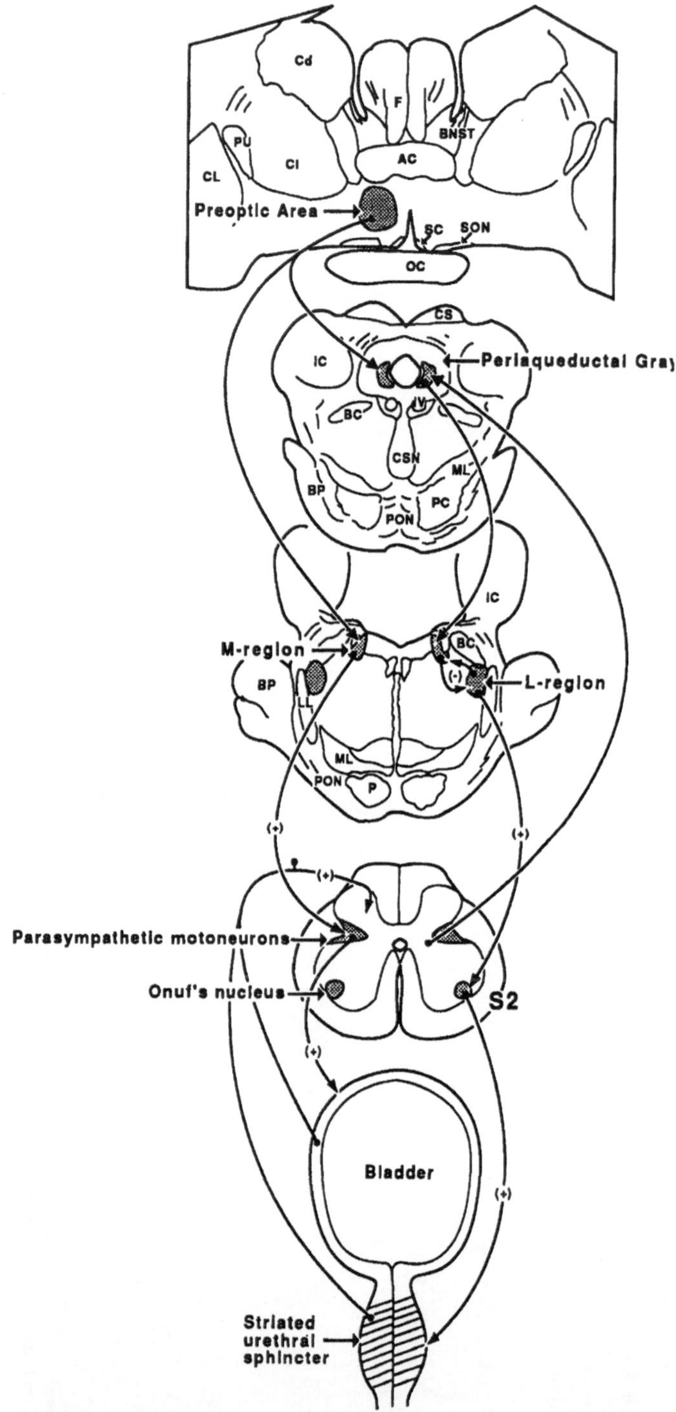

Fig. 13.13. Schematic representation of the spinal and supraspinal structures involved in micturition control. Excitatory pathways are indicated by "(+)", inhibitory projections by "(−)".

ons belong to the medial column. Motoneurons in the lateral motor column innervate the distal extremity muscles, i.e. the fore- and hindpaws in the cat (hands and feet in primates) and the distal portions of the fore- and hindlegs. Motoneurons in the medial column innervate proximal and axial musculature, such as muscles of neck, shoulder, trunk, hip and back. A similar mediolateral organization appears to exist in the propriospinal pathways (p 269) and in the descending pathways belonging to the somatic motor system. The medial motor column receives afferents mainly from cell groups in the brainstem, which project via the ventral funiculus of the spinal cord (Petras 1967; Holstege G. and Kuypers 1982; Holstege 1988b). They form the medial descending system. The lateral motor column receives its supraspinal afferent projections mainly from red nucleus and cerebral motor cortex via the dorsolateral funiculus (lateral descending system) (Nyberg-Hansen and Brodal 1964; Petras 1967; Kuypers and Brinkman, 1970; Armand et al. 1985; Holstege 1987a,b; Holstege and Tan 1988). They represent the lateral descending system.

The Medial Descending System

The function of the medial system is maintenance of erect posture (antigravity movements), integration of body and limbs, synergy of the whole limb and orientation of body and head (Kuypers 1981). Within the medial system most of the proximal and axial muscles are simultaneously active, which explains why they are mutually connected via long propriospinal systems (p 269) and why supraspinal structures, projecting to inter- and motoneurons of the proximal and axial muscles are not clearly somatotopically organized. In order to control orientation of body and head, the medial system also determines the position of the eyes in space, which includes the position of the head on the trunk and the position of the eyes in the orbit. The following brainstem cell groups belong to the medial system: rostral interstitial nucleus of the medial longitudinal fasciculus (rostral iMLF; Fig. 13.14 F,G), the interstitial nucleus of Cajal (Fig. 13.14 H,I) and surrounding reticular formation (INC–RF), the intermediate and deep layers of the superior colliculus (Fig. 14 J–M), a cell group in the lateral PAG and adjacent tegmentum (Fig. 13.14 L,M), the pontine (Fig. 13.14 O–Q) and upper medullary medial tegmentum (Fig. 13.14 R–U), a cell group in the contralateral medullary medial tegmental field (Fig. 13.14 T–U) and the lateral, medial and inferior vestibular nuclei (Fig. 13.14 R–V). They all project (directly or indirectly) to the oculomotor nuclei in the brainstem (p 265) (see Büttner-Ennever and Büttner 1988 for review) and to the neck muscle inter- and motoneurons in the first five cervical segments of the spinal cord (Holstege 1988b).

Pathways Involved in Regulating Axial and Proximal Body Movements

Neurons in the pontine and upper medullary medial tegmentum and in the lateral vestibular nucleus (LVN) send a large number of fibers via the ventral funiculus to laminae VIII and the adjoining part of VII throughout the length of the spinal cord (Jones and Yang (1985) in the rat; Martin et al. (1979c) in the opossum; Nyberg-Hansen and Masicitti (1964),Nyberg-Hansen (1965), Petras (1967); Holstege G. and Kuypers (1982) and Holstege (1988b) in the cat; Fig. 13.15).

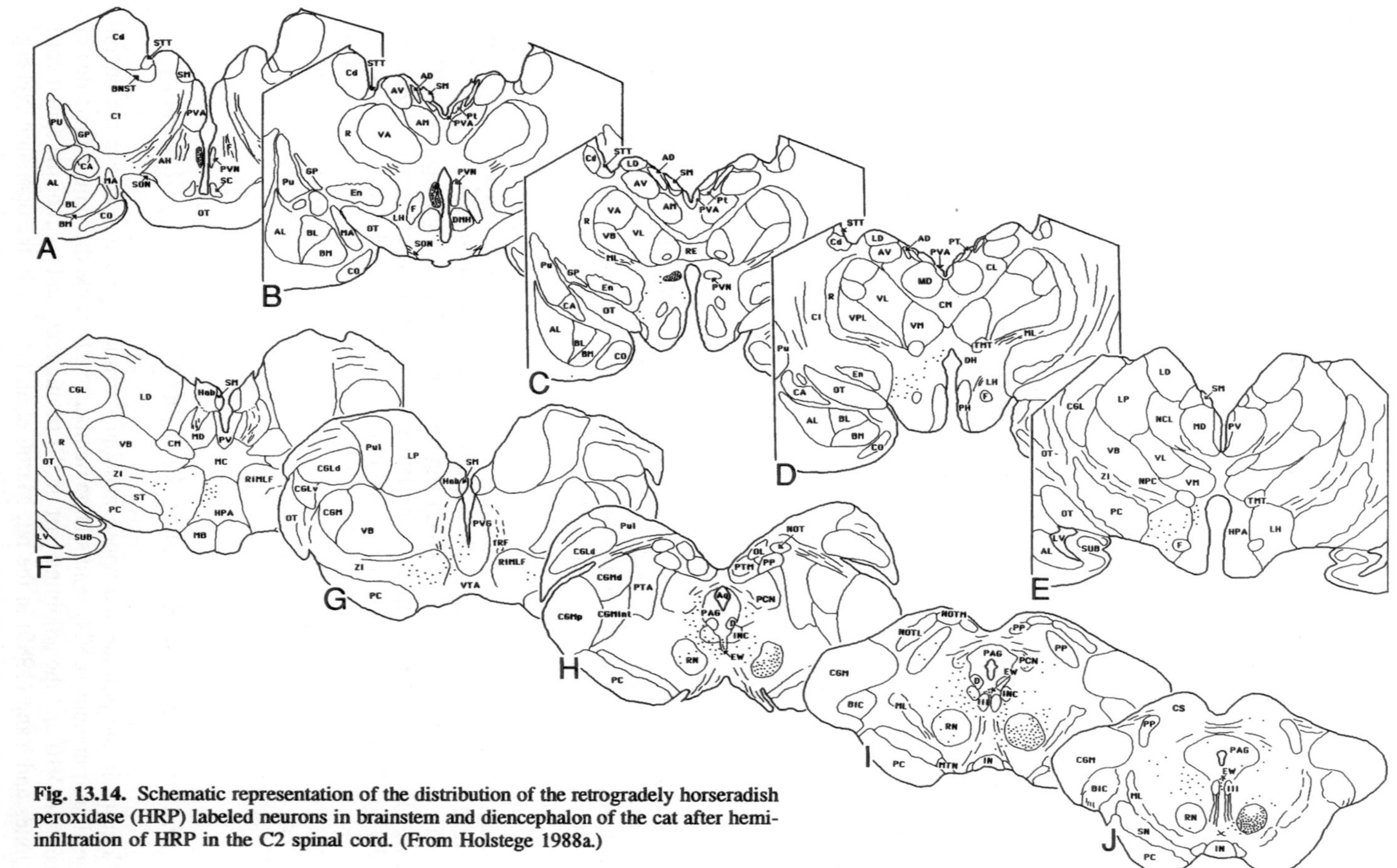

Fig. 13.14. Schematic representation of the distribution of the retrogradely horseradish peroxidase (HRP) labeled neurons in brainstem and diencephalon of the cat after hemi-infiltration of HRP in the C2 spinal cord. (From Holstege 1988a.)

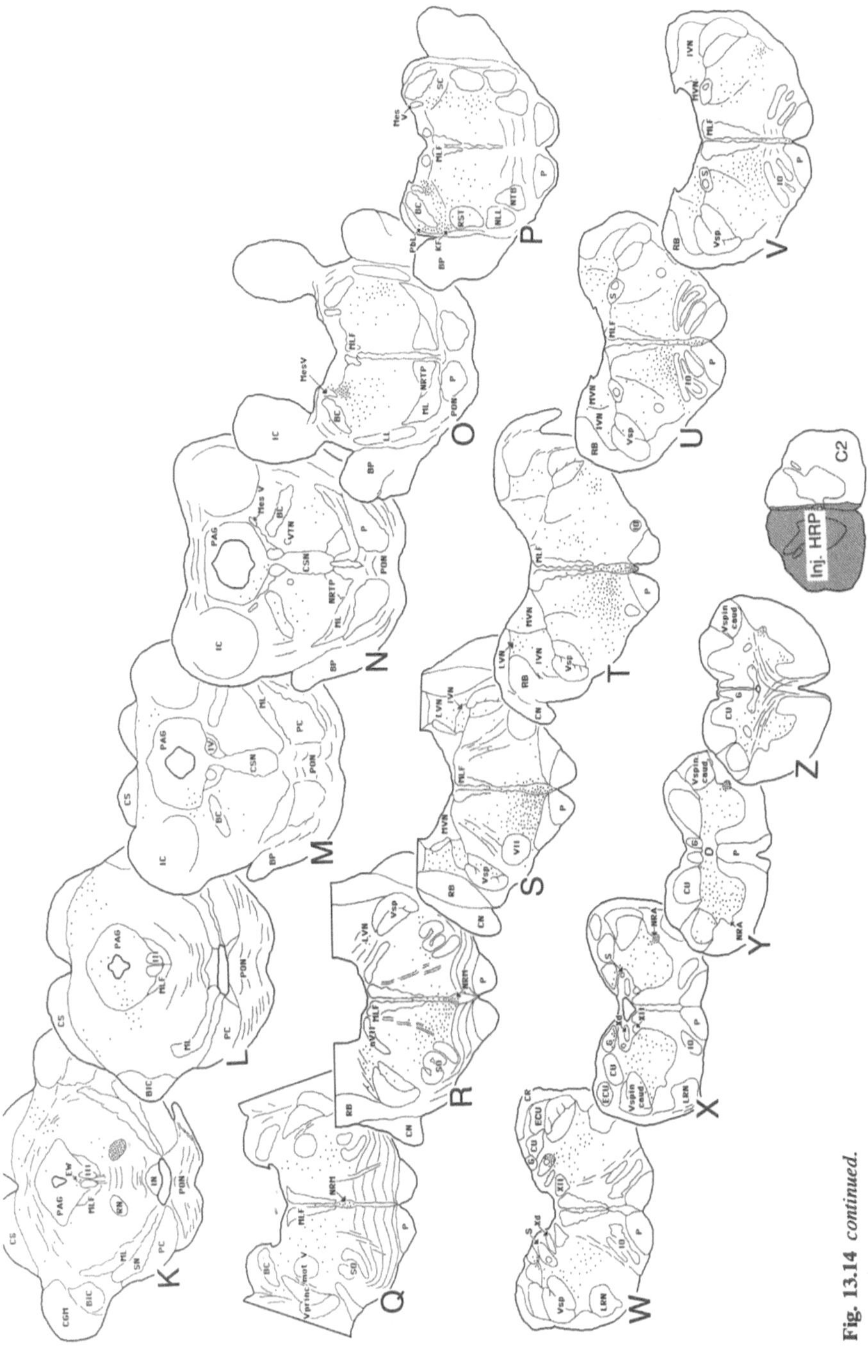

Fig. 13.14 continued.

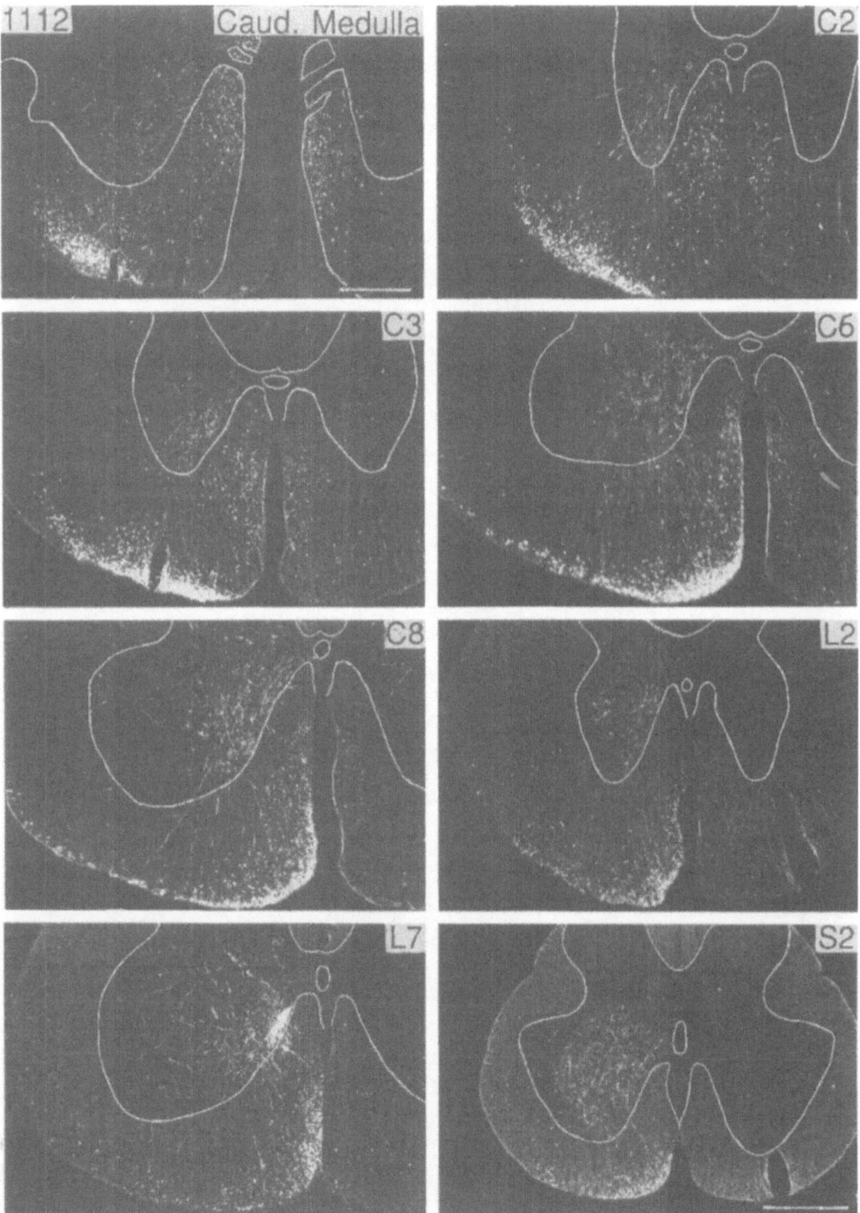

Fig. 13.15. Darkfield photomicrographs of the caudal medulla and seven different levels of the spinal cord in a cat with a ^{3}H-leucine injection in the vestibular complex (lateral vestibular nucleus, rostrodorsal portion of the inferior vestibular nucleus and cell group y). Note the heavily labeled lateral vestibulospinal tract fibers in the ventral part of the ipsilateral ventral funiculus gradually passing medially. Note also the medial vestibulospinal tract fibers on both sides in the dorsal part of the ventral funiculus of the cervical cord. Note further the dense projections to the medial part of the ipsilateral ventral horn throughout the length of the spinal cord. (From Holstege 1988b.)

The function of the long medially descending systems was nicely illustrated by experiments of Lawrence and Kuypers (1968a,b) in the monkey. They made, after pyramidotomy (interruption of the corticospinal fibers at the level of the medulla oblongata), a bilateral lesion of the upper medullary medial tegmentum. The lesion not only destroyed the spinal cord projecting neurons in the upper medulla, but also interrupted all the fibers descending medially in the brainstem, e.g. the ponto-, tecto-, interstitio- and vestibulospinal fibers. Such lesions produced monkeys with postural changes of trunk and limbs, inability to right themselves, and a severe deficit in the steering of axial and proximal limb movements. On the other hand picking up pieces of food with the hand was considerably less impaired. Recovery was slow and when the animals were able to walk, they had great difficulty in avoiding obstacles and frequently veered from course. In the examining chair the animals showed no deficits in picking up pieces of food from a board with their hands and bringing them to the mouth. Unlike the animals in which the medial system is intact, they did not orient themselves to the approaching food, but followed the food only with their eyes.

Pathways Involved in Regulating Eye and Head Movements

The medial system brainstem structures can be subdivided into cell groups steering vertical eye and head movements and those steering horizontal eye and head

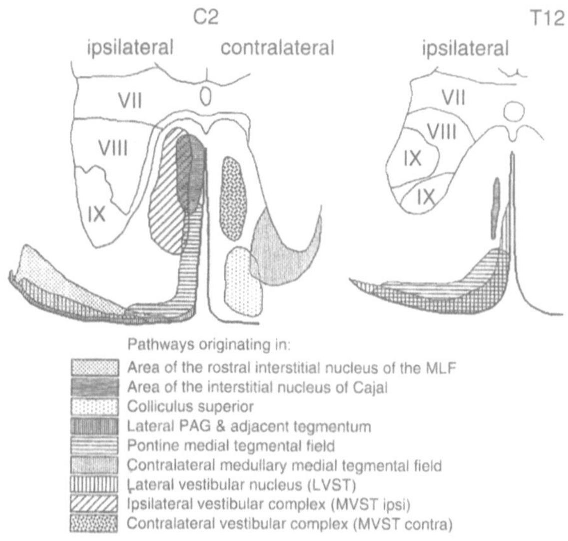

Fig. 13.16. Schematic representation of the spinal white matter location of the various descending pathways, specifically involved in control of neck and axial muscle inter- and motoneurons. On the left a drawing of the C2 spinal segment and on the right a drawing of the T12 spinal segment. It must be emphasized that this scheme does not give any indication about the number of fibers belonging to the different descending pathways. It must also be noted that many other descending fiber systems pass through the same areas as indicated in the drawing (for example, propriospinal, reticulospinal and corticospinal fibers). (From Holstege 1988b.)

movements. Examples of the first group are the interstitial nucleus of Cajal and adjacent reticular formation (INC–RF) and Field H of Forel, which includes the rostral interstitial nucleus of the MLF (Büttner-Ennever et al. 1982; Holstege and Cowie 1989). For the horizontal eye and head movements, the superior colliculus and the pontine and medullary medial tegmental field are most important (Büttner-Ennever and Holstege, 1986).

Concluding Remarks Regarding the Descending Pathways Involved in Regulating Head Movements

Figure 13.16 gives an overview of the white matter location of all the descending pathways belonging to the medial descending system in the upper cervical and low thoracic spinal cord. Only the pontine medial tegmental field and the lateral vestibulospinal tract, and to a limited extent the interstitiospinal tract, descend throughout the length of the spinal cord.

At upper cervical levels, the INC–RF (Fig. 13.17) and the vestibular nuclei project mainly to the *medial* portion of the upper cervical intermediate zone, in which area the prevertebral muscle and some biventer cervicis and complexus muscle motoneurons are located (Abrahams and Keane 1984; Fig. 13.1). These muscles may be specifically involved in head position, although until now such a function has only been described for the biventer cervicis, occipitoscapularis, semispinalis cervicis and rectus capitis muscles (Richmond et al. 1985; Roucoux et al. 1985). In accordance with this concept, both INC–RF and vestibular nuclei are

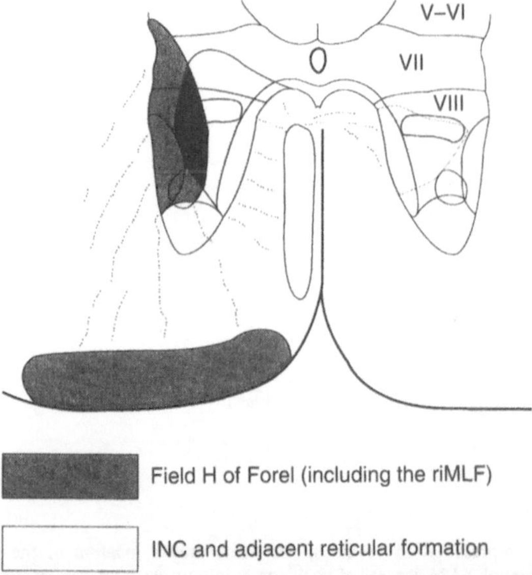

Fig. 13.17. Schematic drawing showing the pathways as well as the termination zones of the projections originating in caudal Field H of Forel and in the INC–RF. Note that the neurons in the caudal Field H of Forel project to more lateral parts of the ventral horn than the neurons of the INC–RF. (From Holstege and Cowie 1989.)

known to be strongly involved in eye position and head posture. On the other hand, the main spinal projection of the caudal Field H of Forel (Fig. 13.17), superior colliculus, lateral PAG and adjacent tegmentum, and pontine medial tegmental field is to the *lateral* parts of the upper cervical ventral horn, which contains motoneurons innervating cleidomastoid, trapezius and splenius capitis muscles. The latter group of muscles appear best suited to produce rapid or phasic torsional movements of the head such as might occur during orienting movements (Callister et al. 1987). This concept corresponds with the observation that stimulation in the caudal Field H of Forel, superior colliculus and pontine medial tegmental field produces eye saccades and fast head movements.

In summary, a concept is put forward (Holstege 1988b) in which the medial somatic system structures are subdivided into two groups; one that controls tonic eye and head position, and one that produces saccadic eye and fast head movements.

The Lateral Descending System

The lateral component of the voluntary motor system produces independent flexion-biased movements of the extremities, in particular of the elbow and hand (Kuypers 1981). The two most important constituents are the rubro- and corticospinal tracts. Vertebrates without extremities, such as snakes and sharks, do not have a rubro- or corticospinal tract, indicating that the presence of such tracts is related to the presence of limbs or limb-like structures (ten Donkelaar 1988). Both red nucleus and motor cortex are somatotopically organized, containing regions such as a face area projecting to the face motor and pre-motor neurons in caudal pons and medulla, an arm or foreleg area projecting to the cervical cord, and a hindleg portion sending fibers to the lumbosacral cord (Kuypers 1981; Armand et al. 1985; Holstege 1987a; Holstege and Tan 1988). There are differences between the organization of the rubro- and corticospinal tract, which depend to a large extent on the species involved.

The Rubrobulbar and Rubrospinal System

There are two different descending pathways from the red nucleus to the caudal brainstem: (1) a mainly contralateral pathway, which sends fibers to the pre-motor interneurons in the lateral tegmental field (p 295), the dorsal column nuclei, pre-cerebellar structures other than the inferior olive, and to the spinal cord; and (2) an ipsilateral fiber system which terminates on neurons in the inferior olive. Many of the neurons in the red nucleus projecting via the contralateral rubrobulbospinal system are of large diameter and are located in the caudal portions of the red nucleus, while the rubro-olivary neurons are of smaller diameter and are located in the rostral parts of the red nucleus. The caudal part of the red nucleus is usually termed the magnocellular red nucleus, while the rostral part of the red nucleus is termed the parvocellular red nucleus. In the cat neurons projecting to both the spinal cord and inferior olive do not exist (Huisman et al. 1982). The subdivision into magno- and parvocellular red nucleus leads to confusion because in the cat the parvocellular (rostral) red nucleus not only contains neurons projecting to the inferior olive, but also neurons projecting to the spinal cord (Holstege and Tan,

1988). Furthermore there exist important species differences regarding the relation between the magnocellular and parvocellular red nucleus. Therefore, Holstege and Tan (1988) proposed a new subdivision of the red nucleus based on the projections of the neurons located in it: a rubrobulbospinal red nucleus and a rubro-olivary red nucleus. It must be emphasized that the rubro-olivary neurons form part of a much larger projection system.

The Rubrobulbospinal Projections. The rubrobulbospinal red nucleus is somatotopically organized in such a way that neurons in its dorsal part project to the bulbar lateral tegmental field and facial nucleus (Kuypers et al. 1962; Martin et al. 1974; Holstege and Tan 1988), neurons in the dorsomedial red nucleus project to the cervical cord and neurons in the ventrolateral red nucleus to the lumbosacral cord (Pompeiano and Brodal 1957; Murray and Gurule 1979; Huisman et al. 1982; Holstege and Tan 1988). In accordance with this somatotopic organization, only very few red nucleus neurons project to both the cervical and lumbar cord (Huisman et al. 1982). All projections are contralateral except for a few ipsilaterally descending fibers, projecting to the intermediate zone of the cervical cord (Holstege 1987a). The red nucleus also projects to the interpositus nucleus in the cerebellum (Huisman et al. 1982) and to some precerebellar structures in the caudal brainstem, such as the nucleus corporis pontobulbaris, lateral reticular nucleus and external cuneate nucleus (Edwards 1972; Martin et al. 1974; Holstege and Tan 1988). Furthermore, neurons in the dorsomedial (foreleg) part of the red nucleus send fibers to the cuneate nucleus, while neurons in the ventrolateral (hindleg) part of the red nucleus project to the gracile nucleus (Edwards 1972; Martin et al. 1974; Holstege and Tan 1988). It has been demonstrated that the projections to the interpositus nucleus are collaterals from rubrobulbospinal fibers (Huisman et al. 1982). In all likelihood, this is also true for the red nucleus projections to the precerebellar structures in the brainstem and dorsal column nuclei (see Anderson 1971), which would correspond with the finding that the latter projections are somatotopically organized (Holstege and Tan 1988).

In the spinal cord the rubrospinal fibers descend via the dorsolateral funiculus and terminate on interneurons in the lateral parts of the intermediate zone (laminae V to VII) and to a limited extent directly to motoneurons. Interneurons receiving rubrospinal fibers receive afferents from other sources also, such as peripheral nerves, propriospinal neurons, and reticulo- and corticospinal tracts. Furthermore, rubrospinal fibers terminate on both first and last order interneurons (Hongo et al. 1969; see also Jankowska 1988 for review). Apparently the red nucleus uses all the interneurons involved in the reflex pathways in the spinal cord.

Although the red nucleus projections to motoneurons are mostly indirect, physiological studies in cat and monkey (Shapovalov et al. 1971; Shapovalov and Kurchavyi 1974; Cheney 1980; Cheney et al. 1988) have demonstrated direct red nucleus projections to spinal motoneurons. Anatomically however, there was only evidence for direct red nucleus projections to motoneurons in the facial nucleus (Courville 1966a,b; Edwards 1972; Martin et al. 1974; Holstege et al. 1984; Holstege and Tan, 1988). Holstege (1987a), Robinson et al. (1987) and McCurdy et al. (1987) demonstrated that the red nucleus in the cat projects directly to a specific group of motoneurons in the dorsolateral part of the C8–TI ventral horn, innervating forelimb digit muscles (see pp 260 and 291). In addition Holstege et al. (1988; Fig. 13.18) at the light microscopical level and Ralston et al. (1988) at the electron microscopical level revealed rubromotoneuronal projections in the monkey,

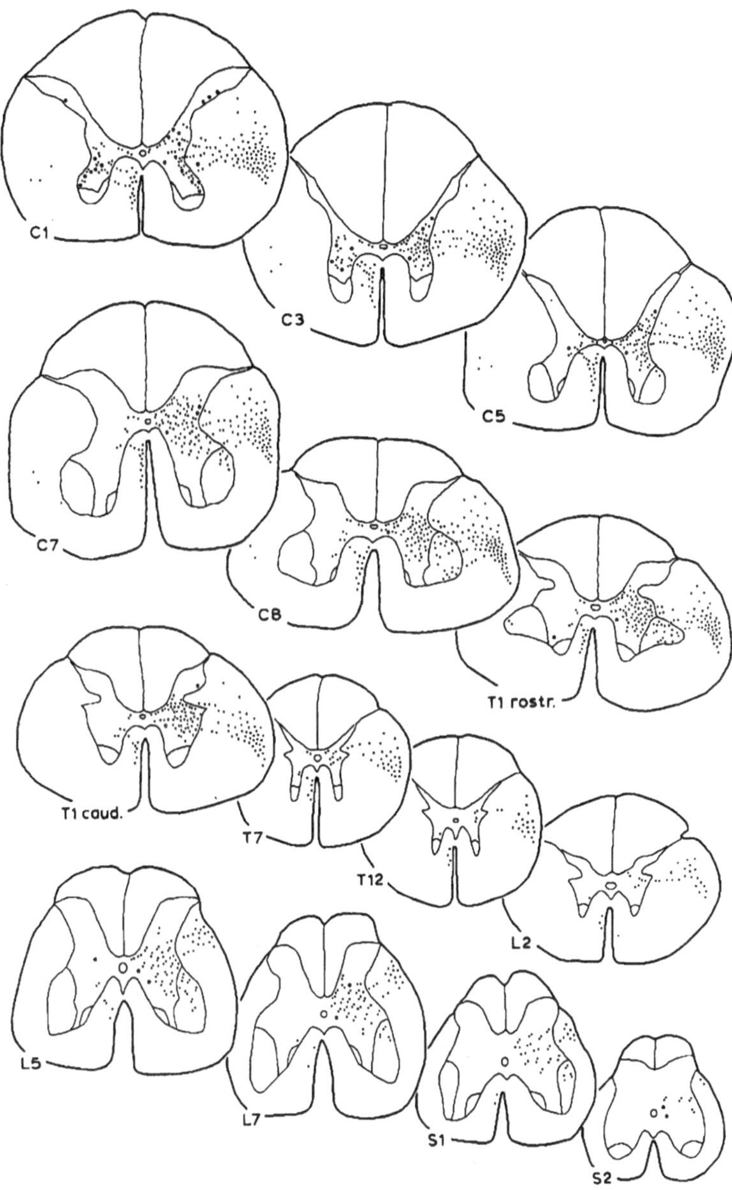

Fig. 13.18. Schematic representation of the labeled fibers (small dots) in the spinal cord of a monkey with an injection of WGA-HRP in the rubrospinal red nucleus. The injection-site extended into the area of the interstitial nucleus of Cajal (INC–RF). The retrogradely labeled neurons are indicated with large dots. Note the contralateral projections to the intermediate zone throughout the length of the spinal cord and to the lateral motoneuronal cell groups in the cervical and lumbosacral enlargements. Note also the ipsilateral (interstitiospinal) fibers in the ventral funiculus on the ipsilateral side. Note further the very few ipsilateral rubrospinal fibers, some of which terminate in the lateral motoneuronal cell groups in rostral T1. (From Holstege et al. 1988.)

which were more extensive than in the cat. These projections involved all distal limb muscle motoneuronal cell groups in the cervical and lumbosacral enlargements. Projections to the axial or proximal muscle motoneurons were never observed.

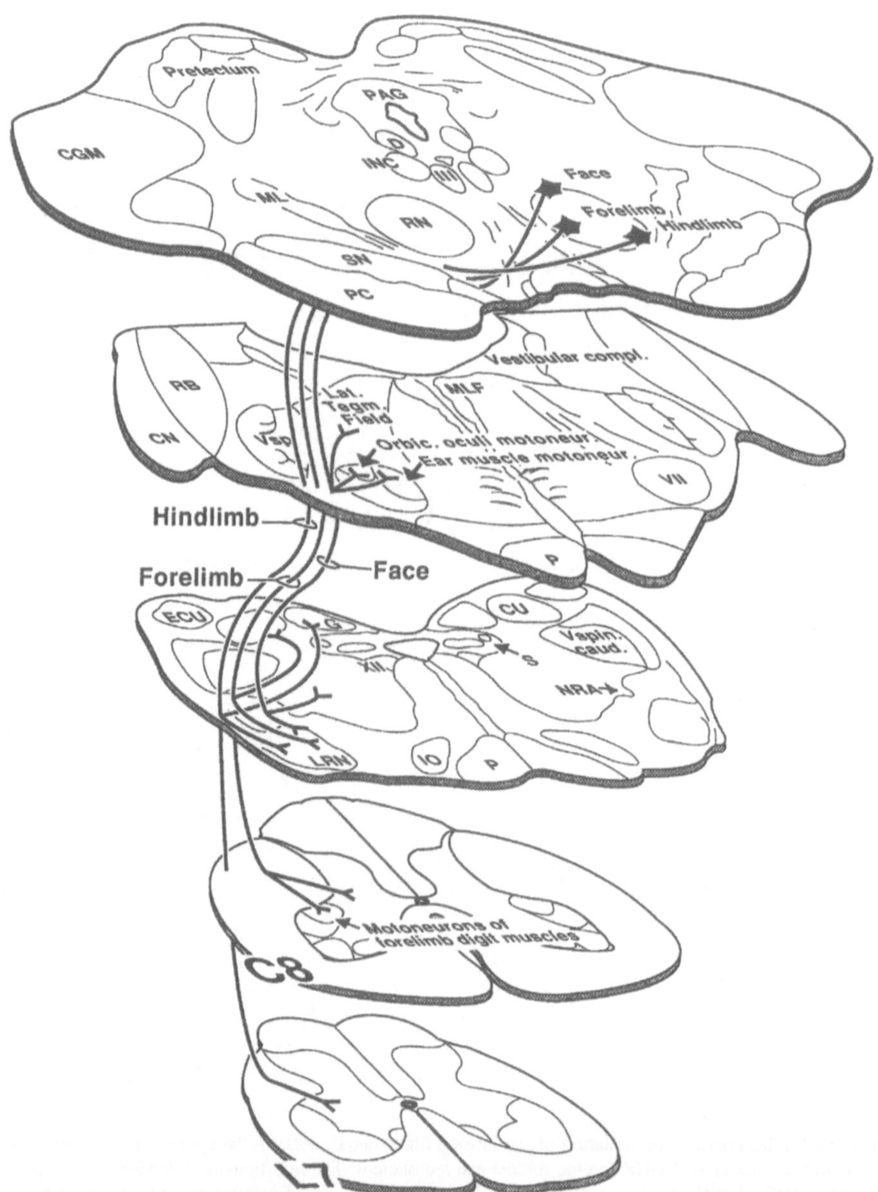

Fig. 13.19. Schematic overview of the rubrobulbospinal projections in the cat. In the monkey the rubrobulbospinal projections are almost identical, with the exception of more extensive projections to the motoneuronal cell groups. (From Holstege 1991.)

Gibson et al. (1985) and Cheney et al. (1988) studied the red nucleus projections to flexor and extensor motoneurons of the wrist and fingers in the monkey and observed a strong preference for facilitation of extensor muscles (see also Cheney et al. 1991). Martin and Ghez (1988), in the cat, studied the differential contributions of the motor cortex and red nucleus neurons to the initiation of a targeted limb response and to the control of trajectory. They concluded that both the motor cortex and the red nucleus contributed to the initiation of the motor responses, but that only the motor cortex is involved in the proper scaling of targeted responses. Fig. 13.19 gives an overview of the rubrobulbospinal projections.

As the evolutionary scale is climbed, the rubrospinal red nucleus becomes smaller, and in humans only very few rubrospinal neurons seem to exist, which do not descend further than C3 (see Nathan and Smith 1982 for review). The most likely reason for such a regression is the development of the corticospinal tract, which is extremely well developed in humans and might render the rubrospinal tract redundant (see Massion 1988). It remains to be determined whether this is also true for the contralateral rubrobulbar projections.

The Corticobulbar and Corticospinal System

The enormous outgrowth of the cerebral cortex in humans, compared to other mammals, is also reflected in the motor cortico-bulbospinal tract, which in primates but especially humans is the most important descending pathway within the somatic motor system. The motor cortex is somatotopically organized with a foreleg area projecting to the cervical cord, a hindleg area projecting to the lumbosacral cord (Armand et al. 1985) and a face area projecting to the lateral tegmental field of caudal pons and medulla (Kuypers 1958c; Holstege unpublished data). In cat, monkey, ape and human the motor cortex not only projects mainly contralaterally to the laterally located interneurons in the spinal cord, but, in contrast to the red nucleus, also bilaterally to more medially located interneurons (lamina VIII). These projections are derived from the so-called common zone. In the cat this area is located in the medial part of the motor cortex next to area 6 and extends caudally between the fore- and hindleg areas (Armand and Kuypers 1980). Stimulation in the area tends to carry the representations of axial movements, i.e. neck, trunk and proximal forelimb movements (Nieollon and Rispal-Padel 1976). Strictly speaking, this cortical projection system belongs to the medial descending system, but is presented together with the other corticospinal projections, because it represents a relatively small portion of the descending corticospinal tract. Not surprisingly, neurons in the common zone, possibly via collaterals of the corticospinal fibers, project to the pontine and upper medullary medial tegmental field, one of the most important parts of the medial descending system (see p 281).

In the monkey (Kuypers 1958b; Ralston and Ralston 1985), but not in the cat (Armand et al 1985), there exist direct cortical projections to motoneurons, innervating the most distal muscles of the extremities. Ralston and Ralston (1985) found that two thirds of the corticomotoneuronal terminals contained round vesicles, suggesting excitatory effects on the motoneuron, and one third pleomorphic or flattened vesicles, suggesting inhibitory effects. It is questionable, however, whether there exist monosynaptic inhibitory corticomotoneuronal connections, but disynaptic inhibitory connections have been demonstrated (Landgren et al. 1962). Jankowska et al. (1975), stimulating the motor cortex in monkeys, observed excitatory postsynaptic potentials (EPSPs) with response latencies of 0.6–1.0 ms, indicating monosynaptic contact. The rubro- and corticospinal tracts in the monkey are very similar, but there are some differences: (1) the motor cortex projects also to more medial parts of the intermediate zone; (2) corticospinal fibers are at least 100 times more numerous than the rubrospinal ones (Holstege et al. 1988); or differences between the corticomotoneuronal and rubromotoneuronal cells, see Cheney et al. (1991). In chimpanzees and humans direct corticomotoneuronal projections are more extensive than in the monkey and terminate also on motoneurons innervating more proximal muscles of the body (Kuypers 1958a,b; Schoen 1964). However, the degeneration findings of Kuypers (1964) and Schoen (1964) do not reveal corticospinal projections to the medial motoneuronal cell column in chimpanzee and human. It is possible that more modern tracing techniques in the chimpanzee would reveal direct cortical projections to medial column motoneurons, but such studies have not yet been done. Since the corticospinal and rubrospinal systems are so similar, it is not surprising that collaterals of the corticospinal tract terminate in the magnocellular red nucleus in a somatotopically organized manner (Kuypers 1981; Holstege unpublished observations).

Behavioural studies on the lateral system of Lawrence and Kuypers (1968a,b) in the monkey have demonstrated that immediately after pyramidotomy (interruption of the corticospinal fibers at the level of the medulla oblongata), the animals can sit, walk, run and climb, but cannot pick up pieces of food with their hands. After a period of recovery they regain this capacity, but individual finger movements such as the thumb-and-index-finger precision grip do not return. In pyramidotomized monkeys, the red nucleus as well as the corticorubral fibers are still intact, and this recovery of hand movements is probably related to the rubrospinal tract taking over many of the functions of the corticospinal tract. Ablation of the precentral motor cortex in adult monkeys (thus lesioning the corticospinal as well as the corticorubral fibers) results in a stronger deficit, i.e. a flaccid paresis of the contralateral extremity muscles. In chimpanzee and humans (patients with stroke or tumour interrupting the corticobulbospinal fibers) this flaccid paresis is more severe than in monkeys, and much more severe than in cats. It is possible that this difference occurs because the rubrospinal neurons in monkey and cat are much more numerous than in chimpanzee and humans. Correspondingly, if in a monkey a bilateral pyramidotomy is combined with an interruption of the rubrospinal tract on one side, the motor deficit on that side is much more pronounced. The monkey is able to sit up, walk and climb, but in the examining chair the fingers and wrist of the arm ipsilateral to the side of the rubrospinal lesion are noticeably limp. In reaching for food, the hand is brought to the food by turning the arm in the shoulder.

Fig. 13.20 gives a summary on the rubro- and corticospinal pathways, based on the findings of several studies (Kuypers 1964, 1973, 1981; Schoen 1964; Kuypers and Brinkman, 1970; Ralston and Ralston, 1985; Armand et al. 1985; Holstege 1987a; Holstege et al. 1988) The corticospinal fibers become more and more numerous and control larger parts of the spinal gray, ascending from cat, via monkey to human, but that is not so for the rubrospinal tract. The enormous predominance of the corticospinal tract over the rubrospinal tract in humans leads to great clinical problems in stroke patients with interruption of the corticospinal tract in the internal capsule. Recovery from such a lesion is much more difficult than in monkeys or cats with similar lesions, because humans do not have the capacity to recruit functionally direct connexions through a well-developed rubrospinal tract.

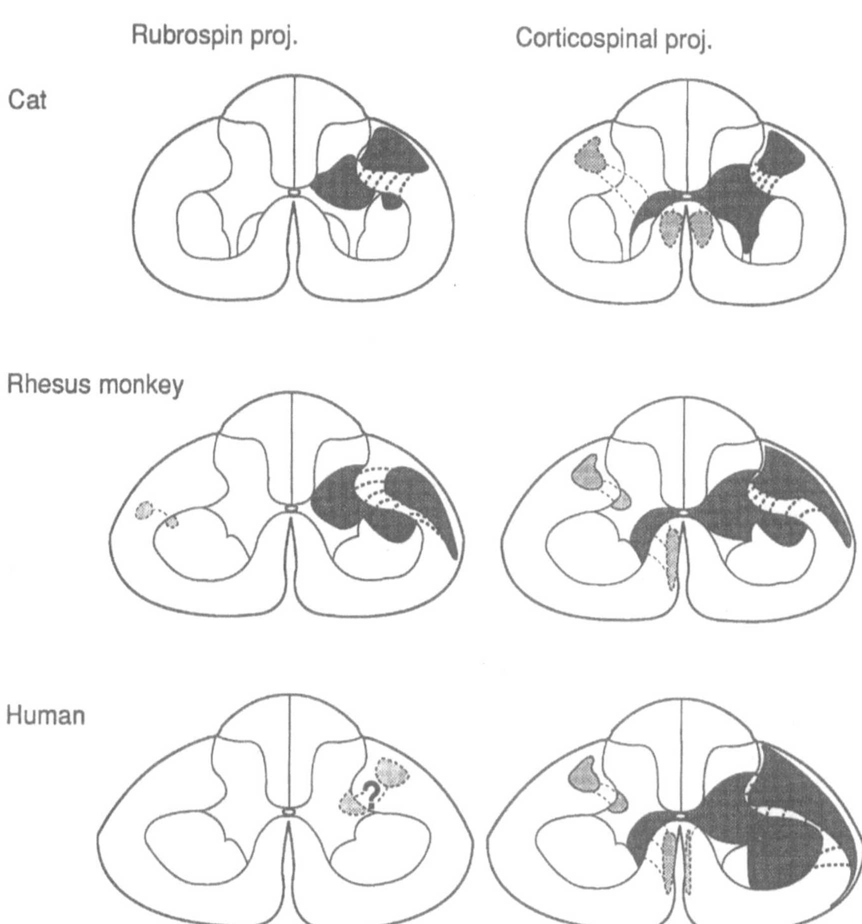

Fig. 13.20. Schematic representation of the rubrospinal and corticospinal projections in cat, rhesus monkey and human at the level of C8. The gray areas in the white matter represent the descending pathways, those in the gray matter represent termination zones. Dark gray areas represent strong projections, lighter gray areas represent light projections. (From Holstege 1991.)

Descending Pathways Involved in Limbic Motor Control Systems

Introduction

It is well known that hemiplegic patients with damage to corticobulbar fibers, resulting in a complete central paresis of the lower face on one side, are able to smile spontaneously, for example, when they enjoy a joke. On the other hand, in cases with postencephalitic parkinsonism, patients are able to show their teeth, whistle, frown, i.e. there is no facial palsy, but the patients' emotions are not reflected in their countenance and they have a stiff, mask-like facial expression (poker face). Patients with irritative pontine lesions sometimes suffer from non-emotional laughter and crying, and patients with pseudobulbar palsy (for example with lesions in the mesencephalon) often suffer from uncontrollable fits of crying or laughter. Such fits are usually devoid of feeling of grief, joy or amusement; they may even be accompanied by entirely incompatible emotions. Fits of crying and laughter may occur in the same patients, other patients show only one of them (Poeck 1969; Rinn 1984). Crying and laughter belong to an expressive behaviour, which in animals is called vocalization. It has been shown in many different species that stimulation in the caudal part of the periaqueductal gray (PAG) produces vocalization. Recently Holstege (1989) has demonstrated that vocalization is based on a specific final common pathway, originating from a distinct group of neurons in the PAG that project to the nucleus retro-ambiguus, which in turn has direct access to all vocalization motoneurons. In all likelihood, in humans this projection forms the anatomical framework for laughing and crying. The vocalization neurons in the PAG receive their afferents from structures belonging to the limbic system, but not from the voluntary system. All this clinical and experimental evidence shows that there exists a complete dissociation between the voluntary and emotional or limbic innervation of motoneurons.

The limbic system is closely involved in the elaboration of emotional experience and expression (MacLean 1952) and is associated with a wide variety of autonomic, visceral and endocrine functions. The limbic system consists of several cortical and subcortical structures, although there is no agreement on exactly which structures belong to it. Some authors argue that the use of the term limbic system should be abandoned (for example Brodal 1981). Nevertheless, many clinicians and scientists still use it and they consider the cingulate, insular, entorhinal, piriform, hippocampal, retrosplenial and orbitofrontal cortex to belong to the limbic system. Subcortical regions usually included in the limbic system are the hypothalamus and the pre-optic region, the amygdala, the bed nucleus of the stria terminalis, the septal nuclei, and the anterior and mediodorsal thalamic nuclei. As early as 1958, Nauta pointed out that the limbic system has extremely strong reciprocal connections with mesencephalic structures such as the periaqueductal gray (PAG) and the laterally and ventrally adjoining tegmentum (Nauta's limbic system–midbrain circuit). More recent findings strongly support Nauta's concept and has led Holstege (1990) to consider the mesencephalic periaqueductal gray (PAG) and large parts of the lateral and ventral mesencephalic tegmentum to belong to the limbic system.

The new tracing techniques of the last 15 years have revealed many new limbic system pathways to caudal brainstem and spinal cord. Some of the most interesting are the projections to the nucleus raphe magnus (nucleus raphe magnus) and pallidus (nucleus raphe pallidus) as well as to the adjacent ventral part of the caudal pontine and medullary reticular formation. These findings are important, because nucleus raphe magnus, nucleus raphe pallidus and adjoining reticular formation in turn project diffusely, but very strongly to all parts of the gray matter throughout the length of the spinal cord. Therefore, the diffuse brainstem–spinal projections will be discussed in the framework of the descending limbic motor control systems.

Pathways Projecting Diffusely to the Spinal Gray Matter

Projections from the Nuclei Raphe Magnus, Pallidus and Obscurus and the Ventral Part of the Caudal Pontine and Medullary Medial Reticular Formation

Retrograde tracing results (Kuypers and Maisky 1975; Tohyama et al. 1979; Holstege G. and Kuypers 1982; Holstege 1988b) have shown that a great number of neurons in the nuclei raphe magnus and pallidus and ventral part of the caudal pontine and medullary medial reticular formation project to the spinal cord (Fig. 13.14). Many of these neurons project to cervical as well as lumbar levels of the spinal cord and to the caudal spinal trigeminal nucleus (Martin et al. 1981b; Hayes and Rustioni 1981; Huisman et al. 1982; Lovick and Robinson 1983).

Anterograde (autoradiographic) tracing techniques demonstrated in the cat that the nucleus raphe magnus and the adjacent reticular formation projects to the marginal layer of the caudal spinal trigeminal nucleus and in the spinal cord to laminae I, II, V, VI and VII, and to the thoracolumbar intermediolateral cell column (Basbaum et al. 1978; Martin et al. 1981a, 1985; Holstege et al. 1979; Holstege and Kuypers 1982; Fig. 13.21 left). Moreover, Holstege G. and Kuypers (1982) demonstrated that the nucleus raphe magnus and adjoining tegmentum project to the sacral intermedial and intermediolateral cell column. Another very important finding was that the nucleus raphe pallidus and its adjoining reticular formation does not project to the dorsal horn of caudal medulla and spinal cord, but to all other parts of the spinal gray matter, i.e. the intermediate zone and the somatic and autonomic motoneuronal cell groups of the spinal cord (Fig. 13.21 right) and to the motoneuronal cell groups V, VII, X and XII in the caudal brainstem (Martin et al. 1979a 1981a; Holstege et al. 1979; Holstege G. and Kuypers 1982). Such projections have also been shown in the monkey (Holstege 1991; Fig. 13.22). The caudal nucleus raphe magnus and rostral nucleus raphe pallidus also project to the thoracolumbar and sacral intermediolateral cell groups (intermediolateral cell group), i.e. the autonomic (sympathetic and parasympathetic) preganglionic motoneuronal cell groups (Fig. 13.21).

Physiological studies are consistent with the anatomy of the descending pathways outlined above. The diffuse organization of nucleus raphe pallidus, nucleus raphe obscurus and ventromedial medulla projections to the motoneuronal cell groups suggests that they do not steer specific motor activities such as movements of distal (arm, hand or leg) or axial parts of the body, but have a more global effect on the level of activity of the motoneurons. Stimulation of the raphe nuclei has a facilitatory effect on motoneurons (Cardona and Rudomin 1983). There exist many different neurotransmitter substances in this area, of which serotonin is the best known. In mammals, there are many serotonergic fibers around the motoneurons (Steinbusch 1981 and Kojima et al. 1983b in the rat; Kojima et al. 1982 in the dog; Kojima et al. 1983a in the monkey). The cell bodies of these serotonergic fibers are mainly located in the nucleus raphe pallidus (Alstermark et al. 1987a,b). Serotonin plays a role in the facilitation of motoneurons, probably directly by acting on the Ca^{2+} conductance or indirectly by reduction of K^+ conductance of the membrane of the motoneuron (McCall and Aghajanian 1979; White and Neuman 1980; VanderMaelen and Aghajanian 1982; Hounsgaard et al. 1986). Thus serotonin enhances the

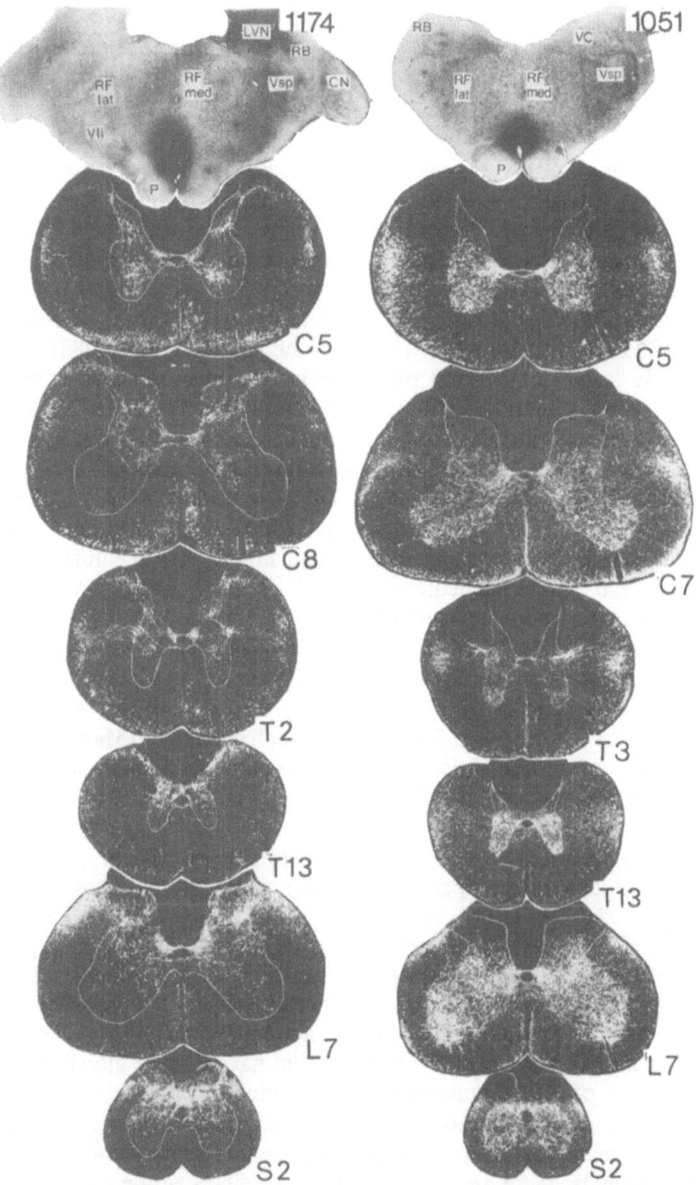

Fig. 13.21. Brightfield photomicrographs of autoradiographs showing tritiated leucine injection sites in the raphe nuclei and darkfield photomicrographs showing the distributions of the labeled fibers in the spinal cord. On the left an injection is shown in the caudal nucleus raphe magnus and adjoining reticular formation. Note that labeled fibers are distributed mainly to the dorsal horn (laminae I, the upper part of II and V), the intermediate zone and the autonomic motoneuronal cell groups. On the right the injection is placed in the nucleus raphe pallidus and immediately adjoining tegmentum. Note that the labeled fibers are not distributed to the dorsal horn, but very strongly to the ventral horn (intermediate zone and autonomic and somatic motoneuronal cell groups. (From Holstege G and Kuypers 1982.)

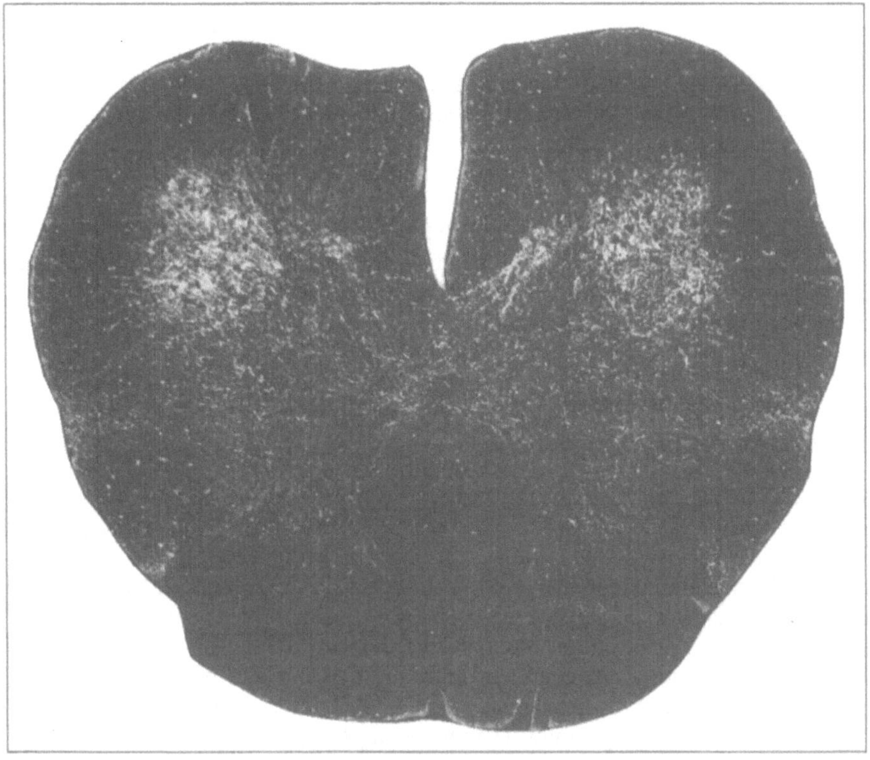

Fig. 13.22. Darkfield photomicrograph of a section through the lumbar spinal cord in the monkey, after injection of ^{3}H-leucine in the ventral part of the medullary medial tegmental field. Note the diffuse projections to the motoneuronal cell groups. (From Holstege 1991.)

excitability of the motoneurons for inputs from other sources, such as red nucleus or motor cortex (McCall and Aghajanian, 1979).

Not only serotonin, but several peptides are also present in the spinally projecting neurons in the ventromedial medulla and nucleus raphe pallidus and obscurus, such as substance P, thyrotropin releasing hormone (TRH), somatostatin, methionine (M-ENK) and leucine-enkephalin (L-ENK), while a relatively small number contains vasoactive intestinal peptide (VIP) and cholecystokinin (CCK). It has been demonstrated that most of these peptides coexist to a variable extent with serotonin in the same neuron (Chan Palay et al. 1978; Hökfelt et al. 1978, 1979; Johansson et al. 1981; Hunt and Lovick, 1982; Bowker et al. 1983, 1988; Mantyh and Hunt 1984; Taber-Pierce et al. 1985; Helke et al., 1986; Léger et al., 1986). Johansson et al. (1981) have even demonstrated the coexistence of serotonin, substance P and TRH in one and the same neuron. This coexistence of serotonin with different peptides not only occurs in the neuronal cell bodies, but also in their terminals in the ventral horn (Pelletier et al. 1981; Bowker 1986; Wessendorf and Elde 1987).

It must be emphasized that a major portion of the diffuse descending pathways to the dorsal horn and the motoneuronal cell groups is not derived from serotonergic neurons (Bowker et al. 1982; Johannessen et al. 1984). This suggests that non-serotonergic neurons terminate differently in the motoneuronal cell groups than the fibers of the serotonergic neurons, which may or may not contain other peptides as well. Spinal motoneurons display a bistable behaviour, i.e. they can switch back and forth to a higher excitable level (Hounsgaard et al. 1984, 1986, 1988; Crone et al. 1988). Bistable behaviour disappears after spinal transection, but reappears after subsequent intravenous injection of the serotonin precursor 5-hydroxytryptophan. Thus, intact descending pathways are essential for this bistable behaviour of motoneurons and serotonin is one of the neurotransmitters involved in switching to a higher level of excitation.

In summary, the diffuse descending pathways originating in the ventromedial medulla, including the nucleus raphe pallidus and obscurus, have very general and diffuse facilitatory or inhibitory effects on motoneurons and probably also on interneurons in the intermediate zone.

Projections from the Dorsolateral Pontine Tegmental Field (A7 Cell Group)

Retrograde and anterograde tracing studies (Martin et al. 1979b; Holstege et al 1979; Westlund and Coulter 1980; Holstege G. and Kuypers, 1982; Holstege J.C. and Kuypers, 1982) show that a large number of neurons in the locus coeruleus in the rat or the nucleus subcoeruleus and ventral part of the parabrachial nuclei in the cat project diffusely to all parts of the spinal gray matter, which projection has also been demonstrated at the ultrastructural level (Holstege J.C. and Kuypers 1987). Many neurons in the locus coeruleus, subcoeruleus and the parabrachial nuclei contain noradrenaline (Westlund and Coulter 1980; Jones and Friedman 1983; Jones and Beaudet 1987) or acetylcholine (Kimura et al., 1981; Jones and Beaudet, 1987). The diffuse projection from this area to the spinal cord is at least in part noradrenergic, since lesioning the dorsolateral pontine tegmental field reduced the number of noradrenergic terminals in the spinal gray matter by 25%–50% in the dorsal horn and by 95% in the ventral horn (Nygren and Olson 1977). Electrical stimulation in the area of the locus coeruleus/subcoeruleus in rat (Chan et al. 1986) and cat (Fung and Barnes 1987) produces a decrease in input resistance and a concurrent non-selective enhancement in motoneuron excitability, indicative of an overall facilitation of motoneurons.

Projections from the Rostral Mesencephalon/Caudal Hypothalamus (A11 Cell Group)

Skagerberg and Lindvall (1985) in the rat demonstrated that dopamine-containing neurons in the so-called A11 cell group projected throughout the length of the spinal cord. The A11 cell group is located in the border region of rostral mesencephalon and dorsal and posterior hypothalamus, extending dorsally along the paraventricular gray of the caudal thalamus. Skagerberg and Lindvall (1985) were not able to determine in which specific parts of the spinal gray matter the A11 dopaminergic fibers terminated. Yoshida and Tanaka (1988), using anti-dopamine serum, found dopamine-immunoreactive fibers throughout the whole gray matter at any level of the spinal cord. The distribution of the dopaminergic fibers in the spinal gray strongly resembles that of the noradrenergic fibers in the spinal cord. Therefore the possibility of labeling dopamine as a precursor of noradrenalin must be kept in mind (for discussion see Yoshida and Tanaka 1988). Functionally there is also a resemblance between noradrenergic and dopaminergic fiber projections to the spinal cord. Infusion of dopamine in the spinal cord increases motoneuron activity (Simon and Schramm 1983).

Projections from the Mesencephalon to Caudal Brainstem and Spinal Cord

In recent years specific information has become available about the anatomy and function of the descending projections of the mesencephalon in relation to emotional behavior. Stimulation in the mesencephalon has been shown to result in pain inhibition, vocalization, aggressive behavior, blood pressure changes, lordosis and locomotion. Many of the neurons involved in these functions are located in the PAG, but neurons in the mesencephalic tegmentum lateral and ventral to the PAG also play a role.

Descending Projections to the Nucleus Raphe Magnus, Nucleus Raphe Pallidus and Ventral Part of the Caudal Pontine and Medullary Medial Tegmentum

Retrograde tracing studies (Abols and Basbaum 1981; Holstege 1988a) indicate that an enormous number of labeled neurons in the PAG and laterally and ventrolaterally adjoining areas project to nucleus raphe magnus, nucleus raphe pallidus and ventral part of the caudal pontine and medullary medial tegmentum (Fig. 13.23). Anterograde (autoradiographic) tracing studies (Jürgens and Pratt, 1979; Mantyh 1983; Holstege 1988a; Fig. 13.24) show that different parts of the PAG and adjacent tegmentum project in the same basic pattern to the caudal brainstem. The descending mesencephalic fibers pass ipsilaterally through the mesencephalic and pontine lateral tegmental field, but gradually shift ventrally and medially at caudal pontine levels. They terminate mainly ipsilaterally in the ventral part of the caudal pontine and medullary medial reticular formation and in the nucleus raphe magnus (Fig. 13.24). Neurons in the ventrolateral portion of the caudal PAG and the ventrally adjoining mesencephalic tegmentum send fibers to the nucleus raphe pallidus (Fig. 13.24 left).

Involvement of the Descending Mesencephalic Projections in Control of Nociception. In animals (see Besson and Chaouch 1987 and Willis 1988 for reviews) as well as in humans (Hosobuchi 1988; Meyerson 1988) the PAG is well known for its involvement in the supraspinal control of nociception. The strong impact on nociception is partly mediated via its projections to the nucleus raphe magnus and adjacent reticular formation, because in cases with reversible blocks of the nucleus raphe magnus and adjacent tegmentum, PAG stimulation results in reduced analgesic effects (Gebhart et al. 1983; Sandkuhler and Gebhart 1984).

Involvement of the Descending Mesencephalic Projections in the Lordosis Reflex. Stimulation in the PAG also facilitates the lordosis reflex (Sakuma and Pfaff 1979a,b). Lordosis, a curvature of the vertebral column with ventral convexity, is an

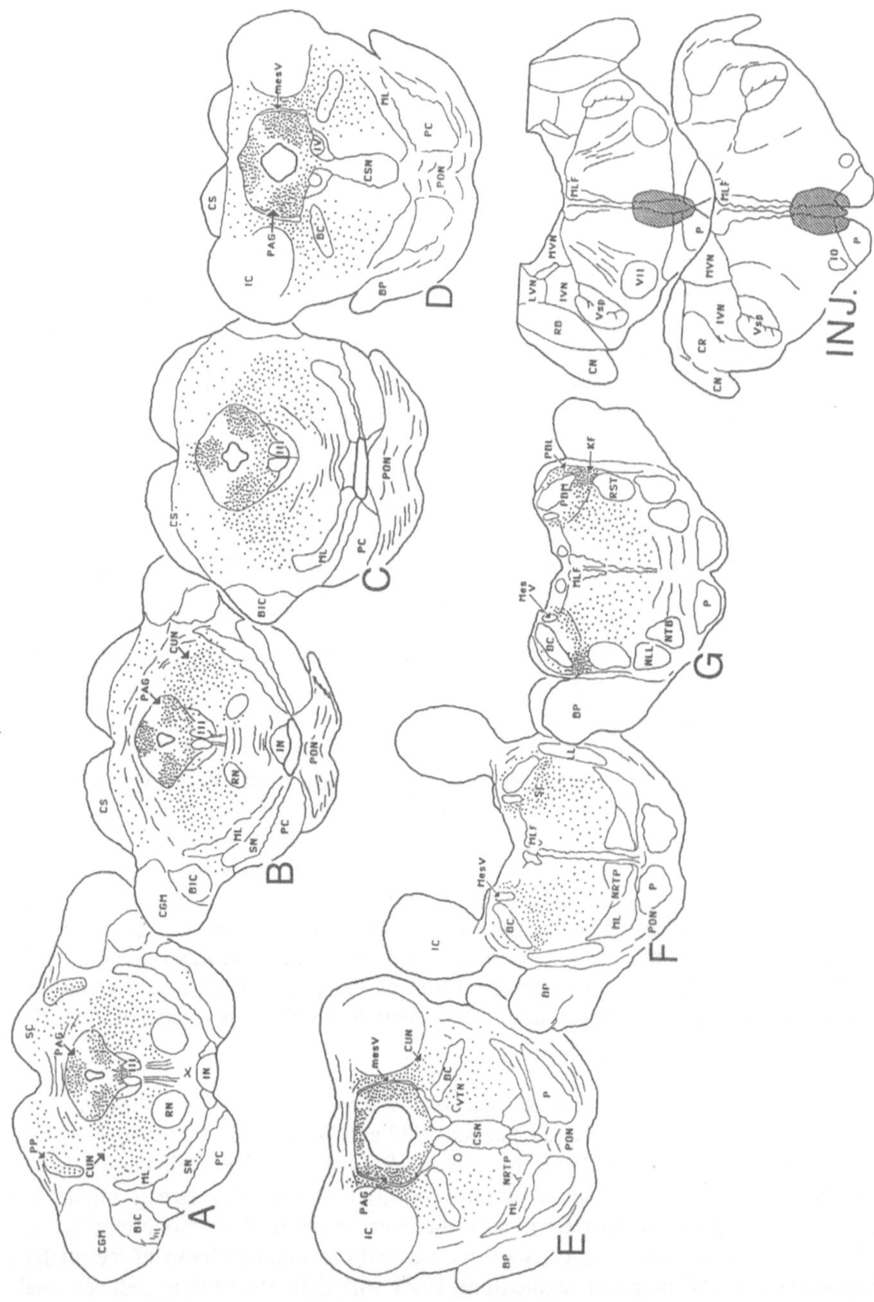

Fig. 13.23. Schematic drawings of HRP-labeled neurons in mesencephalon and pons after injection of HRP in the nucleus raphe magnus/nucleus raphe pallidus region. Note the dense distribution of labeled neurons in the PAG (except its dorsolateral part) and the tegmentum ventrolateral to it. Note also the distribution of labeled neurons in the area of the ventral parabrachial nuclei and the nucleus Kölliker-Fuse. (From Holstege 1988a.)

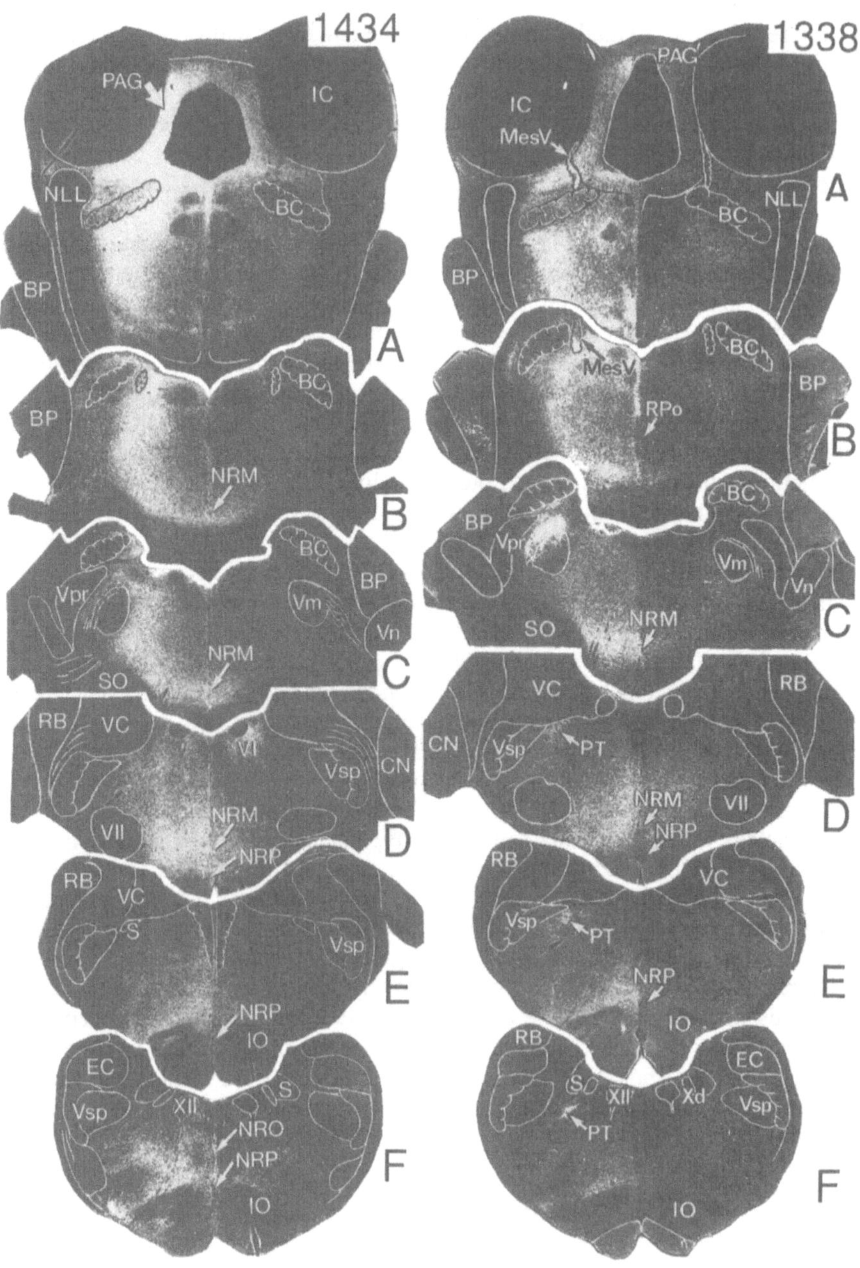

Fig. 13.24. Darkfield photomicrographs of the brainstem in the cases 1434 and 1338 with injections in respectively the ventrolateral PAG and more rostrally in the lateral PAG. Note the strong projections to the nucleus raphe magnus and the ventral part of the medial tegmentum of caudal pons and medulla in both cases. Note that in case 1434, but not in case 1338, labeled fibers were also distributed to the nucleus raphe pallidus. (From Holstege 1988a.)

essential element of female copulatory behavior in rodents. Stimulation of the L1 through S1 dermatomes is sufficient for eliciting the lordosis reflex, but several studies suggest that it is oestrogen dependent, i.e. only occurs when copulation can result in fertilization. Actually, during oestrus the female rat shows several forms of stressful behavior, characterized by frequent locomotion and other stress-like phenomena (Pfaff 1980). It is well known in mammals that various forms of stress, whether it is aggression, fear or sexual arousal, set the motor system at a "high" level. In such circumstances spinal reflexes such as the lordosis reflex can easily be elicited.

Involvement of the Descending Mesencephalic Projections in Locomotion. Just lateral to the brachium conjunctivum, just ventral to the cuneiform nucleus and just rostral to the parabrachial nuclei is located the so-called pedunculopontine nucleus. Stimulation in the pedunculopontine nucleus induces locomotion in cats (Shik et al. 1966), which is the reason that this area is also termed the mesencephalic locomotor region. This region not only comprises the pedunculopontine nucleus, but extends into the cuneiform nucleus, which is located just dorsal to the pedunculopontine nucleus. Garcia-Rill and Skinner (1988) found that during locomotion neurons in the cuneiform nucleus were related preferentially to rhythmic (bursting) activity, while neurons in the pedunculopontine nucleus are preferentially related to the onset or termination of cyclic episodes (on/off cells).

Anatomical studies (Moon Edley and Graybiel 1983; Holstege unpublished results) revealed that the descending projections from the mesencephalic locomotor region are organized similarly to those from the PAG and adjacent tegmentum. Findings of Garcia-Rill and Skinner (1987a,b) indicate that locomotion, elicited in the mesencephalic locomotor region, is based on the projections from this area to the medial part of the ventral medullary medial tegmentum and on the diffuse projections from the latter area to the rhythm generators in the spinal cord.

The afferent connections of the mesencephalic locomotor area are derived from lateral parts of the limbic system, such as the bed nucleus of the stria terminalis, central nucleus of the amygdala and lateral hypothalamus. Strong projections are also derived from the entopeduncular nucleus, subthalamic nucleus and the substantia nigra pars reticulata, but motor cortex projections to the mesencephalic locomotor region are very scarce (Moon Edley and Graybiel, 1983). These findings indicate that the mesencephalic locomotor region is influenced by extrapyramidal and lateral limbic structures, and virtually not by somatic motor structures. This corresponds with the fact that the descending projections from the mesencephalic locomotor region terminate in the ventromedial part of the caudal pontine and medullary tegmental field, which area receives afferents from many other limbic system related areas, but not from the somatic motor structures.

PAG Projections to the Ventrolateral Medulla: Involvement in Blood Pressure Control

Neurons in the rostral part of the ventrolateral tegmental field of the medulla (subretrofacial nucleus) are essential for the maintenance of the vasomotor tone and reflex regulation of the systemic arterial blood pressure. Neurons in the subretrofacial nucleus project specifically to the sympathetic motoneurons in the intermediolateral cell group of the thoracolumbar spinal cord (Lovick 1987; Dampney and McAllen 1988). Carrive et al. (1989) have shown that neurons in the dorsal portions of the caudal half of the PAG have an excitatory effect on the neurons in the subretrofacial nucleus (increase of blood pressure), while neurons in the ventral part of the PAG have an inhibitory effect (decrease of blood pressure). For an extensive review of this control system see Bandler et al. (1991).

PAG Projections to the Nucleus Retroambiguus: Involvement in Vocalization

In many different species, from leopard frog to chimpanzee, stimulation in the caudal PAG results in vocalization (see Holstege 1989 for review), i.e. the non-verbal production of sound. In humans laughing and crying are probably examples of vocalization (see p 294). Holstege (1989) has demonstrated that a specific group of neurons in the lateral and to a limited extent in the dorsal part of the caudal PAG send fibers to the NRA in the caudal medulla (Fig. 13.25). The cell group in the PAG differs from the smaller cells projecting to the raphe nuclei and adjacent tegmentum or the larger cells projecting to the spinal cord. The NRA in turn projects to the somatic motoneurons innervating the pharynx, soft palate, intercostal and abdominal muscles and probably the larynx (Fig. 13.26). Direct PAG projections to these somatic motoneurons do not exist (Holstege 1989). In all likelihood, the projection from the PAG to the NRA forms the final common pathway for vocalization, because DeRosier et al. (1988) found that during vocalization the NRA neurons were more closely related to the vocalization muscle EMG than the PAG. This finding is important, because it shows that a specific expressive motor activity (fixed action pattern) such as vocalization is based on a distinct descending pathway, suggesting that all the other specific motor activities displayed during expressive behavior are based on separate descending pathways.

Projections to the Spinal Cord

Only limited PAG projections to the spinal cord exist (Fig. 13.14 L–N). Some neurons in the lateral PAG and laterally adjacent tegmentum send fibers through the ipsilateral ventral funiculus of the cervical spinal cord to terminate in laminae VIII and the adjoining part of VII (Martin et al. 1979c; Holstege 1988a,b). A very few fibers descend ipsilaterally in the lateral funiculus to terminate in the T1–T2 intermediolateral cell group (Holstege 1988a,b). The projections to the spinal cord may play a role in the defensive behavior observed by Bandler and Carrive (1988), stimulating the PAG. For example, the projection to the medial part of the intermediate zone of the cervical cord may be involved in the contralateral head turning movements as part of defensive behavior, while the projection to the T1–T2 intermediolateral cell group may produce the pupil dilation described by Bandler and Carrive (1988).

Fig. 13.27 gives a schematic overview of the descending projections from the PAG and pedunculopontine and cuneiform nuclei to the caudal brainstem and spinal cord, including the functions in which these projections might be involved.

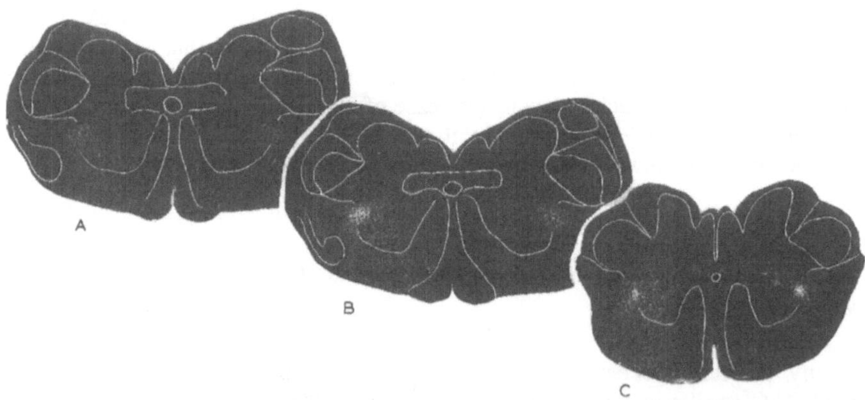

Fig. 13.25. Darkfield photomicrographs of the caudal medulla in a cat (1434, see also Fig. 13.24 left) with an injection of ^{3}H-leucine in the ventrolateral part of the caudal PAG. Note the strong bilateral projections to the nucleus retroambiguus. (From Holstege 1989.)

Projections from the Hypothalamus to Caudal Brainstem and Spinal Cord

The descending hypothalamic projection systems differ greatly, depending on which part of the hypothalamus is considered. In this section the hypothalamus will be subdivided into the anterior hypothalamus, the paraventricular hypothalamic nucleus, the posterior hypothalamus and the lateral hypothalamus.

Projections from the Anterior Hypothalamus/Pre-optic Area

According to retrograde tracing studies (Kuypers and Maisky 1975; Saper et al. 1976; Crutcher et al. 1978; Basbaum et al. 1978 and Holstege 1987b) neurons in the anterior hypothalamus/pre-optic area project strongly to the caudal brainstem, but not to the spinal cord (Fig. 13.28). Neurons in the medial part of the anterior hypothalamus project to the PAG and to the ventromedial tegmentum of caudal pons and medulla, including the nucleus raphe magnus and nucleus raphe pallidus (Fig 13.29). Application of cholinergic drugs in the anterior hypothalamus results in an emotional aversive response, which includes defense posture and autonomic (e.g. cardiovascular) manifestations (Brudzynski and Eckersdorf, 1984; Tashiro et al. 1985).

Projections from the Paraventricular Nucleus of the Hypothalamus

Kuypers and Maisky (1975) were the first to demonstrate paraventricular nucleus projections to the spinal cord (Fig. 13.28). Their findings were later confirmed in

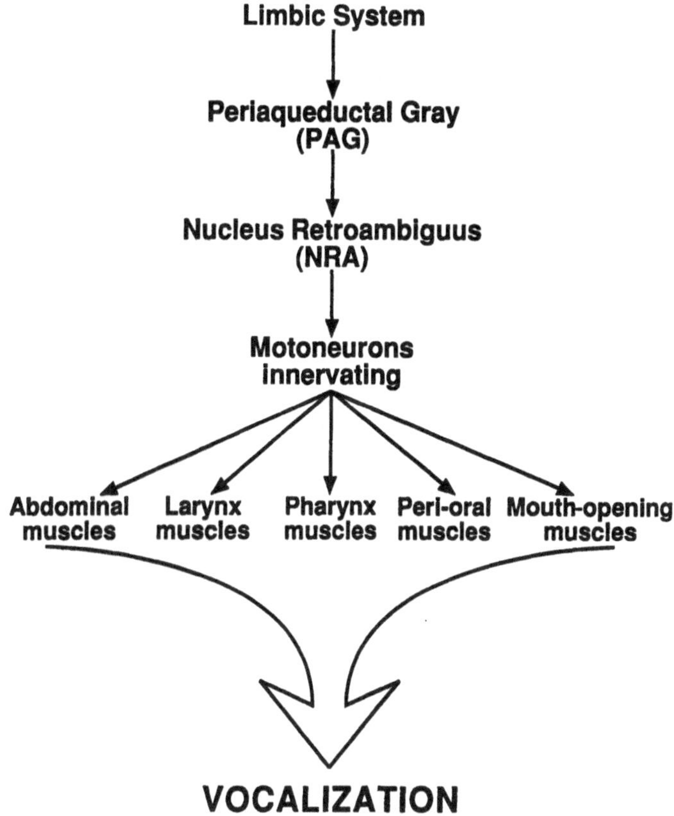

Fig 13.26. Schematic representation of the pathways for vocalization from the limbic system to the vocalization muscles. (From Holstege 1989.)

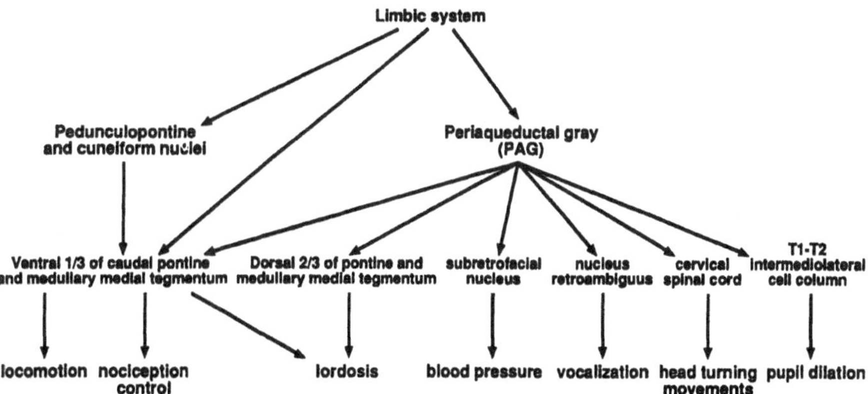

Fig. 13.27. Schematic overview of the descending projections from the PAG and pedunculopontine and cuneiform nuclei to different regions of the caudal brainstem and spinal cord. The functions in which each of the projections might be involved are also indicated. It should be emphasized that these functional interpretations are only tentative. (From Holstege 1991.)

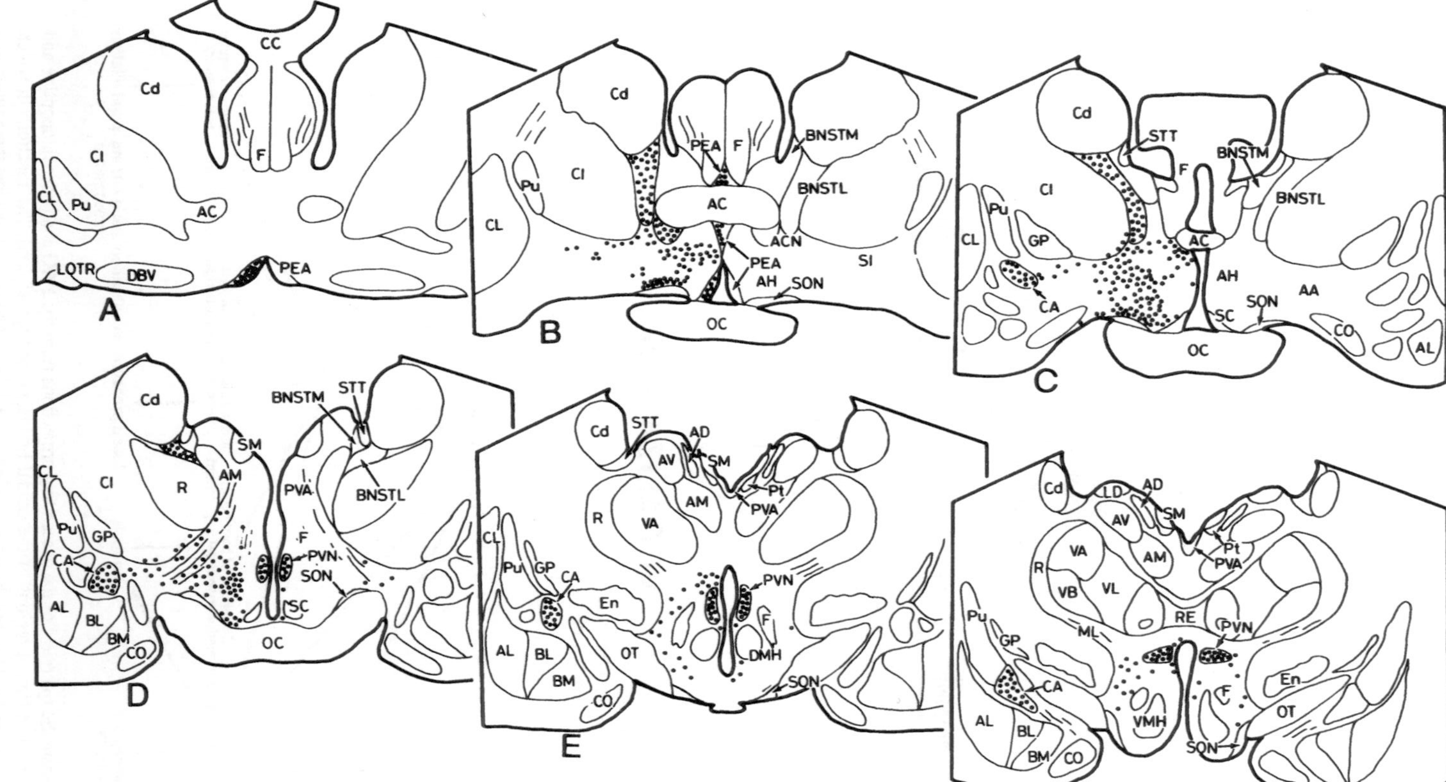

Fig. 13.28. Schematic drawing of HRP neurons in the hypothalamus, amygdala and bed nucleus of the stria terminalis. On the left the pattern of distribution of labeled neurons after a large injection of HRP in the nucleus raphe magnus, rostral nucleus raphe pallidus and adjoining tegmentum is indicated. On the right the pattern of distribution of HRP-labeled neurons after hemi-infiltration of HRP in the C2 spinal segment is shown. (From Holstege 1987.)

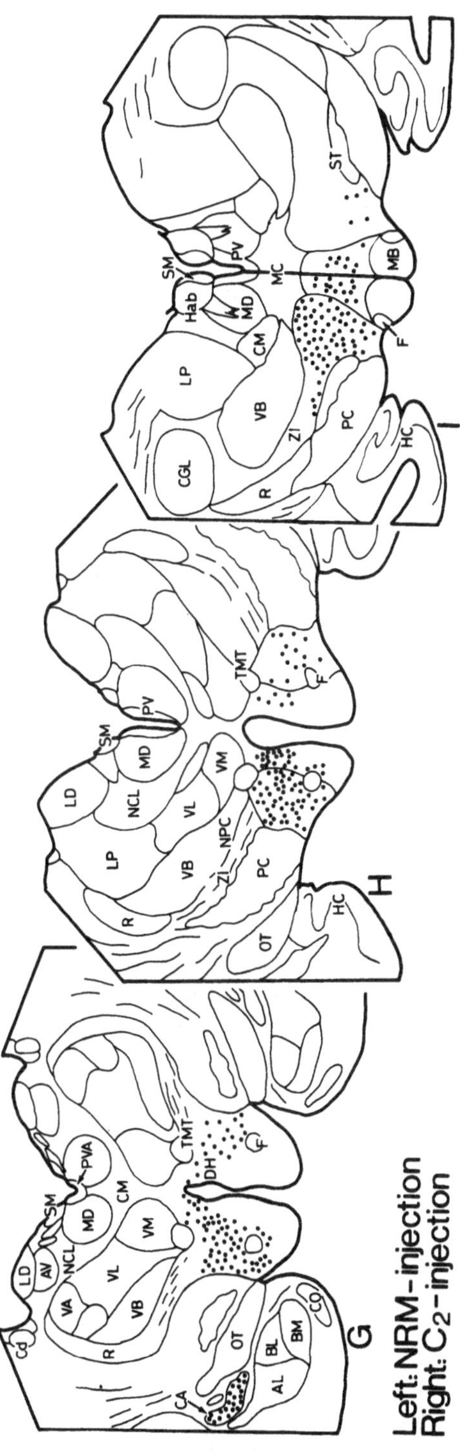

Fig. 13.28 *(continued)*

many other species (Hancock 1976; Crutcher et al. 1978; Kneisley et al. 1978; Blessing and Chalmers 1979; Holstege 1987b). The paraventricular nucleus is best known for its projections to the hypophysis, but Hosoya and Matsushita (1979) and Swanson and Kuypers (1980) have shown that the neurons projecting to the hypophysis differ from the ones projecting to the spinal cord. According to Holstege (1987b), the paraventricular nucleus neurons in the cat send their fibers to the caudal brainstem and spinal cord via the medial forebrain bundle and more caudally via a well-defined pathway through the lateral part of the mesencephalon and upper pons and just lateral to the pyramidal tract. The paraventricular fibers descend further into the lateral and dorsolateral funiculus of the spinal cord throughout its total length (Fig. 13.30). Via this pathway paraventricular nucleus sends fibers to the nucleus raphe magnus, rostral nucleus raphe pallidus and adjoining reticular formation, and to the caudal brainstem parasympathetic motoneurons and possibly the noradrenergic brainstem nuclei A1 and A2. Furthermore, paraventricular nucleus fibers terminate in mainly the rostral half of the solitary nucleus and in the area postrema (Hosoya and Matsushita 1981; Holstege 1987b). In the spinal cord the paraventricular nucleus projects bilaterally, but mainly ipsilaterally, to lamina X next to the central canal, the thoracolumbar (T1–L4) intermediolateral (sympathetic) motoneuronal cell group, and to the sacral intermediomedial and intermediolateral (parasympathetic) motoneuronal cell groups. The projections to the sympathetic intermediolateral cell column at the levels L2, L3 and upper L4 are especially strong and extensive. One might speculate, in view of the strong PVN projections to the L2–L4 intermediolateral sympathetic motoneurons, the sacral intermediolateral parasympathetic motoneurons and the nucleus of Onuf (Fig. 13.30 bottom left), that the PVN might play a role in sexual activity and/or control of the uterus contractions in pregnant women (see Holstege and Tan 1987 for a review). On the other hand, the PVN projects to all pre-ganglionic motoneurons (sympathetic and parasympathetic), which suggests a more general function, for example a similar function to the hormone ACTH. Finally, the paraventricular nucleus projects to the nucleus of Onuf (Holstege 1987b; Holstege and Tan, 1987). The paraventricular nucleus sends also fibers to lamina I of the caudal spinal trigeminal nucleus and throughout the length of the spinal cord (Holstege 1987b; Fig. 13.30). This suggests a role of the paraventricular nucleus in nociception control mechanisms. The paraventricular nucleus contains a large number of transmitter substances such as oxytocin, vasopressin, somatostatin, dopamine, methionine-enkephalin, leucine-enkephalin, neurotensin, cholecystokinin, dynorphin, substance P, glucogen, renin, and corticotropin releasing factor (see Swanson and Sawchenko 1983 for a review). Swanson (1977) and Nilaver et al. (1980) traced a pathway containing neurophysin (a carrier protein for oxytocin and vasopressin) from the paraventricular nucleus through the medial forebrain bundle to the caudal brainstem and spinal cord, distributing fibers to the nucleus of the solitary tract, the dorsal vagal nucleus and the thoracic intermediolateral cell column and Rexed's laminae I and X. Similar oxytocinergic brainstem projections were found by Hermes et al. (1988) in the garden dormouse, but they also reported oxytocinergic fibers terminating in the nuclei raphe magnus, pallidus and obscurus. Furthermore, Holstege and Van Leeuwen in the cat (unpublished observations) observed oxytocinergic and vasopressinergic fibers in the nucleus of Onuf and the sacral intermediolateral (parasympathetic) cell group. Oxytocin and vasopressin in the spinal cord are only derived from the paraventricular nucleus (Hawthorn et al. 1985), but according to Sawchenko and Swanson (1982) only 20% of the paraventricular nucleus neurons projecting to the

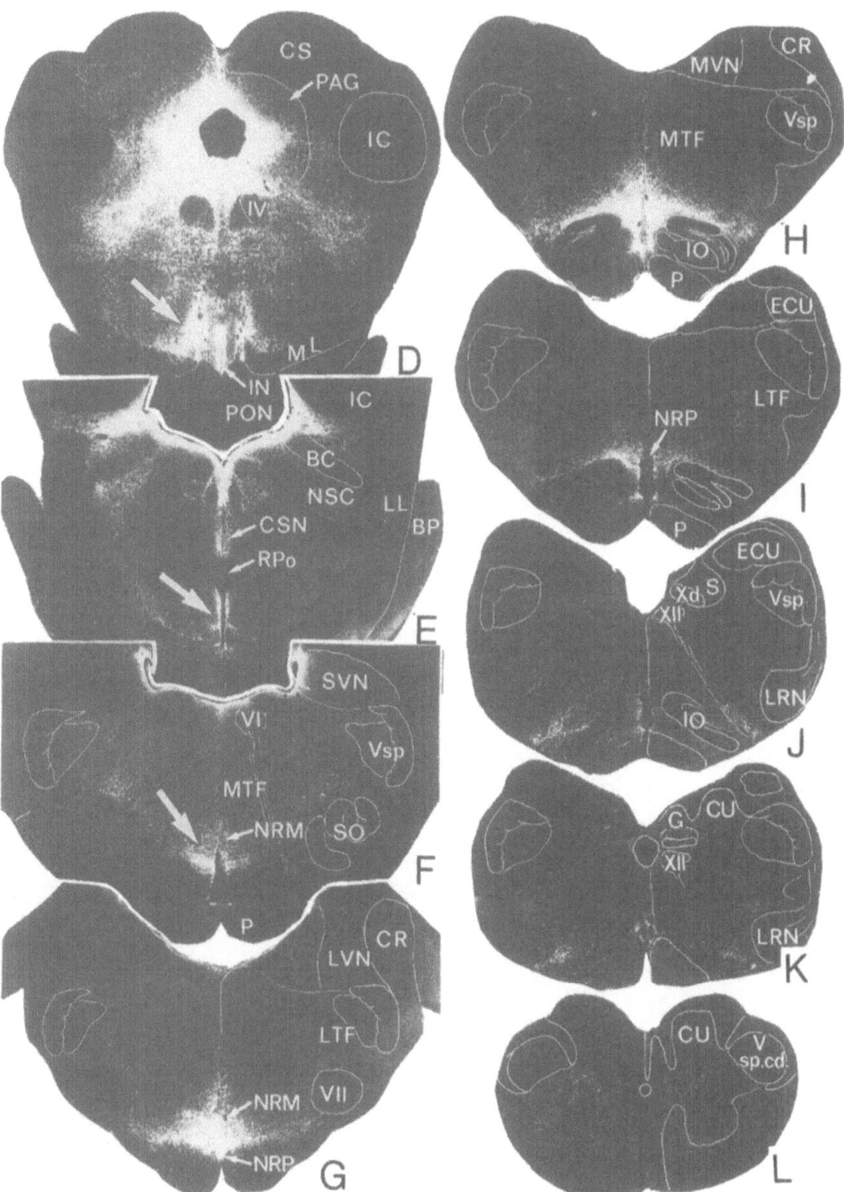

Fig. 13.29. Darkfield photomicrographs of the brainstem in a case with a ^{3}H-leucine injection in the medial part of the anterior hypothalamic area. Note the strong projections, via a medial fiberstream (see large arrows in D to F) to the medially located nucleus raphe magnus/nucleus raphe pallidus and to the ventral part of the caudal pontine and upper medullary medial tegmentum. Note also that only the most rostral part of the nucleus raphe pallidus receives labeled fibers. (From Holstege 1987a,b.)

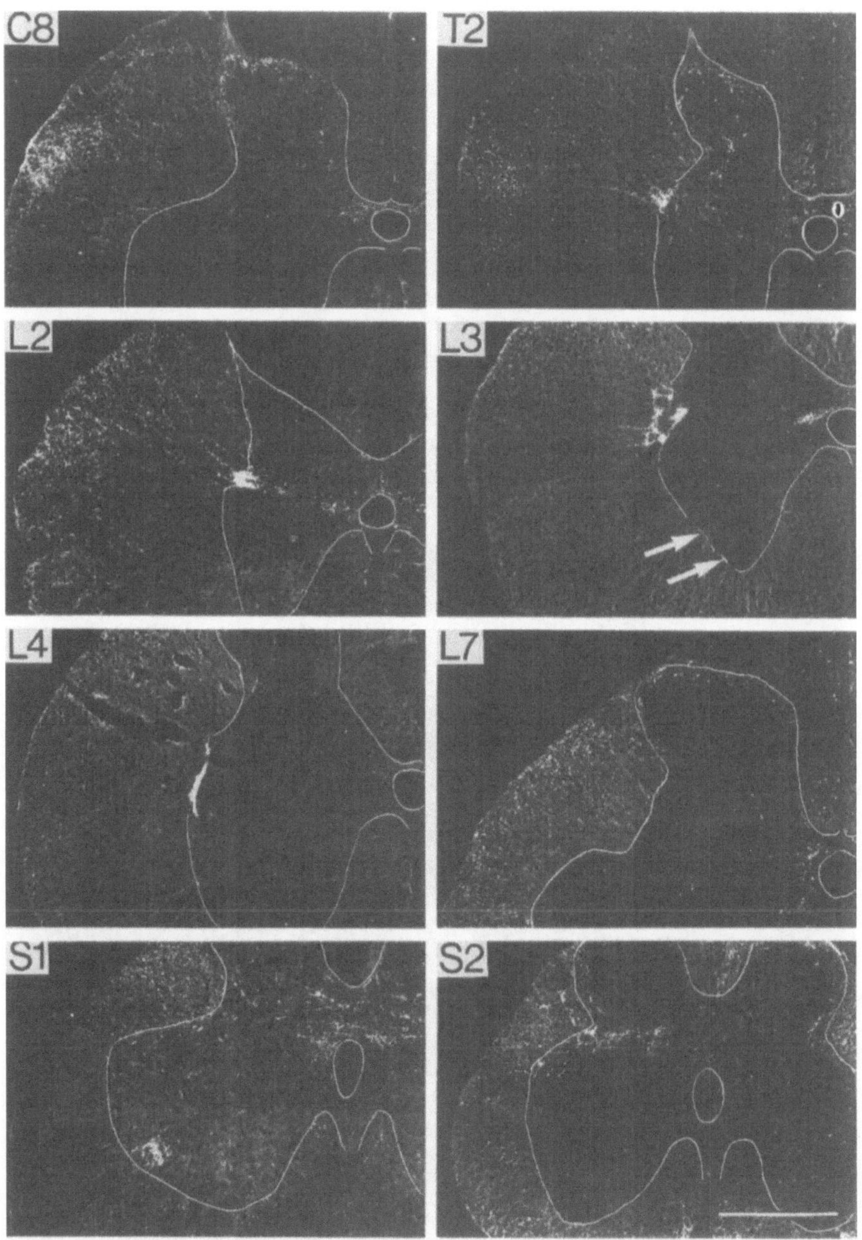

Fig. 13.30. Darkfield photomicrographs of the spinal cord of a cat with a ^{3}H-leucine injection in the area of the paraventricular nucleus of the hypothalamus. Note the projection to lamina I (C8, T2 and L7), the sympathetic intermediolateral cell group (T2, L2, L3 and L4), the nucleus of Onuf (S1) and the parasympathetic intermediomedial and intermediolateral cell group (S2). The arrows in L3 probably indicate projections to distal dendrites of the motoneurons located in the sympathetic intermediolateral cell group. (From Holstege 1987a.)

spinal cord contain oxytocin or vasopressin and another 5% contain tyrosine hydroxylase (presumably dopamine) and met-enkephalin. Therefore, other neuroactive substances must be involved in this paraventricular nucleus-caudal brainstem/spinal pathway.

Projections from the Medial Part of the Posterior Hypothalamic Area

The posterior hypothalamus projects to the spinal cord as well as to the nucleus raphe magnus and nucleus raphe pallidus (Holstege 1987b). Anterograde tracing studies have revealed that the posterior hypothalamic area sends fibers mainly to the nucleus raphe pallidus and to only a limited extent to the nucleus raphe magnus (Hosoya 1985; Holstege 1987b). The medial part of the posterior hypothalamus sends fibers into the lateral funiculus of the spinal cord, where they terminate in the upper thoracic intermediolateral cell column and in lamina X throughout the length of the spinal cord (Holstege 1987b).

There seems to exist a rostrocaudal organization in the medial hypothalamic projections to the raphe nuclei and spinal cord, in which the rostral portion of the hypothalamus projects to the rostral parts of the raphe nuclei (i.e. the nucleus raphe magnus and the rostral nucleus raphe pallidus), while the caudal hypothalamus projects to all parts of the nucleus raphe pallidus and to the spinal cord. Functionally, such differences in projections may be important, because nucleus raphe magnus and nucleus raphe pallidus project to different parts of the spinal gray.

Projections from the Lateral Hypothalamic Area

Functional and anatomical studies on the lateral hypothalamus have always been difficult, because the fibers of the medial forebrain bundle pass through it. This important fiber bundle not only contains fibers originating in the lateral hypothalamus, but also in many other areas, and stimulation or lesions in this area not only affect lateral hypothalamic neurons, but also fibers derived from many other limbic structures (see Nieuwenhuys et al. 1982 for a review). Retrograde tracing studies (Saper et al. 1976; Hosoya 1980; Holstege 1987b) reveal that many neurons in the more caudal portions of the lateral hypothalamus project to the spinal cord. Anterograde autoradiographic tracing studies (Hosoya and Matsushita 1981; Berk and Finkelstein 1982; Holstege 1987b; Berk 1987), which do not label fibers of passage (Lasek et al 1968; Cowan et al. 1972), show that the lateral hypothalamus sends fibers to the PAG, the nucleus subcoeruleus, the caudal pontine and medullary lateral tegmental field, (as defined by Holstege et al. 1977; see Fig 13.8), and to the ventral part of the caudal pontine and medullary medial tegmentum. Some fibers terminate in the periphery of the dorsal vagal nucleus and in the rostral half of the solitary nucleus. The rostral portion of the lateral hypothalamus also projects strongly to the area just ventral and medial to the mesencephalic trigeminal tract, probably representing Barrington's (1925) nucleus or the M-region of Holstege et al. (1986c). This last area is strongly involved in micturition control (p 275), and an anterior hypothalamic projection to it corresponds with the observation of Grossman and Wang (1956) that stimulation of the pre-optic area, which, according to Bleier (1961) is the same as the anterior part of the anterior hypothalamic area, produces micturition-like bladder contractions. Only the caudal portion of the lateral hypothalamus sends fibers throughout the length of the spinal cord via the lateral and dorsolateral funiculi to the intermediate zone, lamina X and the thoracolumbar sympathetic intermediolateral cell column.

The lateral hypothalamic projection to the caudal pontine and medullary lateral tegmental field and to the intermediate zone throughout the length of the spinal cord is interesting, since the caudal brainstem lateral tegmentum can be considered as the rostral continuation of the spinal intermediate zone (p 269). No direct lateral hypothalamic projections exist to the oculomotor, trochlear, trigeminal, abducens, facial and hypoglossal nerve motor nuclei, nor to the nucleus ambiguus or the interneurons in the nucleus retroambiguus. On the other hand, the many parasympathetic motoneurons located in the caudal brainstem lateral tegmentum, such as those innervating the salivatory glands (Hosoya et al. 1983), receive lateral hypothalamic afferents. In summary, the lateral hypothalamus has direct access to autonomic motoneurons in brainstem and spinal cord, and indirect access, via premotor interneurons, to the somatic motoneurons of brainstem and spinal cord.

Many of the brainstem motoneurons are involved in activities such as swallowing, chewing and licking. It is interesting that the lateral hypothalamus is involved in feeding and drinking behavior (Grossman et al. 1978) as well as in salivation (Epstein 1971). It is probably also involved in cardiovascular control and defense behavior (see next section).

Projections from Amygdala and Bed Nucleus of the Stria Terminalis to Caudal Brainstem and Spinal Cord

Retrograde tracing studies have shown that a continuum of neurons from the central nucleus of the amygdala dorsomedially along the medial border of the internal capsule into the lateral portion of the bed nucleus of the stria terminalis (BNST) project to the caudal brainstem. Such a distribution pattern is suggestive of a nucleus split into two different parts by the fibers of the internal capsule in the same way as the caudate nucleus and the putamen. As early as 1923 Johnston considered the central and medial amygdaloid nuclei and the BNST as a single anatomical entity, and many others have accepted this concept (see De Olmos et al. 1984; Holstege et al. 1985; Heimer et al. 1991). Other arguments in favor of this concept are that both areas: (1) contain neurons with the same neuropeptides, for example neurotensin, substance P, cholecystokinin, vasoactive intestinal polypeptide, enkephalin, somatostatin and dynorphin; (2) receive afferents from the same brainstem structures; and (3) have identical projections to the caudal brainstem (Hopkins and Holstege 1978; Holstege et al. 1985). Both structures send many fibers to the lateral hypothalamic area, and via the medial forebrain bundle, into the lateral part of the mesencephalon, pons, and medulla oblongata (Fig. 13.31). At mesencephalic levels fibers terminate in the PAG (except its dorsolateral part), the ventrolaterally adjoining nucleus cuneiformis and pedunculopontine nucleus, and the mesencephalic lateral tegmental field. In the pons, fibers terminate laterally in the tegmentum. At the level of the motor trigeminal nucleus some fibers branch off from the lateral descending fiber bundle, passing ventrally and medially to terminate in the ventral part of the caudal pontine and upper medullary medial tegmentum. At medullary levels many fibers terminate in the lateral tegmental field as defined by Holstege et al. (1977) (p 280) as well as in the rostral and caudal parts of the solitary nucleus and the peripheral parts of the dorsal vagal nucleus. No direct projections exist from the central nucleus

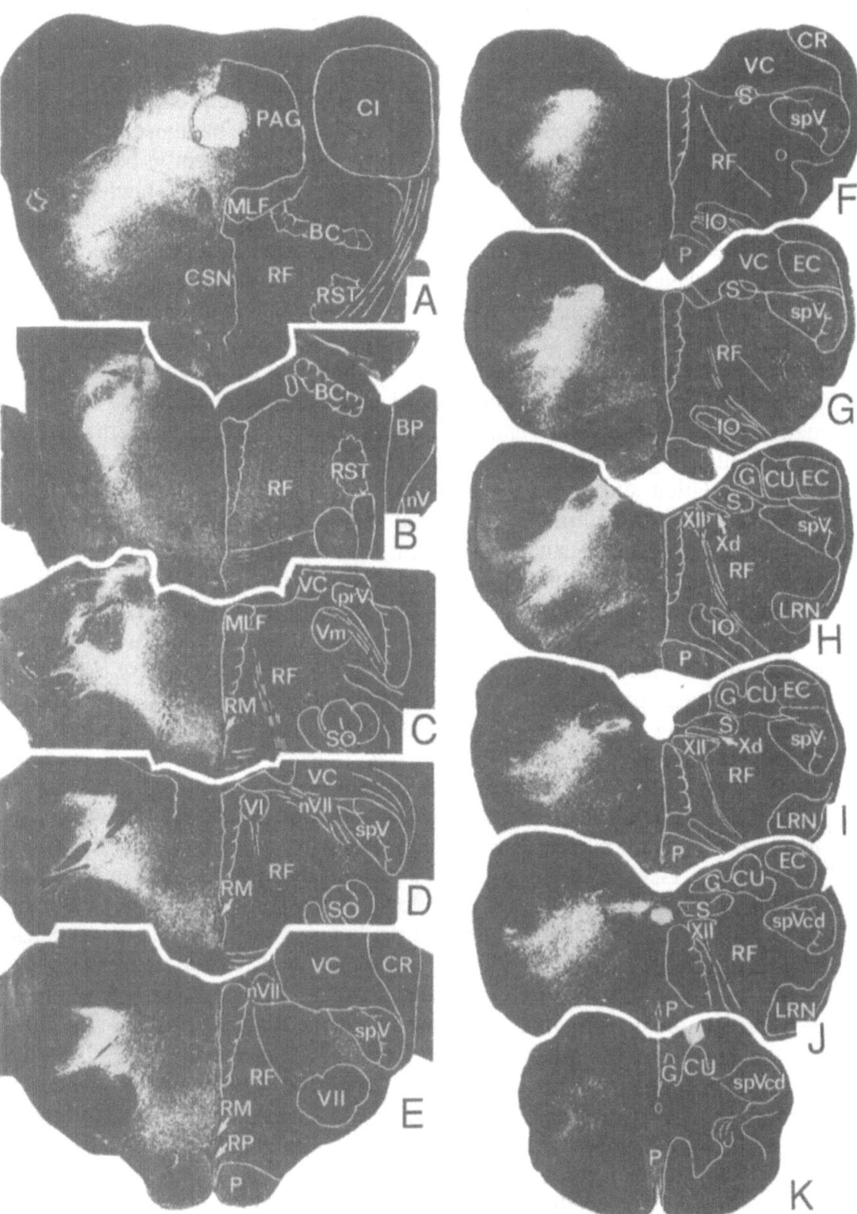

Fig 13.31. Darkfield photomicrographs of 11 brainstem sections of a cat with an injection of ^{3}H-leucine in the bed nucleus of the stria terminalis. Note the strong projection to the PAG, with the exception of its dorsolateral part. Note also the strong projection to the bulbar lateral tegmental field and the projection to the ventral part of the caudal pontine and upper medullary medial tegmentum. (From Holstege et al. 1985.)

of the amygdala and BNST to the somatic motor nuclei in the caudal brainstem. Both structures send a few fibers to the intermediate zone of the C1 spinal cord, but not beyond that level. A great similarity exists between the caudal brainstem projections originating in the central nucleus of the amygdala and BNST on the one hand and the lateral hypothalamic area on the other. All three areas have very strong mutual connections.

Neurons in CA and BNST receive many afferent fibers from other (basolateral and basomedial) amygdaloid nuclei (Krettek and Price, 1978), but these connections are not reciprocal (see also Price et al. 1987 for review). Apparently, both CA and BNST serve as "output nuclei" for other parts of the amygdala/bed nucleus of the stria terminalis complex to reach the caudal brainstem. The lateral hypothalamus also may have this function, although its afferent connections are less clearly defined, mainly because of the many fibers of passage in the medial forebrain bundle. The direct projections from CA, BNST and the lateral hypothalamus to the caudal brainstem lateral tegmental field may form the anatomical framework of the final output of the defense response of the animal. Electrical stimulation in the amygdala (especially the basal and central nuclei), bed nucleus of the stria terminalis, lateral hypothalamus, and PAG elicits defensive behavior (Fernandez de Molina and Hunsperger 1962; Bandler et al. 1991). In fact, there exists a column of electrical stimulation sites from central nucleus of the amygdala, BNST, lateral hypothalamus, and PAG through the lateral mesencephalic tegmentum into the lateral tegmentum of the caudal brainstem, which elicits defensive behavior (Abrahams et al. 1960; Coote et al. 1973). Kaada (1972) gives an excellent description of the defense response in cats. The initial phase of such a response is arrest of all spontaneous ongoing activities, and the whole attitude of the animal changes to one of attention. This arousal is followed by orienting or searching movements towards the contralateral side, frequently accompanied by sniffing, swallowing, chewing, and by twitching of the ipsilateral facial musculature. Later in the defense reaction the cat retracts its head and crouches with the ears flattened to a posterior position. The cat growls or hisses, the pupils are dilated and there is piloerection, elevation of blood pressure with bradycardia, increased rate of breathing, alteration of gastric motility and secretion. On stronger stimulation an "affective" attack may take place, in which the cat strikes with its paw with claws unsheathed, in a series of swift, accurate blows. If the stimulus continues, the cat will bite savagely. Many of the activities in the beginning of the defense response are co-ordinated in the caudal brainstem lateral tegmental field. The observation that part of this behavior appears to be ipsilateral corresponds with the predominantly ipsilateral projection of central nucleus of the amygdala, BNST and lateral hypothalamus to the caudal brainstem lateral tegmentum. Edwards and Flynn (1972) have shown that during the strike movement in the "affective" attack, a pure facilitation of the pyramidal tract neurons of the ipsilateral motor cortex takes place. In addition there are mainly facilitatory influences at the motoneuronal level in the spinal cord, which might be the result of the central nucleus of the amygdala, BNST, lateral hypothalamus, and mesencephalic projections to the ventral part of the medullary medial reticular formation, which in turn projects diffusely to all motoneuronal cell groups in the spinal cord.

Fig. 13.32 gives a schematic overview of the descending projections to the caudal brainstem from hypothalamus, amygdala and BNST. There exists a mediolateral organization within this descending system in which the medial hypothalamus forms the medial, and the lateral hypothalamus, amygdala and BNST the lateral

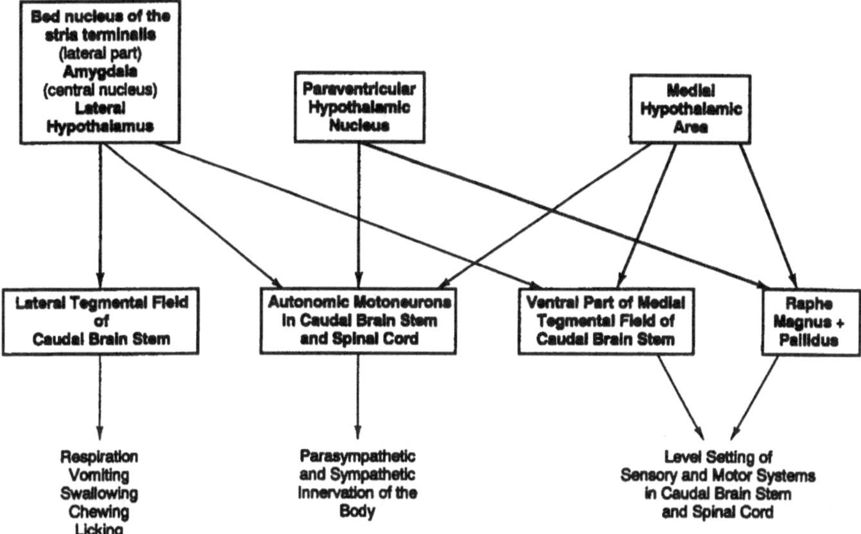

Fig. 13.32. Schematic overview of the mediolateral organization of the limbic system pathways to brainstem and spinal cord and its possible functional implications. The strongest projections are indicated by thick arrows. (From Holstege 1988a.)

component. The paraventricular nucleus, with its direct projections to all preganglionic (sympathetic and parasympathetic) motoneurons in brainstem and spinal cord, occupies a separate position within this framework.

Projections from the Prefrontal Cortex to Caudal Brainstem and Spinal Cord

In recent years it has been shown that the prefrontal cortex projects directly to the caudal brainstem. Most of these studies are done in the rat, in which the medial frontal cortex sends fibers to the solitary nuclei and the PAG (Van der Kooy et al. 1984; Neafsy et al. 1986; Terreberry and Neafsy 1987). The insular cortex projects also to the solitary nuclei and PAG (Ruggiero et al. 1987; Neafsy et al. 1986). Ruggiero et al. (1987) also found that electrical stimulation of the rat's insular cortex leads to elevation of arterial pressure and cardioacceleration. In animals other than the rat studies on the prefrontal cortical projections to the brainstem are extremely scarce. In the cat, Willett et al. (1986), found that the orbital gyrus, anterior insular cortex and infralimbic cortex project to the solitary nuclei. In the light of the findings in rat and cat, it is extremely unlikely that the frontal cortex in the monkey and human does not project to the caudal brainstem. In the section on the lateral descending system (p 280) it is pointed out that there exist major differences between cat, monkey and human in respect of the projections and the functions of the motor cortex. The motor cortex in primates has taken over many of the "motor tasks", performed by brainstem structures in rat and cat. This might also be the case for the fronto-orbital cortical projections in primates.

Conclusions

An enormous number of new studies have been published in the last 10 years on the descending motor pathways to caudal brainstem and spinal cord and about the physiological and pharmacological properties of them. Nevertheless, all the new pathways seem to belong to one of three major motor systems in the central nervous system, which determine the activity of the somatic and autonomic motoneurons. In this concept the motoneuronal cell columns themselves are not considered a central motor system, but the beginning of the peripheral motor system (motoneuronal cell body–motor nerve–muscle).

The First Motor System

The first system (Fig. 13.33) is formed by the pre-motor interneuronal projections to the motoneurons. These neurons receive direct or indirect afferent information from the periphery via peripheral afferent nerves and from the second and/or third motor system. They are of paramount importance for determining the final output of the motoneurons. It is not always true that these interneurons are located close to their target motoneurons. For example those involved in back-musculature control travel over large distances through the spinal cord.

As has been pointed out previously (p 267) the bulbar lateral tegmental field can be considered as the rostral extent of the spinal intermediate zone. Correspondingly this area contains interneurons, not only for the brainstem motoneurons of the cranial nerves V, VII, X and XII, but also for some motoneuronal cell groups in the spinal cord. Also these interneurons belong to the first system. Examples are the medullary interneurons projecting to the phrenic and other respiratory-related motoneuronal cell groups (Fig. 13.10). Since most of the afferent information from the respiratory organs does not enter the central nervous system via the spinal cord, but via the brainstem (vagal nerve), it is natural that the interneurons involved are located in the region of entrance. The author of this paper regards the micturition-related interneurons in the dorsolateral pons as also part of this system. These interneurons are of enormous importance for micturition, because via their long descending pathways they determine whether bladder and bladder-sphincter function synergistically. The question arises why these neurons are located so far from their target motoneurons? In that respect it is important to realize that micturition is strongly correlated with the emotional state of the individual. Therefore, the micturition interneurons need to receive afferent information from the limbic system, which is available in the dorsolateral pons, but not in the sacral cord.

In summary, the first motor system is formed by the interneuronal projections to the motoneurons. They are present in the caudal brainstem, the spinal cord, and between brainstem and spinal cord.

The Second Motor System

The second motor system (Fig. 13.33) is discussed under the section on descending pathways of somatic central systems. The projections of this system have been studied for some time, mainly because they consist of thick fibers, which could be

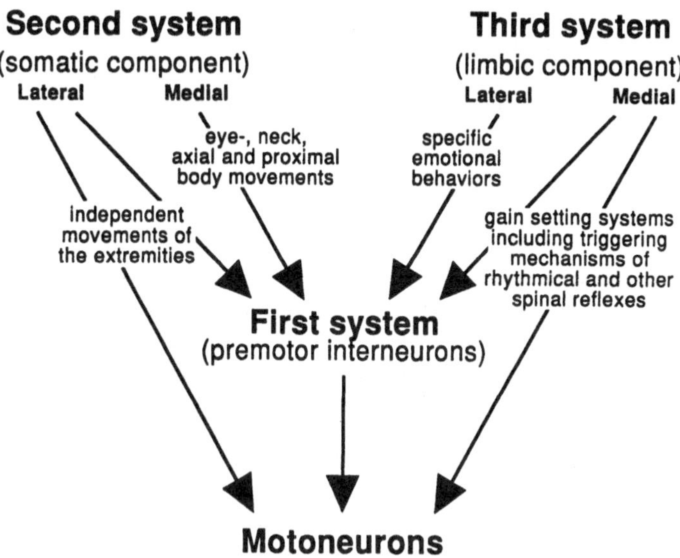

Fig. 13.33. Schematic overview of the three subdivisions of the motor system.

detected with the lesion-degeneration techniques in the 1950s and 1960s. The fibers of this system terminate to only a limited extent directly on motoneurons, but for the most part on the interneurons of the first motor system. Kuypers was the first to point out the mediolateral organization within this system. The medial component originates mainly in the brainstem (dorsal two thirds of the pontine and medullary medial tegmentum, vestibular nuclei, superior colliculus, interstitial nucleus of Cajal and caudal Field H of Forel), descends medially in the ventral funiculus of the spinal cord and terminates on inter- and to a lesser extent motoneurons of the medial motor column in the spinal cord. The medial motor column controls eye and neck movements and axial and proximal body movements. The function of the medial system is maintenance of erect posture (antigravity movements), integration of body and limbs, synergy of the whole limb and orientation of body and head (Kuypers, 1981). On the other hand, the lateral component of this second motor system is formed by laterally descending fiber systems, terminating in laterally located inter- and to a more limited extent motoneurons in caudal brainstem and spinal cord (the lateral motor column). These systems are represented by the rubrospinal tract (in humans of minor importance) and the lateral corticospinal tract (in humans extremely well developed). The lateral motor column in the spinal cord innervates the distal body musculature, i.e. those of the distal limbs. The lateral component of the voluntary motor system produces independent flexion-biased movements of the extremities, in particular of the elbow and hand (Kuypers 1981).

The Third Motor System

The third motor system (Fig. 13.33) was discovered only recently. Although there was clinical evidence that this separate motor system existed, anatomical studies did not find any evidence for such a system. In the last 15 years that position has changed drastically. It appeared that modern tracing techniques were able to demonstrate a large number of new pathways. They all consisted of thin fibers, which was the reason that the lesion-degeneration techniques were not able to demonstrate them earlier. The development of immunohistochemical techniques has revealed a large number of neurotransmitters or neuromodulators within the central nervous system. Interestingly, with the exception of acetylcholine, glutamate and aspartate, all these new monoamines and peptides were found in the third motor system.

The third motor system is strongly connected with the limbic system and systematically skips the areas belonging to the second motor system, such as red nucleus, interstitial nucleus of Cajal and other perioculomotor areas, the dorsal two thirds of the caudal brainstem medial tegmentum, vestibular nuclei or precerebellar structures as pontine nuclei, inferior olive or lateral reticular nucleus. Similarly, the second motor system does not overlap in its projections with the third motor system. An exception to this rule is the monaminergic projections originating in the raphe nuclei and locus coeruleus/subcoeruleus complex. These structures, which belong to the third system, send fibers to many structures in the central nervous system, including some belonging to the second system (e.g. the inferior olive and cerebellum).

A mediolateral organization is present within the third motor system. The medial component originates in the medial portions of hypothalamus and in the mesencephalon and terminates in the area of locus coeruleus/subcoeruleus and in the ventral part of the caudal pontine and medullary medial tegmental field. The latter structures determine the final output of this system. The lateral component originates laterally in the limbic system, i.e. in the lateral hypothalamus, central nucleus of the amygdala and bed nucleus of the stria terminalis. These structures project to the lateral tegmental field of caudal pons and medulla, but not to the somatic motoneurons in this area. How far the prefrontal cortex plays a role within these systems remains to be determined. There are some exceptions to this general subdivision into medial and lateral third motor systems: 1. Within the PAG and lateral adjacent tegmentum, some specific groups of neurons exist, projecting to areas outside the caudal brainstem ventromedial tegmental field, such as the nucleus retroambiguus, cervical spinal cord or subretrofacial nucleus. These neurons are probably related to specific functions, such as vocalization, head movements involved in emotional behavior or blood pressure control. They may serve as final common pathway, especially for the lateral component of the third motor system. 2. Some of the fibers of the lateral component of the third descending system terminate in the ventromedial tegmentum at levels around the facial nucleus. Neurons in this area, in turn, project diffusely to the dorsal horn of the spinal cord. Via these fibers the lateral component structures may have some control over nociception.

The functional implications of the third system motor pathways differ depending on whether they belong to the medial or lateral system. The medial system, via its projections to the locus coeruleus/nucleus subcoeruleus and nucleus raphe magnus and nucleus raphe pallidus/nucleus raphe obscurus and the diffuse coeruleo- and raphe-spinal pathways, has a global effect on the level of activity of the somatosensory and motoneurons in general by changing their membrane excitability. In other words, the emotional brain has a great impact on the sensory as well as on the motor system. In both systems it sets the gain or level of functioning of the neurons. The emotional state of the individual determines this level. For example it is well known that many forms of stress, such as aggression, fear and sexual arousal, induce analgesia, while at the same time the motor system is set at a "high" level and motoneurons can easily be excited by the second motor system. In this concept the brainstem structures, which project diffusely to all parts of the spinal cord, can be considered as tools for the limbic system controlling spinal cord activity. The diffuse descending system is also used to trigger rhythmical (locomotion) or other (lordosis) spinal reflexes in certain behavior. Whether functions such as locomotion and lordosis use different or the same diffuse pathways from the caudal brainstem to the spinal cord is not yet clear. If they use the same pathways, the differences lie in the output of the spinal generators for each of these functions.

The lateral component of the third motor system projects the caudal brainstem lateral tegmental field, which contains first motor system interneurons involved in specific functions such as respiration, vomiting, swallowing, chewing, and licking. These activities are displayed in the beginning of flight or defense responses and can easily be elicited by stimulation of the lateral parts of the limbic system. Therefore, it seems that the lateral component of the third motor system is involved in more specific activities, related to emotional behavior.

It is intriguing that both the medial and lateral components of the second and third motor systems are involved in similar activities. The medial components are involved in general activities such as in integration of body and limbs and orientation of body and head for the second system and level setting of neurons for the third system. On the other hand, the lateral components are involved in specific activities such as independent movements of the extremities for the second motor system and blood pressure and respiration control, vocalization, vomiting, swallowing, chewing, and licking for the third motor system.

References

Abols IA, Basbaum AI (1981) Afferent connections of the rostral medulla of the cat: a neural substrate for midbrain-medullary interactions in the modulation of pain. J Comp Neurol 201:285–297

Abrahams VC, Hilton SM, Zbrozyna A (1960) Active muscle vasodilation produced by stimulation of the brain stem: its significance in the defence reaction. J Physiol 154:491–513

Abrahams VC, Keane J (1984) Contralateral, midline and commissural motoneurons of neck muscles: a retrograde HRP study in the cat. J Comp Neurol 223:448–456

Alstermark B, Lundberg A, Norsell U, Sybirska E (1981) Integration in descending motor pathways controlling the forelimb in the cat. 9. Differential behavioural defects after spinal cord lesions interrupting defined pathways from higher centers to motoneurones. Exp Brain Res 42:299–318

Alstermark B, Kimmel H, Tantisira B (1987a) Monosynaptic raphespinal and reticulospinal projection to forelimb motoneurons in cats. Neurosci Lett 74: 286–290

Alstermark B, Kimmel H, Pinter MJ, Tantisira B (1987b) Branching and termination of C3–C4 propriospinal neurones in the cervical spinal cord of the cat. Neurosci Lett 74:291–296

Anderson ME (1971) Cerebellar and cerebral inputs to physiologically identified efferent cell groups in the red nucleus of the cat. Brain Res 30:49–66

Armand J, Kuypers, HGJM (1980) Cells of origin of crossed and uncrossed corticospinal fibers. A quantitative horseradish peroxidase study. Exp Brain Res 42:299–318

Armand J, Holstege G, Kuypers HGJM (1985) Differential corticospinal projections in the cat. An autoradiographical tracing study. Brain Res. 343:351–355

Baldissera F, Hultborn H, Illert M (1981) Integration in spinal neuronal systems. In: Burke RE (ed)

Handbook of physiology, section I, the nervous system, vol. II, motor systems. American Physiological Society, Washington DC, pp 509–595
Bandler R, Carrive P (1988) Integrated defence reaction elicited by excitatory amino acid microinjection in the midbrain periaqueductal grey region of the unrestrained cat. Brain Res 439:95–106
Bandler R, Carrive P, Zhang SP (1991) Integration of somatic and autonomic reactions within the midbrain periaqueductal grey: Viscerotopic, somatotopic and functional organization. In: Holstege G (ed) Role of the forebrain in sensation and behavior. Elsevier, Amsterdam, pp 269–305 (Progress in brain research 87)
Barrington FJF (1925) The effect of lesions of the hind- and mid-brain on micturition in the cat. J Exp Physiol Cogn Med 15:81–102
Basbaum AI, Clanton CH, Fields HL (1978) Three bulbospinal pathways from the rostral medulla of the cat: an autoradiographic study of pain modulating systems. J Comp Neurol 178:209–224
Batini C, Buisseret-Delmas C, Corvisier J (1976) Horseradish peroxidase localization of masticatory muscle motoneurons in cat. J Physiol 72:301–309
Berger AJ (1980) The distribution of the cat's carotid sinus nerve afferent and efferent cell bodies using the horseradish peroxidase technique. Brain Res 190:309–320
Berk ML (1987) Projections of the lateral hypothalamus and bed nucleus of the stria terminalis to the dorsal vagal complex in the pigeon. J Comp Neurol 260:140–156
Berk ML, Finkelstein JA (1982) Efferent connections of the lateral hypothalamic area of the rat: an autoradiographic investigation. Brain Res Bull 8:511–526
Berthier NE, Moore JW (1983) The nictitating membrane response: an electrophysiological study of the abducens nerve and nucleus and the accessory abducens in rabbit. Brain Res 258:201–211
Besson J-M, Chaouch A (1987) Peripheral and spinal mechanisms of nociception. Physiol Rev 67:67–186
Blaivas JG (1982) The neurophysiology of micturition: a clinical study of 550 patients. J Urol 127:958–963
Bleier R (1961) The hypothalamus of the cat: a cytoarchitectonic atlas in the Horsley-Clarke co-ordinate system. Johns Hopkins University Press, Baltimore
Blessing WW, Chalmers JP (1979) Direct projection of catecholamine (presumably dopamine) containing neurons from hypothalamus to spinal cord. Neurosci Lett 11:35–41
Bolk L, Groppert E, Kallius EE, Lubosch W (1938) Handbuch der vergleichende Anatomie den Wirbeltiere. Urban & Schwarzenberg, Berlin
Bowker RM (1986) Serotonergic and peptidergic inputs to the primate ventral spinal cord as visualized with multiple chromagens on the same tissue section. Brain Res 375:345–350
Bowker RM, Westlund KN, Sullivan MC, Coulter JD (1982) Organization of serotonergic projections to the spinal cord. Progr Brain Res 57:239–265
Bowker RM, Westlund KN, Sullivan MC, Wilber JF, Coulter JD (1983) Descending serotonergic, peptidergic and cholinergic pathways from the raphe nuclei: a multiple transmitter complex. Brain Res 288:33–48
Bowker RM, Abbott LC, Dilts RP (1988) Peptidergic neurons in the nucleus raphe magnus and the nucleus gigantocellularis: their distributions, interrelationships, and projections to the spinal cord. In: Besson JM, Fields HL (eds) Pain modulation. Elsevier, Amsterdam, pp 95–127 (Progress in brain research 77)
Brink EE, Morrell JI, Pfaff, DW (1979) Localization of lumbar epaxial motoneurons in the rat. Brain Res 170:23–43
Brodal A (1981) Neurological anatomy in relation to clinical medicine, 3rd edn. Oxford University Press, Oxford
Brudzynski SM, Eckersdorf B (1984) Inhibition of locomotor activity during cholinergically induced emotional-aversive response in the cat. Behav Brain Res 14:247–253
Burke RE, Strick PL, Kanda K, Kim CC, Walmsley B (1977) Anatomy of medial gastrocnemius and soleus motor nuclei in cat spinal cord. J Neurophysiol 40:667–680
Büttner-Ennever J, Holstege G (1986) Anatomy of premotor centers in the reticular formation controlling oculomotor, skeletomotor and autonomic motor systems. Progr Brain Res 64:89–98
Büttner-Ennever JA, Büttner U, Cohen B, Baumgarter G (1982) Vertical gaze paralysis and the rostral interstitial nucleus of the medial longitudinal fasciculus. Brain 105:125–149
Büttner-Ennever JA, Büttner U (1988) The reticular formation. In: Büttner-Ennever JA (ed) Neuroanatomy of the oculomotor system. Elsevier, Amsterdam pp 119–176
Callister RJ, Brichta AM, Peterson EH (1987) Quantitative analysis of cervical musculature in rats: histochemical composition and motor pool organization. II. Deep dorsal muscles. J Comp Neurol 255:369–385

Cardona A, Rudomin P (1983) Activation of brainstem serotoninergic pathways decreases homosynaptic depression of monosynaptic responses of frog spinal motoneurons. Brain Res 280:373–378

Carrive P, Bandler R, Dampney RAL (1989) Viscerotopic control of regional vascular beds by discrete groups of neurons within the midbrain periaqueductal gray. Brain Res 493:385–390

Chan JYH, Fung SJ, Chan SHH, Barnes CD (1986) Facilitation of lumbar monosynaptic reflexes by locus coeruleus in the rat. Brain Res 369:103–109

Chan Palay V, Jonsson G, Palay SL (1978) Serotonin and substance P coexist in neurons of the rat's central nervous system. Proc Natl Acad Sci USA 75:1582–1586

Cheney PD (1980) Response of rubromotoneuronal cells identified by spike triggered averaging of EMG activity in awake monkeys. Neurosci Lett 17:137–143

Cheney PD, Mewes K and Fetz EE (1988) Encoding of motor parameters by corticomotoneuronal (CM) and rubromotoneuronal (RM) cells producing postspike facilitation of forelimb muscles in the behaving monkey. Behav Brain Res 28:181–191

Cheney PD, Fetz EE, Mewes K (1991) Neural mechanisms underlying corticospinal and rubrospinal control of limb movements. In: Holstege G (ed) Role of the forebrain in sensation and behavior. Elsevier, Amsterdam, pp 213–252 (Progress in brain research 87)

Coote JH, Hilton SM, Zbrozyna AW (1973) The ponto-medullary area integrating the defence reaction in the cat and its influence on muscle blood flow. J Physiol 229:257–274

Courville J (1966a) The nucleus of the facial nerve; the relation between cellular groups and peripheral branches of the nerve. Brain Res 1:338–354

Courville J (1966b) Rubrobulbar fibres to the facial nucleus and the lateral reticular nucleus (nucleus of the lateral funiculus). An experimental study in the cat with silver impregnation methods. Brain Res 1:317–337

Cowan WM, Gottlieb DI, Hendrickson AE, Price JL, Woolsey TA (1972) The autoradiographic demonstration of axonal connections in the central nervous system. Brain Res 37:21–51

Crone C, Hultborn H, Kiehn O, Mazieres L, Wigstrom H (1988) Maintained changes in motoneuronal excitability by short-lasting synaptic inputs in the decerebrate cat. J Physiol 405:321–343

Crouch JE (1969) Text-atlas of cat anatomy. Lea and Febiger, Philadelphia

Crutcher KA, Humbertson AO jr, Martin GF (1978) The origin of brainstem spinal pathways in the North American Opossum (*Didelphis virginiana*). Studies using the horseradish peroxidase. J Comp Neurol 179:169–194

Cullheim S, Kellerth J-O (1978) A morphological study of the axons and recurrent axon collaterals of cat α-motoneurons supplying different functional types of muscle unit. J Physiol 281:301–314

Czarkowska J, Jankowsk E, Sybirska E (1976) Axonal projections of spinal interneurones excited by group I afferents in the cat, revealed by intracellular staining with horseradish peroxidase. Brain Res 118:115–118

Dampney RAL, McAllen RM (1988) Differential control of sympathetic fibres supplying hindlimb skin and muscle by retrofacial neurones in the cat. J Physiol 395:41–56

Davis PJ, Nail BS (1984) On the location and size of laryngeal motoneurons in the cat and rabbit. J Comp Neurol 230:13–32

Dekker JJ, Lawrence DG, Kuypers HGJM (1973) The location of longitudinally running dendrites in the ventral horn of the cat spinal cord. Brain Res 51:319–325

De Olmos JS, Alheid GF, Beltramino CS (1984) Amygdala. In Paxinos G (ed) The rat nervous system. Academic Press, Sydney, pp 223–334

DeRosier EA, West RA, Larson CR (1988) Comparison of single unit discharge properties in the periaqueductal gray and nucleus retroambiguus during vocalization in monkeys. Soc Neurosci Abstr 14:1237

ten Donkelaar HJ (1988) Evolution of the red nucleus and rubrospinal tract. Behav Brain Res 28:9–20

Donoghue S, Garcia M, Jordan D, Spyer KM (1982) The brain-stem projections of pulmonary stretch afferent neurons in cats and rabbits. J Physiol 322:352–364

Duron B, Marlot D, Larnicol N, Jung-Caillol MC, Macron JM (1979) Somatotopy in the phrenic motor nucleus of the cat as revealed by retrograde transport of horseradish peroxidase (HPP). Neurosci Lett 14:159–163

Edwards SB (1972) Descending projections of the midbrain reticular formation of the cat: an experimental study using a "protein transport", tracing method. Anat Rec 172:305

Edwards SB, Flynn JP (1972) Corticospinal control of striking in centrally elicited attack behavior. Brain Res 41:51–65

Epstein AN (1971) The lateral hypothalamic syndrome: its implications for the physiological psychology of hunger and thirst. In: Stellar E, Sprague JM (eds) Progress in physiological psychology, vol. 4 Academic Press, New York pp 263—317

Evinger C (1988) Extraocular motor nuclei: location, morphology and afferents. In: Büttner-Ennever JA (ed) Neuroanatomy of the the oculomotor system. Elsevier, Amsterdam, pp 81–117
Feldman JL (1986) Neurophysiology of breathing in mammals. In: Bloom FE (ed) Handbook of physiology, section 1: the nervous system, vol. IV. Intrinsic regulatory systems of the brain. American Physiological Society, Bethesda, pp 463–524
Fernandez de Molina A, Hunsperger RW (1962) Organization of the subcortical system governing defense and flight reactions in the cat. J Physiol 160:200–213
Fetcho JR (1987) A review of the organization and evolution of motoneurons innervating the axial musculature of vertebrates. Brain Res Dev 12:243–280
Friauf E, Baker R (1985) An intracellular HRP-study of cat tensor tympani motoneurons. Exp Brain Res 57:499–511
Fritz N, Illert M, Saggau P (1986a) Location of motoneurones projecting to the cat distal forelimb. I. Deep radial motornuclei. J Comp Neurol 244:286–301
Fritz N, Illert M, Reeh P (1986b) Location of motoneurones projecting to the cat distal forelimb. II. Median and ulnar motornuclei. J Comp Neurol 244:302–312
Fritz N, Illert M, Saggau P (1978) Monosynaptic convergence of group I muscle afferents from the forelimb onto interosseus motoneurones. Neurosci Lett (Suppl)I:S95
Fritz N, Illert M, de la Motte S, Reeh P (1984) Pattern of monosynaptic Ia connections from forelimb nerves onto median and ulnar motoneurones. Neurosci Lett 18:S264
Fung SJ, Barnes CD (1987) Membrane excitability changes in hindlimb motoneurons induced by stimulation of the locus coeruleus in cats. Brain Res 402:230–242
Garcia-Rill E, Skinner RD (1987a) The mesencephalic locomotor region. I. Activation of a medullary projection site. Brain Res 411:1–12
Garcia-Rill E, Skinner RD (1987b) The mesencephalic locomotor region. II. Projections to reticulospinal neurons. Brain Res 411–13–20
Garcia-Rill E, Skinner RD (1988) Modulation of rhythmic function in the posterior midbrain. Neurosci 27:639–654
Gebhart GF, Sandkuhler J, Thalhammer JG, Zimmermann M (1983) Quantitative comparison of inhibition in spinal cord of nociceptive information by stimulation in periaqueductal gray or nucleus raphe magnus of the cat. J Neurophysiol 50:1433–1445
Gelfand IM, Orlovsky GN, Shik ML (1988) Locomotion and scratching in tetrapods. In: Cohen AH et al. (eds) Neural control of rhythmic movements in vertebrates. Wiley, New York, pp 167–199
Gibson AR, Houk, JC, Kohlerman NJ (1985) Magnocellular red nucleus activity during different types of limb movement in the macaque monkey. J Physiol 358:527–549
Giovanelli Barilari MS Kuypers HGJM (1969) Propriospinal fibers interconnecting the spinal enlargements in the cat. Brain Res 14:321–330
Griffiths D, Holstege G, Dalm E, de Wall H (1990) Control and coordination of bladder and urethral function in the brain stem of the cat. Neurourol Urodynam 9:63–82
Grillner S (1981) Control of locomotion in bipeds, tetrapods, and fish. In: Burke RE (ed) Handbook of physiology, section I, the nervous system, vol II, motor systems. Washington, American Physiological Society, Washington DC pp 1179–1236
Grossman RG, Wang SC (1956) Diencephalic mechanism of control of the urinary bladder of the cat. Yale J Biol Med 28:285–297
Grossman SP, Dacey D, Halaris AE, Collier T, Routtenberg A (1978) Aphagia and adipsia after preferential destruction of nerve cell bodies in hypothalamus. Science 202:537–539
Hahne M, Illert M, Wietelmann D (1988) Recurrent inhibition in the cat distal forelimb. Brain Res 456:188–192
Hancock MB (1976) Cells of origin of hypothalamo-spinal projections in the rat. Neurosci Lett 3:179–184
Hawthorn J, Ang, VTY, Jenkins JS (1985) Effects of lesions in the hypothalamic paraventricular, supraoptic and suprachiasmatic nuclei on vasopressin and oxytocin in the rat brain and spinal cord. Brain Res 346:51–57
Hayes NL Rustioni A (1981) Descending projections from brainstem and sensorimotor cortex to spinal enlargements in the cat. Exp Brain Res 41:89–107
Heimer L, de Olmos J, Alheid GF, Zaborsky L (1991) "Perestroika" in the basal forebrain: Opening the border between neurology and psychiatry. In: Holstege G (ed) Role of the forebrain in sensation and behavior. Elsevier, Amsterdam, pp 109–165 (Progress in brain research 87)
Helke CJ, Sayson SC, Keeler JR, Charlton CG (1986) Thyrotropin-releasing hormone-immunoreactive neurons project from the ventral medulla to the intermediolateral cell column: partial coexistence with serotonin. Brain Res 381:1–7
Henneman E, Mendell LM (1981) Functional organization of motoneuron pool and its inputs. In: Burke

RE (ed) Handbook of physiology, section I, the nervous system, vol II, motor systems. American Physiology Society, Washington DC pp 423–507

Hermes ML, Buijs RM; Masson-Pevet M, Pevet P (1988) Oxytocinergic innervation of the brain of the garden dormouse (*Eliomys quercinus* L.). J Comp Neurol 273:252–262

Hiraoka M, Shimamura M (1977) Neural mechanisms of the corneal blinking reflex in cats. Brain Res 125:265–275

Hökfelt T, Ljungdahl A, Steinbusch H et al. (1978) Immunohistochemical evidence of substance P-like immunoreactivity in some 5-hydroxytryptamine-containing neurons in the rat central nervous system. Neuroscience 3:517–538

Hökfelt T, Terenius T, Kuypers HGJM, Dann O (1979) Evidence for enkephalin immunoreactivity neurons in the medulla oblongata projecting to the spinal cord. Neurosci Lett 14:55–61

Holstege G (1987a) Anatomical evidence for an ipsilateral rubrospinal pathway and for direct rubrospinal projections to motoneurons in the cat. Neurosci Lett 74:269–274

Holstege G (1987b) Some anatomical observations on the projections from the hypothalamus to brainstem and spinal cord: an HRP and autoradiographic tracing study in the cat. J Comp Neurol 260:98–126

Holstege G (1988a) Direct and indirect pathways to lamina I in the medulla oblongata and spinal cord of the cat. In: Fields HL, Besson JM (eds) Descending brainstem controls of nociceptive transmission. Elsevier, Amsterdam, pp 47–94 (Progress in brain research 77)

Holstege G (1988b) Brainstem-spinal cord projections in the cat, related to control of head and axial movements. In: Büttner-Ennever JA (ed) Neuroanatomy of the oculomotor system. Elsevier, Amsterdam, pp 429–468

Holstege G (1989) An anatomical study on the final common pathway for vocalization in the cat. J Comp Neurol 284:242–252

Holstege G (1990) Subcortical limbic system projections to caudal brainstem and spinal cord. In: Paxinos G (ed) The human nervous system. Academic Press, Sydney

Holstege G (1991) Descending motor pathways and the spinal motor system. Limbic and non-limbic components. In Holstege G (ed) Role of the forebrain in sensation and behavior. Elsevier, Amsterdam, pp 307–421 (Progress in brain research)

Holstege G, Cowie RJ (1989) Projections from the rostral mesencephalic reticular formation to the spinal cord. Exp Brain Res 75:265–279

Holstege G, Kuypers HGJM (1977) Propriobulbar fibre connections to the trigeminal, facial and hypoglossal motor nuclei I. An anterograde degeneration study in the cat. Brain 100:239–264

Holstege G, Kuypers HGJM (1982) The anatomy of brain stem pathways to the spinal cord in the cat. A labeled amino acid tracing study. Progr Brain Res 57:145–175

Holstege G, Tan J (1987) Supraspinal control of motoneurons innervating the striated muscles of the pelvic floor including urethral and anal sphincters in the cat. Brain 110:1323–1344

Holstege G, Tan J (1988) Projections from the red nucleus and surrounding areas to the brainstem and spinal cord in the cat. An HRP and autoradiographical tracing study. Behav Brain Res 28:33–57

Holstege G, Kuypers HGJM, Dekker JJ (1977) The organization of the bulbar fibre connections to the trigeminal, facial and hypoglossal motor nuclei. II. An autoradiographic tracing study in cat. Brain 100:265–286

Holstege G, Kuypers HGJM, Boer RC (1979) Anatomical evidence for direct brain stem projections to the somatic motoneuronal cell groups and autonomic preganglionic cell groups in cat spinal cord. Brain Res 171:329–333

Holstege G, Graveland G, Bijker-Biemond C, Schuddeboom I (1983) Location of motoneurons innervating soft palate, pharynx and upper esophagus. Anatomical evidence for a possible swallowing center in the pontine reticular formation. An HRP and autoradiographical tracing study. Brain Behav Evol 23:47–62

Holstege G, Tan J, van Ham J, Bos A (1984) Mesencephalic projection to the facial nucleus in the cat. An autoradiographic tracing study. Brain Res 311:7–22

Holstege G, Meiners L, Tan K (1985) Projections of the bed nucleus of the stria terminalis to the mesencephalon, pons, and medulla oblongata in the cat. Exp Brain Res 58:379–391

Holstege G, van Ham JJ, Tan J (1986a) Afferent projections to the orbicularis oculi motoneuronal cell group. An autoradiographical tracing study in the cat. Brain Res 374:306–320

Holstege G, Tan J, van Ham JJ, Graveland GA (1986b) Anatomical observations on the afferent projections to the retractor bulbi motoneuronal cell group and other pathways possibly related to the blink reflex in the cat. Brain Res 374:321–334

Holstege G, Griffiths D, De Wall H, Dalm E (1986c) Anatomical and physiological observations on supraspinal control of bladder and urethral sphincter muscles in the cat. J Comp Neurol 250:449–461

Holstege G, Van Neerven J, Evertse F (1987) Spinal cord location of the motoneurons innervating the abdominal cutaneous maximus, latissimus dorsi and longissimus dorsi muscles in the cat. Exp Brain Res 67:179–194

Holstege G, Blok BF, Ralston DD (1988) Anatomical evidence for red nucleus projections to motoneuronal cell groups in the spinal cord of the monkey. Neurosci Lett 95:97–101

Holstege JC, Kuypers HGJM (1982) Brain stem projections to spinal motoneuronal cell groups in rat studied by means of electron microscopy autoradiography. Progr Brain Res 57:177–183

Holstege JC, Kuypers HGJM (1987) Brainstem projections to lumbar motoneurons in rat. I. An ultrastructural study using autoradiography and the combination of autoradiography and horseradish peroxidase histochemistry. Neuroscience 21:345–367

Hongo T, Jankowska E, Lundberg A (1969) The rubrospinal tract. I. Effects on alpha-motoneurons innervating hindlimb muscles in cats. Exp Brain Res 7:344–364

Hopkins DA Holstege G (1978) Amygdaloid projections to the mesencephalon, pons, and medulla oblongata in the cat. Exp Brain Res 32:529–547

Hosobuchi Y (1988) Current issues regarding subcortical electrical stimulation for pain control in humans. In: Fields HL, Besson JM (eds) Nociception control. Elsevier, Amsterdam, pp 189–192 (Progress in brain research 77)

Hosoya Y (1980) The distribution of spinal projection neurons in the hypothalamus of the rat, studied with the HRP method. Exp Brain Res 40:79–87

Hosoya Y (1985) Hypothalamic projections to the ventral medulla oblongata in the rat, with special reference to the nucleus raphe pallidus: a study using autoradiographic and HRP techniques. Brain Res 344:338–350

Hosoya Y, Matsushita M (1979) Identification and distribution of the spinal and hypophyseal projection neurones in the paraventricular nucleus of the rat. A light and electron microscopic study with the HRP method. Exp Brain Res 35:315–331

Hosoya Y, Matsushita, M (1981) Brainstem projections from the lateral hypothalamic area in the rat, as studied with autoradiography. Neurosci Lett 24:111–116

Hosoya Y, Matsushita M, Sugiura Y (1983) A direct hypothalamic projection to the superior salivatory nucleus neurons in the rat. A study using anterograde autoradiographic and retrograde HRP methods. Brain Res 266:329–334

Hounsgaard J, Hultborn H, Jespersen B, Kiehn O (1984) Intrinsic membrane properties causing a bistable behavior of α-motoneurons. Exp Brain Res 55:391–394

Hounsgaard J, Hultborn H, Kiehn O (1986) Transmitter-controlled properties of α-motoneurones causing long-lasting motor discharge to brief excitatory inputs. Progr Brain Res 64:39–49

Hounsgaard J, Hulborn H, Jespersen B, Kiehn O (1988) Bistability of alpha-motoneurons in the decerebrate cat and in the acute spinal cat after intravenous 5-hydroxytryptophan. J Physiol 405:345–367

Hughes JT (1982) Pathology of amyotrophic lateral sclerosis. In: Rowland LP (ed) Human motor neuron diseases. Raven Press, New York, pp 61–74

Huisman AM, Kuypers HGJM, Verburgh CA (1982) Differences in collateralization of the descending spinal pathways from red nucleus and other brain stem cell groups in cat and monkey. Progr Brain Res 57:185–217

Hunt SP, Lovick TA (1982) The distribution of serotonin, met-enkephalin and b-lipotropin-like immunoreactivity in neuronal perikarya of the cat brainstem. Neurosci Lett 30:139–145

Illert M, Lundberg A, Padel Y, Tanaka R (1978) Integration in descending motor pathways controlling the forelimb in the cat. 5 Properties of and monosynaptic excitatory convergence on C3–C4 propriospinal neurones. Exp Brain Res 33:101–130

Ishizuka N, Mannen H, Hongo T, Sasaki S (1979) Trajectory of group 1a afferent fibers stained with horseradish peroxidase in the lumbosacral spinal cord of the cat: three dimensional reconstructions from serial sections. J Comp Neurol 186:189–213

Jankowska E (1988) Target cells of rubrospinal tract fibres within the lumbar spinal cord. Behav Brain Res 28:91–96

Jankowska E, Lindström S (1971) Morphological identification of Renshaw cells. Acta Physiol Scand 81:428–430

Jankowska E, Lindström S (1972) Morphology of interneurones mediating Ia reciprocal inhibition of motoneurones in the spinal cord of the cat. J Physiol 226:805–823

Jankowska E, Roberts WJ (1972) An electrophysiological demonstration of the axonal projections of single spinal interneurones in the cat. J Physiol 222:597–622

Jankowska E, Padel Y, Tanaka R (1975) Projections of pyramidal tract cells to alpha-motoneurons innervating hindlimb muscles in the monkey. J Physiol 249:637–667

Jankowska E, McCrea D, Mackel R (1981) Oligosynaptic excitation of motoneurones by impulses in group Ia muscle spindle afferents in the cat. J Physiol 316:411–425
Johannessen JN, Watkins LR, Mayer DJ (1984) Non-serotonergic origins of the dorsolateral funiculus in the rat ventral medulla. J Neuroscience 4:757–766
Johansson, O, Hökfelt T, Pernow B et al. (1981) Immunohistochemical support for three putative transmitters in one neuron: coexistence of 5-hydroxytryptamine, substance P- and thyrotropin-releasing hormone-like immunoreactivity in medullary neurons projecting to the spinal cord. Neuroscience 6:1857–1881
Johnston JB (1923) Further contributions to the study of the evolution of the forebrain. J Comp Neurol 35:337–481
Jones BE, Beaudet A (1987) Distribution of acetylcholine and catecholamine neurons in the cat brainstem: A choline acetyltransferase and tyrosine hydroxylase immunohistochemical study. J Comp Neurol 261:15–32
Jones BE, Friedman L (1983) Atlas of catecholamine perikarya, varicosities and pathways in the brainstem of the cat. J Comp Neurol 215:382–396
Jones BE, Yang T-Z (1985) The efferent projections from the reticular formation and the locus coeruleus studied by anterograde and retrograde axonal transport in the rat. J Comp Neurol 242:56–92
Joseph MP, Guinan JJ Jr, Fullerton BC, Norris BE, Kiang NYS (1985) Number and distribution of stapedius motoneurons in cats. J Comp Neurol 232:43–54
Jürgens U, Pratt R (1979) The cingular vocalization pathway in the squirrel monkey. Exp Brain Res 34:499–510
Kaada B (1972) Stimulation and regional ablation of the amygdaloid complex with reference to functional representation. In: Eleftheriou BE (ed): The neurobiology of the amygdala. Plenum Press, New York, pp 145–204
Keller JT, Saunders MC, Ongkiko CM et al. (1983) Identification of motoneurons innervating the tensor tympani and tensor veli palatini muscles in the cat. Brain Res 270:209–215
van Keulen LCM (1979) Axon trajectories of Renshaw cells in the lumbar spinal cord of the cat as reconstructed after intracellular staining with horseradish peroxidase. Brain Res 167:157–163
Kimura H, McGeer PL, Peng JH, McGeer EG (1981) The central cholinergic system studied by choline acetyltransferase immunohistochemistry in the cat. J Comp Neurol 200:151–201
Kneisley LW, Biber MP, LaVail JH (1978) A study of the origin of brain stem projections to monkey spinal cord using retrograde transport method. Exp Neurol 60:116–139
Kojima M, Takeuchi Y, Goto M, Sano Y (1982) Immunohistochemical study on the distribution of serotonin fibers in the spinal cord of the dog. Cell Tissue Res 226:477–491
Kojima M, Takeuchi Y, Goto M, Sano Y (1983a) Immunohistochemical study on the localization of serotonin fibers and terminals in the spinal cord of the monkey (*Macaca fuscata*). Cell Tissue Res 229:23–36
Kojima M, Takeuchi Y, Kawata M, Sano Y (1983b) Motoneurons innervating the cremaster muscle of the rat are characteristically densely innervated by serotonergic fibers as revealed by combined immunohistochemistry and retrograde fluorescence DAPI-labelling. Anat Embryol 168:41–49
Krettek JE, Price JL (1978) A description of the amygdaloid complex in the rat and cat with observations on intra-amygdaloid axonal connections. J Comp Neurol 178:225–280
Kugelberg E (1952) Facial reflexes. Brain 75:385–396
Kume M, Uemura M, Matsuda K, Matsushima K, Mizuno N (1978) Topographical representation of peripheral branches of the facial nerve within the facial nucleus: An HRP study in the cat. Neurosci Lett 8:5–8
Kuypers HGJM (1958a) Some projections from the peri-central cortex to the pons and lower brain stem in monkey and chimpanzee. J Comp Neurol 110:221–255
Kuypers HGJM (1958b) Corticobulbar connections to the pons and lower brain stem in man. An anatomical study. Brain 81:364–388
Kuypers HGJM (1958c) An anatomical analysis of cortico-bulbar connections to the pons and lower brain stem in the cat. J Anat 92:198–218
Kuypers HGJM (1964) The descending pathways to the spinal cord, their anatomy and function. In: Eccles JC, Schadé JP (eds) Organization of the spinal cord, pp 178–200 (Progress in brain research 11)
Kuypers HGJM (1973) The anatomical organization of the descending pathways and their contributions to motor control especially in primates. New Dev EMG Clin Neurophysiol 3:38–68
Kuypers HGJM (1981) Anatomy of the descending pathways. In: Burke RE (ed) Handbook of physiology, section I, the nervous system, vol II. Motor systems. American Physiological Society, Washington DC, pp 597–666

Kuypers HGJM, Brinkman J (1970) Precentral projections to different parts of the spinal intermediate zone in the rhesus monkey. Brain Res 24:29–48

Kuypers HGJM, Maisky VA (1975) Retrograde axonal transport of horseradish peroxidase from spinal cord to brain stem cell groups in the cat. Neurosci Lett 1:9–14

Kuypers HGJM, Fleming WR, Farinholt JM (1962) Subcorticospinal projections in the rhesus monkey. J Comp Neurol 118:107–137

Kuzuhara S, Kanazawa I, Nakanishi T (1980) Topographical localization of the Onuf's nuclear neurons innervating the rectal and vesical striated sphincter muscles: a retrograde fluorescent double labeling in cat and dog. Neurosci Lett 16:125–130

Landgren S, Phillips CG, Porter R (1962) Minimal synaptic actions of pyramidal impulses on some alpha-motoneurons of the baboon's hand and forearm. J Physiol 161:9–111

Lasek R, Joseph BS, Whitlock DG (1968) Evaluation of a radioautographic neuroanatomical tracing method. Brain Res 8:319–336

Lawn AM (1966) The nucleus ambiguus of the rabbit. J Comp Neurol 127:307–320

Lawrence DG, Kuypers HGJM (1968a) The functional organization of the motor system in the monkey. I. The effects of bilateral pyramidal lesions. Brain 91:1–14

Lawrence DG, Kuypers HGJM (1968b) The functional organization of the motor system in the monkey. II. The effects of lesions of the descending brainstem pathways. Brain 91:15–36

Lawyer T, Netsky MG (1953) Amyotrophic lateral sclerosis. A clinicoanatomic study of 53 cases. Arch Neurol Psychiatry 69:171–192

Léger L, Charnay Y, Dubois PM, Jouvet M (1986) Distribution of enkephalin-immunoreactive cell bodies in relation to serotonin-containing neurons in the raphe nuclei of the cat: immunohistochemical evidence for the coexistence of enkephalins and serotonin in certain cells. Brain Res 362:63–73

Lindqvist C, Martensson A (1970) Mechanisms involved in cat's blink reflex. Acta Physiol Scand 80:149–159

Lipski J, Martin-Body RL (1987) Morphological properties of respiratory intercostal motoneurons in cats as revealed by intracellular injection of horseradish peroxidase. J Comp Neurol 260:423–434

Loewy AD, Saper CB, Baker RP (1979) Descending projections from the pontine micturition center. Brain Res 172:533–539

Lovick TA (1987) Differential control of cardiac and vasomotor activity by neurones in nucleus paragigantocellularis lateralis in the cat. J Physiol 389:23–35

Lovick TA, Robinson JP (1983) Bulbar raphe neurones with projections to the trigeminal nucleus caudalis and the lumbar cord in the rat: A fluorescence double-labelling study. Exp Brain Res 50:299–309

Lyon MJ (1975) Localization of the efferent neurons of the tensor tympani muscle of the newborn kitten using horseradish peroxidase. Exp Neurol 49:439–455

Lyon MJ (1978) The central location of the motor neurons to the stapedius muscle in the cat. Brain Res 143:437–444

Mackel R (1979) Segmental and descending control of the external urethral and anal sphincters in the cat. J Physiol 294:105–123

MacLean PD (1952) Some psychiatric implications of physiological studies on frontotemporal portion of limbic system. EEG Clin Neurophysiol 4:407–418

Mannen T, Iwata M, Toyokura Y, Nagashima K (1977) Preservation of a certain motoneurone group of the sacral cord in amyotrophic lateral sclerosis: its clinical significance. J Neurol Neurosurg Psychiatry 40:464–469

Mannen T, Iwata M, Toyokura Y, Nagashima K (1982) The Onuf's nucleus and the external anal sphincter muscles in amyotrophic lateral sclerosis and Shy–Drager syndrome. Acta Neuropathol 58:255–260

Mantyh PW (1983) Connections of midbrain periaqueductal gray in the monkey. II. Descending efferent projections. J Neurophysiol 49:582–595

Mantyh PW, Hunt SP (1984) Evidence for cholecystokinin-like immunoreactive neurons in the rat medulla oblongata which project to the spinal cord. Brain Res 291:49–54

Martin GF, Dom R (1970) Rubrobulbar projections of the Opossum (*Didelphis virginiana*). J Comp Neurol 139:199–214

Martin GF, Dom R, Katz S, King JS (1974) The organization of projection neurons in the opossum red nucleus. Brain Res 78:17–34

Martin GF, Humbertson AO Jr, Laxson C, Panneton WM (1979a) Evidence for direct bulbospinal projections to laminae IX, X and the intermediolateral cell column. Studies using axonal transport techniques in the North American opossum. Brain Res 170:165–171

Martin GF, Humbertson AO Jr, Laxson C, Panneton WM (1979b) Dorsolateral pontospinal systems. Possible routes for catecholamine modulation of nociception. Brain Res 163:333–339

Martin GF, Humbertson AO Jr, Laxson LC, Panneton WM, Tschismadia I (1979c) Spinal projections from the mesencephalic and pontine reticular formation in the North American opossum: a study using axonal transport techniques. J Comp Neurol 187:373–401

Martin GF, Cabana T, Humbertson AO Jr, Laxson LC, Panneton WM (1981a) Spinal projections from the medullary reticular formation of the North American opossum: evidence for connectional heterogeneity. J Comp Neurol 196:663–682

Martin GF, Cabana T, Humbertson AO Jr (1981b) Evidence for collateral innervation of the cervical and lumbar enlargements of the spinal cord by single reticular and raphe neurons. Studies using fluorescent markers in double-labelling experiments on the North American opossum. Neurosci Lett 24:1–6

Martin GF, Vertes RP, Waltzer R (1985) Spinal projections of the gigantocellular reticular formation in the rat. Evidence for projections from different areas to laminae I and II and lamina IX. Exp Brain Res 58:154–162

Martin JH, Ghez C (1988) Red nucleus and motor cortex: parallel motor systems for the initiation and control of skilled movement. Behav Brain Res 28:217–223

Massion J (1988) Red nucleus: past and future. Behav Brain Res 28:1–8

McCall RB, Aghajanian GK (1979) Serotonergic facilitation of facial motoneuron excitation. Brain Res 169:11–29

McCurdy ML, Hansma DI, Houk JC, Gibson AR (1987) Selective projections from the cat red nucleus to digit motor neurons. J Comp Neurol 265:367–379

McKenna K.E, and Nadelhaft I (1986) The organization of the pudendal nerve in the male and female rat. J Comp Neurol 248:532–549

Meessen H, Olszewski J (1949) A cytoarchitectonic atlas of the rhombencephalon of the rabbit. Karger, Basel

Mendell LM, Henneman E (1971) Terminals of single 1a fibers: location, density, and distribution within a pool of 300 homonymous motoneurons. J Neurophysiol 34:171–187

Meyerson BA (1988) Problems and controversies in PAG and sensory thalamic stimulation as treatment for pain. In: Fields HL, Besson JM (eds) Nociception and control. Elsevier, Amsterdam, pp 175–188 (Progress in brain research 77)

Miller AD (1987) Localization of motoneurons innervating individual abdominal muscles of the cat. J Comp Neurol 256:600–606

Miyazaki T, Yoshida Y, Hirano M, Shin T, Kanaseki T (1981) Central location of the motoneurons supplying the thyrohyoid and the geniohyoid muscles as demonstrated by horseradish peroxidase method. Brain Res 219:423–427

Mizuno N, Konishi A, Sato M (1975) Localization of masticatory motoneurons in the cat and rat by means of retrograde axonal transport of horseradish peroxidase. J Comp Neurol 164:105–116

Mizuno N, Nomura S, Konishi A et al. (1982) Localization of motoneurons innervating the tensor tympani muscles: an horseradish peroxidase study in the guinea pig and cat. Neurosci Lett 31:205–208

Molenaar I (1978) The distribution of propriospinal neurons projecting to different motoneuronal cell groups in the cat's brachial cord. Brain Res 158:203–206

Molenaar I, Kuypers HGJM (1978) Cells of origin of propriospinal, ascending supraspinal and medullospinal fibers. A HRP study in cat and Rhesus monkey. Brain Res 152:429–450

Molenaar I, Rustioni A, Kuypers HGJM (1974) The location of cells of origin of the fibers in the ventral and the lateral funiculus of the cat's lumbosacral cord. Brain Res 78:239–254

Moon Edley S, Graybiel AM (1983) The afferent and efferent connections of the feline nucleus tegmenti pedunculopontine, pars compacta. J Comp Neurol 217:187–216

Murray HM, Gurule ME (1979) Origin of the rubrospinal tract of the rat. Neurosci Lett 14:19–25

Nadelhaft I, De Groat WC, Morgan C (1980) Location and morphology of parasympathetic preganglionic neurons in the sacral spinal cord of the cat revealed by retrograde axonal transport of horseradish peroxidase. J Comp Neurol 193:265–281

Nathan PW, Smith MC (1958) The centrifugal pathway for micturition within the spinal cord. J Neurol Neurosurg Psychiatry 21:177–189

Nathan PW, Smith MC (1982) The rubrospinal and central tegmental tracts in man. Brain 105:223–269

Nauta WJH (1958) Hippocampal projections and related neural pathways to the mid-brain in the cat. Brain 80:319–341

Neafsy EJ, Hurley-Gius KM, Arvanitis D (1986) The topographical organization of neurons in the rat medial frontal, insular and olfactory cortex projecting to the solitary nucleus, olfactory bulb, periaqueductal gray and superior colliculus. Brain Res 377:261–270

Nieollon A, Rispal-Padel L (1976) Somatotopic localization in cat motor cortex. Brain Res 105:405–422

Nieuwenhuys R, Geeraedts LMG, Veening J (1982) The medial forebrain bundle of the rat. I. General introduction. J Comp Neurol 206:49–81
Nilaver G, Zimmerman EA, Wilkins J, Michaels J, Hoffman D, Silverman A (1980) Magnocellular hypothalamic projections to the lower brain stem and spinal cord of the rat. Neuroendocrinology 30:150–158
Nyberg-Hansen R (1965) Sites and mode of termination of reticulo-spinal fibers in the cat. An experimental study with silver impregnation methods. J Comp Neurol 124:71–100
Nyberg-Hansen R, Brodal A (1964) Sites and mode of termination of rubrospinal fibres in the cat. J Anat Lond 98:235–253
Nyberg-Hansen R, Masicitti TA (1964) Sites and mode of termination of fibres of the vestibulospinal tract in the cat. An experimental study with silver impregnation methods. J Comp Neurol 122:369–387
Nygren LG, Olson L (1977) A new major projection from locus coeruleus: the main source of noradrenergic nerve terminals in the ventral and dorsal columns of the spinal cord. Brain Res 132:85–93
Ongerboer de Visser BW, Kuypers HGJM (1978) Late blink reflex changes in lateral medullary lesions. Brain 101:285–295
Onufrowicz B (1899) Notes on the arrangement and function of the cell groups in the sacral region of the spinal cord. J Nervous Mental Dis 26:498–504
Panneton WM, Burton H (1981) Corneal and periocular representation within the trigeminal sensory complex in the cat studied with transganglionic transport of horseradish peroxidase. J Comp Neurol 199:327–344
Panneton WM, Martin GF (1983) Brainstem projections to the facial nucleus of the opossum. Brain Res 267:19–33
Papez JW (1927) Subdivisions of the facial nucleus. J Comp Neurol 43:159–191
Paxinos G, Törk I, Halliday G, Mehler WR (1990) Human homologs to brainstem nuclei identified in other animals as revealed by acetylcholinesterase activity. In: Paxinos G (ed) The human nervous system. Academic Press, New York, pp 149–202
Pelletier G, Steinbusch HWM, Verhofstad AAJ (1981) Immunoreactive substance P and serotonin present in the same dense core vesicles. Nature 293:71–72
Petras JM (1967) Cortical, tectal and tegmental fiber connections in the spinal cord of the cat. Brain Res 6:275–324
Pfaff DW (1980) Estrogens and brain function. Neuronal analysis of a hormone-controlled mammalian reproductive behavior. Springer, Heidelberg New York Berlin, p 281
Poeck K (1969) Pathophysiology of emotional disorders associated with brain damage. In Vinken PJ, Bruyn GW (eds) Handbook of clinical neurology, vol. 3. North-Holland, Amsterdam, pp 343–367
Pompeiano O, Brodal A (1957) Experimental demonstration of a somatotopical origin of rubrospinal fibers in the cat. J Comp Neurol 108:225–252
Price JL, Russchen FT, Amaral DG (1987) The limbic region. II. The amygdaloid complex. In Bjorkund et al. (eds) Handbook of chemical neuroanatomy, vol. 5. Integrated systems of the CNS, part I. Hypothalamus, hippocampus, amygdala, retina. Elsevier, Amsterdam, pp 279–388
Pullen AH (1988) Quantitative synaptology of feline motoneurones to external anal sphincter muscle. J Comp Neurol 269:414–424
Ralston DD, Ralston HJ III (1985) The terminations of corticospinal tract axons in the macaque monkey. J Comp Neurol 242:325–337
Ralston DD, Milroy AM, Holstege G (1988) Ultrastructural evidence for direct monosynaptic rubrospinal connections to motoneurons in *Macaca mulatta*. Neurosci Lett 95:102–106
Rexed B (1954) A cytoarchitectonic atlas of the spinal cord in the cat. J Comp Neurol 100:297–380
Richmond FJR, Loeb GE, Reesor D (1985) Electromyographic activity in neck muscles during head movements in the alert, unrestrained cat. Soc Neurosci Abstr 11:83
Rinn WE (1984) The neurophysiology of facial expression: a review of the neurological and psychological mechanisms for producing facial expressions. Psychol Bull 95:52–77
Robinson FR, Houk JC, Gibson AR (1987) Limb-specific connections of the cat magnocellular red nucleus. J Comp Neurol 257:553–577
Romanes GJ (1951) The motor cell columns of the lumbo-sacral spinal cord of the cat. J Comp Neurol 94:313–363
Roppolo JR, Nadelhaft I, de Groat WC (1985) The organization of pudendal motoneurons and primary afferent projections in the spinal cord of the rhesus monkey revealed by horseradish peroxidase. J Comp Neurol 234:475–488
Roucoux A, Crommelinck M, Decostre MF, Crémieux J (1985) Gaze shift related neck muscle activity in trained cats. Soc Neurosci Abstr 11:83

Ruggiero DA, Mraovitch S, Granata AR, Anwar M, Reis DJ (1987) A role of insular cortex in cardiovascular function. J Comp Neurol 257:189–207

Rustioni A, Kuypers HGJM, Holstege G (1971) Propriospinal projections from the ventral and lateral funiculi to the motoneurons in the lumbosacral cord of the cat. Brain Res 34:255–275

Sakuma Y, Pfaff DW (1979a) Mesencephalic mechanisms for integration of female reproductive behavior in the rat. Am J Physiol 237:R285–R290

Sakuma Y, Pfaff DW (1979b) Facilitation of female reproductive behavior from mesencephalic central gray in the rat. Am J Physiol 237:R278–R284

Sandkuhler J, Gebhart GF (1984) Characterization of inhibition of a spinal nociceptive reflex by stimulation medially and laterally in the midbrain and medulla in the pentobarbital-anesthetized rat. Brain Res 305:67–76

Saper CB, Loewy AD, Swanson LW, Cowan WM (1976) Direct hypothalamo-autonomic connections. Brain Res 117:305–312

Sato M, Mizuno N, Konishi A (1978) Localization of motoneurons innervating perineal muscles: a HRP study in cat. Brain Res 140:149–154

Sawchenko PE, Swanson LW (1982) Immunohistochemical identification of neurons in the paraventricular nucleus of the hypothalamus that project to the medulla or to the spinal cord in the rat. J Comp Neurol 205:260–272

Schoen JHR (1964) Comparative aspects of the descending fibre systems in the spinal cord. In: Eccles JC, Schadé JP (eds) Organization of the spinal cord. Progr Brain Res 11:203–222

Schrøder HD (1980) Organization of the motoneurons innervating the pelvic muscles of the rat. J Comp Neurol 192:567–587

Schrøder HD (1981) Onuf's nucleus X: a morphological study of a human spinal nucleus. Anat Embryol 162:443–453.

Shahani BT, Young RR (1972) Human orbicularis oculi reflexes. Neurology 22:149–154

Shapovalov AI, Karamyan OA, Kurchavyi GG, Repina ZA (1971) Synaptic actions evoked from the red nucleus on the spinal alpha-motoneurons in the rhesus monkey. Brain Res 32:325–348

Shapovalov AI, Kurchavyi GG (1974) Effects of trans-membrane polarization and TEA injection on monosynaptic actions from motor cortex, red nucleus and group Ia afferents on lumbar motoneurons in the monkey. Brain Res 82:49–67

Shaw MD, Baker R (1983) The locations of stapedius and tensor tympani motoneurons in the cat. J Comp Neurol 216:10–19

Shik ML, Severin FV, Oriovski GN (1966) Control of walking and running by means of electrical stimulation of the mid-brain. Biophysics 11:756–765

Simon OR, Schramm LP (1983) Spinal superfusion of dopamine excites renal sympathetic nerve activity. Neuropharmacology 22:287–293

Skagerberg G, Lindvall O (1985) Organization of diencephalic dopamine neurones projecting to the spinal cord in the rat. Brain Res 342:340–351

Smith CL, Hollyday M (1983) The development and postnatal organization of motor nuclei in the rat thoracic spinal cord. J Comp Neurol 220:16–28

Sprague JM (1948) A study of motor cell localization in the spinal cord of the rhesus monkey. Am J Anat 82:1–26

Steinbusch HWM (1981) Distribution of serotonin-immunoreactivity in the central nervous system of the rat: cell-bodies and terminals. Neuroscience 6:557–618

Sterling P, Kuypers HGJM (1967) Anatomical organization of the brachial spinal cord of the cat. II. The motoneuron plexus. Brain Res 4:16–32

Sterling P, Kuypers HGJM (1968) Anatomical organization of the brachial spinal cord of the cat. III. The propriospinal connections. Brain Res. 7:419–443

St John WM, Barltlett Jr D, Knuth KV, Hwang J-C (1981) Brain stem genesis of automatic ventilatory patterns independent of spinal mechanisms. J Appl Physiol 51:204–210

Sung JH (1979) Autonomic neurons affected by lipid storage in the spinal cord in Fabry's disease: distribution of autonomic neurons in the sacral spinal cord. J Neuropathol Exp Neurol 28:87–98

Sung JH, Mastri AR (1980) Spinal autonomic neurons in Werdnig–Hoffmann disease, mannosidosis, and Hurler's syndrome: distribution of autonomic neurons in the sacral spinal cord. J Neuropathol Exp Neurol 39:441–451

Sung JH, Mastri AR, Segal E (1979) Pathology of Shy–Drager syndrome. J Neuropathol Neurol 38:353–368

Swanson LW (1977) Immunohistochemical evidence for a neurophysin-containing pathway arising in the paraventricular nucleus of the hypothalamus. Brain Res 128:346–353

Swanson LW, Kuypers HGJM (1980) The paraventricular nucleus of the hypothalamus: cytoarchitectonic subdivisions and organization of projections to the pituitary, dorsal vagal complex, and spinal cord as demonstrated by retrograde fluorescence double labeling methods. J Comp Neurol 194:555–570

Swanson LW, Sawchenko PE (1983) Hypothalamic integration: organization of the paraventricular and supraoptic nuclei. Ann Rev Neurosci 6:269–324

Taber-Pierce E, Lichtenstein E, Feldman SC (1985) The somatostatin systems of the guinea-pig brainstem. Neuroscience 15:215–235

Tabira T, Namikawa T, Goto I, Kuroiwa Y (1980) Fabry's disease with special reference to EMG findings of external sphincters. Onuf's nucleus, anhydrosis and sweat gland. Autonom Nerv Syst (Tokyo) 17:346–351

Takeuchi Y, Nakano K, Uemura M, Matsuda K, Matsushima R, Mizuno N (1979) Mesencephalic and pontine afferent fiber system to the facial nucleus in the cat: a study using the horseradish peroxidase and silver impregnation techniques. Exp Neurol 66:330–343

Tashiro N, Tanaka T, Fukumoto T, Hirata K, Nakao H (1985) Emotional behavior and arrhythmias induced in cats by hypothalamic stimulation. Life Sci 36:1087–1094

Terreberry RR, Neafsy EJ (1987) The rat medial frontal cortex projects directly to autonomic regions of the brainstem. Brain Res Bull 19:639–649

Tohyama M, Sakai K, Salvert D, Touret M, Jouvet M (1979) Spinal projections from the lower brain stem in the cat as demonstrated by the horseradish peroxidase technique. I: Origins of the reticulospinal tracts and their funicular trajectories. Brain Res 173:383–405

Uemura M, Matsuda K, Kume M, Takeuchi Y, Matsushima R, Mizuno N (1979) Topographical arrangement of hypoglossal motoneurons: An HRP study in the cat. Neurosci Lett 13:99–104

Ueyama T, Mizuno N, Nomura S, Konishi A, Itoh K Arakawa H (1984) Central distribution of afferent and efferent components of the pudendal nerve in cat. J Comp Neurol 222:38–46

Van der Kooy, D Koda LY, McGinty JF, Gerfen CR, Bloom FE (1984) The organization of projections from the cortex, amygdala, and hypothalamus to the nucleus of the solitary tract in rat. J Comp Neurol 224:1–24

VanderMaelen CP, Aghajanian GK (1982) Serotonin-induced depolarization of rat facial motoneurons in vivo: comparison with amino acid transmitters. Brain Res 239:139–152

Wessendorf MW, Elde R (1987) The coexistence of serotonin- and substance P-like immunoreactivity in the spinal cord of the rat as shown by immunofluorescent double labelling. J Neurosci 7:2352–2363

Westlund KN, Coulter JD (1980) Descending projections of the locus coeruleus and subcoeruleus/medial parabrachial nuclei in monkey: axonal transport studies and dopaminehydroxylase immunocytochemistry. Brain Res Rev 2:235–264

White SR, Neuman RS (1980) Facilitation of spinal motoneurone excitability by 5-hydroxytryptamine and noradrenaline. Brain Res 188:119–127

Willett CJ Gwyn DG, Rutherford JG, Leslie RA (1986) Cortical projections to the nucleus of the tractus solitarius: An HRP study in the cat. Brain Res Bull 16:497–505

Willis WD (1988) Anatomy and physiology of descending control of nociceptive responses of dorsal horn neurons: comprehensive review. In: Fields HL, Besson JM (eds), Pain modulation. Elsevier, Amsterdam, pp 1–29 (Progress in brain research 77)

Yoshida M, Tanaka M (1988) Existence of new dopaminergic terminal plexus in the rat spinal cord: assessment by immunohistochemistry using anti-dopamine serum. Neurosci Lett 94:5–9

Yoshida Y, Miyazaki O, Hirano M, Shin T, Totoki T, Kanaseki T (1981) Localization of efferent neurons innervating the pharyngeal constrictor muscles and the cervical esophagus muscle in the cat by means of the horseradish peroxidase method. Neurosci Lett 22:91–95

14 Neurophysiological Changes in Motor Neuron Disease

M.S. Schwartz and M. Swash

Motor neuron disease is characterised by features of both lower and upper motor neuron dysfunction (Swash and Schwartz 1992). Thus, there is atrophy, weakness and fasciculation, with hyperreflexia, extensor responses and, in some cases, spasticity. These clinical features vary in their distribution at different stages of the disease and in many patients the disease seems to commence asymmetrically, often involving only one limb, or predominantly to involve bulbar muscles or limb muscles. Fasciculation or muscle cramp sometimes precedes the development of more florid upper and lower motor neuron features and in some patients fatigue is a prominent symptom. In about 10% of patients sensory symptoms, mainly consisting of paraesthesiae, are part of the clinical syndrome. These major clinical features of the disease are accompanied by characteristic neurophysiological abnormalities and the latter may be used to assess the extent, severity and rate of progression of the disease (Swash and Schwartz 1984; Stalberg and Sanders 1984).

Denervation of muscle, the major manifestation of lower motor neuron involvement in the disease, is accompanied by compensatory reinnervation, a process that initially enables muscle strength to be preserved in the face of the continuing loss of functioning anterior horn cells (Wohlfart 1951). Wohlfart (1958) estimated that 30% of motor neurons may be lost before weakness becomes apparent. Electromyography is therefore capable of detecting neurogenic change even in apparently normal muscles. This investigation is especially valuable in diagnosis since it reveals the generalised abnormality, even when the clinical presentation is limited in distribution. Furthermore, in the absence of a specific immunological, biochemical or radiological abnormality, the correlation of neurophysiological abnormalities with the clinical features is important in establishing diagnostic criteria, and can be used to establish staging and prognosis (Swash and Schwartz 1984). At present, definitive diagnosis can only be made by post-mortem examination of the central nervous system.

Electromyography (EMG)

EMG is the most useful investigation in the assessment of patients with motor neuron disease. Lambert (1969) suggested the following EMG criteria for the diagnosis of motor neuron disease:

1. Fibrillation and fasciculation in muscles of the upper and lower extremities; cranial musculature may also be involved.
2. Motor unit action potentials of increased amplitude and duration, and reduced in number as represented by a reduced interference pattern.
3. Motor conduction velocity is normal in nerves innervating relatively unaffected muscles, and not less than 10% of normal in severely affected, atrophic muscles.
4. Normal amplitude of sensory action potential, and normal sensory conduction velocity in peripheral nerves, even in severely affected muscles.

Further to these criteria, EMG features of chronic partial denervation found in three extremities, counting cranial muscles as one extremity, in three different muscles with different radicular or peripheral nerve distributions, are required to fulfil the first of Lambert's (1969) criteria. These more rigorous criteria are important in excluding cervical spondylosis, multiple entrapment neuropathies, sensorimotor neuropathy and syringomyelia. Entrapment neuropathies are common in patients in the age range of those with motor neuron disease, and this coincident disorder should be considered in assessing patients since it does not necessarily exclude the diagnosis.

Fasciculations in MND

Fasciculations consist of spontaneous firing of motor units, and therefore of potentials that are identical in configuration to voluntarily activated motor unit potentials. In motor neuron disease fasciculation potentials as recorded by concentric needle or monopolar EMG electrodes, almost always consist of abnormal potentials of increased complexity, whereas in normal subjects fasciculation potentials consist of normal motor unit potentials. In normal subjects (Reed and Kurland 1963) fasciculations usually occur in calf or intrinsic hand muscles, often after strong contraction, but are not associated with weakness or muscle atrophy, and are not accompanied by EMG features of chronic partial denervation. In motor neuron disease fasciculations are characteristically accompanied by EMG evidence of chronic partial denervation, although atrophy is not necessarily a feature. Further, in motor neuron disease fasciculations are often widespread, involving muscles not clinically involved at the time of the EMG examination. Visible fasciculations occur in superficial motor units; needle EMG allows fasciculations occurring more deeply in the muscle to be recorded. Surface electrode recordings, using multiple channels, is a useful technique for ascertaining fasciculation potentials in clinical practice.

In MND fasciculations occur repetitively in individual motor units at a rate of about 1 per 3 seconds (Trojaborg and Buchthal 1965). These slow rate fasciculations were termed "malignant" fasciculations to indicate their association with motor neuron disease (Trojaborg and Buchthal 1965). Fasciculations occurring in root disorders, such as cervical spondylosis or in peripheral neuropathy, tend to occur at a more rapid rate, e.g. about 1 per second. The fasciculations of MND are clinically larger and more obvious, and are more polyphasic in EMG recordings than those found in root or peripheral nerve disorders, indicating more marked enlargement of remaining motor units in muscles partially denervated and in which collateral sprouting is developing.

Fasciculations in MND were at first thought to arise in the soma of diseased anterior horn cells (Denny-Brown 1953). However, fasciculations may persist after more proximal nerve block (Roth 1984) and it is probable that they may arise at any point in the lower motor neuron (Wettstein 1979). Most fasciculations arise in the distal part of the motor nerve from abnormal sites of excitation (ectopic generators) within the motor axon arborisation, or even at the end-plate synaptic terminals themselves (Conradi et al. 1982; Roth 1984). The fasciculating motor units themselves can be voluntarily activated, and are mostly low threshold units with moderately slow conduction velocities in the range 3–45 m/s (Conradi et al. 1982). The clinical impression that abundant fasciculations imply rapid progression and few fasciculations a more benign prognosis is not proven; indeed these two concepts may not be converse.

Fibrillation Potentials and Positive Sharp Waves

Fibrillation potentials consist of biphasic potentials 20–300 μV in amplitude and 1–5 ms in duration. They originate from single denervated muscle fibres and can be recorded only with needle electrodes (Ekstedt 1964). Fibrillations occur at frequencies of 1–10 Hz (Conrad et al. 1972) and often occur as doublets (paired discharges). They occur characteristically in neurogenic diseases but may also be recorded in myopathies, especially in acute polymyositis.

Positive sharp waves consist of potentials with a sharp, positive initial deflection of about 100 μV amplitude followed by a slow negative decay toward the baseline (Buchthal and Pirelli 1953), of about 4 ms duration. Their firing rate is similar to that of fibrillation potentials and can be recorded before fibrillation potentials following acute denervation, e.g. after nerve injury. Their presence in MND therefore implies acute denervation, and like fibrillation potentials are associated with denervation rather than reinnervation.

Other Spontaneous Discharges in MND

Complex repetitive discharges (bizarre high frequency potentials) occur in MND, as in other neurogenic disorders. They are sometimes recorded also in myopathies. They consist of continuous trains of spikes, of simple or complex patterns, repeated at a regular frequency in the range 5–150 Hz, with an abrupt onset and termination. They are often triggered by electrode movement, and probably originate from multiple pacemaker sites on denervated muscle fibres that drive adjacent muscle fibres ephaptically through excitatory loops (Trontelj and Stalberg 1983). The morphology of the potential varies according to the number of fibres activated, and thus varies during the course of the discharge. However, since end-plate activation is not involved, the discharges show a very low interpotential jitter (Stalberg and

Trontelj 1982). Complex repetitive discharges in MND are therefore a non-specific feature of denervation in the disease, but have no prognostic significance.

Motor Unit Potentials in MND

In concentric needle or single fibre EMG recordings motor unit potentials are polyphasic units of increased duration. Concentric needle recordings characteristically reveal units of increased amplitude, recruited early in voluntary contraction. With maximal voluntary contraction the interference pattern is reduced. In atrophic muscles giant units of amplitude greater than 12 mV may be recorded, representing relatively compact motor units containing many muscle fibers formed by collateral sprouting close to the uptake area of the electrode. In single fibre EMG recordings these giant units are found to consist of many separate components distributed within a unit of relatively short duration. The use of the trigger-delay line, and increased low frequency filters (up to 1000 Hz), allows recognition of complex units of increased duration in both concentric needle and single fibre EMG recordings. These complex potentials have been termed potentials with satellite potentials or late units, and can only be recognised with the trigger delay line (Stalberg et al. 1975).

In muscles of normal strength and bulk routine concentric needle EMG studies may be normal. However, single fibre EMG recordings in such muscles often reveal slightly increased fibre density and an increased neuromuscular jitter in some potentials (Stalberg et al. 1975; Schwartz and Swash 1982), indicating the presence of early reinnervative change in these muscles representing a well-compensated stage of the disease. As the disorder progresses muscle weakness and atrophy develop and this is correlated with EMG features of chronic partial denervation with fibrillation potentials, fasciculation potentials, large polyphasic motor units on voluntary activation and a reduced interference pattern. In the very advanced stage the muscle is markedly atrophic and EMG recordings reveal areas of electrical silence or fibrillation with single, large, polyphasic motor units that have late components with a prominent neuromuscular jitter, and which fire at slow rates. At all stages of the disease there are marked differences in the electrophysiological features of different muscles, although certain patterns of involvement are common; for example, hand muscles are frequently more severely involved than more proximal upper limb muscles (Swash 1980). Turns analysis of the interference pattern shows typical features of neurogenic change, with increased amplitude/turns relationship (Hayward and Willison 1977).

Multi-electrode studies, and scanning EMG investigations (Schwartz et al. 1976; Stalberg and Sanders 1984), show that there is no increase in the territorial extent of the reinnervated motor unit in MND, but that there is an increase in the packing density of fibres innervated by the same motor unit within this territory, giving a motor unit potential of larger amplitude, and a higher fibre density in single fibre EMG studies. Macro EMG illustrates the total activity of all the muscle fibres innervated by a motor unit, and in MND this technique reveals the large potential for reinnervation in the preterminal phases of the disease. However, a much greater potential for reinnervation is found in poliomyelitis, Type III spinal muscular atrophy and Charcot–Marie–Tooth syndrome (HMSN) than in MND, a finding suggesting that the anterior horn cell is less capable of sustaining collateral sprouting in MND than in other neurogenic disorders (Stalberg 1982; Swash and Schwartz 1984). This electrophysiological evidence is consistent with the suggestion made by McComas (1977) that motoneurons in MND may enter a phase of "sickness" that precedes their ultimate death. Indeed, as the disease progresses the fibre density may fall in the preterminal atrophic phase (Swash and Schwartz 1982, 1984).

Fatigue, Decrement and Neuromuscular Jitter

During the course of MND patients often complain of fatiguability and this may respond to treatment with neostigmine (Mulder et al. 1959). Repetitive nerve stimulation at slow rates (2 Hz) may show a decrement, and at high rates (20 Hz) a potentiation in about 30% of patients with MND. However, these changes are mild and not as prominent as in patients with myasthenia gravis. A decremental response to stimulation at 2 Hz is associated with a more rapidly progressive course (Bernstein and Antel 1981). Increased jitter and blocking on single fibre EMG studies is associated with abnormal decremental responses to repetitive stimulation, indicating reduced safety factor for transmission at the neuromuscular junction, and in the terminal axonal tree. During continuous activation of a muscle unit for a minute or more the neuromuscular jitter may increase and neuromuscular blocking may develop (Fig. 14.1), indicating that the safety factor is low in these abnormal, reinnervated and unstable motor units in MND (Ingram et al 1985).

Motor Conduction in MND

Loss of anterior horn cells results in axonal degeneration; this is accompanied by axonal sprouting in the terminal axonal tree during the process of compensatory reinnervation. This might therefore be expected to cause changes in motor conduction in peripheral nerves.

The magnitude of the disturbance in motor conduction is related to loss of motor units and to atrophy of the muscle innervated by the motor nerve under investigation. Thus, in weak, atrophic muscles, motor conduction studies show an increased distal motor latency, slowed motor nerve conduction velocity and reduced amplitude of the evoked muscle action potential. The distal motor latency consists of components including the motor axons in the peripheral nerve, the terminal arborization, motor end-plate function and propagation velocity of muscle fibres. All these components may be abnormal to some degree, but the major component of slowing of distal motor conduction is probably related to loss of large, fast conducting motor nerve fibres with relative preservation of smaller, slower conducting fibres. However, Nakanishi et al. (1989) found that maximal motor conduction and minimal motor conduction velocity were both slowed. In the ulnar nerve Nakanishi et al. (1989) found that maximal motor conduction was 62 m/s in controls and 54 m/s in MND patients, and minimal motor conduction was 53 m/s in controls and 43 m/s in MND patients. Thus, the difference between maximal and minimal

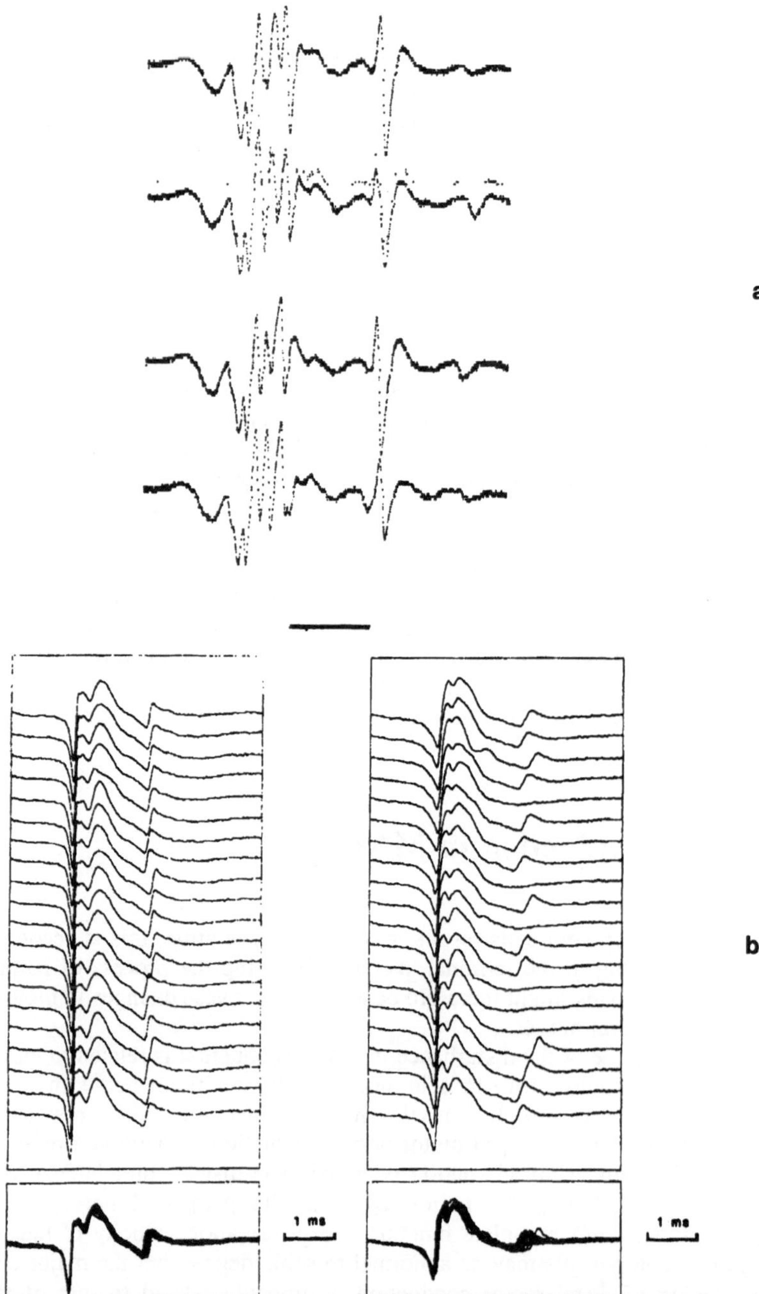

Fig. 14.1. **a** A complex motor unit action potential recorded by single fibre EMG. The waveform is more complex than normal and there is a series of late components. The morphology of the waveform is stable, indicating that the innervation of this complex unit is functionally stable during repetitive action. **b** Falling leaf display of consecutive discharges of a single motor unit action potential showing the development of increasing jitter of the late component with respect to the early component, and of impulse blocking in the later sequence shown on the right. Superimposed traces are shown on the bottom of the figures.

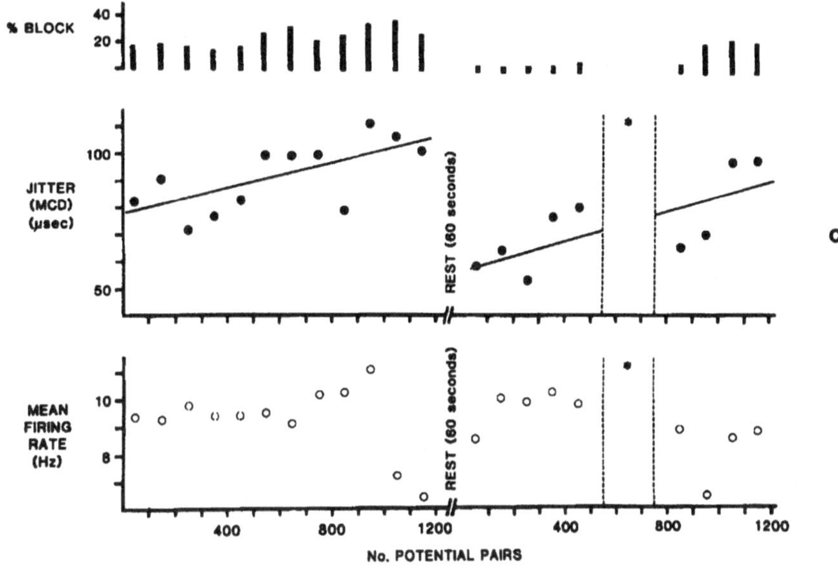

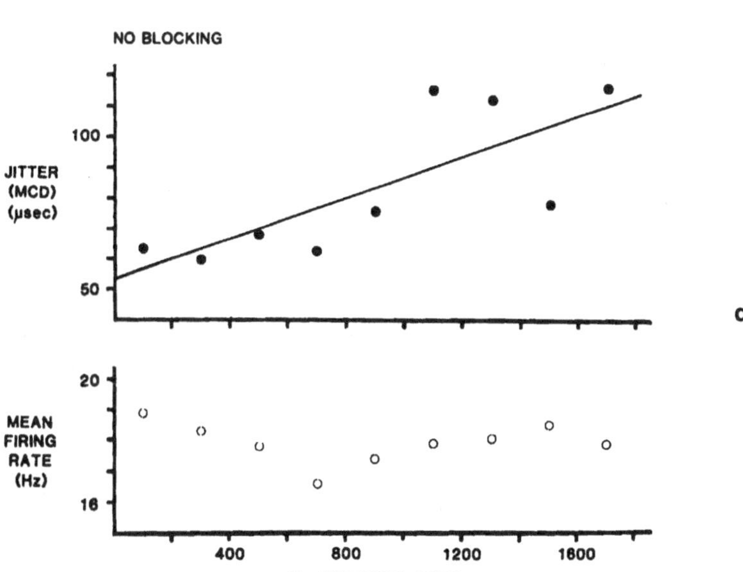

Fig. 14.1 (*continued*) **c** Type 3 spinal muscular atrophy. Single fibre EMG recording from forearm extensor muscle, showing increasing jitter, and blocking (upper bars), related to firing rate (lower box). A period of rest of 60 seconds was followed by recovery of neuromuscular function, but jitter and blocking reappeared after a further period of contraction. Impulse blocking is accompanied by fatigue when sufficient muscle fibres cease to contract because of failure of transmission at the neuromuscular junction. **d** Motor neuron disease. In this sequential recording the jitter increased, but there was no blocking during the period of the recording.

motor nerve conduction velocity was similar in controls and patients, suggesting that both faster and slower conducting motor nerve fibres were similarly involved. This finding is consistent with the histological finding of involvement of somatic extrafusal and intrafusal (gamma) motor nerve fibres in the disease (Swash and Fox 1974), and the finding that the slowest motor conduction velocity in MND is found in nerves innervating the most atrophic muscles (Hansen and Ballantyne 1978). The electrophysiological findings show no evidence of preferential involvement of large, fast conducting motor units (Hausmanowa-Petrusewicz and Kopec 1970; Hansen and Ballantyne 1978; Ingram et al. 1988; Nakanishi et al. 1989). Motor nerve conduction (maximal conduction) velocity is not slower than 75% of normal fastest motor conduction.

The muscle action potential duration is not increased (Hansen and Ballantyne 1978), indicating that there are no very slowly conducting motor units in peripheral nerves in MND. Since the disorder is axonal, focal slowing of motor conduction at points of potential entrapment is not a feature of the disease.

Late Responses in Motor Conduction Studies

F-wave responses depend on excitability of the anterior horn cell and on the number of surviving motor units in the nerve investigated. In motor neuron disease there is a combined upper and motor neuron lesion with an increased central excitatory state and loss of anterior horn cells, and therefore of motor axons. Pieroglou-Harmoussi et al. (1987) found that the F-wave responses were less complex than in normal subjects. However, there was a tendency for the same F responses to appear repeatedly in recordings, indicating both a smaller than normal pool of anterior horn cells and an increased excitability of remaining anterior horn cells. The reduction in F-wave response frequency was explained by loss of anterior horn cells, so that fewer cells were available to respond, and the less complex configuration of F responses was also related to the smaller number of motor units making up each response in atrophic muscles in recordings made with surface electrodes. Petajan (1985) found that the number of repeated F responses was only slightly increased in a study of the extensor digitorum brevis muscle in amyotrophic lateral sclerosis.

H reflex studies in MND are useful for showing the upper motor neuron component of the disease. There is an increased incidence of H reflex potentials in recordings made in soleus muscle after stimulating the posterior tibial nerve, indicating increased excitability of the anterior horn cell pool. Similarly, H reflexes can be recorded in small hand and foot muscles after stimulating ulnar, median, tibial and peroneal nerves (Norris 1975).

Sensory Conduction in MND

In the spinal cord in MND there is loss of neurons in Clarke's column, pallor of spinocerebellar pathways and damage to axons in these ascending tracts and, to a lesser extent, in the median portion of the posterior columns, in addition to the well-defined motor disorder (Feller et al. 1966; Swash et al. 1988). Dayan et al. (1969) showed that there was damage to Schwann cells in peripheral sensory nerves, and Shahani et al. (1971) reported that these nerves showed abnormal resistance to ischaemia. Mulder et al. (1983) found that there were increased thresholds to detection of vibration in patients with MND. However, conventional clinical and electrophysiological tests of sensation are normal, although Jamal et al. (1985) reported abnormalities of thermal thresholds in 80% of patients. Sensory action potentials are normal, suggesting that these subclinical sensory features are associated with changes in the spinal cord itself, rather than abnormalities in the peripheral nerves. However, Kawamura et al. (1981) found abnormalities in the peripheral sensory neuron, as assessed by morphometry of dorsal root and posterior root ganglia.

Studies of somatosensory evoked potentials (SSEP) in motor neuron disease have yielded variable results, perhaps reflecting the severity and clinical distribution of the disease. Most studies have revealed abnormalities in the N9-N13, or N13-N19 latencies, reflecting abnormalities in conduction in the root entry zone, or in the cord itself (Radtke et al. 1986; Matheson et al. 1986; Ghezzi et al. 1989). Brain stem and visual evoked potentials are normal.

Central Motor Conduction in MND

Electrical or magnetic stimulation of the cerebral cortex and spinal cord can be used to measure central motor conduction times, and thus to search for electrophysiological evidence of disturbed function in the major descending motor pathways in the central nervous system. In MND prominent and often asymmetrical slowing of central motor conduction, most marked in the spinal cord, has been found (Ingram and Swash 1987; Berardelli et al. 1987). This can be correlated with the severity of clinical signs of upper motor neuron involvement in the disease (Fig. 14.2), and with the presence or absence of extensor plantar responses (Ingram and Swash 1987). This finding illustrates the major abnormality in the corticospinal tracts in the disease, a feature that is often marked clinically by the overwhelming evidence of lower motor neuron disease. Motor conduction in central pathways may be less than 20 m/s, compared with a normal value of >45 m/s. It is presumed that this marked slowing of central motor conduction, which resembles the abnormality found in multiple sclerosis (Mills and Murray 1985; Ingram et al. 1988), is due to loss of large, heavily myelinated, fast-conducting corticospinal fibres.

Stages of the Disease: EMG Criteria

EMG studies have shown that even when the disease begins focally in a single limb, there are neurogenic changes distributed widely in other muscles and in other limbs. Single fibre EMG recordings reveal an increased fibre density in muscles that are

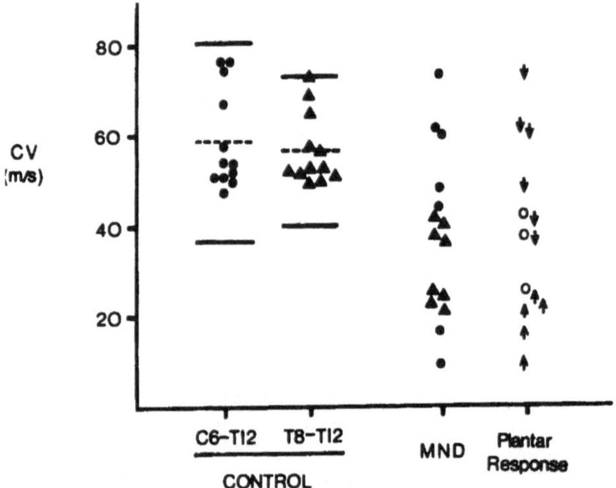

Fig. 14.2. Motor neuron disease. In this experiment motor conduction velocity in the spinal cord was measured between C6–T12, or T8–T12, using transcutaneous electrical stimulation and recording from tibialis anterior. In many of the patients with motor neuron disease the spinal motor conduction velocity was slowed, and this abnormality broadly correlated with the presence of extensor plantar responses. Closed circles, C6–T12 cord motor conduction velocity; closed triangles, T8–T12 cord motor conduction velocity. Arrows indicate plantar responses; circles, equivocal responses.

apparently unaffected (Swash 1980; Swash and Schwartz 1984). The disease is therefore widespread in the motor system at least from the time of diagnosis in most cases (Stalberg 1982). Hansen and Ballantyne (1978) found that the drop-out of motor units was effectively compensated by increased motor unit size until less than 50% of motor units remained, and that reinnervation ceased when fewer than 5% of motor units in a muscle remained functional. The rate of progression of the disease varies from muscle to muscle, and does not follow a consistent pattern from patient to patient. Indeed, homologous muscle on the two sides of the body may show a markedly different extent of involvement (Swash and Schwartz 1984) (Fig. 14.3).

In the early stages of the disease strength is normal and there is no detectable wasting, but the single fibre EMG fibre density is slightly increased. This is the phase of early reinnervation. In the second stage, strength and muscle bulk remain normal but the fibre density is increased and there may be some abnormal jitter and blocking. This is the first electrophysiological indication of the impending failure of functional reinnervation. This is the phase of relatively well compensated reinnervation. In this stage the macro EMG potentials reach their largest amplitude. In the third stage of the disease the muscle is weak and easily fatiguable; wasting is often present. The fibre density is markedly increased (>3.0) but impulse blocking is now a prominent feature. This is the phase of early decompensation with loss of anterior horn cells and, perhaps, impairment of the capacity for reinnervation in the remaining units. In the fourth stage the muscle is very weak and wasted, the fibre density is decreased, and jitter and blocking may be prominent in the few potentials recorded. Macro EMG shows small potentials. In this phase reinnervation has failed and the muscle is irreversibly wasted. During the second well-compensated stage of the disease macro EMG recordings of motor units may be as much as 10 times normal in amplitude. Macro EMG potentials as an index of reinnervation, are of

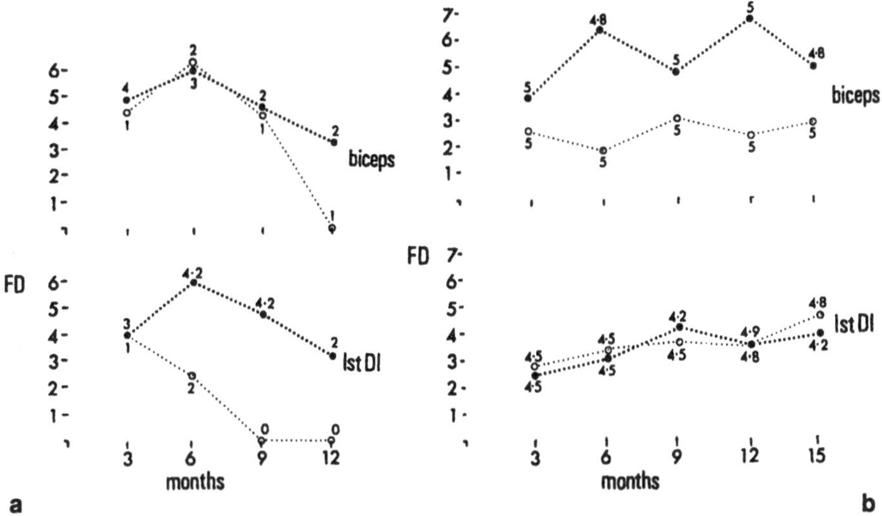

Fig. 14.3 a,b. The fibre density was measured in biceps brachii and 1st dorsal interosseous muscles, on the two sides, in two patients with motor neuron disease at intervals (abscissa) from the time of diagnosis during a period of a year or more. Note that the rate of progression, as shown by this measure of reinnervation, varied from muscle to muscle, and was different on the two sides of the body (cord). Indeed, in **a** there was effective reinnervation in one 1st dorsal interosseous muscle while the other was undergoing irreversible denervation. Open symbols, left; closed symbols, right. Fibre density noted at each data point.

interest in relation to other, more chronic neurogenic disorders. For example, in motor neuron disease the macro EMG potential is four times larger than the amplitude of that normal in normal subjects, in Kugelberg–Welander disease it is about 15 times the normal value, and in old poliomyelitis it may be as much as 40 times larger than normal (Stalberg 1982). Thus it appears that the capacity for reinnervation in MND is not as great as in some other anterior horn cell disorders, an observation that may reflect the underlying rates of progression of these disorders, and the underlying biological factors involved in determining the capacity for functional reinnervation.

Electrophysiological Assessment in Treatment Trials

Quantitative EMG techniques can be used in assessing new treatments in patients with MND. For example, neostigmine was shown by Mulder et al. (1959) to increase muscular strength and to improve the decremental response to repetitive stimulation in patients in whom fatiguability was a prominent symptom. Similarly, a period of rest will reduce the neuromuscular jitter and continuous activation may increase it in patients with fatiguability (Ingram et al. 1985). Macro EMG and single fibre EMG studies have been used to assess the effect of thyrotropin-releasing hormone (TRH) and various analogues in short-term experiments. The macro EMG recordings showed no change (Guiloff et al. 1988) and a single fibre EMG study suggested that the neuromuscular jitter increased (Massey et al. 1989). Clinical trials of this treatment have shown no benefit (Brooke et al. 1986). Objective studies of this type are important.

Differential Diagnosis of MND by its Electrophysiological Features

Other anterior horn cell disorders also produce the features of chronic partial denervation (Swash and Schwartz 1988). In *spinal muscular atrophy* (Kugelberg–Welander type) the motor unit action potentials are very complex, with a markedly increased fibre density, and jitter and blocking are not prominent features. In patients with *old poliomyelitis*, if the disorder is stable the motor unit action potentials are highly complex, of increased duration and stable, without increased jitter or blocking. Secondary myopathic features may be evident, consisting of low amplitude, polyphasic motor units. In the post-polio syndrome (Sever 1986), when there is increasing disability later in life, increased jitter and blocking are found, indicating age-related damage to the reinnervated sprouts in unusually large motor units, together with secondary myopathic change (Weichers and Hubble 1981). This late complication is slowly progressive, but is not a form of motor neuron disease (Dalakas et al. 1986). Syringomyelia often causes neurogenic atrophy in the upper extremities, but there is a constant pattern of involvement, the C8/T1 segments being most affected. The motor units are very complex and the fibre density is very high; the legs are usually unaffected (Schwartz et al. 1980). *Cervical spondylosis*, with radiculopathy, produces a syndrome in which the EMG evidence of chronic partial denervation is clearly radicular in distribution. The root lesion may be confirmed by increased F-wave latencies. Axonal neuropathies, such as HMSN Type II, are difficult to distinguish from MND on EMG criteria. The sensory action potentials are usually of reduced amplitude in the legs in neuropathies, but are normal in MND. Hereditary distal spinal muscular atrophy is virtually indistinguishable on electrophysiological criteria from MND, the distinction depending on clinical and especially genetic features (Harding and Thomas 1980).

Three lower motor neuron syndromes with clinical resemblance to motor neuron disease have been described (Pestronk et al. 1990). Each is associated with raised levels of anti-GM1 antibodies in the serum. In one there is a multifocal motor neuropathy with weakness of distal arm muscles, in which there is electrophysiological evidence of multifocal conduction block. In the second there is an asymmetrical clinical syndrome without conduction block, and in the third there is an asymmetrical proximal muscular syndrome without conduction block. In electrophysiological diagnosis the presence of conduction block is a key feature that delineates the presence of a neuropathy, and that indicates that this is a disorder that is distinct from motor neuron disease itself.

References

Berardelli A, Inghilleri M, Formisano R, Accornero N, Manfredi M (1987) Stimulation of motor tracts in motor neuron disease. J Neurol Neurosurg Psychiatry 50:732–737

Bernstein LP, Antel JP (1981) Motor neuron disease: decremental responses to repetitive nerve stimulation. Neurology 29:627–631

Brooke MH, Florence JM, Heller LS et al. (1986) Controlled trial of thyrotropin-releasing hormone in amyotrophic lateral sclerosis. Neurology 36:146–151

Buchthal F, Pirelli L (1953) Analysis of muscle action potentials as a diagnostic aid in neuromuscular disorders. Acta Med Scand 142 (Suppl 226):315–327

Conrad B, Sindermann F, Prochazka VJ (1972) Interval analysis of repetitive denervation potentials of human skeletal muscle. J Neurol Neurosurg Psychiatry 35:834–840

Conradi S, Grimby L, Lundemo G (1982) Pathophysiology of fasciculations studied by electromyography of single motor units. Muscle Nerve 5:202–208

Dalakas MC, Elder G, Hallett M et al. (1986) A long-term follow-up study of patients with post-poliomyelitis neuromuscular symptoms. N Eng J Med 314:959–963

Dayan AD, Graveson GS, Robinson PK (1969) Schwann cell damage in motor neuron disease. Neurology 19:242–246

Denny-Brown D (1953) Clinical problems in neuromuscular physiology. Am J Med 15:368–390

Ekstedt J (1964) Human single muscle fibre action potentials. Acta Physiol Scand 61 (Suppl 226):1–96

Feller TG, Jones RE, Netsky MG (1966) Amyotrophic lateral sclerosis and sensory changes. Va Med 93:328–335

Ghezzi A, Mazzalovo E, Locatelli C, Zibetti A, Zaffaroni M, Montanini R (1989) Multimodality evoked potentials in amyotrophic lateral sclerosis. Acta Neurol Scand 79:353–356

Guiloff RJ, Modarres Sadeghi H, Stalberg E, Rogers H (1988) Short-term stability of single motor unit recordings in motor neuron disease: a macro EMG study. J Neurol Neurosurg Psychiatry 51:671–676

Hansen S, Ballantyne JP (1978) A quantitative electrophysiological study of motor neuron disease. J Neurol Neurosurg Psychiatry 41:773–783

Harding A, Thomas PK (1980) Hereditary distal spinal muscular atrophy. J Neurol Sci 45:337–348

Hausmanowa-Petrusewicz I, Kopec J (1970) Motor nerve conduction velocity in anterior horn lesions. Electromyography 3:227–237

Hayward M, Willison RG (1977) Automatic analysis of the electromyogram in patients with chronic partial denervation. J Neurol Sci 33:415–423

Ingram DA, Swash M (1987) Central motor conduction is abnormal in motor neuron disease. J Neurol Neurosurg Psychiatry 50:159–166

Ingram DA, Davis GR, Schwartz MS, Swash M (1985) The effect of continuous voluntary activation on neuromuscular transmission: an SF EMG study of myasthenia gravis and anterior horn cell disorders. Electroencephalogr Clin Neurophysiol 60:207–213

Ingram DA, Thompson AJ, Swash M (1988) Central motor conduction in multiple sclerosis: evaluation of abnormalities revealed by transcutaneous magnetic stimulation of the brain. J Neurol Neurosurg Psychiatry 51:487–494

Jamal GA, Weir AI, Hansen S, Ballantyne JP (1985) Sensory involvement in motor neuron disease: further evidence from automated thermal threshold determinations. J Neurol Neurosurg Psychiatry 48:906–910

Kawamura Y, Dyck PJ, Shimura M, Okazaki H, Tateishi J, Doi H (1981) Morphometric comparison of the vulnerability of peripheral motor and sensory neurons in amyotrophic lateral sclerosis. J Neuropathol Exp Neurol 40:667–675

Lambert EH (1969) Electromyography in amyotrophic lateral sclerosis. In: Norris FH Jr, Kurland LT (eds) Motor neuron diseases: research on amyotrophic lateral sclerosis and related disorders. Grune and Stratton, New York, pp 135–153

Massey JM, Nandedkar SD, Sanders DB (1989) EMG studies in patients with amyotrophic lateral sclerosis receiving thyrotropin-releasing hormone. Electroencephalogr Clin Neurophysiol 73:19–20

Matheson JK, Harrington HJ, Hallett M (1986) Abnormality of multi-modality evoked potentials in amyotrophic lateral sclerosis. Arch Neurol 43:338–340

McComas AJ (1977) Neuromuscular function and disorders. Butterworth, London

Mills KR, Murray NMF (1985) Corticospinal tract conduction time in multiple sclerosis. Ann Neurol 18:601–605

Mulder DW, Lambert EH, Eaton LM (1959) Myasthenic syndrome in patients with amyotrophic lateral sclerosis. Neurology 9:627–631

Mulder DW, Bushek W, Spring E, Karnes J, Dyck PJ (1983) Motor neuron disease (ALS): evaluation of detection thresholds of cutaneous sensation. Neurology 33:1625–1627

Nakanishi T, Tamaki M, Arasaki K (1989) Maximal and minimal motor nerve conduction velocities in amyotrophic lateral sclerosis. Neurology 39:580–583

Norris FH Jr (1975) Adult spinal motor neuron disease. Progressive muscular atrophy (Aran's disease) in relation to amyotrophic lateral sclerosis. In: Vinken PJ, Bruyn GW (eds) Handbook of clinical neurology, vol 22. North-Holland, Amsterdam, pp 1–56

Pestronk A, Chaudhury V, Feldman EL et al. (1990) Lower motor neuron syndrome defined by patterns of weakness, nerve conduction, abnormalities and high fibers of anti-glycolipid antibodies. Ann Neurol 27:216–236

Petajan J (1985) F waves in neurogenic atrophy. Muscle Nerve 8:690–696

Pieroglou-Harmoussi S, Fawcett PRW, Howel D, Barwick DD (1987) F response frequency in motor neuron disease and cervical spondylosis. J Neurol Neurosurg Psychiatry 50:593–599

Radtke RA, Erwin A, Erwin GW (1986) Abnormal sensory evoked potentials in amyotrophic lateral sclerosis. Neurology 36:796–801

Reed DM, Kurland LT (1963) Muscle fasciculations in a healthy population. Arch Neurol 9:353–367

Roth G (1984) Fasciculations and their F response. J Neurol Sci 63:299–306

Schwartz MS, Stalberg E, Schiller HH, Thiele B (1976) The reinnervated motor unit in man. J Neurol Sci 27:303–312

Schwartz MS, Stalberg E, Swash M (1980) Pattern of segmental motor involvement in syringomyelia: a single fibre EMG study. J Neurol Neurosurg Psychiatry 43:150–155

Schwartz MS, Swash M (1982) Pattern of involvement in the cervical segments in the early stage of motor neuron disease – a single fibre EMG study. Acta Neurol Scand 65:424–431

Sever J (1986) A long-term follow-up study of patients with post-poliomyelitis neuromuscular symptoms. N Eng J Med 314:959–963

Shahani B, Davies-Jones GAB, Russell WR (1971) Motor neuron disease. Further disease evidence for an abnormality of nerve metabolism. J Neurol Neurosurg Psychiatry 34:185–191

Stalberg E (1982) Electrophysiological studies of reinnervation in ALS. In: Rowland LP (ed) Human motor neuron diseases. Raven Press, New York, pp 47–59

Stalberg E, Sanders DB (1984) The motor unit in ALS studies with different neurophysiological techniques. In: Rose FC (ed) Research progress in motor neuron diseases. Pitman, London, pp 105–122

Stalberg E, Schwartz MS, Trontelj JV (1975) Single fibre electromyography in various processes affecting the anterior horn cell. J Neurol Sci 24:403–415

Stalberg E, Trontelj JV (1982) Abnormal discharges generated within the motor unit as observed with single fibre EMG. In: Culp WJ, Ochoa J (eds) Abnormal nerves and muscles as impulse generations. Oxford University Press, Oxford, pp 443–474

Swash M (1980) Vulnerability of lower brachial myotomes in motor neuron disease. J Neurol Sci 47:59–68

Swash M, Fox KP (1974) Pathology of the muscle spindle – effect of denervation. J Neurol Sci 22:1–24

Swash M, Scholtz CL, Vowles G, Ingram DA (1988) Selective and asymmetric vulnerability of corticospinal and spinocerebellar tracts in motor neuron disease. J Neurol Neurosurg Psychiatry 51:785–789

Swash M, Schwartz MS (1982) A longitudinal study of changes in motor units in motor neuron disease. J Neurol Sci 56:185–197

Swash M, Schwartz MS (1984) Staging motor neuron disease: single fibre EMG studies of asymmetry, progression and compensatory reinnervation. In: Rose FC (ed) Research progress in motor neuron disease. Pitman, London, pp 123–140

Swash M, Schwartz MS (1988) Neuromuscular diseases: a practical approach to diagnosis and management, 2nd edn. Springer, London, Berlin, Heidelberg, p 456

Swash M, Schwartz MS (1992) What do we really know about ALS? J Neurol Sci 113:4–16

Trojaborg W, Buchthal F (1965) Malignant and benign fasciculations. Acta Neurol Scand 41(Suppl 13):251–254

Trontelj JV, Stalberg E (1983) Bizarre repetitive discharges recorded with single fibre EMG. J Neurol Neurosurg Psychiatry 46:305–309

Weichers DO, Hubble SL (1981) Late changes in the motor unit after acute poliomyelitis. Muscle Nerve 4:524–528

Wettstein A (1979) The origin of fasciculation in motor neuron disease. Ann Neurol 5:295

Wohlfart G (1951) Collateral regeneration from residual motor nerve fibres in amyotrophic lateral sclerosis. Neurology 7:124–134

Wohlfart G (1958) Collateral reinnervation in partially denervated muscle. Neurology 8:175–180

15 Clinical Pharmacology of Motor Neurons

R.J. Guiloff

Advances in neurobiology, neuropharmacology and in the physiology of the human motor system in recent years have been such that it is now clear that a new field of clinical pharmacology of motor neurons is emerging. This chapter brings together information on this subject, that is dispersed in the literature, in order to identify the main themes which are relevant to motor neuron disease.

Introduction

The development of substances that are active in human upper and lower motor neuron systems is of importance in attempting to treat patients with motor neuron disease (MND). That development requires knowledge both of the disease and of human motor neurons. Unfortunately, the aetiology of the condition is not known and the mechanisms of motor neuron death are not yet well understood.

Drugs that have no serious side effects may be considered useful in the treatment of MND if one or more of the following effects can be demonstrated: (1) prolonged symptomatic improvement; (2) reduction in the rate of deterioration in some or all affected functions; (3) prolongation of survival time; (4) reduction in mortality. These effects could be achieved by a drug action on the, so far unknown, cause of the disease, on the mechanisms leading to motor neuron loss or dysfunction, or on the mechanisms mediating the symptomatic effects of motoneuron loss or dysfunction. Anatomically, the actions of such drugs may be exerted on cortical, bulbar or spinal motor neurons, or on their connections (Modarres-Sadeghi et al. 1988). These connections include sensory afferents and interneurons. Further, such actions may be exerted either directly on any of these neuronal populations, or indirectly through, for example, an effect on supporting structures such as glia or blood vessels.

Unless a drug that arrests completely disease progression is found it will be necessary to consider the points mentioned in the preceding paragraph when developing strategies to treat the disease. If a drug does arrest the disease, it does not follow that symptomatic improvement will be seen for functions already affected or lost. Conversely, symptomatic improvements may be of little benefit if they are short-lived. There is no reason to assume that a given drug will necessarily have a similar action on all sets of motor neurons (Guiloff et al. 1987a). It is known, for example, that the anatomical location and synaptic connectivity of anterior horn cells in lamina IX of Rexed is different for proximal and distal muscles (Kuypers 1973). Drugs with an action on spinal motor neurons might help disability but would probably not influence survival time or mortality. Drugs with a more prominent action on bulbar and high cervical cord motor neurons might increase survival time but somatic disability would remain unchanged.

There are several important problems facing researchers in this area:

Delivery of Drugs to Motor Neurons

The major obstacle to drug delivery is the blood–brain barrier. For example, most neuropeptides are lipid-insoluble and highly polar (Ermisch et al. 1985). Only up to 0.20% of thyrotrophin-releasing hormone (TRH) was detected in mouse brain after 5–60 min of various parenteral administrations (Mitsuma and Nogimori 1983). Enkephalins, arginine vasopressin (AVP) and substance P also penetrate poorly (Cornford et al. 1978; Mens et al. 1983; Ermisch et al. 1985). In humans, only 2.7% of a bolus of intrathecal TRH passed to the blood (Munsat et al. 1987) and only 1.4% of the blood concentration of the TRH analogue RX77368 was found in CSF after a 2-hour intravenous infusion of 0.3 mg/kg (Guiloff 1989).

Assessment of Drug Effects on Motor Neurons

There is currently no clinically applicable method that allows *direct* measurement of the action of a drug on human motor neurons. Clinical measurements of force, bulbar and respiratory function, or activities of daily living are all indirect ways of indexing complex motor pathways, including motor neurons. Laboratory measurements of motor unit function (such as electromyography), or spasticity also measure motor pathway function or dysfunction.

The Clinical Variability of MND

This poses serious problems when studying a given drug effect in a population of patients. Weakness and wasting are patchy, with variable degree and distribution, from patient to patient. Spasticity, bulbar involvement and pyramidal signs are also quite variable. The course of the disease may be as short as 6 months or as long as 10 or more years. In order to find a suitable control group of MND patients against which to compare a certain drug effect, careful matching is required and may be quite difficult. Alternatively, very large numbers of patients are required so that the standard deviation of the parameters measured becomes small enough not to obscure a useful, but not major, drug effect (Guiloff et al. 1990).

Lack of a Marker for "Sick" Motor Neurons

Both clinical and laboratory methods have traditionally addressed the degree of involvement of muscles by the amount of motor unit loss (wasting, weakness, interference pattern, motor unit counts) or the degree of chronic partial reinnervation (motor unit territory, SFEMG fibre density, motor unit potential parameters such as duration, amplitude, area or number of phases). A more direct approach, with obvious benefit for a clinical pharmacology of motor neurons, would be the characterization and quantification of sick motor units within a muscle. No information on this point exists. Motor units with small electrical signals for the level of recruitment, and with reduced twitch force (compared to normals) have been suggested as representing sick motor units. Loss of terminal axonal branches, previously supporting a larger territory, might account for these physiological features (Dengler et al 1990).

Lack of a Satisfactory Experimental Model of MND

An experimental model is needed to test the effect of drugs (Sillevis Smitt and Delong. 1989). Differences in the anatomy and physiology of motor systems in different species preclude direct extrapolation of drug effects seen in animals to humans.

Tophic Factors and Motor Neurons

The possibility of influencing motor neuron survival is relevant to developing a clinical pharmacology of motor neurons which can be applied to possible treatment strategies in MND. No clinically useful drug with a proven action on motor neuron survival has yet been found.

The influence of target muscle on motor neuron in normal development and growth has been widely studied (Oppenheim 1989). Although a variety of muscle trophic activities on motor neurons have been described in vitro and, recently, in vivo models (McManaman et al. 1990) during development, the relevance of this work to adult motor neurons and MND seems questionable. For example, there is no evidence of motor neuronal loss in patients with severe chronic muscle disease. Further, axotomy studies in mammals suggest that muscle may not be critical for motor neuron survival during adult life (Crews and Wigston 1990). Since both upper and lower motor neurons are affected in MND, any growth factor produced in the periphery would have to be shown to have an action beyond those neurons directly linked to the target tissue, or be produced both in the periphery and by cells within the CNS, unless trans-synaptic degeneration can be unequivocally demonstrated.

There is in vitro evidence that neurotransmitters may regulate neuronal survival, outgrowth and plasticity (Mattson 1988). This regulation may involve second messenger systems; both intracellular calcium (voltage activated calcium channels and receptor operated calcium channels) and other second messengers (such as cAMP) may be activated by neurotransmitters (Lipton and Kater 1989). In the developing nervous systems impulse activity appears to play a role in the formation of connections. Experiments that use tetrodotoxin (a blocker of sodium-dependent calcium channels) result in changes in axonal organisation and synaptic connections in the peripheral and central nervous system (Frank 1987). It may be that the pattern of electrical activity is critical for the formation of normal synaptic patterns. In neuronal cultures low frequency stimulation produces a reversible cessation of neurite extension (Cohan and Kater 1986).

There have been several criticisms of the use of dissociated neuronal culture systems for the in vitro study of the action of putative drugs or trophic factors active on motor neurons. An important criticism is that pure cultures of motor neurons have not always been used. It has been difficult to ensure that any of the effects described, such as increase in survival, neurite extension or in the production of choline acetyltransferase are the result of an action on motor neurons or on other neuronal sets within the ventral horn (Schmidt-Achert et al. 1984; Jehanli et al. 1987). It should be remembered, for example, that in the ventral horn of the cat there are about seven interneurons for each motor neuron and, if the intermediate grey matter and the base of the dorsal horn are included, the ratio is 13 : 1 (Aitken and Bridger 1961). Further, neurons in culture die relatively quickly (at the most they survive a few weeks) and a number of factors added to the culture medium may promote neuronal survival, including nutritive components of a normal tissue culture medium or agents that can destroy toxic substances accumulating in the medium (Barde 1988). The technique of organotypic roller tube culture has been recently applied to the spinal cord of neonatal rats to study the effect of toxins on spinal cord neurons (Delfs et al. 1989). This technique has the advantage of preserving anatomical relationships on slices that are just one to three cell layers thick, thus facilitating morphological, histochemical and biochemical studies in regions of the ventral horn known to contain motor neurons.

Many factors appear to promote survival of cultured motor neurons. Ciliary neurotrophic factor (CNTF), basic fibroblast growth factor and leukaemia-inhibiting factor are particularly potent. Insulin-like growth factor, transforming growth factor B and other fibroblast growth factors have less activity. CNTF is expressed exclusively from the 4th post-natal day in rat Schwann cells and some subpopulations of glial cells and prevents motor neuron degeneration following experimental axotomy in early post-natal life. This suggests that CNTF has a role more as a lesion factor than as a target-derived neurotrophin. There is some evidence that CNTF may prolong survival, improve motor function and reduce loss of motor neurons in mouse with mutant progressive motor neuropathy, an autosomal recessive condition with motor neuron degeneration (Sendtner 1992). No abnormalities have been demonstrated to date in CNTF levels in MND, but therapeutic trials with CNTF are in progress.

Several experimental studies have indicated a role for androgens in the regulation of the number and size of motor neurons innervating perineal muscles during development. Such regulation appears to be exerted on normally occurring neuronal cell death during development, and is relevant to the differences in number of motor neurons in the spinal nucleus of the bulbocavernosus and in the dorsolateral nucleus in the lumbar spinal cord of male and female rats (Breedlove 1986; Sengelaub and Arnold 1989; Sengelaub et al. 1989a,b). Again, the relevance of motor neuron death during development to the processes leading to motor neuron death in MND is unknown. Castration in adult rats, producing a decrease in androgen levels, leads to structural changes in the motor neurons of such nuclei including reductions in dendritic length and in the size of the soma (Kurz et al. 1986). The expression of calcitonin gene-related peptide in these motor neurons is also reduced (Popper and Micevych 1989). These changes are reversed by androgen replacement. In adult female rats, neuronal loss in the brain stem nuclei following transection of the hypoglossal and facial nerves can be reduced by administration of testosterone, progesterone or by lactation (Yu 1989). Such works hold promise that, in the future, manipulation of defined humoral factors may result in both functional and anatomical changes in other motor neuronal populations.

Interestingly, X-linked spinal and bulbar muscular atrophy, an adult onset form of lower motor neuron disease with gynaecomastia and reduced fertility, suggesting androgen insensitivity, is caused by a mutation in the androgen receptor gene (La Spada et al. 1991). Androgens may also be important for adult lower motor neuron survival, but their role in MND has yet to be determined. The trophic effects associated with the action of thyrotrophin-releasing hormone (TRH) are considered below.

Neurotransmitters, Neuropeptides and Receptors in the Ventral Horn

A great deal of data, mostly from animal work, on neurotransmitters and neuropeptides has accumulated in recent years. The salient and broadly agreed views are given below; detailed discussions can be found in the reviews by Krnjevic (1981) and Schoenen (1988).

Spinal motor neurons release acetylcholine, which has an excitatory action at the neuromuscular junction, onto nicotinic receptors in Renshaw cells (recurrent inhibition), and, in the cat, onto other motor neurons via short direct excitatory connections (Cullheim et al. 1977; Gogan et al. 1977). Indirect evidence from F-wave studies suggests that the latter may also exist in man (Guiloff and Modarres-Sadeghi 1991). Some spinal interneurons may exert excitatory effects on motor neurons by releasing aspartate or acetylcholine, the latter acting on muscarinic receptors. Corticofugal projections to motor neurons probably release glutamate. Primary sensory afferents may release glutamate at their synaptic site on motor neurons but small diameter nociceptive fibres probably release substance P onto dorsal horn cells. Renshaw cells and Ia inhibitory motor neurons release glycine at their synapse with motor neurons, although taurine may play a minor role. Other interneurons exert inhibitory effects on motoneurons mediated by gamma aminobutyric acid (GABA). GABA is released by neurons in the substantia gelatinosa onto primary sensory afferents via axo-axonal synapses, causing presynaptic inhibition (Krnjevic 1981; Schoenen 1988). The projections to motor neurons from the raphe nuclei release TRH, 5-hydroxytryptamine (5-HT) and substance P but their action may be mediated by interneurons, with net excitatory effects (White et al. 1989). Projections from the locus coeruleus and brain stem to motor neurons may release norepinephrine (Westlund et al. 1984). Other substances found in the human ventral horn include enkephalins, cholecystokinin, oxytocin, calcitonin gene-related peptide vasoactive intraspinal peptide (VIP), neuropeptide Y, neurokinin and somatostatin (Gibson et al. 1988; Schoenen 1988).

Advances in molecular biology, autoradiography, immunocytochemistry and single ion channel recording techniques, have resulted in much information about the pharmacology of neurotransmitter receptors in the central nervous system. This has not yet been matched by a knowledge of the relevance of peptides and neurotransmitters to the pathogenesis of the motor neuron diseases. Advances in understanding of the molecular mechanisms underlying the function at receptor sites are likely to stimulate the development of drugs acting on specific sites which may then be applied to human disease. The receptors most relevant to the topics considered later in this chapter are briefly listed below:

Excitatory Amino Acid Receptors

L-Glutamate and L-aspartate are probably the major excitatory neurotransmitters in the CNS. Their actions are mediated by at least three types of receptors, named after the agonists that activate them: N-methyl-D-aspartate (NMDA) and kainate and quisqualate (non-NMDA receptors). Both types of receptors are widely distributed in the CNS (Cotman et al. 1987), including the spinal cord, and can be blocked by a number of specific competitive and non-competitive antagonists (Watkins and Olverman 1987). The NMDA receptors have characteristically high Ca^{2+} permeability, are blocked by Mg^{2+} in a voltage-dependent way and have an allosteric modulatory site activated by glycine that potentiates the effect of NMDA or glutamate (Ascher and Nowak 1987). The action of glycine may relate to an increase in NMDA receptor affinity; glycine may also depress receptor desensitization (Thomson 1989; Thomson et al. 1989). The NMDA receptor complex contains other modulatory sites which can also be targeted for pharmacological action (McGeer 1990). Non-NMDA receptors mediate fast synaptic potentials whilst those mediated by NMDA receptors have a longer duration of activity (MacDermott and Dale 1987).

GABA Receptors

GABA, a major inhibitory neurotransmitter in the spinal cord, binds to $GABA_A$ and $GABA_B$ receptors. GABA receptors have sequence similarities with glycine and nicotinic cholinergic receptors (Schofield et al. 1987; Grenningloh et al. 1987). $GABA_A$ is composed of alpha, beta, gamma and delta subunits. Combinations of these subunits may give rise to many forms of GABA receptors in the CNS. The binding site for GABA is located in the beta subunit. $GABA_A$ receptors contain chloride channels and binding sites for benzodiazepines and barbiturates. They are inhibitory because by increasing intracellular Cl^- they stabilise the neuron resting potential during the activation produced by excitatory receptors (Bormann 1988; Macdonald and Twyman 1990; Seeburg et al. 1990). $GABA_B$ receptors do not contain ion channels but regulate calcium and potassium channels via pertussis toxin-sensitive G proteins. The inhibition of Ca^{2+} currents and the inhibitory outwardly directed K^+ currents produced by $GABA_B$ receptor activation suggest that inhibition of transmitter release at presynaptic terminals may be one mechanism by which they control synaptic transmission (Bormann 1988). $GABA_B$ receptors are selectively activated by levobaclofen (see below).

Glycine Receptors

Glycine is another major inhibitory neurotransmitter in the spinal cord (van den Pol and Gorcs 1988). Glycine receptors have been cloned (Grenningloh et al. 1987) and contain three polypeptide subunits. One of them binds the glycine receptor antagonist strychnine and another is localized to the cytoplasmic side of the post-synaptic membrane (Betz and Becker 1988). There is evidence that glycine receptors are heterogeneous. According to the type of mRNA, or the affinity for strychnine, two types have been shown, in the neonatal and in the adult rat spinal cord respectively; topographical differences between adult spinal cord and brain also exist (Akagi and Miledi 1988; Lewis et al. 1989). The strychnine-sensitive glycine receptor, similarly to the GABA receptor complex, has a chloride channel. However, glycine can also act at a strychnine-insensitive allosteric site of the NMDA receptor potentiating the excitatory effects of glutamate (Johnson and Ascher 1987; Thomson 1989; Thomson et al. 1989; Mayer et al. 1989).

5-HT receptors

5-Hydroxytryptamine has been shown to have both excitatory and inhibitory actions in the CNS. The main categories of receptor subtypes include 5-HT_1, 5-HT_3. 5-HT_1 receptors are further subdivided in four types (A, B, C and D). 5-HT_1B sites are not present in man. 5-HT_1C receptors have been cloned and sequenced. Specific agonists and antagonists exist for these subtypes which also have different effects on ion conductances. At least three other subtypes exist but are presently poorly defined (5-HT I_1, I_m, I_k). 5-HT_1A receptors are found in the hypoglossi nuclei and spinal cord, 5-HT_2 in facial motoneurons and 5-HT I_1 in spinal motoneurons (Peroutka 1988; Bobker and Williams 1990).

Motor Neuron Disease

The content of TRH is reduced when expressed as a fraction of wet weight but not when expressed relative to protein content (Jackson et al. 1986). Reduced glutamate in the cerebral and cerebellar cortex and in the spinal cord, and reduced aspartate in the spinal cord of MND patients have also been reported (Plaitakis et al. 1988a; Rothstein et al. 1990). Gibson et al. (1988) found that, in the ventral horns of MND patients, fibres immunoreactive for neurokinin, enkephalin and TRH, normally associated with motor neurons, were absent and calcitonin gene-related peptide immunoreactive motor neurons were not seen. There is also a reduction in fibres containing substance P.

In the ventral horn of MND patients reduction in acetylcholinesterase, choline acetyltransferase and in muscarinic, benzodiazepine, glycinergic, TRH and substance P receptors have been described (Whitehouse et al. 1983; Manaker et al. 1988; Dietl et al. 1989). However, there is no change in the concentration of β-adrenergic receptors or in the norepinephrine sites. 5-HT_1A receptor densities are increased in lamina IX. The preservation of some receptors in MND may indicate that not all the changes seen are non-specific findings associated with motor neuron loss (Manaker et al. 1988).

Excitotoxins and Motor Neurons

Since the description of the neurotoxic action of systemically administered glutamate, in areas of the brain without blood–brain barrier in newborn mice, rats and monkeys (Olney and Sharpe 1969; Olney 1971), the concept that excitatory amino acids may play a role in the pathogenesis of a number of neurological diseases has been intensively investigated. The neuronal loss induced by glutamate may be mainly mediated by the NMDA receptor but excitatory amino acids activate at least two other sites; the kainate and the quisqualate receptors. Whilst brief NMDA receptor activation can induce rapid damage to cortical neurons in culture, the kainate and quisqualate receptors require several hours of exposure to non-NMDA receptor agonists to mediate such neurotoxic effects (Choi et al. 1989; Koh et al. 1990). NMDA-mediated neuronal death includes neuronal swelling induced by depolarization, influx of chloride and water and a slower intracellular accumulation of Ca^{2+} leading to necrosis (Rothman and Olney 1987).

Glutamate

The notion that excitatory amino acids may be involved in the pathogenesis of ALS is speculative. However, it has led already to a therapeutic trial with branched chain amino acids (Plaitakis et al. 1988b). Based on the findings in MND of elevated fasting plasma levels of glutamate, abnormal elevation in plasma glutamate with an oral loading test with monosodium glutamate, and decreased spinal cord levels of glutamate and aspartate, Plaitakis has suggested an abnormal distribution of glutamate between the intracellular and extracellular pools in motor neuron disease. The increased amount of glutamate at the synaptic (extracellular) site would result in motor neuron damage. A number of possibilities are offered for the abnormal distribution of glutamate including a defect in the high affinity uptake system, or in the metabolism of glutamate in the synapses by glial cells, or an increase in the release, or leakage, of glutamate from the nerve terminal. In order to account for the selective involvement of motor neurons, and in view of the reduced content of glutamate in nerve tissue, it is proposed that glutamatergic excitatory transmission is potentiated by glycine at the NMDA receptor site, either via glycinergic interneurons or direct glycinergic innervation of motoneurons (Plaitakis and Caroscio 1987; Plaitakis et al. 1988a; Plaitakis 1990a,b). High glycine levels in CSF (de Belleroche et al. 1984) and evidence of defective removal of glycine from plasma (de Belleroche et al. 1990) have been described in MND patients.

According to Plaitakis (1990a,b) the presence of high levels of glycine with glutamatergic potentiation may prevent the receptor desensitization which normally protects post-synaptic neurons against glutamate neurotoxicity. However, others report normal concentrations of glutamate in fasting plasma, and of glutamate aspartate and glycine in the CSF, of most patients with MND (Perry et al. 1990).

A number of pharmacological approaches to the treatment of MND follow from Plaitakis's hypothesis. One is to increase the activity of glutamate dehydrogenase (GDH), a predominantly glial enzyme that metabolizes glutamate to alpha-ketoglutarate, with the aim of reducing the hypothesized excess of glutamate at the synaptic site. To this end branched chain amino acids (1-leucine, 1-isoleucine and 1-valine) were administered for 1 year in a controlled pilot trial to MND patients. Seven treated and five placebo patients completed the trial. A significant difference in spinal scores ($P=0.02$) was found at 6 months but not at 3, 9 or 12 months and there was no difference in bulbar scores (Plaitakis et al. 1988b,c). The small number of patients and the multiple statistical comparisons made make the results inconclusive. The results of a large European multi-centre trial of branched chain amino acids therapy in MND are awaited. Further possible approaches include the use of NMDA receptor antagonists acting specifically on glutamatergic excitation or on the strychnine-insensitive allosteric site activated by glycine (Plaitakis 1990b). Drugs that inhibit the release of glutamic acid presynaptically are also being tested (Bensimon et al. 1994).

β-*N-Oxalylamino-L-Alanine (L-BOAA)*

Lathyrism, a condition characterised by a spastic paraparesis without low motor neuron features, is known to be caused by consumption of *Lathyrus sativus* (chickling or grass pea). This pea contains a non-protein amino acid, L-BOAA, which, when administered orally to macaques, produces a motor syndrome similar to that seen at the early stage of the human disease (Spencer et al. 1986; Ludolph et al. 1987). L-BOAA behaves like an excitotoxic amino acid. It induces death of cultured mouse cortical neurons and appears to activate non-NMDA glutamate receptors (Weiss et al. 1989; MacDonald and Morris 1984; Bridges et al. 1989). Receptor binding studies in mouse synaptic membranes, organolytic cultures of mouse cortex, and electrophysiological work on hippocampal pyramidal neurons suggest that BOAA has the highest affinity for the quisqualate receptor, can act as a low affinity agonist at the kainate receptor, produces post-synaptic vacuolation in the superficial layers of cortical explants and an acute neuronal degeneration in deeper layers (Allen et al. 1990).

β-*N-Methylamino-L-alanine (L-BMAA)*

This non-protein amino acid is found in small amounts in cycads, which have been suspected of playing a role in the aetiology of the Western Pacific amyotrophic lateral sclerosis/parkinsonism-dementia (ALS/PD) (Spencer et al. 1987). That L-BMAA itself is the cause of this disease complex seems unlikely (Duncan et al. 1990; Spencer 1990). L-BMAA appears to be an atypical excitatory amino acid, with neurotoxic and excitatory effects on cortical neuronal cultures, when applied at high doses, which depend on the simultaneous presence of bicarbonate ions. Its actions may be exerted via non-NMDA receptors (Allen et al. 1990; Weiss and Choi 1990).

The work with L-BOAA and L-BMAA indicates that non-NMDA receptors may also be capable of mediating motor neuron damage. Experiments with intrathecal injection of kainic acid in rats, resulting in neuronal degeneration changes which were more marked in the ventral horn, support this concept (Hugon et al. 1989). A neurotoxic effect of MND CSF on rat embryo cortical neuronal cultures was blocked by an AMPA/kainate receptor antagonist but not by the NMDA receptor antagonist (Couratier et al. 1993). If such mechanisms of motor neuron damage are found to operate in MND then pharmacological intervention with specific non-NMDA receptor antagonists, or with agents acting pre-synaptically or in the post-synaptic neuron, may become feasible.

Assessment Tools for Drug Effects on Human Motor Neurons

Clinical Measurements

There has been an increasing interest in recent years in developing clinical measurements that can reflect accurately the effect of drugs in therapeutic trials in MND. A detailed review of this area is beyond the scope of this chapter and only the salient points will be mentioned.

Use of Global Scales

The first widely used scale for MND was described by Norris et al. (1974). It is an ordinal scale of 34 items, combining self-reported and observed activities and the neurological examination. It has been criticised for being insensitive to small changes, for excluding quantification of lung function and for not giving an overall balanced assessment of the items included. In addition its reliability is not well known. An interval scale derived from the Tufts quantitative neurological examination has been reported (Andres et al. 1986). This scale contains items that assess pulmonary function, timed motor activities, maximal isometric strength in upper and lower limbs and assessment of bulbar function with timed measurements of sound repetition. Another recently proposed scale, which included assessments of respiration, strength and function in upper and lower limbs, swallowing and speech, was found to increase roughly linearly with progression of the disease over 1 year (Appel et al. 1987). A severity scale comprising ordinal staging of functional assessments of speech, swallowing, lower limbs (walking) and upper limbs (dressing and hygiene) has been reported to be reliable and can be combined with vital capacity measurements (Hillel et al. 1989).

Assessment of Muscle Force

Manual strength testing applying the Medical Research Council (1976) five-point scale, is a rapid and simple way of assessing strength in a large number of muscles. A limited number of muscles may be selected in upper and lower limbs and a weakness score may be obtained for upper or lower limbs, or for both, by adding the scores of all muscles measured (Plaitakis et al. 1988b). The main problem of the MRC scale, however, is uncertainty in Grade 4, which covers a wide range of weakness; in addition it is essentially subjective. For this reason methods to measure maximal isometric voluntary contraction, using manual and fixed dynamometry, have been developed (Andres et al. 1987; Guiloff and Eckland 1987; van der Ploeg et al. 1991; Goonetilleke et al. 1994). Torque at various velocities has also been measured using isokinetic dynamometry (Brooks et al. 1989). These force assessments cannot distinguish between the relative contributions of upper and lower motor neuron involvement to the weakness documented.

Assessment of Respiratory Function

Forced vital capacity, volume expired in the first second (FEV_1), peak flow, maximal voluntary ventilation, and maximal inspiratory and expiratory pressures have all been used (Andres et al. 1986; Braun 1987; Guiloff et al. 1987a). The last three may be more sensitive in neuromuscular disorders (Braun 1987; Griggs et al. 1981). These parameters also measure the net result of upper motor neuron drive and segmental, high cervical cord function. Weakness in facial muscles in MND is common and may be an important source of error since significant amounts of air may leak through the mouthpiece.

Assessment of Bulbar Function

Since in most patients with MND death is related to bulbar and higher cervical cord dysfunction, with respiratory failure (Braun 1987), prolongation of life will be critically dependent on a drug effect on these functions. This had led to the development of timed tests of tongue, jaw and palatal movement, of swallowing and of word repetition (Modarres-Sadeghi et al. 1988). Learning effects are seen for many of these timed tests and need to be considered when establishing baselines (Guiloff et al. 1990). Methods to test maximal isometric contraction in lips, tongue and jaw muscles have also been described (DePaul et al. 1988).

Functional Assessments and Activities of Daily Living

A large number of tests such as time walking, handling a variety of objects, measurements of leg function, complex activities such as dressing or eating can be standardised and used as measures of outcome in therapeutic interventions. Excellent accounts and discussions can be found in two recent monographs (Potvin and Tourtellotte 1985; Munsat 1989).

Assessment of Spasticity

The Ashworth scale (Ashworth, 1964), an ordinal 5 point clinical scale ranked between no increase in tone (scale = 1) and a rigid limb in flexion or extension (scale = 5), has been widely used and appears to be reliable and reproducible (Bohannon and Smith 1987; Lee et al. 1989). Laboratory-based methods to quantify spasticity that can be applied to drug evaluations may be difficult to adapt to a clinical setting.

A number of factors operate in the resistance detected clinically as spasticity. These include the mechanical properties of the joint, the elasticity of the muscle, the viscosity of muscle and connective tissues, which is velocity dependent, and the gravitational forces acting on the segment of the limb examined. To this is added the abnormal reflex response to stretch seen in spasticity, which is also velocity dependent. Adequate evaluation of these various factors requires systems that incorporate mechanical and electromyographic recordings (Meyer and Adorjani 1983; Knutsson 1983, 1985; Ashby et al. 1987).

Measurements of integrated EMG activity and its relation to the angular velocity of stretch and to the threshold velocity triggering such activity have been used to document the effects of "antispastic drugs" such as diazepam or baclofen (Jones et al. 1970; Ashby et al. 1972). However, integrated EMG activity may vary widely, even with relatively similar resistance to passive movement as assessed by the torque measured by an isokinetic dynamometer (Knutsson 1985).

Electrophysiology

It has been proposed that electrophysiological tests in humans need to meet a number of conditions before they can be used to assess reflex pathways. They should comply with ethical requirements, they should be easy and relatively quick to perform, the need for active collaboration from the subject should be minimized, and measurements should be stable, sensitive and specific. In fact, many of the procedures described fulfil these requirements only partially (Delwaide et al. 1983). A number of tests can be used to study the effect of drugs on the human anterior horn cells (Delwaide et al. 1983; Eisen 1987; Delwaide 1988).

Alpha-motor neuron excitability may be measured by the ratio between the maximal amplitudes of the H-reflex and the M response (H_{max}/M_{max}) (Angel and Hoffmann 1963). This ratio is a measure of the proportion of motor neurons that can be activated via the monosynaptic H-reflex and is increased, for example, in spastic muscles. Certain F-wave parameters have also been used as an index of motor neuron excitability. The F-wave causes a motoneuronal response to antidromic stimulation but complex segmental events, including a reflex component, need to be considered (Guiloff and Modarres-Sadeghi 1991). Microneurography has been used to study the gamma system (Prochazka and Hulliger 1983). Presynaptic inhibition, acting upon the terminals of Ia afferent fibres, can be assessed by studying the effect on the H_{max} of a muscle of vibration applied to its tendon (Iles and Roberts 1986). Reciprocal inhibition, mediated by interneurons receiving Ia afferents and projecting to the motor neurons of the antagonist muscle may be assessed by conditioning the H-reflex of a given muscle to stimuli delivered to the Ia fibres of the antagonist muscle (Tanaka 1983). Techniques involving double collision and measurements of H amplitude have been used to test recurrent inhibition by Renshaw cells (Pierrot-Deseilligny et al. 1976, 1983; Mazzocchio and Rossi 1989). It is also possible to test Ib or autogenic inhibition, i.e. inhibition mediated by afferences from the Golgi receptors of the same muscle (Delwaide 1988). The reflex response to electrical stimulation of exteroceptive (cutaneous) fibres has been used as a measure of the interneurons mediating the flexor reflex (Young 1973). Excitability curves of the monosynaptic H-reflex and tonic voluntary EMG activity conditioned by stimulation of a sensory nerve can also be used to test interneurons involved in the flexion reflex (Meinck et al. 1983).

A variety of electromyographic techniques have been used to quantify chronic partial reinnervation, including concentric needle EMG, single fibre EMG and macro EMG. The latter has also been used in acute studies on the effect of a TRH analogue in MND (Guiloff et al. 1987b, 1988). Methods for estimating the number of motor units in a muscle have been described using (a) surface EMG recordings and graded stimulation of the appropriate motor nerve (McComas et al. 1971), (b) concentric needle recording and a spike-triggered averaging technique (Brown et al. 1988; Strong et al. 1988) and (c) macro EMG (de Koning et al. 1988).

Excluding the neuromuscular junction, there are two main examples in which drugs with clear actions in upper and/or lower motor neuron systems have been used clinically. They are the experimental use of TRH and analogues in motor neuron disease and the use of drugs to treat spasticity. In both these examples the relevant animal and clinical data, which exemplify well the development of a clinical pharmacology of motor neurons as a field of science are reviewed below.

TRH and TRH Analogues

Thyrotropin-releasing hormone (TRH) is a tripeptide (pyroglutamyl-histidyl-proline amide) which stimulates the release of thyrotropin (TSH) and prolactin in the anterior pituitary. Around two thirds of total TRH in the CNS are present in extrahypothalamic sites. It has a variety of actions in the CNS, most of which are not mediated by the pituitary–thyroid axis. TRH is present in nerve terminals around motor neurons in the spinal cord and brain stem but is also seen in Rexed's laminae II, III, VII and X as well as in the intermediolateral column. TRH is co-localized with serotonin and probably with substance P in the nerve terminals in the ventral horns. This TRH appears to originate from neurons in the medullary raphe nuclei (n. pallidus, n. obscurus, n. magnus). The descending fibres travel via the dorsolateral, ventral and ventrolateral funiculi. This pathway may be involved in motor function but its precise physiological role is not known (Guiloff 1987). Much of the TRH in the intermediolateral columns may originate from the nucleus interfascicularis hypoglossi and the nucleus paragigantocellularis lateralis (Helke et al. 1986). This pathway may be important in the modulation of sympathetic activity. Most of the TRH in the dorsal horns is part of an intrinsic system and may play a role in the processing of pain (Winokur et al. 1989). The distribution of TRH receptors in the CNS is wide. In human spinal cord, receptor density is highest in the substantia gelatinosa and moderate in lamina IX (containing motor neurons) (Manaker et al. 1985). It is not clear whether TRH receptors in brain and spinal cord are identical.

Trophic and Protective Effects of TRH on Motor Neurons

In rat embryo ventral spinal cord cultures treated with TRH for 2–5 weeks, significant increase in choline acetyltransferase activity (ChAT), creatine kinase activity and in the growth of neurites (Schmidt-Achert et al. 1984) was seen. The results with the TRH analogue RGH2202, but not with another (DN1417), were comparable and electron microscopic observations suggested that there was an increase in the number of axosomatic synapses (Askanas et al. 1989). Treatment with TRH for 21 days induced increased soma size and heightened neurite complexity in murine ventral cord cultures (Banda et al. 1989). A more modest increase in ChAT, and an increase in antineurofilament antibody binding were seen with TRH in another spinal cord culture preparation; the TRH analogue RX77368 had similar but more marked effects (Jehanli et al. 1987). Although the data from these culture studies support a trophic action for TRH there is as yet no convincing evidence that this action is of relevance for motor neuronal survival in adult animals, in normal man or in motor neuron disease.

In infant rats treated with TRH, proximal section of the sciatic nerve resulted in less marked reduction in spinal cord neurons than in untreated rats (Banda et al. 1987). In neonatal rats treated with 5,7 dihydroxytryptamine (5,7 DHT), an agent that destroys the raphe spinal system, there was a slow recovery of the compound muscle action potential of plantar muscles after botulinum injections in the foot (Van den Bergh et al. 1987). However, in adult rats depletion of TRH from the spinal cord with 5,7 DHT did not result in motor or electrophysiological abnormalities, nor did it produce loss of motor neurons in the lumbar cord or denervation in the gastrocnemius muscle (Van den Bergh et al. 1987). In the Wobbler mouse, a genetic lower motor neuron degeneration, TRH administration did not reduce mortality but a few parameters of neurological deficit improved (Kozachuk et al. 1987). A protective effect of TRH in cat spinal cord trauma has been reported (Faden et al. 1981). TRH also prevents necrosis and motor neuron loss produced by intrathecal injection of a substance P antagonist (Spantide) in rats; it is not clear whether such protection relates to the vasodilating effect of TRH or to an interaction at the level of the substance P receptor (Freedman et al. 1989). These studies suggest that TRH may contribute to motor neuron growth or survival during development or when damaged by trauma or by certain specific peptide antagonists, but a relation of these phenomena to motor neuron death in MND is not apparent.

Neurophysiological Data

Animals

There is a body of experimental data showing a number of effects of TRH and its analogues on upper and lower motor neurons and interneurons. TRH usually increases motor neuron excitability by interacting with serotonin and substance P, but tachyphylaxis occurs (White et al. 1989).

In isolated spinal cord preparations of frog and toad, TRH increased excitability and depolarisation of anterior horn cells (Nicoll 1977, 1978; Phillis and Kirkpatrick 1979). Ventral root depolarisation and an increase in mono and disynaptic reflexes with TRH and the analogue DN1417, facilitation of glutamate and aspartate-induced lumbar motor neuron activity by iontophoretic application of TRH, an increase in the amplitude of motor neuron field potentials evoked by antidromic ventral root stimulation with an increase in the size of the test response in recovery cycle experiments with the analogue RX7768, have all been described in the rat (Ono and Fukuda 1982; White 1985; Clarke and Stirk 1983, 1984). In spinal rats TRH also enhances the flexor reflex (Pawlowski et al. 1980) and, when applied iontophoretically, enhances glutamate- and aspartate-induced firing in rat hypoglossal neurons (Ross and White 1986). An acute increase in EMG activity is seen with TRH in chronic spinally transected rats and is prevented by the 5-HT antagonist cyproheptadine. DN1417, an analogue with low affinity for spinal cord TRH receptors, produces a similar but delayed and more marked effect on EMG activity. It was suggested that TRH-induced lower motor neuron excitation is mediated by non-TRH receptors (Barbeau and Bédard 1981; Hawkins and Engel 1985; Hawkins et al. 1986). In rat cerebral cortex the effects of TRH have been less consistent; both potentiation of the excitatory action of acetylcholine and lack of it, have been reported (Yarbrough 1976; Winokur and Beckman 1978). Both non-specific depression of firing in most cortical neurons, including pyramidal tract neurons, and excitation in a few, were described (Phillis and Kirkpatrick 1980).

TRH-induced increased EMG activity in flexor and extensor groups, increased monosynaptic and polysynaptic evoked reflex potentials, tremor, shivering and increased muscle tone have been reported in spinal cats (Cooper and Boyer 1978). Some pyramidal neurons of the cat were also directly excited by TRH, or the excitatory action of acetylcholine was enhanced by this drug (Braitman et al. 1980).

Using a preparation of human muscle innervated by co-cultured foetal rat spinal cord with spinal root ganglia Askanas et al. (1985) showed that TRH increased the number of spontaneous muscle fibre contractions observed. No effect of TRH on neuromuscular transmission was found during stimulation of the rat sciatic nerve (Pawlowski et al. 1980). In vitro experiments in rat diaphragm documented a TRH-induced increase in the frequency and amplitude of miniature end plate potentials and, in vivo, an increase in fibrillations in denervated rat muscle was reported with TRH (Uchida et al. 1986).

It has been found that TRH has a more pronounced and more prolonged effect on spinal cord neurons when there is chronic damage to descending pathways than in the normal spinal cord. This had led to the suggestion that, in MND, administration of TRH may enhance the excitability of abnormal motor neurons (White et al. 1989).

Humans

The report of acute effects of TRH on muscle force and spasticity in patients with MND (Engel et al. 1983) triggered a number of electrophysiological studies to investigate its mechanism of action in man. With the exception of drugs used in spasticity (see below), and of agents active at the neuromuscular junction, more information exists for TRH and in analogues than for any other drug in this respect.

With small doses of intravenous TRH a short-term increase in mean amplitude of the rectified averaged F-wave response in flexor hallucis and a decrease in the antagonist extensor digitorum brevis muscle was found, suggestive of an inhibitory action on interneurons mediating the flexor reflex (Delwaide and Schoenen 1985). A TRH-induced increase in vibratory inhibition of soleus and quadriceps H-reflex and of flexor carpi radialis and soleus H-reflex threshold in MND patients suggested that an increase in pre-synaptic inhibition acting on Ia fibres plays a role in reducing spasticity (Pierrot-Deseilligny et al. 1985; Morin and Pierrot-Deseilligny 1988).

An excitatory effect of TRH and analogues on motor neurons is suggested by the observed increase in fasciculation potentials and by spontaneous discharges of groups of motor units, sometimes simultaneously in agonist and antagonist muscles. Such spontaneous discharges precede the onset of clinical "shivering" and occur randomly in many muscles for many seconds at a time (Guiloff 1989). During acute administration of the TRH analogue RX77368 there is a 25%–30% increase in macro EMG median amplitude and area and in mean corrected fibre density (Guiloff et al. 1987b). A similar acute increase in fibre density was confirmed with high dose subcutaneous TRH but not when a low dose was used (Biletch et al. 1989). The changes in macro EMG and fibre density were interpreted as related to a central, direct or indirect, action on motor neurons. The lack of change in the compound muscle action with peripheral nerve stimulation, absence of increase in the percentage of individual macro EMG potentials with amplitude above the upper limit or normal, lack of change in macro EMG amplitude and area of single motor units followed during the period of infusion, and similar changes in fibre density and macro EMG in weak and strong muscles all supported such interpretation (Guiloff et al. 1987b, 1988). The findings with RX77368 were consistent with a drug-induced selective reduction in the excitation threshold of pathological motor units or with a change in recruitment order (Guiloff 1989).

No evidence of an acute effect of TRH or RX77368 in the human neuromuscular junction has been found using repetitive stimulation (Carlo et al. 1984; Guiloff et al. 1987a; Guiloff 1989). Although a decrease in jitter was reported with TRH the number of measurements per patient was too small to draw useful conclusions (Tahmoush et al. 1985).

In humans the plasma half-life of TRH is about 35 minutes, in CSF about 54 minutes, and in brain tissue 18 minutes. The tripeptide is rapidly cleared by various enzymes. The main degradation products are the free amino acids, TRH-OH and cyclo(His-Pro). TRH analogues have been synthesized in order to enhance neuropharmacological actions, prolong drug effects or dissociate endocrine and neuropharmacological actions. For RX77368 (which has a dimethyl substitution in the proline amide end of the molecule) half-life in plasma is 1080 min and in brain 168 min; 80% of each dose is excreted unchanged in the urine. Oral absorption of RX77368 is rapid but poor (Guiloff 1987; Modarres-Sadeghi and Guiloff 1990).

Although transient, focal and modest changes in muscle force and spasticity have been documented in clinical trials with TRH in MND the results have not been conclusive and TRH is no longer under active clinical investigation (Brooks et al. 1989). In spite of the reported experimental trophic effects of TRH, long-term treatment in MND patients failed to stop the progression of the disease (Engel 1989). The TRH analogue RX77368 has more prominent actions on bulbar functions, in addition to effects on spasticity and cramps. Mild to moderate beneficial effects have been observed with repeated intravenous and oral administration for several weeks and long-term administration is currently under investigation (Guiloff et al. 1987a; Modarres-Sadeghi et al. 1988; Modarres-Sadeghi and Guiloff 1990).

Spasticity

The use of drugs, such as baclofen, diazepam, tizanidine and dantrolene, in the treatment of spasticity probably provides the current best example of a clinical pharmacological approach to abnormal motor neuron function, if disorders of the neuromuscular function are excluded. Spasticity has been defined as a velocity sensitive increase in resistance to passive stretch of muscles associated with exaggerated deep tendon reflexes (Davidoff 1985). It occurs with lesions in the hemispheres, brain stem or spinal cord caused by many different diseases. It is therefore not specific to MND and indeed, may be absent from the clinical picture in a number of patients.

Understanding of the actions of drugs on human spasticity and their development has been limited by several problems. First, there is no good animal model for chronic human spasticity (Young 1987; Wright and Rang 1990). Second, spasticity is often associated with other manifestations of upper motor neuron dysfunction. Third, the mechanisms leading to spasticity may be different with different lesions (Mazzocchio and Rossi 1989). Fourth, assessment of spasticity both clinically and instrumentally, may be difficult (Young 1987; Wright and Rang 1990).

Spasticity in humans appears to be the result of directly or indirectly heightened segmental motor neuron excitability. The precise mechanisms of such increase in excitability are not completely understood. It does not seem to depend on over-activity of the fusimotor system (Burke 1983). There is evidence that Ia presynaptic inhibition of motor neurons may be reduced or absent (Iles and Roberts 1986). Recurrent inhibition via the Renshaw cells may be reduced in some patients but not in others (Katz and Pierrot-Deseilligny 1982; Mazzocchio and Rossi 1989). The inhibitory effect of Ib interneurons on motor neurons has been reported to be absent (Delwaide 1988). Reductions in reciprocal inhibition have been reported in hemiplegics with and without spasticity, suggesting that spasticity does not correlate with such abnormalities (Nakashima et al. 1989).

It has been suggested that the stretch reflex threshold is reduced in spasticity with no significant increase in reflex gain (Lee et al. 1987; Powers et al. 1988, 1989). However, experiments that closely reproduce clinical testing of tone in man indicate that the spastic biceps shows late bursts of EMG activity, in addition to the early one also seen in normal subjects. The level of this late activity, named the spastic stretch reflex, correlates with the displacement velocity used, and its duration with the duration of the applied displacement, suggesting that a pathological increase in stretch reflex gain is a major factor in the pathogenesis of spasticity in man. The late response has a latency consistent with it being mediated via a predominantly polysynaptic pathway. The early reflex EMG response appears at displacement velocities which are lower in spastic than in normal subjects, indicating that there is also some reduction in the reflex threshold, but confined to this early component (Thilmann et al. 1991).

Drugs Used in Spasticity

Four drugs, currently and generally accepted to be useful in spasticity, are briefly considered below.

Baclofen (Lioresal)

This drug is a lipophilic derivative of GABA (beta-(4-chlorophenyl) GABA) and can slowly penetrate the blood–brain barrier. It seems to reduce the excitability of alpha-motor neurons mainly at spinal level.

Experiments with genetically "spastic" and normal rats suggest that in these animals the substantia nigra pars reticulata is an important supraspinal site of action for its action on tone (Turski et al. 1990). Levo-baclofen is a specific agonist of $GABA_B$ receptors in brain and spinal cord. $GABA_B$ receptors may mediate prolonged inhibitory processes in the CNS. In rat hippocampal neurons they mediate late inhibitory post-synaptic potential (Dutar and Nicoll 1988). Because of baclofen's inhibitory GABA-like action, monosynaptic and polysynaptic excitation of motor neurons is reduced.

There is experimental evidence that, at therapeutic doses, the inhibitory effects of baclofen are mainly pre-synaptic, by diminishing the influx of calcium into the pre-synaptic terminals. Depressed reflex activity may then result from a reduction in the release of excitatory neurotransmitters by afferent fibres, and interneurons, making contact with motor neurons. Baclofen depresses the effect of nociceptive stimuli on dorsal horn neurons, reduces fusimotor activity and stimulates Renshaw cell inhibitory activity, but the role of these various actions in its therapeutic effects is not known (Davidoff 1985). Phaclofen, a putative weakly selective $GABA_B$ receptor antagonist can block the pre-synaptic effects of baclofen on cat dorsal horn afferent fibres but not the post-synaptic effects in the spinal cord (Kerr et al 1987).

However, in spastic patients intramuscular baclofen did not reinforce vibratory inhibition of the H reflex but the recovery curve of the H reflex was affected, with lessening of the facilitation seen between 70 and 200 ms. These changes were interpreted as a modification of motor neuron excitability secondary to segmental and suprasegmental mechanisms, possibly by an action on interneurons (Delwaide and Schoenen 1985). Intrathecal baclofen suppresses tonic EMG activity in genetically "spastic" rats and H and spinal flexor reflexes in normal rats (Klockgether et al. 1989); in patients with severe spasticity it reduces both the amplitude of the H-reflex and the H_{max}/M_{max} (Latash et al. 1989; MacDonnel et al. 1989).

Baclofen is well absorbed from the gastrointestinal tract. Peak plasma concentrations are reached about 2 hours after oral administration (Faigle and Keberle 1972; Wuis et al. 1990). The half-life in plasma is about 3 hours; in the cerebrospinal fluid it is 5 hours (Penn and Kroin 1987). Most of the drug is excreted unchanged by the kidney (Wuis et al. 1990). The drug is usually given orally, but an increasing number of reports show that intrathecal administration may be helpful in cases that are refractory to oral administration (Penn and Kroin 1987; Siegfried and Rea 1987; Zierski et al. 1988; Ochs et al. 1989; Lazorthes et al. 1990).

Benzodiazepines

Diazepam increases pre-synaptic inhibition in the spinal cord of the cat (Schmidt et al. 1967; Stratten and Barnes 1971) and in man (Verrier et al. 1975a,b), probably by increasing the post-synaptic actions of GABA (Costa and Guidotti 1979). The effects are apparently related to an increase in the affinity of the binding of GABA to $GABA_A$ receptors (Skerritt and Johnston 1983). Benzodiazepine receptors are closely related to $GABA_A$ receptors in many GABAergic synapses. The gamma subunit of the $GABA_A$ receptor is activated by benzodiazepines and the alpha subunits may determine the type of benzodiazepine receptor (BZ_1 and BZ_2) (Pritchett et al. 1989). GABA binding increases chloride conductance across the membrane (Study and Barker 1982). The action of diazepam requires GABAergic transmission to be intact; the drug only makes it more efficient by increasing the binding to the receptor and increasing the frequency of chloride channel opening (Twyman et al. 1989). This efflux of chloride from afferent terminals may lead to the depolarisation which is probably responsible for pre-synaptic inhibition.

Although diazepam has widespread effects in the central nervous system its antispastic action is exerted mainly in the spinal cord since it is also seen in patients with spinal cord transection (Cook and Nathan 1967). There is experimental evidence that polysynaptic spinal reflexes may also be depressed, but not by an increase in pre-synaptic inhibition, and that fusimotor activity may be reduced

(Brausch et al. 1973). Intrathecal midazolam suppresses EMG activity in genetically "spastic" rats and H and flexor reflexes in normal rats (Klockgether et al. 1989). However, in spastic patients, intramuscular diazepam produced significant reinforcement of vibratory inhibition, but had no clear effect on the recovery curve of the H-reflex. These findings suggested a significant effect on pre-synaptic inhibition exerted on Ia fibres (Delwaide and Schoenen 1985). Intravenous diazepam can often reduce substantially, or even abolish, pathological flexion reflexes in spastic patients (Meinck et al. 1985).

Diazepam crosses the blood–brain barrier quickly (Marcucci et al. 1971). After oral administration peak plasma concentrations are reached in 2–4 hours; the half-life shows two peaks, one at about 8 hours, the other at 2–6 days (Van Der Kleijn 1969; Zbinden and Randall 1967). The drug is metabolized mainly by demethylation and hydroxylation (Schwartz et al. 1965) and most of the products are excreted in the urine.

Tizanidine

Tizanidine (5-chloro-4-(2-imidazolin-2-gamma 1-amino)-2,1,3 benzothiodazole) is a clonidine derivative with prominent $alpha_2$-adrenergic properties.

Noradrenergic fibres project from the locus coeruleus to the spinal cord and terminate both in the dorsal and ventral horns (Westlund et al. 1984). Stimulation of the locus coeruleus neurons results in a facilitation of spinal reflexes (Fung and Barnes 1987). Tizanidine appears to affect mainly interneuronal transmission as judged from its effects on spinal polysynaptic reflexes. A depression of the flexor response was seen in normal and decerebrated rats but in spinal rats the response was enhanced by tizanidine. The first effect was antagonized by an $alpha_2$ blocker (yohimbine) and the second by an $alpha_1$ blocker (prazosin) (Chen et al. 1987). Intrathecal tizanidine failed to suppress H-reflexes and flexor reflexes in genetically "spastic" rats (Klockgether et al. 1989). This evidence suggests that descending projections from the locus coeruleus and interneuronal pathways are required for the action of tizanidine. Recent work showing that the flexor reflex reduction produced by tizanidine in the rat is associated with, and has a similar temporal profile to, a reduction in the firing rate of locus coeruleus neurons supports this hypothesis (Palmeri and Wiesendanger 1990). Tizanidine also reduces spasticity in genetically "spastic" rats and H-reflexes in normal rats when injected in the substantia nigra pars reticulata (Turski et al. 1990). However, tizanidine depresses dorsal horn neuron responses to noxious stimuli and spontaneous firing (Davies 1989); this depression is also mediated by $alpha_2$ adrenoceptors.

In spastic patients modest, non-significant, reduction in the H_{max}/M_{max} ratio and in reinforcement of vibratory inhibition with no change in the H-reflex recovery curve were reported with oral tizanidine (Delwaide and Schoenen 1985) and intravenous tizanidine produced marked decrease in flexion reflexes (Meinck et al. 1985).

Tizanidine is rapidly absorbed with oral administration with a biovailability of around 20%. It is nearly completely metabolized before excretion by the kidney. Administration of 4 mg every 8 hours achieves a steady state after 2 or 3 doses (Tse et al. 1987).

A number of trials suggest that tizanidine is clinically as effective as baclofen or diazepam (Hassan and McLennan 1980; Lapierre et al. 1987; Bes et al. 1988; Bass et al. 1988; Hoogstraten et al. 1988; Pellkofer and Paulig 1989).

Dantrolene Sodium

Dantrolene (1-{[5-(p-paraphenyl) furfurylidene]amino} hydantoin) is the only widely used antispastic agent that has a peripheral action.

It is a hydantoin, with no proven effect on reflex pathways, that exerts its action by uncoupling the excitation–contraction coupling mechanisms in skeletal muscle (Ellis and Bryant 1972). The effect is seen both in the muscle fibres of the muscle spindles and in extrafusal muscle fibres (Leslie and Part 1981). Dantrolene produces a partial block of calcium ion release from the sarcoplasmic reticulum (Hainaut and Desmedt 1975) but effects on signal transmission along the transverse (T) tubules and on the coupling between T tubules and the terminal cisternae of the sarcoplasmic reticulum have also been suggested (Morgan and Bryant 1977). This action is more marked in the muscle fibres of fast- than of slow-twitch motor units (Jami et al. 1983; Everts and Van Hardeveld 1987). No direct effect on the discharge of gamma nerve fibres to the soleus muscle was found in the decerebrate rat nor is there evidence of a direct action on alpha motoneuron discharge rate (Farquhar and Part 1988). The inhibitory effect of dantrolene on the muscle cell is markedly reduced when the stimulus frequency is increased so that fused tetanic contractions are less inhibited than subtetanic and twitch contractions (Ellis and Bryant 1972; Leslie and Part 1981). Using a rat phrenic nerve diaphragm preparation evidence was presented that the inhibitory effect is related to Ca^{2+} dependent K^+ currents (Røed 1989). Effects on central nervous system neurons (hippocampal) have been described (Obenaus et al. 1989; Krnjevic and Xu 1989) but are probably not relevant to its action on spasticity.

Although dantrolene is a lipophilic compound, expected to cross the blood–brain barrier, whole body autoradiography in the marmoset monkey showed only traces of the molecule in the brain. The possibility exists that its two main metabolites (5-hydroxydantrolene and 7-aminodantrolene) do cross the barrier (Wuis et al. 1989). After oral administration about 70% of dantrolene is absorbed with variable peak plasma concentrations at 6 hours. After intravenous administration the elimination half-life is about 12 hours. Dantrolene is metabolised in liver microsomes by hydroxylation, reduction and acetylation. It is excreted in the urine and bile, 4% unchanged and 86% as metabolic products (Harrison 1988).

Conclusion

The field of clinical pharmacology of motor neurons aims at developing drugs that can influence motor function through a direct action on upper and lower motor neurons or an indirect effect upon the neuronal circuits or the supporting tissues related to them. It also aims at influencing motor neuron survival in diseases in which motoneuron death is a major feature, such as MND. Advances in our knowledge of the basic mechanisms underlying human motor neuron function and survival in normal and diseased states, as well as in our ability to assess reliably and accurately, by clinical and laboratory-based methods, the effects of such drugs, are required for this area of neuropharmacology to flourish. Drugs to treat spasticity remain, so far, the only ones with established therapeutic efficacy but recent work with excitatory amino acids and neuropeptides, such as TRH and TRH analogues, has contributed much to expand the field. The rapid progress in neurobiological techniques suggests that trophic factors influencing motor neuron survival will be increasingly tried in a clinical setting in the future.

Acknowledgements. I thank the Special Trustees of Westminster and Roehampton Hospitals, the Motor Neurone Disease Association and the North-West Thames Regional Health Authority for financial assistance to the author's work quoted in this review. I am also grateful to Mr R. Wentz and Miss J. Roberts for help with the bibliography and typing.

References

Aitken JT, Bridger JE (1961) Neuron size and neuron population density in the lumbosacral region of the cat's spinal cord. J Anat 95:38–53

Akagi H, Miledi R (1988) Heterogeneity of glycine receptors and their messenger RNAs in rat brain and spinal cord. Science 242:270–273

Allen CN, Ross SM, Spencer PS (1990) Properties of the neurotoxic nonprotein amino acids, Beta-N-methylamino-L-alanine (BMAA) and Beta-N-oxalylamino-L-alanine (BOAA). In: Clifford Rose F, Norris FH (eds) Amyotrophic lateral sclerosis: new advances in toxicology and epidemiology. Smith-Gordon, London, pp 41–48

Andres PL, Hedlund W, Finison L, Conlon P, Felmus M, Munsat TL (1986) Quantitative motor assessment in amyotrophic lateral sclerosis. Neurology 36:937–941

Andres PL, Thibodeau LM, Finison LJ, Munsat TL (1987) Quantitative assessment of neuromuscular deficit in ALS. Neurol Clin 5:125–141

Angel RW, Hoffmann WW (1963) The H reflex in normal, spastic and rigid subjects. Arch Neurol 8:591–596

Appel V, Stewart SS, Smith G, Appel SH (1987) A rating scale for amyotrophic lateral sclerosis: description and preliminary experience. Ann Neurol 22:328–333

Ascher P, Nowak L (1987) Electrophysiological studies of NMDA receptors. Trends Neurosci 10:284–288

Ashby P, Burke D, Rao S, Jones RF (1972) Assessment of cyclobenzaprine in the treatment of spasticity. J Neurol Neurosurg Psychiatry 35:599–605

Ashby P, Mailis A, Hunter J (1987) The evaluation of "spasticity". J Neurol Sci 14:497–500

Ashworth B (1964) Preliminary trial of Carisoprodol in multiple sclerosis. Practitioner 192:540–542

Askanas V, Engel WK, Kobayashi T (1985) Thyrotropin-releasing hormone enhances motor neuron-evoked contractions of cultured human muscle. Ann Neurol 18:716–719

Askanas V, Engel WK, Eagleson K, Micaglio G (1989) Influence of TRH and TRH analogues RGH-2202 and DN-1417 on cultured ventral spinal cord neurons. In: Metcalf G, Jackson IMD (eds) Thyrotropin-releasing hormone: biomedical significance. Ann NY Acad Sci 553:325–336

Banda RW, Means ED, Fitzgerald M (1987) Thyrotropin-releasing hormone decreases neuronal loss induced by axotomy in infant rats. Neurology 37:285

Banda RW, Means ED, Scherch HM (1989) Trophic effect of thyrotropin-releasing hormone in murine ventral horn neuronal cultures. In: Metcalf G, Jackson IMD (eds) Thyrotropin-releasing hormone: biomedical significance. Ann NY Acad Sci 553:588–589

Barbeau H, Bédard PJ (1981) Similar motor effects of 5HT and TRH in rats following chronic spinal transection and 5-7-dihydroxytryptamine injection. Neuropharmacology 20:611–616

Barde YA (1988) What, if anything, is a neurotrophic factor? Trends Neurosci 11:343–346

Bass B, Weinshenker B, Rice GP et al. (1988) Tizanidine versus baclofen in the treatment of spasticity in patients with multiple sclerosis. Can J Neurol Sci 15:15–19

Bensimon G, Lacomblez L, Meininger V et al (1994) A controlled trial of riluzole in amyotrophic lateral sclerosis. N Engl J Med 330:585–592

Bes A, Eyssette M, Pierrot-Deseilligny E, Rohmer F, Warter JM (1988) A multi-centre, double-blind trial of tizanidine, a new antispastic agent, in spasticity associated with hemiplegia. Curr Med Res Opin 10:709–718

Betz H, Becker CM (1988) The mammalian glycine receptor: biology and structure of a neuronal chloride channel protein. Neurochem Int 13:137–146

Biletch M, Eichman P, Sufit R, Turner J, Brooks BR (1989) Increased fiber density after subcutaneous TRH in amyotrophic lateral sclerosis (ALS) patients. A placebo and low-dose controlled study. In: Metcalf G, Jackson IMD (eds) Thyrotropin-releasing hormone: biomedical significance. Ann NY Sci 553:614–617

Bobker DH, Williams JT (1990) Ion conductances affected by 5-HT receptor subtypes in mammalian neurons. Trends Neurosci 13:169–173

Bohannon RW, Smith MB (1987) Interrater reliability of a modified Ashworth scale of muscle spasticity. Phys Ther 67:206–207

Bormann J (1988) Electrophysiology of $GABA_A$ and $GABA_B$ receptor subtypes. Trends Neurosci 11:112–116

Braitman DJ, Auker CR, Carpenter DO (1980) Thyrotropin-releasing hormone has multiple actions in cortex. Brain Res 194:244–248

Braun SR (1987) Respiratory system in amyotrophic lateral sclerosis. Neurol Clin 5:9–31

Brausch U, Henatsch HD, Student C, Takano K (1973) In: Gorattini S, Mussini E, Randall LO (eds) The benzodiazepines. Raven Press, New York

Breedlove SM (1986) Cellular analyses of hormone influence on motoneuronal development and function. J Neurobiol 17:157–176

Bridges RJ, Stevens DR, Kahle JS, Nunn PB, Kadri M, Cotman CW (1989) Structure-function studies on N-oxalyl-diamino-dicarboxylic acids and excitatory amino acid receptors: evidence that Beta-L-ODAP is a selective non-NMDA agonist. J Neurosci 9:2073–2079

Brooks BR, Sufit RL, Clough JA et al. (1989) Isokinetic and functional evaluation of muscle strength over time in amyotrophic lateral sclerosis. In: Munsat TL (ed) Quantification of neurologic deficit. Butterworth, London, p 143

Brown WF, Strong MJ, Snow R (1988) Methods for estimating numbers of motor units in biceps-brachialis muscles and losses of motor units with aging. Muscle Nerve 11:423–432

Burke D (1983) Critical examination of the case for or against fusimotor involvement in disorders of muscle tone. In: Desmedt JE (ed) Motor control mechanisms in health and disease. Raven Press, New York, pp 133–150

Carlo JR, Engel WK, Van den Bergh P (1984) Thyrotropin-releasing hormone (TRH) does not affect neuromuscular transmission in patients with amyotrophic lateral sclerosis (ALS). Neurology 34(Suppl 1):146

Chen D-F, Bianchetti M, Wiesendanger M (1987) The adrenergic agonist tizanidine has differential effects on flexor reflexes of intact and spinalized rat. Neuroscience 23:641–647

Choi DW, Visekul V, Amirthanayagam M, Monyer H (1989) Aspartate neurotoxicity on cultured cortical neurons. J Neurosci Res 23:116–121

Clarke KA, Stirk G (1983) Motor neurone excitability after administration of a thyrotropin-releasing hormone analogue. Br J Pharmacol 80:561–565

Clarke KA, Parker AJ, Stirk CG (1984) Motor neurone excitability during antidromically evoked inhibition after administration of a thyrotropin releasing hormone (TRH) analogue. Neuropeptides 4:403–411

Cohan CS, Kater SB (1986) Suppression of neurite elongation and growth cone motility by electrical activity. Science 232:1638–1640

Cook JB, Nathan PW (1967) On the site of action of diazepam in spasticity in man. J Neurol Sci 5:33–37

Cooper BR, Boyer CE (1978) Stimulant action of thyrotropin-releasing hormone on cat spinal cord. Neuropharmacology 17:153–156

Cornford EM, Braun LD, Crane PD, Oldendorf WH (1978) Blood–brain barrier restriction of peptides and the low uptake of enkephalins. Endocrinology 103:1297–1303

Costa E, Guidotti A (1979) Molecular mechanisms in the receptor action of benzodiazepines. Ann Rev Pharmacol Toxicol 19:531–545

Cotman CW, Monaghan DT, Ottersen OP, Storm-Mathisen J (1987) Anatomical organization of excitatory amino acid receptors and their pathways. Trends Neurosci 10:273–280

Couratier P, Hugon J, Sindon P, Vallet JM, Dumas M (1993) Cell culture evidence for neuronal degeneration in amyotrophic lateral sclerosis being linked to glutamate AMPA/kainate receptors. Lancet 341:265–268

Crews LL, Wigston DJ (1990) The dependence of motoneurons on their target muscle during postnatal development of the mouse. J Neurosci 10:1643–1653

Cullheim S, Kellerth JO, Conradi S (1977) Evidence for direct synaptic interactions between cat spinal alpha motoneurons via the recurrent axon collaterals; a morphological study using intracellular injection of horseradish peroxidase. Brain Res 132:1–10

Davidoff RA (1985) Antispasticity drugs: mechanisms of action. Ann Neurol 17:107–116
Davies J (1989) Effects of tizanidine, eperisone and afloqualone on feline dorsal horn neuronal responses to peripheral cutaneous noxious and innocuous stimuli. Neuropharmacology 28:1357–1362
de Belleroche J, Recordati A, Rose FC (1984) Elevated levels of amino acids in the CSF of motor neurone disease patients. Neurochem Pathol 2:106–111
de Belleroche J, Lane RJM, Bandopadhyay R, Clifford Rose F (1990) Abnormalities in amino acid metabolism in amyotrophic lateral sclerosis. In; Clifford Rose F, Norris FH (eds) Amyotrophic lateral sclerosis. New advances in toxicology and epidemiology. Smith-Gordon, London, pp 261–264
de Koning P, Wieneke GH, van der Most van Spijk D, et al (1988) Estimation of the number of motor units based on macro-EMG. J Neurol Neurosurg Psychiatry 51:403–411
Delfs J, Friend J, Ishimoto S, Saroff D (1989) Ventral and dorsal horn acetylcholinesterase neurons are maintained in organotypic cultures of postnatal rat spinal cord explants. Brain Res 488:31–42
Delwaide PJ (1988) Electrophysiological exploration of the human anterior horn. Clinical implications. Rev Neurol 144:656–659
Delwaide PJ, Schoenen J (1985) The effects of TRH on F-waves recorded from antagonistic muscles in human subjects. Ann Neurol 18:366–367
Delwaide PJ, Schoenen J, Burton L (1983) Central actions of neurotrophic drugs assessed by reflex studies in man. In: Desmedt JE (ed) Motor control mechanisms in health and disease. Raven Press, New York, pp 977–996
Dengler R, Konstanzer A, Küther G, Hesse S, Wolf W, Struppler A (1990) Amyotrophic lateral sclerosis: macro-EMG and twitch forces of single motor units. Muscle Nerve 13:545–550
DePaul R, Abbs JH, Caligiuri M, Gracco VL, Brooks BR (1988) Hypoglossal, trigeminal, and facial motoneuron involvement in amyotrophic lateral sclerosis. Neurology 38:281–283
Dietl MM, Sanchez M, Probst A, Palacios JM (1989) Substance P receptors in the human spinal cord: decrease in amyotrophic lateral sclerosis. Brain Res 483:39–49
Duncan MW, Steele JC, Kopin IJ, Markey SP (1990) 2-amino-3-(methylamino)-propanoic acid (BMAA) in cycad flour: an unlikely cause of amyotrophic lateral sclerosis and parkinsonism-dementia of Guam. Neurology 40:767–772
Dutar P, Nicoll RA (1988) A physiological role for $GABA_B$ receptors in the central nervous system. Nature 332:156–158
Eisen A (1987) Electromyography in disorders of muscle tone. Can J Neurol Sci 14:501–505
Ellis KO, Bryant SH (1972) Excitation contraction uncoupling in skeletal muscle by dantrolene sodium. Naunyn Schmiedebergs Arch Pharmacol 274:107–109
Engel WK, Siddique T, Nicoloff JT (1983) Effect on weakness and spasticity in amyotrophic lateral sclerosis of thyrotropin-releasing hormone. Lancet ii:73–75
Engel WK (1989) High-dose TRH treatment of neuromuscular diseases: summary of mechanisms and critique of clinical studies. Summary of section IX. In: Metcalf G, Jackson IMD (eds) Thyrotrophin-releasing hormone: biomedical significance. Ann NY Acad Sci 553:462–472
Ermisch A, Ruhle HJ, Landgraf R, Hess J (1985) Blood–brain barrier and peptides. J Cereb Blood Flow Metab 5:350–357
Everts ME, Van Hardeveld C (1987) Effects of dantrolene on force development in slow- and fast-twitch muscle of euthyroid, hypothyroid and hyperthyroid rats. Br J Pharmacol 92:47–54
Faden AI, Jacobs TP, Holaday JW (1981) Thyrotropin-releasing hormone improves neurologic recovery after spinal trauma in cats. N Engl J Med 305:1063–1067
Faigle JW, Keberle H (1972) The chemistry and kinetics of Lioresal. Postgrad Med J 48:9–13
Farquhar R, Part NJ (1988) The effect of dantrolene sodium on the discharge of alpha and gamma motor neurones to the soleus muscle in the decerebrate rat. Br J Pharmacol 93:257–266
Frank E (1987) The influence of neuronal activity on patterns of synaptic connections. Trends Neurosci 10:188–189
Freedman J, Hokfelt T, Post C et al. (1989) Immunohistochemical and behavioral analysis of spinal lesions induced by a substance P antagonist and protection by thyrotropin-releasing hormone. Exp Brain Res 74:279–292
Fung SJ, Barnes CD (1987) Membrane excitability changes in hindlimb motoneurons induced by stimulation of the locus coeruleus in cats. Brain Res 402:230–242
Gibson SJ, Polak JM, Katagiri T et al. (1988) A comparison of the distributions of eight peptides in spinal cord from normal controls and cases of motor neurone disease with special reference to Onuf's nucleus. Brain Res 474:255–278
Gogan P, Gueritaud JP, Horcholle-Bossavit G, Tyc-Dumont S (1977) Direct excitatory interactions between spinal motoneurons of the cat. J Physiol 272:755–767
Goonetilleke A, Modarres-Sadeghi H, Guiloff RJ (1994) Accuracy, reproductibility and variability of hand-held dynamometry in motor neuron disease. J Neurol Neurosurg Psychiatr 57:326–332

Grenningloh G, Rienitz A, Schmitt B et al. (1987) The strychnine-binding subunit of the glycine receptor shows homology with nicotinic acetylcholine receptors. Nature 328:215–220

Griggs RC, Donahue KM, Utell MJ et al. (1981) Evaluation of pulmonary function in neuromuscular disease. Arch Neurol 38:9–12

Guiloff RJ (1987) Thyrotropin-releasing hormone and motor neurone disease. Rev Neurosci 1:201–219

Guiloff RJ (1989) Use of TRH analogues in motor neurone disease. In: Metcalf G, Jackson IMD (eds) Thyrotropin-releasing hormone: biomedical significance. Annals of the New York Academy of Sciences, New York, pp 399–421

Guiloff RJ, Eckland DJ (1987) Observations on the clinical assessment of patients with motor neuron disease. Experience with a TRH analogue. Neurol Clin 5:171–192

Guiloff RJ, Modarres-Sadeghi H (1991) Preferential generation of recurrent responses by groups of motor neurons in man. Conventional and single unit F-wave studies. Brain 114:1771–1801

Guiloff RJ, Eckland DJ, Demaine C, Hoare RC, MacRae KD, Lightman SL (1987a) Controlled acute trial of a thyrotrophin-releasing hormone analogue (RX77368) in motor neuron disease. J Neurol Neurosurg Psychiatry 50:1359–1370

Guiloff RJ, Stälberg E, Eckland DJ, Lightman SL (1987b) Electrophysiological observations in patients with motor neuron disease receiving a thyrotropin-releasing hormone analogue (RX77368). J Neurol Neurosurg Psychiatry 50:1633–1640

Guiloff RJ, Modarres-Sadeghi H, Stälberg E, Rogers H (1988) Short-term stability of single motor unit recordings in motor neuron disease: a macro EMG study. J Neurol Neurosurg Psychiatry 51:671–676

Guiloff RJ, Modarres-Sadeghi H, Rogers H (1990) Motor neurone disease: aims and assessment methods in trial design. In: Clifford Rose F (ed) Methodologic problems in clinical neurologic trials: amyotrophic lateral sclerosis, vol 1. Demos, New York, pp 19–31

Hainaut K, Desmedt JE (1975) Effect of dantrolene sodium on calcium movements in single muscle fibres. Nature 252:728–729

Harrison GG (1988) Malignant hyperthermia. Dantrolene-dynamics and kinetics. Br J Anaesth 60:279–286

Hassan N, McLellan DL (1980) Double-blind comparison of single doses of DS 103–282, baclofen and placebo for suppression of spasticity. J Neurol Neurosurg Psychiatry 43:1132–1136

Hawkins EF, Beydoun SR, Haun CK, Engel WK (1986) Analogs of thyrotropin-releasing hormone: hypotheses relating receptor binding to net excitation of spinal lower motor neurons. Biochem Biophys Res Commun 138:1184–1190

Hawkins EF, Engel WK (1985) Analog specificity of the thyrotropin-releasing hormone receptor in the central nervous system: possible clinical implications. Life Sci 36:601–611

Helke CJ, Sayson SC, Keeler JR, Charlton CG (1986) Thyrotropin-releasing hormone-immunoreactive neurons project from the ventral medulla to the intermediolateral cell column: partial coexistence with serotonin. Brain Res 38:1–7

Hillel AD, Miller RM, Yorkston K, McDonald E, Norris FH, Konikow N (1989) Amyotrophic lateral sclerosis severity scale. Neuroepidemiology 8:142–150

Hoogstraten MC, van der Ploeg RJ, van der Burg W, Vreeling A, van Marle S, Minderhoud JM (1988) Tizanidine versus baclofen in the treatment of spasticity in multiple sclerosis patients. Acta Neurol Scand 77:224–230

Hugon J, Vallat JM, Spencer PS, Leboutet MJ, Barthe D (1989) Kainic acid induces early and delayed degenerative neuronal changes in rat spinal cord. Neurosci Lett 104:258–262

Iles JF, Roberts RC (1986) Presynaptic inhibition of monosynaptic reflexes in the lower limbs of subjects with upper motoneuron disease. J Neurol Neurosurg Psychiatry 49:937–944

Jackson IMD, Adelman LS, Munsat TL, Forte S, Lechan RM (1986) Amyotrophic lateral sclerosis: thyrotropin-releasing hormone and histidyl proline diketopiperazine in the spinal cord and cerebrospinal fluid. Neurology 36:1218–1223

Jami L, Murthy KSK, Petit J et al. (1983) Action of dantrolene sodium on single motor units of cat muscle in vivo. Brain Res 261:285–294

Jehanli A, Harrison R, Lunt GG, Guiloff RJ (1987) Effect of TRH analogue RX77368 on spinal cord neurons in culture. J Neurol Neurosurg Psychiatry 51:946

Johnson JW, Ascher P (1987) Glycine potentiates the N-MDA response in cultured mouse brain neurons. Nature 325:529–531

Jones RF, Burke D, Marosszeky JF, Gillies JD (1970) A new agent for the control of spasticity. J Neurol Neurosurg Psychiatry 33:464–468

Katz R, Pierrot-Deseilligny E (1982) Recurrent inhibition of alpha-motoneurons in patients with upper motor neuron lesions. Brain 105:103–124

Kerr DIB, Ong J, Prager RH, Gynther BD, Curtis DR (1987) Phaclofen: a peripheral and central baclofen antagonist. Brain Res 405:150–154

Kleijn E van der (1969) Protein binding and lipophilic nature of ataractics of the meprobamate – and diazepine – group. Arch Int Pharmacodyn Ther 179:225–250

Klockgether T, Schwartz M, Wullner U, Turski L, Sontag K-H (1989) Myorelaxant effect after intrathecal injection of antispastic drugs in rats. Neurosci Lett 97:221–226

Knutsson E (1983) Analysis of gait and isokinetic movement for evaluation of antispastic drugs or physical therapies. In: Desmedt JE (ed) Motor control mechanisms in health and disease. Raven Press, New York, pp 1013–1034

Knutsson E (1985) Quantification of spasticity. In: Struppler A, Weindl A (eds) Electromyography and evoked potentials. Theories and applications. Springer, Berlin Heidelberg New York, pp 84–91

Koh J, Goldberg MP, Hartley DM, Choi DW (1990) Non-NMDA receptor mediated neurotoxicity in cortical culture. J Neurosci 10:693–705

Kozachuk WE, Mitsumoto H, Salanga VD, Beck GJ, Wilber JF (1987) Thyrotropin-releasing hormone (TRH) in murine motor neuron disease (the wobbler mouse). J Neurol Sci 78:253–260

Krnjevic K (1981) Transmitters in motor systems. In: Handbook of physiology, section 1: the nervous system, vol. II. Motor control, part 1. American Physiological Society, Bethesda, pp 107–154

Krnjevic K, Xu YZ (1989) Dantrolene suppresses the hyperpolarization or outward current observed during anoxia in hippocampal neurons. Can J Physiol Pharmacol 67:1602–1604

Kurz EM, Sengelaub DR, Arnold AP (1986) Androgens regulate the dentritic length of mammalian motor neurons in adulthood. Science 232:395–398

Kuypers HGJM (1973) The anatomical organisation of the descending pathways and their contributions to motor control especially in primates. In: Desmedt S (ed) New developments in electromyography and clinical neurophysiology. Karger, Basel, pp 38–68

Lapierre Y, Bouchard S, Tansey C, Gendron D, Barkas WJ, Francis GS (1987) Treatment of spasticity with tizanidine in multiple sclerosis. Can J Neurol Sci 14:513–517

La Spada AR, Wilson EM, Lubahn DB et al. (1991) Androgen receptor mutations in X-linked spinal and bulbar muscular atrophy. Nature 352:77–79

Latash ML, Penn RD, Corcos DM, Gottlieb GL (1989) Short-term effects of intrathecal baclofen in spasticity. Exp Neurol 103:165–172

Lazorthes Y, Sallerin-Caute B, Verdie JC, Bastide R (1990) Chronic intrathecal baclofen administration for control of severe spasticity. J Neurosurg 72:393–402

Lee KC, Carson L, Kinnin E, Patterson V (1989) The Ashworth scale: a reliable and reproducible method of measuring spasticity. Neurology 39:143–143

Lee WA, Boughton A, Rymer WZ (1987) Absence of stretch reflex gain enhancement in voluntarily activated spastic muscle. Exp Neurol 98:317–335

Leslie GC, Part NJ (1981) The effect of dantrolene sodium on intrafusal muscle fibres in the rat soleus muscle. J Physiol 318:73–83

Lewis CA, Ahmed Z, Faber DS (1989) Characteristics of glycine-activated conductances in cultured medullary neurons from embryonic rat. Neurosci Lett 96:185–190

Lipton SA, Kater SB (1989) Neurotransmitter regulation of neuronal outgrowth plasticity and survival. Trends Neurosci 12:265–270

Ludolph AC, Hugon J, Dwivedi MP, Schaumburg HH, Spencer PS (1987) Studies on the etiology and pathogenesis of motor neuron diseases. I. Clinical findings in established cases of lathyrism. Brain 110:149–165

MacDermott AB, Dale N (1987) Receptors, ion channels and synaptic potentials underlying the integrative actions of excitatory amino acids. Trends Neurosci 10:280–284

MacDonald JF, Morris ME (1984) Lathyrus excitotoxin: mechanism of neuronal excitation by L-2-oxalylamino-3-amino- and L-3-oxalylamino-propionic acid. Exp Brain Res 57:158–166

Macdonald RL, Twyman RE (1990) GABA/benzodiazepine receptors and glycine receptors. In: Walton J (ed) Current opinion in neurology and neurosurgery. Current Science, London, pp 538–543

MacDonnel RA, Talalla A, Swash M, Grundy D (1989) Intrathecal baclofen and the H-reflex. J Neurol Neurosurg Psychiatry 52:1110–1112

Manaker S, Caine SB, Winokur A (1988) Alterations in receptors for thyrotropin-releasing hormone, serotonin, and acetylcholine in amyotrophic lateral sclerosis. Neurology 38:1464–1474

Manaker S, Winokur A, Rhodes CH, Rainbow TC (1985) Autoradiographic localization of thyrotropin-releasing hormone (TRH) receptors in human spinal cord. Neurology 35:328–332

Marcucci F, Guaitani A, Fanelli R, Mussini E, Garattini S (1971) Metabolism and anticonvulsant activity of diazepam in guinea pigs. Biochem Pharmacol 20:1711–1713

Mattson MP (1988) Neurotransmitters in the regulation of neuronal cytoarchitecture. Brain Res 472:179–212

Mayer ML, Vyklicky L, Clements J (1989) Regulation of NMDA receptor desensitization in mouse hippocampal neurons by glycine. Nature 338:425–427

Mazzocchio R, Rossi A (1989) Recurrent inhibition in human spinal spasticity. Ital J Neurol Sci 10:337–347

McComas AJ, Fawcett PRW, Campbell MJ, Sica REP (1971) Electrophysiological estimation of the number of motor units within a human muscle. J Neurol Neurosurg Psychiatry 34:121–131

McGeer EG (1990) Neurotransmitters. In: Walton J (ed) Current opinion in neurology and neurosurgery. Current Science, London, pp 530–537

McManaman JL, Oppenheim RW, Prevette D, Marchetti D (1990) Rescue of motoneurons from cell death by a purified skeletal muscle polypeptide: effects of the ChAT development factor, CDF. Neuron 4:891–898

Medical Research Council (1976) Aids to the examination of the peripheral nervous system. London (Memorandum No 45)

Meinck HM, Benecke R, Kuster S et al. (1983) Cutaneomuscular (flexor) reflex organization in normal men and in patients with motor disorders. In: Desmedt JE (ed) Motor control mechanisms in health and disease. Raven Press, New York, pp 787–796

Meinck HM, Benecke R, Conrad B (1985) Cutaneo-muscular control mechanisms in health and disease: possible implications on spasticity. In: Struppler A, Weindl A (eds) Electromyography and evoked potentials. Theories and applications. Springer, Berlin Heidelberg New York, pp 75–83

Mens WBJ, Witter A, Greidanus TBVW (1983) Penetration of neurohypophyseal hormones from plasma into cerebrospinal fluid (CSF): half-times of disappearance of these neuropeptides from CSF. Brain 262:143–149

Meyer M, Adorjani C (1983) Quantification of the effects of muscle relaxant drugs in man by tonic stretch reflex. In: Desmedt JE (ed) Motor control mechanisms in health and disease. Raven Press, New York, pp 997–1011

Mitsuma T, Nogimori T (1983) Influence of the route of administration on thyrotropin-releasing hormone concentration in the mouse brain. Experientia 39:620–622

Modarres-Sadeghi H, Rogers H, Emami J, Guiloff RJ (1988) Subacute administration of a TRH analogue (RX77368) in motoneuron disease: an open study. J Neurol Neurosurg Psychiatry 51:1146–1157

Modarres-Sadeghi H, Guiloff RJ (1990) Comparative efficacy and safety of intravenous and oral administration of a TRH analogue (RX77368) in motor neuron disease. J Neurol Neurosurg Psychiatry 53:944–947

Morgan KG, Bryant SH (1977) The mechanism of action of dantrolene sodium. J Pharmacol Exp Ther 201:138–147

Morin C, Pierrot-Deseilligny E (1988) Spinal mechanism of the antispastic action of TRH in patients with amyotrophic lateral sclerosis. Rev Neurol 144:701–703

Munsat TL (1989) Quantification of neurologic deficit. Butterworth, London

Munsat TL, Taft J, Jackson IMD (1987) Pharmacokinetics of intrathecal thyrotropin-releasing hormone. Neurology 37:597–601

Nakashima K, Rothwell JC, Day BL, Thompson PD, Shannon K, Marsden CD (1989) Reciprocal inhibition between forearm muscles in patients with writer's cramp and other occupational cramps, symptomatic hemidystonia and hemiparesis due to stroke. Brain 112:681–697

Nicoll RA (1977) Excitatory action of TRH on spinal motor neurones. Nature 265:242–243

Nicoll RA (1978) The action of thyrotropin-releasing hormone, substance P and related peptides on frog spinal motor neurones. J Pharmacol Exp Ther 207:817–824

Norris FH, Callachini PR, Fallat RJ et al. (1974) Administration of guanidine in amyotrophic lateral sclerosis. Neurology 24:721–728

Obenaus A, Mody I, Baimbridge KG (1989) Dantrolene-Na (Dantrium) blocks induction of long-term potentiation in hippocampal slices. Neurosci Lett 98:172–178

Ochs G, Struppler A, Meyerson BA et al. (1989) Intrathecal baclofen for long-term treatment of spasticity: a multi-centre study. J Neurol Neurosurg Psychiatry 52:933–939

Olney JW (1971) Glutamate-induced neuronal necrosis in the infant mouse hypothalamus: an electron microscopic study. J Neuropathol Exp Neurol 30:75–90

Olney JW, Sharpe LG (1969) Brain lesion in an infant rhesus monkey treated with monosodium glutamate. Science 166:386–388

Ono H, Fukuda H (1982) Ventral root depolarization and spinal reflex augmentation by a TRH analog in rat spinal cord. Neuropharmacology 21:739–744

Oppenheim RW (1989) The neurotrophic theory and naturally occurring motoneuron death. Trends Neurosci 12:252–255

Palmeri A, Wiesendanger M (1990) Concomitant depression of locus coeruleus neurons and of flexor reflexes by an $alpha_2$-adrenergic agonist in rats: a possible mechanism for an $alpha_2$-mediated muscle relaxation. Neuroscience 34:177–187

Pawlowski L, Ruczynska J, Przegalinski E (1980) The effect of thyroliberin and some of its analogues on the hind limb flexor reflex in the spinal rat. Pol J Pharmacol Pharmacy 32:539–550
Pellkofer M, Paulig M (1989) Comparative double-blind study of the effectiveness and tolerance of baclofen, tetrazepam and tizanidine in spastic movement disorders of the lower extremities. Med Klin 84:5–8
Penn RD, Kroin JS (1987) Continuous intrathecal baclofen infusion for treatment of spasticity. J Neurosurg 66:181–185
Peroutka SJ (1988) 5-Hydroxytryptamine receptor subtypes: molecular, biochemical and physiological characterization. Trends Neurosci 11:496–500
Perry TL, Krieger C, Hansen S, Eisen A (1990) Amyotrophic lateral sclerosis: amino acid levels in plasma and cerebrospinal fluid. Ann Neurol 28:12–17
Phillis JW, Kirkpatrick JR (1979) Actions of various gastrointestinal peptides on the isolated amphibian spinal cord. Can J Physiol Pharmacol 57:887–889
Phillis JW, Kirkpatrick JR (1980) The actions of motolin, luteinizing hormone-releasing hormone, cholecystokinin, somatostatin, vasoactive intestinal peptide, and other peptides on rat cerebral cortical neurons. Can J Physiol Pharmacol 58:612–623
Pierrot-Deseilligny E, Bussel B, Held JP, Katz R (1976) Excitability of human motoneurones after discharge in a conditioning reflex. Electroencephalogr Clin Neurophysiol 40:279–287
Pierrot-Deseilligny E, Engel WK, Fardeau M (1985) Effect of high dose TRH (HD-TRH) on H-reflex vibratory inhibition and H-reflex threshold in amyotrophic lateral sclerosis (ALS) patients. Neurology 35(Suppl 1):128
Pierrot-Deseilligny E, Katz R, Hultborn H (1983) Functional organization of recurrent inhibition in man: changes preceding and accompanying voluntary movements. In: Desmedt JE (ed) Motor control mechanisms in health and disease, advances in neurology. Raven Press, New York, pp 443–458
Plaitakis A (1990a) Glutamate dysfunction and selective motor neuron degeneration in amyotrophic lateral sclerosis: a hypothesis. Ann Neurol 28:3–8
Plaitakis A (1990b) Glutamate alterations in amyotrophic lateral sclerosis. In: Clifford Rose F, Norris FH (eds) Amyotrophic lateral sclerosis. New advances in toxicology and epidemiology. Smith-Gordon, London, pp 265
Plaitakis A, Caroscio JT (1987) Abnormal glutamate metabolism in amyotrophic lateral sclerosis. Ann Neurol 22:575–579
Plaitakis A, Constantakakis E, Smith J (1988a) The neuroexcitotoxic amino acids glutamate and aspartate are altered in the spinal cord and brain in amyotrophic lateral sclerosis. Ann Neurol 24:446–449
Plaitakis A, Smith J, Mandeli J, Yahr MD (1988b) Pilot trial of branched-chain amino acids in amyotrophic lateral sclerosis. Lancet i:1015–1018
Plaitakis A, Mandeli J, Smith J, Yahr MD (1988c) Branched-chain aminoacids in amyotrophic lateral sclerosis. Lancet ii:680–681
Popper P, Micevych PE (1989) The effect of castration on calcitonin gene-related peptide in spinal motor neurons. Neuroendocrinology 50:338–343
Potvin AR, Tourtellotte WW (1985) Quantitative examination of neurologic functions, vols I and II. CRC Press, Boca Raton
Powers RK, Marder-Meyer J, Rymer WZ (1988) Quantiative relations between hypertonia and stretch reflex threshold in spastic hemiparesis. Ann Neurol 23:115–124
Powers RK, Campbell DL, Rymer WZ (1989) Stretch reflex dynamics in spastic elbow flexor muscles. Ann Neurol 25:32–42
Pritchett DB, Luddens H, Seeburg PH (1989) Type I and Type II $GABA_A$-benzodiazepine receptors produced in transfected cells. Science 245:1389–1392
Prochazka A, Hulliger M (1983) Muscle afferent function and its significance for motor control mechanisms during voluntary movements in cat, monkey and man. In: Desmedt JE (ed) Motor control mechanisms in health and disease. Raven Press, New York, pp 93–132
Røed A (1989) Effects of dantrolene on twitch and tetanic contractions of the rat phrenic nerve-diaphragm preparation. Arch Int Pharmacodyn Ther 297:260–271
Ross J, White SR (1986) Modulation of hypoglossal motor neurone excitability by thyrotropin releasing hormone (TRH) and serotonin. Soc Neurosci Abstr 12: 153
Rothman SM, Olney JW (1987) Excitotoxicity and the MDA receptor. Trend Neurosci 10:244–302
Rothstein JD, Tsai G, Kuncl RW et al. (1990) Abnormal excitatory amino acid metabolism in amyotrophic lateral sclerosis. Ann Neurol 28:18–25
Schmidt RF, Vogel ME, Zimmermann M (1967) Die Wirkung von Diazepam auf die prasynaptische Hemmung und andere Ruckenmarksreflexe. Naunyn Schmiedebergs Arch Pharmacol 258:69–82
Schmidt-Achert KM, Askanas V, Engel WK (1984) Thyrotropin-releasing hormone enhances choline acetyltransferase and creatine kinase in cultured spinal ventral horn neurons. J Neurochem 43:586–589

Schoenen J (1988) Neuroanatomie chimique de la moelle épinière humaine. Rev Neurol 144:630–642

Schofield PR, Darlison MG, Fujita N et al. (1987) Sequence and functional expression of the $GABA_A$ receptor show a ligand-gated receptor super-family. Nature 328:221–227

Schwartz MA, Koechlin BA, Postma E, Palmer S, Krol G (1965) Metabolism of diazepam in rat, dog and man. J Pharmacol Exp Ther 149:423–435

Seeburg PH, Pritchett DB, Luddens H, Shivers BD (1990) Structural and functional heterogeneity of $GABA_A$ receptors. Neurochem Abst 21:191

Sendtner M, Schmalbruch H, Stockli KA et al. (1992) Ciliary neurotrophic factor prevents degeneration of motor neurons in mouse mutant progressive neuropathy. Nature 358:502–504

Sengelaub DR, Arnold AP (1989) Hormonal control of neuron number in sexually dimorphic spinal nuclei of the rat. I. Testosterone-regulated death in the dorsolateral nucleus. J Comp Neurol 280:622–629

Sengelaub DR, Jordan CL, Kurz EM, Arnold AP (1989a) Hormonal control of neuron number in sexually dimorphic spinal nuclei of the rat. II. Development of the spinal nucleus of the bulbocavernosus in androgen-insensitive (Tfm) rats. J Comp Neurol 280:630–636

Sengelaub DR, Nordeen EJ, Nordeen KW, Arnold AP (1989b) Hormonal control of neuron number in sexually dimorphic spinal nuclei of the rat. III. Differential effects of the androgen dihydrotestosterone. J Comp Neurol 280:637–644

Siegfried J, Rea GL (1987) Intrathecal application of baclofen in the treatment of spasticity. Acta Neurochir (Suppl) 39:121–123

Sillevis Smitt PAE, DeJong JMBV (1989) Animal models of amyotrophic lateral sclerosis and the spinal muscular atrophies. J Neurol Sci 91:231–258

Skerritt JH, Johnston GAR (1983) Enhancement of GABA binding by benzodiazepines and related anxiolytics. Eur J Pharmacol 89:139–198

Spencer PS (1990) Linking cycad to the etiology of Western Pacific amyotrophic lateral sclerosis. In: Clifford Rose F, Norris FH (eds) Amyotrophic lateral sclerosis: new advances in toxicology and epidemiology. Smith-Gordon, London, p 29–34

Spencer PS, Ludolph A Dwivedi MP, Roy DN, Hugon J, Schaumburg HH (1986) Lathyrism: evidence for role of the neuroexcitatory amino acid BOAA. Lancet i:1066–1067

Spencer PS, Nunn PB, Hugon J, Ludolph AC, Ross SM, Roy DN, Robertson RC (1987) Guam amyotrophic lateral sclerosis-parkinsonism-dementia linked to a plant excitant neurotoxin. Science 237:517–522

Stratten WP, Barnes CD (1971) Diazepam and presynaptic inhibition. Neuropharmacology 10:685–696

Strong MJ, Brown WF, Hudson AJ, Snow R (1988) Motor unit estimates in the biceps-brachialis in amyotrophic lateral sclerosis. Muscle Nerve 11:415–422

Study RE, Barker JL (1982) Diazepam and (–)pentobarbital: fluctuation analysis reveals different mechanism for potentiation of gamma-aminobutyric acid responses in cultured central neurons. Proc Natl Acad Sci USA 78:7180–7184

Tahmoush AJ, Heiman-Patterson TD, Tahmoush GP, Francis ME (1985) Single fibre electromyography (SFEMG) studies in patients with amyotrophic lateral sclerosis before and during TRH infusion. Muscle Nerve 8:613–614

Tanaka R (1983) Reciprocal Ia inhibitory pathway in normal man and in patients with motor disorders. In: Desmedt JE (ed) Motor control mechanisms in health and disease. Raven Press, New York, pp 433–441

Thilmann AF, Fellows SJ, Garms E (1991) The mechanism of spastic muscle hypertonus. Variation in reflex gain over the time course of spasticity. Brain 114:233–244

Thomson AM (1989) Glycine modulation of the NMDA receptor/channel complex. Trends in Neurosci 12:349–353

Thomson AM, Walker Ve, Flynn DM (1989) Glycine enhances NMDA-receptor mediated synaptic potentials in neocortical slices. Nature 338:422–424

Tse FL, Jaffe JM, Bhuta S (1987) Pharmacokinetics of orally administered tizanidine in healthy volunteers. Fundam Clin Pharmacol 1:479–488

Turski L, Klockgether T, Schwartz M, Turski WA, Sontag K-H (1990) Substantia nigra: a site of action of muscle-relaxant drugs. Ann Neurol 28:341–348

Twyman RE, Rogers CJ, Macdonald RL (1989) Differential regulation of γ-aminobutyric acid receptor channels by diazepam and phenobarbital. Ann Neurol 25:213–220

Uchida H, Nemoto H, Kinoshita M (1986) Action of thyrotropin-releasing hormone (TRH) on the occurrence of fibrillation potentials and miniature end plate potentials (MEPPs) – an experimental study. J Neurol Sci 76:125–130

Van den Bergh P, Kelly JJ Jr, Adelman L, Munsat TL, Jackson IMD, Lechan RM (1987) Effect of spinal cord TRH deficiency on lower motor neuron function in the rat. Muscle Nerve 10:397–405

van den Pol AN, Gorcs T (1988) Glycine and glycine receptor immunoreactivity in brain and spinal cord. J Neurosci 8:472–492

van der Ploeg RJO, Fidler V, Oosterhuis HJGH (1991) Hand-held myometry: reference values. J Neurol Neurosurg Psychiatry 54:244–247

Verrier M, MacLeod S, Ashby P (1975a) The effects of diazepam and phenobarbital. Ann Neurology 25:213–220

Verrier M, MacLeod S, Ashby P (1975b) The effects of diazepam on presynaptic inhibition in patients with complete and incomplete spinal cord lesions. Can J Neurol Sci 2:179–184

Watkins JC, Olverman HJ (1987) Agonists and antagonists for excitatory amino acid receptors. Trends Neurosci 10:265–272

Weiss JH, Koh J, Choi DW (1989) Neurotoxicity of β-N-methylamino-L-alanine (BMAA) and β-N-oxalylamino-L-alanine (BOAA) on cultured cortical neurons. Brain Res 497:64–71

Weiss JH, Choi DW (1990) Non N-methyl-D-aspartate receptors and amyotrophic lateral sclerosis. In: Clifford Rose F, Norris FH (eds) Amyotrophic lateral sclerosis: new advances in toxicology and epidemiology. Smith-Gordon, London, pp 283–286

Westlund KN, Bowker RM, Ziegler MG, Coulter JD (1984) Origins and terminations of descending noradrenergic projections to the spinal cord of monkey. Brain Res 292:1–16

White SR (1985) A comparison of the effects of serotonin, substance P and thyrotropin-releasing hormone on excitability of rat spinal motoneurons in vivo. Brain Res 335:63–70

White SR, Crane GK, Jackson DA (1989) Thyrotropin-releasing hormone (TRH) effects on spinal cord neuronal excitability. In: Metcalf G, Jackson IMD (eds) Thyrotropin-releasing hormone: biomedical significance. Ann NY Sci 553:337–350

Whitehouse PJ, Wamsley JK, Zarbin MA, Price DL, Tourtellotte WW, Kuhar MJ (1983) Amyotrophic lateral sclerosis: alterations in neurotransmitter receptors. Ann Neurol 14:8–16

Winokur A, Beckman AL (1978) Effects of thyrotropin-releasing hormone, norepinephrine and acetylcholine on the activity of neurons in the hypothalamus, septum and cerebral cortex of the rat. Brain Res 150:205–209

Winokur A, Manaker S, Kreider MS (1989) TRH and TRH receptors in the spinal cord. In: Metcalf G, Jackson IMD (eds) Thyrotropin-releasing hormone: biomedical significance. Ann NY Acad Sci 553:314–324

Wright J, Rang M (1990) The spastic mouse. And the search for an animal model of spasticity in human beings. Clin Orthop 12–19

Wuis EW, Rijntjes NV, Kleijn van der E (1989) Whole-body autoradiography of 14C-dantrolene in the marmoset monkey. Pharmacol Toxicol 64:156–158

Wuis EW, Dirks MJ, Vree TB, Van der Kleijn E (1990) Pharmacokinetics of baclofen in spastic patients receiving multiple oral doses. Pharmaceutisch Weekblad: Scientific Edition 12:71–74

Yarbrough GG (1976) TRH potentiates excitatory actions of acetylcholine on cerebral cortical neurones. Nature 263:523–524

Young RR (1973) The clinical significance of exteroceptive reflexes. In: Desmedt JE (ed) New developments in electromyography and clinical neurophysiology, vol 3. Karger, Basel, pp 697–712

Young RR (1987) Physiologic and pharmacologic approaches to spasticity. Neurol Clin 5:529–539

Yu W-HA (1989) Survival of motoneurons following axotomy is enhanced by lactation or by progesterone treatment. Brain Res 491:379–382

Zbinden G, Randall LO (1967) Pharmacology of benzodiazepines: laboratory and clinical correlations. Adv Pharmacol 5:213–291

Zierski J, Mueller H, Dralle D, Wurdinger T (1988) Implanted pump systems for treatment of spasticity. Acta Eurochir (Suppl) 43:94–99

16 The Management of Motor Neuron Disease

R. Langton Hewer

Introduction

Motor neuron disease is a condition that most health service professionals hope that they themselves will never suffer. Its relentless course, the lack of effective therapy, and the frightening nature of some of the symptoms, underlie this fear.

The disorder poses considerable ethical, logistical and educational problems. The ethical problems (Carey 1986) involve such matters as the use of ventilatory support, artificial methods of feeding, and of narcotic drugs during the last phases of the disorder. The logistical and educational problems arise from the relative rarity of the disorder, and the fact that many doctors and nurses have little experience of dealing with chronic and progressive bulbar and respiratory paralysis. Possible solutions include more education of medical and paramedical staff and the concentration of patients in one place, such as has occurred in some neurological centres in the USA and in a few hospices in the UK.

Patients and their relatives should expect to be able to obtain efficient help throughout the course of the disease – from diagnosis to death – wherever they live. This includes the competent management of the many symptoms and problems that occur. There is currently much discussion of minimum standards of care and of audit. These matters are discussed later in this chapter.

Background

Patients and relatives have described their experience from time to time (Carus 1980; Wilson 1982). One of the most well known was Roger Carus, who wrote an account in the *British Medical Journal* entitled "Motor Neuron Disease: a Demeaning Illness" (Carus 1980). He said "I have the highest respect for my consultant neurologist as a man who can label and diagnose neurological illnesses, but in the context of my illness he is no more use to me than the milkman". Carus was particularly bitter that he was not told the diagnosis in a straightforward way. However, he expressed gratitude to his general practitioner: "To his eternal credit, my G.P. has never stopped coming to see me, no matter what the circumstances, even though he can do nothing constructive".

A study of 42 motor neuron disease patients in Bristol (Newrick and Langton Hewer 1984) found considerable room for improvement. There were particular problems with symptom control, co-ordination, access to neurological expertise, and the supply of aids and equipment. There do not appear to be any other published attempts to audit the quality of services for MND patients either in the UK or in other parts of the world. However, discussion within the Motor Neurone Disease Association (MNDA) of Great Britain, and with others, indicates that major problems are still occurring. Nevertheless, interest in the management of motor neuron disease has increased considerably during the last few years, and there have been a number of reviews (Mulder 1979; Norris et al. 1985; Cochrane 1987).

Epidemiology

If an incidence rate of 1.5–2 per 100 000 is accepted, there will be between 3 and 5 new cases per quarter of a million population per year (an "average" health district in the United Kingdom). If an average survival of 3 years is accepted the approximate prevalence per quarter of a million population is 13–15 cases. Our experience (Newrick and Langton Hewer 1984) is that 8 or 9 of these will have a bulbar problem and that 6 or 7 will be using a wheelchair. The "average" general practitioner in the UK (having 2000 persons on his list) will see a "new" patient with MND about once every 25 years.

It is worth recalling that a notional population of a quarter of a million will contain 250–300 patients with multiple sclerosis; 400 with Parkinson's disease; and about 1200 who have survived a stroke, of whom about 800 will be permanently disabled (Wade and Langton Hewer 1987). MND is thus a relatively small contributor to the overall pool of neurological disability. These data have important implications for the management of MND, and are discussed below.

Who Looks After MND Patients?

MND is usually thought of as being a disorder that is managed by neurologists. The reality is, however, different. An unpublished analysis of 1276 discharges and deaths from hospital beds in four regions of England in 1988 and 1989 shows that 26% were discharged from, or died in, neurological beds. The comparable figure for general medicine/geriatrics was 46%. There are no published figures, and the position elsewhere in the world is unknown. As far as England is concerned, it is clear that any attempt to improve the care of MND patients in hospital must be targeted, at least in part, on general physicians and geriatricians.

The Organisation of Services

The precise way in which services for MND patients are organised will depend on many factors. The most important is probably the existence of someone with interest, knowledge and commitment to organise the local services. In the UK this could be a consultant in rehabilitation/disability.

It is suggested that the following principles should underlie a "model" plan for MND patients:

1. *Overall* the objective is to achieve the best possible service at the lowest possible cost.
2. *Where practical* existing services should be used, shared, and supplemented. Supplanting existing services with new ones is not usually appropriate. For example, many of the resources needed by MND patients are also required by those with multiple sclerosis, Parkinson's disease, and stroke. Thus, the district nursing service will be used by all these groups of patients.
3. Adequate training of relevant personnel, medical, nursing, and social work is of great importance.
4. The numbers of medical and paramedical staff involved at any one time, during the last stages of the disorder, should be reduced to a minimum, preferably not more than three or four professionals.
5. The problem of co-ordination and continuity of care must be solved. This may be best done by ensuring that one person has responsibility for co-ordinating activities in any given case. The possible role for a "keyworker" is discussed below.
6. As far as possible care and help should be provided *within* the health district. Travelling far afield should, if possible, be avoided. In some instances it may be appropriate for adjacent health districts to pool resources. Much will depend on local circumstances and geography.
7. The patient must have access to appropriate medical expertise for such matters as diagnosis, advice on prognosis, and the prescription of appropriate medication (e.g. opiates, later in the course of the disorder). Expert help regarding gastrostomy and ventilatory assistance must be available.
8. Equipment: there should be, locally, the ability to obtain advice about, and supply of, equipment, including wheelchairs, splints, beds, and hoists.
9. Home alterations: it should be possible to undertake home alterations, such as widening of doors to take a wheelchair, ramping of steps, etc. without delay.
10. There should be the appropriate support services for the carer. These will include respite care beds (see later) and some form of home support system, such as "Crossroads".
11. Finances: it is essential to ensure that the patient and his family receive all the financial allowances to which they are entitled by the State. In the UK, this would include mobility and invalidity benefit.
12. There should be a district service for the management of bulbar and respiratory problems. This should be available for patients in their own homes and in nursing homes, as well as in hospitals.
13. Full use should be made of the local section of the Motor Neurone Disease Association, or any similar organisation. These bodies frequently provide much support and factual information, for example, booklets and other written advice, as well as facilitating the provision of mechanical aids to daily living and of nursing and social support.

The Hospital

Initially, the patient may need to be admitted to hospital for investigation. Electrophysiological tests, MRI scanning of the spine, and other investigations may be required, and sometimes these are best performed in hospital, and possibly in a regional neurosciences centre.

Later in the course of their disease the local district general hospital (DGH) will probably function as a base and as a resource for the management of patients with motor neuron disease.

Hospital Care

Admission to hospital or hospice may be needed during the last stages of the disorder in the following situations:

1. Crisis admission. Thus, if there is a sudden deterioration in the patient's health or if there is a domestic crisis such as the carer becoming suddenly ill, the patient may need to be admitted to hospital without delay.
2. Intermittent admission/respite care. This may be required in order to give the carer a planned "break" from her/his responsibilities.
3. Terminal care. In many cases the terminal hours and days are spent in hospital.

As mentioned elsewhere in this section, it is essential that the nursing staff on the wards are fully trained and knowledgeable about the particular problems that occur in MND problems.

Outpatient Departments

Outpatients will usually be seen in the hospital. Both the medical and rehabilitation needs can be assessed by a multidisciplinary team at this time. Patients should not be supervised by untrained staff whether medical or otherwise.

Keyworker

The concept of the keyworker (KW) has been in existence for some years. The need for such a person arose because of the multiplicity of problems that can occur during the course of a disorder such as motor neuron disease. There is a clear need to co-ordinate the patient's care and to ensure that he/she has access to all appropriate services. It is *not* the duty of the keyworker to take over all the duties of other professionals involved.

Background

The keyworker can come from a wide variety of backgrounds. Remedial therapy, nursing, or social work may be appropriate. In some instances, the keyworker may be a doctor. The person concerned would usually be working within the health service, and it is not usually suggested that an extra salary would be payable. Occasionally, a relative of the patient may fulfil part of the role of a keyworker.

Training

It is essential that the keyworker should be thoroughly familiar with all the common problems experienced by MND patients. He/she must also be familiar with the rehabilitation facilities available within the locality.

Role of the Keyworker

The precise role of the keyworkers will depend on what is available within the locality. The co-ordinating role has already been mentioned, and is clearly important. They may also be involved with giving advice on many areas of care, including mobility, dysphagia, etc.

Time

The amount of time needed for keyworkers to discharge their responsibility in relation to MND patients is at the moment uncertain, and work studies are needed to clarify this point. Similarly, it is unclear as to what the caseload might be. However, a given keyworker will probably not look after more than a few patients at any one time.

The keyworker should have a close link with the local Motor Neurone Disease Association, or equivalent organisation.

Much more work needs to be done to define the role of keyworkers and other staff in the efficient and sympathetic management of MND patients.

Appendix I consists of a suggested model for the management of an MND patient and identifies, particularly, the role that might be filled by the consultant, general practitioner, and keyworker.

Phase I: The Stage of Diagnosis

The Diagnosis

In most instances – probably about 80% of cases – the diagnosis of MND is easy. Indeed, an experienced neurologist can frequently make the diagnosis within minutes of seeing the patient. In about 10% of cases diagnosis is more difficult, and in a further 10% the diagnosis may not be made for several months. Presentations which may cause particular difficulty include the following:

1. Monoparesis. An upper motor neuron lesion, without wasting, confined to one limb, can pose particular problems.
2. Paraparesis with minimal sensory loss. Cervical myelopathy associated with cervical spondylosis can be confusing, particularly if there is wasting of the hand muscles. The coexistence of two conditions may need to be considered.
3. Wasting of the muscles of one limb. Three common examples are wasting of one hand, which may be caused by cervical rib; a wrist drop possibly caused by a

radial nerve lesion; and a foot drop caused by a common peroneal lesion, or compression of the L5 nerve root.

4. Miscellaneous group. Confusion may be caused in other situations, including dysarthria of uncertain cause, where the patient is very young, or where there is an exceptionally long history extending over many years.

Persistent diagnostic uncertainty makes management difficult. Full investigation of such cases is essential, and a second neurological opinion may be desirable. Paradoxically, a straightforward diagnosis also poses problems. Thus, distress may be caused if there is no need for investigation and the diagnosis is given at the first consultation. Most neurologists avoid doing this, and indeed, many would admit the patient to hospital for investigation, even if this is not strictly speaking essential.

At this early stage the general practitioner and consultant will share the "lead" role.

"Treatment"

No medical treatment has been shown to influence the course of the disorder. Older textbooks record a wide variety of medications, including vitamin B_{12} injections. The majority of physicians today would not indulge in what is, in reality, subterfuge.

However, randomized trials of a variety of putative therapeutic agents have been undertaken during the last few years and some trials are currently in process. These include trials of TRH analogues, amino acid mixtures, and various antiviral agents.

A practical problem is that drug trials are sometimes inevitably conducted in centres which may be situated at a distance from the patient's home. This may involve considerable expense, discomfort, and inconvenience, with repeated long journeys. These facts need to be borne in mind when the patient is considering participation in such clinical trials. However, many patients desperate for a remedy will do anything to obtain one, however slim the chance and most are enthusiastic supporters of the need for research into the disease and its treatment.

Prognosis

It is not possible to give a firm prognosis early in the course of the disease. However, the early development of a pseudobulbar palsy is usually a sign of a poor prognosis. Lower motor neuron signs confined to the limbs may imply a more favourable outlook. Other features that need to be considered include the patient's pre-existing physical and mental health, the presence of coexisting disease, and the patient's age. Other important non-medical factors include educational background, financial position, housing, and the health of the partner. All the above factors may be important in drawing up a rough plan of action which may need to cover the next 2–3 years. Some 15%–20% of patients may survive 5 years or more, but the mean survival from the time of diagnosis is about 3 years.

Telling the Diagnosis

In general, the patient and partner should be told the diagnosis once this is reasonably well established. However, certain points need to be borne in mind:

Wishing to Know. A few patients clearly do not wish to be told the diagnosis, or to have the prognosis spelled out. Obviously, their wishes should be respected, although these may change with time.

Public Knowledge. MND was once a disorder totally unknown to the public. Today, however, various well known personalities, such as the late David Niven, are known to have had the disorder, and it is now easier to establish a reference point. Furthermore, the MNDA has advertised widely on hoardings and in the press, and many people have at least heard the name.

Literature. The existence of high-quality lay literature (as supplied by the MNDA), to which both patients and professionals have contributed, is useful. Leaflets, suitable for patients in both the early and later stages of the disorder, are available.

The Interviews. It is well known that patients take in very little information after they have been told a very serious diagnosis. For this reason, only a small amount of information should be given at the first interview. A second interview, 2 weeks later, is usually desirable. At this time, further information can be given, and the patient should be encouraged to ask questions.

Many patients are afraid to ask the doctor a large number of questions for fear of appearing ignorant, or of wasting his time. It is often desirable for the interview to involve someone else on the team, such as the social worker, who can provide further information and support later.

Certain Particular Points. Many patients will have some preconceived ideas as to what is wrong with them. It is sometimes helpful to obtain this information. Certain more specific points can be made:

1. The patient does not suffer with multiple sclerosis or cancer.
2. Various faculties remain intact, generally including intellect, sexual and sphincter function.
3. The disorder is not familial and it is therefore unlikely that the patient's children will inherit the disorder.
4. Pain does occur in some patients but can usually be controlled.

Research. The existence of current research projects in most developed countries should be mentioned so that the patient and his partner realise that efforts are being made to find an effective treatment for the disorder.

Second Opinion. The diagnosis of motor neuron disease can be regarded as a death sentence. In this situation many patients feel the need of a second opinion. In Bristol this service is routinely provided by other neurologists working in the department.

Abandonment. It must be made clear to patients that they will not be abandoned by the consultant and the team. Continued support throughout the course of the disorder must be available, and there should be the offer of rapid advice by telephone in the event of a crisis. Furthermore, it must be indicated that patients will not be required to attend a huge outpatient clinic, where they may be seen by a junior doctor who has little knowledge of the disorder.

Quality of Life. The patients should be encouraged to lead a normal life for as long as possible. They may wish to ask various questions related to such topics as employment and personal finances.

The method and manner by which the diagnosis is told is of the greatest possible importance. Many patients regard this as the single most important event that occurred during the course of their illness. The whole pattern of later management may well be determined by the relationships built up at this early stage. It is well worth investing time and thought at this critical point. At this stage also, it may well be appropriate to introduce patients and their partners to a keyworker and, at the same time, to put them in touch with the MNDA.

Phase II: The Stage of Deterioration Without Major Disability

This stage may last for 12 to 18 months. The consultant plays a major role at this time, and should see the patient as often as seems required, but unnecessary hospital attendances should be avoided. The patient should know that the consultant is available if needed. The following particular points should be noted:

1. The patient may have increasing difficulty with his or her job and may need advice about this.
2. Advice regarding driving may be needed, and in some instances modifications to the car may be required.
3. He or she will ask for information about MND research and may request permission to participate in drug trials.
4. There is a danger that the patient will undertake untried and expensive therapies which may actually do harm. Advice will be needed about this.

Phase III: The Phase of Major Disability with Multiple Problems

The third phase will probably be reached within 18–24 months of onset. It is at this point that major management problems can occur, and that social services are liable to "break down".

The roles of the consultant, keyworker and general practitioner are outlined in Appendix I. The consultant's role particularly involves being available as necessary, advising on the management of various problems such as drooling, swallowing and speech difficulties and pain, and arranging admission to hospital as necessary. The keyworker now becomes extremely important. The various disability services within the health district must be mobilized and efficiently organized.

Bulbar Problems in Motor Neuron Disease

Before discussing the detailed management of dysphagia and related problems (see also Enderby and Langton Hewer 1987), it is perhaps worth making some initial points:

1. Many of the therapies recommended in the literature have not been properly evaluated. This is partly due to the absence, until recently, of objective assessments of bulbar dysfunction.
2. The effective management of bulbar problems requires considerable skill and experience. The best results are undoubtedly obtained by staff who have acquired wide experience. Conversely, poor results will probably be obtained if the staff do not understand the problems and have little experience.
3. Severe bulbar problems are uncommon compared with many of the other disabilities encountered in hospital and elsewhere, such as incontinence and immobility. The result is that patients may be managed by nurses, and others, who have little or no experience. For instance, a patient with MND and bulbar palsy may be admitted to a general medical ward as an emergency because he cannot swallow. He or she is likely to be the *only* patient on that ward with major swallowing problems. Difficulty with breathing and with speech may also be present. Low levels of nursing staffing, particularly at night, are likely to compound the problem. Patients finding themselves in this situation are likely to become extremely frightened. How often needless death occurs is unknown, but clearly the possibility exists.
4. Education of staff at all levels is of the greatest importance. For instance, written guidance should always be available in any ward where there are patients with bulbar problems.
5. As discussed elsewhere in this chapter there should, ideally, be a 24-hour dysphagia (and bulbar) advice service within the health district so that assistance with difficult cases can always be obtained from an expert.

Salivary Dribbling (Drooling)

The normal person produces about 2–3 litres of saliva per day (Crossner 1984). Most people swallow automatically about 600 times per day (Lear et al. 1965). The frequency of swallowing is greatest during eating, and least during sleep, with other activities occupying an intermediate place. Records in normal people made during sleep invariably show periods of 20 minutes or more during which deglutition is absent. No published studies of swallowing frequency in motor neuron disease have been found.

The effective disposal of saliva depends on many factors, including the production of the normal amount of saliva, the ability to produce a good lip-seal, and the ability to swallow normally.

Persistent salivary dribbling can be a major source of distress. The following are some of the results:

1. The problem is sometimes perceived as being associated with mental deficiency and psychiatric disorder. In the present context this is, of course, nonsense. Nevertheless, reassurance is necessary.

2. Constant mopping of the lips and mouth is needed, either by the patient or by the carer.
3. The patient may need to wear some sort of protective "bib" to absorb the saliva. This by itself can be humiliating and invoke associations with early childhood.
4. The most important result is that of social embarrassment and isolation. The patient no longer feels able to be seen in public. Loss of friends and acquaintances can easily occur as a result.
5. The corners of the lips become painful and sore.
6. Drooling is yet another symptom that contributes to exhaustion.
7. The flow of air through a pool of saliva at the back of the throat can cause an unpleasant noise, which is distressing for the patient and his carer.
8. The collection of pools of saliva in the back of the throat is almost certainly an important cause of nocturnal coughing and choking.

Contributory Factors and their Management

Many factors contribute to drooling. The first part of management concerns a general explanation of the physiology and anatomy of what has happened. It is necessary to explain the normal mechanism of salivary production and to emphasize firstly that the problem is not a sign of mental deficiency, and secondly, that patients will not drown in their own saliva.

1. Correct head position. Saliva tends to pool in the floor of the mouth and/or in a vallecular between the teeth/gums and lower lips, which tend to prolapse forwards and downwards. If the head is allowed to fall forwards the saliva will dribble out. Correct head positioning is important in the management of drooling and also of other problems such as dysphagia. Other factors are involved, including finding the correct chair in which to sit, the possible use of a collar, and adjusting the amount of anteroposterior tilt of the body.
2. Infection in the mouth. Monilial infection of the mouth is very common. It is likely that this contributes to the excessive production of saliva. Fungal infections should be treated in the usual way with frequent mouth hygiene, and with antifungal agents such as nystatin or amphotericin.
3. Loss of the automatic swallow reflex. Most normal people swallow automatically once or twice a minute. This may become depressed in MND. The swallow reflex can sometimes be triggered by the chewing of sweets or chewing gum. However, this technique should be used with the greatest possible care; particularly if there is any danger of choking.
4. Weakness of lip closure. Most patients with motor neuron disease are unable to whistle. This is because they cannot get their lips together. The phenomenon may be due to weakness of the muscles, spasticity, or a combination of both. Exercises to improve the awareness of lip closure and to strengthen lip-seal may be of some assistance. For instance, the patient may be encouraged to hold a tongue depressor between the lips as he watches television. Other lip exercises, such as closing the lips against finger resistance or producing bilabial sounds such as "p" or "b" may be used (Enderby and Langton Hewer 1987).

5. Reduction of spasticity. It is probable that bunching of the tongue can impede the automatic swallow and also form a mechanical barrier between the mouth and the pharynx when it is bunched up against the palate. It is possible that tongue spasticity can be reduced either by the local application of ice (either by external application to the neck, or by sucking), or by the use of anti-spasticity drugs such as baclofen).
6. Avoidance of precipitating factors. Various drugs, particularly Mestinon, may increase salivary production. They should be avoided if drooling becomes a problem.
7. Medication. Various anticholinergic drugs can be used to reduce oropharyngeal secretion. The main drugs used orally are amitriptyline, propantheline, and hyoscine. The latter can be given by transdermal application.
8. Dietary change. There is some suggestion that dairy foods (e.g. milk and cream) can increase the volume and tenacity of mouth secretions. An experimental withdrawal of dairy products may be worthwhile (Enderby and Langton Hewer 1987).
9. Cosmetic. In some cases it may be impossible to avoid having some sort of absorptive material on the front of the patient's chest. This can sometimes be discreetly hidden. A bag for disposing of damp tissues should always be available. It is important that the patient should be allowed to look as normal as possible.

   In some instances, it may be helpful to have available a portable suction device which can be used either by the patient or by the carer. In this way pools of saliva can be readily removed. This equipment is usually readily acceptable to patients and their carers.
10. Surgical measures. Occasionally, drooling cannot be controlled by conservative measures. A few patients may benefit by the use of transtympanic neurectomy. This technique involves section of the chorda tympani in the middle ear (Zalin and Cooney 1974). The technique is difficult to evaluate but the few patients who have had the operation, in our experience, do seem to notice less drooling for at least a few months, although some complain of loss of the sense of taste.

    Other techniques can theoretically be used, including transposition of the salivary ducts to the back of the throat, and irradiation of the salivary glands. These procedures are not widely used, and irradiation of the salivary glands appears relatively ineffective in reducing the volume of saliva produced.
11. The problem of thick secretions. Many patients produce very thick, tenacious secretions. These can impair swallowing and talking. Precisely why these secretions occur is uncertain, but, as mentioned above, the problem may be made worse by the use of anticholinergic drugs. Excessive mouth breathing may be another factor. The possibility that increased viscosity of saliva may be related to intake of dairy products has already been mentioned.

One of the major problems in dealing with drooling is that we do not have any adequate method of measuring salivary production. Cannulation of salivary ducts is impractical. This means that it is almost impossible to evaluate either the severity of the symptom, or the efficacy of treatment methods. One's overall impression is that the problem of drooling is not well managed, and that there is scope for considerable improvement in our management techniques. Research is clearly needed in this area.

Dysphagia

Difficulty with swallowing probably affects between a third and a half of all MND patients at any one time. However, most patients will experience these problems at some point during their illness. Associated symptoms include choking, dehydration, dribbling, weight loss, exhaustion, and inhalation pneumonia.

The act of swallowing can be arbitrarily divided into five phases. All phases can be affected in patients with motor neuron disease. In any given patient one or more phases may be involved.

1. Pre-buccal. This involves the act of getting the food and fluid to the mouth. Correct positioning of the relevant body parts is important.
2. Early buccal phase. This involves accepting the food into the mouth, keeping the food and fluid in the mouth by lip-seal and by correct positioning of the head, and mastication of the food. This phase may be interrupted if lip closure is weak or if the head is allowed to fall forwards so that mouth contents escape through open lips. During this phase, the tongue is bunched up at the back of the throat acting as a valve to prevent oral contents prematurely entering the pharynx.
3. Late buccal phase. The tongue contracts in such a way that semi-solid masticated food is propelled in a bolus to the back of the throat. Fluid is moved backwards by the formation of a trough in the middle of the tongue. If the tongue is immobile because of spasticity, weakness or both, mouth contents may remain in the mouth almost indefinitely.
4. Pharyngeal phase. This phase involves the food bolus moving from the back of the mouth to the pharynx. A negative pressure is created by the palate, which seals the oronasal gap, and by the tongue and lips which prevent air getting into the mouth. Weakness or spasticity of the lips, tongue, palate or pharyngeal muscles, will result in food failing to enter the pharynx. The larynx elevates at this point, and the laryngeal inlet becomes sealed. Failure of this mechanism may result in material entering the air passages with resultant choking.
5. Oesophageal phase. In this phase the food and fluid leaves the pharynx, passes the cricopharyngeal (CP) sphincter and then moves down through the oesophagus. Sometimes the CP muscle fails to relax, thus preventing food from moving into the oesophagus. The possibility of surgical section of the CP muscle is discussed below.

Investigation of Dysphagia in Motor Neuron Disease Patients

The following are some of the most important points:

1. A careful history should be taken from the patient and the partner. Information should be collected about, for instance, how long the patient takes to eat an ordinary meal.
2. There should be careful observation of the patient eating and drinking a standard amount of fluid or food, e.g. 50 ml of water, and a biscuit with butter and jam. The activity can be timed and the presence of choking will be carefully noted.
3. Physical examination. This forms an important part of the evaluation of dysphagia and will include an assessment of the following:

lip closure and the ability to whistle
tongue strength, mobility and tone
palatal movement in response to tactile stimulation
evidence of 'palatal escape'
the quality and strength of the cough
the presence or absence of exaggerated jaw and facial reflexes (these invariably accompany bulbar spasticity)
the evidence of salivary pooling
the presence or absence of dysarthria.

4. Assessment of whether or not there has been inhalation of mouth contents. This involves ascertaining how often choking occurs, and whether or not there is any clinical or radiographic evidence of inhalational pneumonia.
5. Sequential measurements of body weight and of urinary output.
6. Cine-videofluoroscopy. This technique involves the swallowing by the patient of barium suspension of varying consistency, fluid and semi-solid. An analysis of the various stages of swallowing is made. Particular note is taken of the level of the cricopharyngeus and any tendency for the barium to enter the larynx. Opinions vary as to the value of this technique in the routine assessment of dysphagia in MND patients, but there is some evidence that fluoroscopy is necessary to reliably detect laryngeal penetration (Linden and Siebens 1983). Our experience is that the technique is of some, but limited, use. Much of the required information can be obtained by careful clinical assessment. An exception is the diagnosis of hold-up at the level of the CP muscle.

Management of Dysphagia

It is important that from the outset the clinician should have a good idea of the severity of the dysphagia, and what components of the swallowing process are affected. The mechanism of swallowing, and what has gone wrong, can be simply explained to the patient and partner.

Environment. The ability to swallow, as to micturate, is much influenced by environment. Slowness of eating, dribbling, and choking, may cause severe embarrassment. As a result anxiety is increased and swallowing becomes impossible. The patient should be encouraged to eat somewhere where he feels relaxed and comfortable. Some people prefer to eat on their own.

Positioning. Faulty positioning of the patient, or of the food, can adversely affect swallowing. It is obviously important that the food should be easily accessible. A non-slip mat may sometimes be useful. The position of the head is important. Thus, swallowing will be difficult if the head falls forward. Some form of head support may be needed, and it is sometimes helpful to tip chair and patient backwards slightly so that the head falls backwards by gravity. Some patients find that swallowing is made easier by turning the head to one side. If a collar is used it will be important to ensure that chewing is not mechanically hindered.

Diet. The composition and texture of the diet is of great importance. Where possible the assistance of a dietician should be sought. The following points are noted:

1. It is advisable to avoid dry foods which are difficult for the patient to swallow. The food should be firm and smooth but not solid or unyielding. As the disorder progresses the consistency of the food can be varied by appropriate use of food blenders/mixers.
2. Certain foods are unsuitable. These include those that are crumbly, or those that contain solid material which floats in fluid (e.g. vegetable soup).
3. It is advisable to avoid food and drink which precipitates choking and coughing. Examples include highly spiced foods and whisky. Alcohol can be highly therapeutic but wine and sweet sherry are probably best and cause the least irritation. In each case experience will dictate what is acceptable.
4. Most patients find that semi-solids are easier to manage than fluids. If dehydration is a risk foods containing a substantial proportion of water should be used e.g. semi-solid soup containing gelatine.

Medication and Other Measures to Assist Swallowing

The following drugs can be of use:

1. Drugs that reduce spasticity. Baclofen in a daily dose of 80–90 mg may be helpful.
2. Pyridostigmine and neostigmine (Campbell and Enderby 1984). Some physicians use anticholinesterase drugs in an attempt to improve dysphagia and other bulbar functions. The author's experience is that these agents are not usually helpful. There is, furthermore, the danger that the problem of drooling will be made worse.
3. Monilial (thrush) infections of the throat, pharynx and oesophagus can affect swallowing adversely. They must be recognized and treated appropriately.
4. Palatal supports. There is evidence that a weak palate can be stimulated by a palatal loop attached to a plate or existing denture. Similarly, a flaccid palate can be mechanically supported by a palatal splint (Enderby et al. 1984).

Cricopharyngeal Myotomy (CPM)

Division of the fibres of the cricopharyngeus muscle has been practised for many years in an attempt to relieve swallowing difficulty and the technique is recommended in many neurological texts and papers. Preoperatively, it is necessary to undertake cine-videofluoroscopy. The technique may be indicated if there is marked hold-up of barium at the level of the CP muscle.

A review of five published papers has recently been undertaken (Langton Hewer and Enderby 1990). The five papers gave details of the results of CPM in 106 patients with motor neuron disease. The post-operative mortality varied between 6% and 30% in different series, the average being 14%. Benefit was reported in 100% of survivors in two studies, and in 64% and 68% in a further two studies, but detailed examination of these publications shows that in none was there any attempt to describe the improvement in meaningful terms such as speed of eating, frequency of choking, or weight gain. Reliance appears to have been placed entirely upon clinical impression. In the remaining study Norris et al. (1985) report a morbidity and treatment failure of less than 5%, but details were not given.

Overall, it appears that the technique of CPM has not been subjected to adequate scientific evaluation, and the value of the operation is not known. However, discussions with neurologists in the UK and elsewhere in the world indicate that many experienced clinicians have largely abandoned the procedure.

Alternative Methods of Feeding

Alternative methods of feeding should be considered when the patient can no longer swallow in the usual way. They may, however, be contraindicated if the patient is clearly dying, with "end-stage" MND, or if he/she expresses a wish against further intervention.

It is important that the medical and nursing staff have the requisite expertise and skills, and that whatever alternative feeding route is chosen the technique is efficiently handled. For instance, the passing of a nasogastric (NG) tube may result in gagging, and even vomiting, which can be highly dangerous when the airway is unprotected. Similarly, the establishment and operation of a gastrostomy requires the availability of an endoscopist with expertise in this area, and of trained nursing staff. Complications can occur and, as a result, a difficult situation can be made even worse. It is important to "do no harm" at this critical point in the course of the disorder.

Indications for Alternative Methods of Feeding. The following are suggested as reasonably firm indications:

1. Exhaustion. This may occur when the patient is taking an unacceptable length of time to eat and drink, and is thus becoming exhausted.
2. Frequent choking.
3. Inadequate fluid intake with resultant severe dehydration. In this connection, it is noted that some patients with MND can "manage" with only 300–400 ml fluid intake per day. There does not appear to be any published data on this point. However, most appear to feel much better when the dehydration is corrected.
4. Evidence of inhalation pneumonia.
5. Severe and continuing weight loss, particularly when the patient's neurological condition is not too severely compromised.

Nasogastric Feeding

No-one wants to have a nasogastric (NG) tube. However, much relief may be provided once the tube is in situ – after weeks and months of attempted oral feeding. Cosmetic aspects are important. All tubes protrude from the nose, although narrow-bore tubes are less obtrusive than wide-bore ones. Research is needed in this area. Sometimes a protruding tube may be hidden by a scarf. It is important that both the patient and his carer become familiar with what is involved in NG feeding. Their initial distaste for the idea can be overcome by well-trained, considerate staff who must be available to provide long-term support, particularly when the patient returns home.

The following points are noted:

1. Temporary feeding. Dysphagia may be made temporarily worse by upper respiratory tract infections, or by monilial infections. In these instances it may be possible to remove the NG tube when the infection has cleared.
2. Intermittent feeding. Intermittent feeding, for 1–2 days at a time, every 2 weeks, can occasionally be acceptable. It can be useful in giving the patient a "rest" and by correcting dehydration and low food intake.
3. All tubes are liable to cause mucosal irritation with an increase in secretions.
4. Narrow-bore tubes involve prolonged drip feeding, which means that the patient has to be attached to his reservoir for many hours during the day and night. Nonetheless, these tubes are usually preferred by most patients, and do not require frequent changing.

Nasogastric tube feeding has recently been supplemented by the advent of the technology of endoscopic gastroscopy.

Gastrostomy

The "old-style" gastrostomy involved the insertion into the stomach, via the abdominal wall, of a large-bore tube—usually under general anaesthetic. The giving of a general anaesthetic can be hazardous to MND patients, particularly when there is weakness of the respiratory muscle. Furthermore, experience has indicated that many patients disliked having a large-bore, weighty and wide tube protruding from the abdominal wall. More recently, percutaneous endoscopic gastrostomy, using a small-bore catheter inserted under local anaesthesia, has proved to be a simple, safe, and effective procedure for malnourished, dysphagic patients (Norris et al. 1985; Gauderer et al. 1980; Russell et al. 1984). It has been suggested that aspiration is less likely to occur with jejunal feeding than with feeding into the stomach, but there is no clear evidence on this point (Lazarus et al. 1990).

Oesophagostomy

Norris et al. (1985) found that a cervical oesophagostomy could be performed easily and safely using local or general anaesthesia. Within a month, a cutaneo-oesophageal fistula is formed, and the tube can be easily changed. A shirt collar or scarf may hide the tube. However, the technique has probably been largely replaced by percutaneous endoscopic gastrostomy, and few surgeons have any experience of the procedure.

Choking

Choking usually accompanies dysphagia and occurs initially mainly with fluids. The severity of choking will depend on many factors, including the strength of the cough and the amount of material that has entered the air passages.

Measures to Prevent Choking

1. Alleviation of anxiety. Anxiety is likely to be present, particularly if the patient has become frightened. The presence of other people who know what to do in an

emergency can be helpful. The patient can be reassured that death as a result of choking is uncommon in MND. However, there is little published work on the nature of death in MND and research is needed in this area.

2. Small frequent meals. It is usually recommended that the patient should take five or six small meals rather than two larger ones. This may do something to avoid exhaustion.
3. Types of food. It is wise to avoid foods and fluid which may precipitate choking, e.g. curry or whisky.
4. Consistency of food. Experience will determine which foods are most suitable in any given case. Semi-solid food prepared in a liquidizer may be appropriate.
5. Regurgitation. Some patients tend to choke shortly after a meal. The patient should remain in a sitting position for 30–40 minutes after eating.
6. Suction. A suction apparatus should be readily available both in hospital and when the patient is at home. Attendants must be trained in its use.

Management of Choking

It is most important that the people concerned should know what to do beforehand.

1. Empty the mouth. A sucker, if available, may be useful.
2. Act calmly and try to get the patient to relax and breathe slowly.
3. The patient should be encouraged to lean forward in his chair.
4. The carer may be taught to exert pressure on the abdomen in time with the patient's attempts at coughing.
5. The Heimlich manoeuvre (Goldblatt 1977) is occasionally needed, although as far as can be ascertained, no formal evaluation of its effectiveness has been undertaken. The patient is bent over a chair or the attendant's arm so that his head is lower than his chest. The attendant, standing behind the patient, and encircling him with his arms, hands interlocked, presses forcefully upwards on the upper abdomen, forcing the diaphragm upward. As a result, residual air in the chest cavity is expelled and any obstructing foreign body is dislodged.

Frequent choking may indicate the need to cease oral feeding. Occasionally, Teflon injections into the vocal folds (Goldblatt 1977) can be used to improve airway protection. The author has no experience of this technique.

Dysarthria and Communication Problems

Several factors impair speech production in MND. The bulbar problems frequently have both lower motor neuron (LMN) and upper motor neuron (UMN) components. The result is a combination of weakness, spasticity, and inco-ordination of movements. The disability gradually progresses so that about 75% of the patients will have major speech problems prior to death (Saunders et al. 1981). Communication aids of varying sophistication may be required if the patient is also unable to write.

The early problems involve palatal weakness, poor lip-seal, and impaired tongue movement. (Enderby and Langton Hewer 1987). As the disorder progresses, the speech becomes slow and slurred. The problem is compounded by impaired control of the vocal cords and associated breathing difficulties.

Dysarthria can be quantitatively analysed using a formal assessment such as the Frenchay Dysarthria Assessment (Enderby 1983).

Management of Dysarthria and Speech Problems

The following may be useful:

1. Decreasing the speed of speech. The use of short phrases and sentences may improve intelligibility.
2. Getting and retaining the attention of the listener. This may be helped by getting the patient to point to the first letter of each word on a letter board as he speaks, thus giving the listener a visual clue. It is also important that the listener should be properly instructed as to the nature of the communication disorder. Impatience only makes the problem worse.
3. Measures to reduce spasticity. To some extent, these have been discussed elsewhere. Occasionally, the use of local ice, and appropriate medication such as baclofen, may be helpful.
4. Dentures. Loose dentures may be an important factor in the genesis of dysarthria and should be dealt with.
5. Palatal lift. There is some evidence that a palatal loop and/or palatal lift may be helpful, particularly when there is marked hypernasality due to nasal escape of air (Enderby and Langton Hewer 1987; Enderby et al. 1984).

Communication Aids

It is worth remembering that the pen is the best communication aid and should obviously be used where possible. It is also important to consider the spouse, who may be deaf or have visual problems (Enderby and Langton Hewer 1987).

Most communication aids are operated by one finger, but other activating techniques can be used when the hands become useless. For example, a switch can be operated by head turning or tilting.

1. *Direct-select aid.* Examples include the Canon Communicator, or the Memowriter. The patient operates a keyboard after identifying the appropriate letter, word or phrase which he wishes to display. The Canon is a small electronic typewriter which can be attached to the wrist. The message is displayed on a thermosensitive tape. The overall size is 10×20 cm. The keys are rather close together and this can pose a problem. The Lightwriter may be preferred, the keys being larger and more widely spaced. The Memowriter has a memory, which can be used to display simple messages comprising several words.
2. *Scanning aids.* Scanning aids may be required if there is severe physical impairment. Each item, letter or symbol, is identified by a cursor light or pointer, which can be stopped by a switch. The Possum Communicator is one example. The use of this apparatus requires considerable patience on the part both of the patient himself and his family.

A very large number of communication aids are now available, and it is essential to obtain informed and up-to-date advice. Speech syntheses are available on a software package that can run on a modern portable or notebook computer at quite moderate cost. Male or female voices can be provided. This development promises to revolutionize speech and communication aids devices in the next few years. Speech Therapists are usually mainly involved with the management of dysarthria. The objective is to achieve some form of communication throughout the time course of the illness. Our experience is that this objective may well be attained in many patients up to the last few weeks. However, there is no good published evidence.

Respiratory Failure

The majority of deaths from MND are probably due to pulmonary complications of respiratory muscle weakness and/or bulbar paralysis (Caroscio et al. 1987; Howard et al. 1989; Fallat et al. 1979).

The Symptoms

The symptoms of respiratory paralysis in motor neuron disease include shortness of breath on mild exertion or when lying flat, difficulty with coughing and talking and insomnia and daytime somnolence.

Respiratory insufficiency is not always easy to diagnose, particularly as some of the symptoms may be due to associated factors. Thus, a poor cough may sometimes be due to weakness of the muscles controlling the vocal cords. Common early symptoms include headache and sleeplessness. Overall, the problem is almost certainly under-diagnosed. The following should be checked:

1. Respiratory rate
2. Chest expansion
3. Presence or absence of diaphragmatic paralysis (usually detected by paradoxical movements of the abdominal wall)
4. Strength of cough
5. Ability to count. Most "normal" persons can count up to 20 in one breath or blow out a match with the mouth open in a single breath
6. Chest radiograph
7. Arterial gases: PO_2 and PCO_2
8. Tests of respiratory function, notably vital capacity in different positions: lying, sitting and, if possible, standing, and possibly the forced expiratory volume (FEV1)
9. Observations of breathing at night to detect underventilation and sleep apnoea (Howard et al. 1989).

Management of Respiratory Insufficiency

General management includes the following:

1. The detection and prevention of inhalation pneumonia
2. The correct positioning of the patient. Many patients can breathe better if they have several pillows in bed. Frequent turning of the patient at night may be helpful
3. The judicious use of antibiotics at the first hint of pulmonary infection. Do not wait for infections to develop first, rather, treat in anticipation
4. Physiotherapy, particularly when the patient has difficulty with expectorating
5. As far as possible, contact with people (including doctors and nurses) who have significant upper respiratory tract infections should be avoided

If breathlessness causes distress during the later stages of the disease the use of small amounts of morphine, particularly at night, may be helpful. Significant respiratory depression does not usually occur if the initial dose is small and subsequent doses are *gradually* increased (Saunders et al. 1981).

Assisted Ventilation

For the majority of patients assisted ventilation is not a reasonable option. However, in a minority of cases, it can be considered:

1. Patients with known MND should not normally be intubated or subjected to intermittent positive pressure ventilation (IPPV). Failure to observe this general rule may lead to permanent ventilator dependency with associated trauma and distress to the patient and his family. Assisted ventilation is an option to be considered only after full discussion of the implications with family, neurologist and pulmonary physiology team.
2. A tracheotomy is almost never indicated for purposes of temporary assisted ventilation alone. However, it may occasionally be helpful if tracheal secretions cannot otherwise be aspirated.
3. Non-invasive negative pressure assisted ventilation may be acceptable in patients in whom symptoms due to respiratory insufficiency constitute a major limitation on the patient's activities. Howard et al. (1989) reported 9 patients who used a cuirass ventilator during sleep. All patients derived considerable subjective improvement in sleep, mobility, exercise tolerance, morning headache, and daytime fatigue. "Iron lungs" have occasionally been used in the past.

Weight and Nutrition

Most patients with MND lose weight during the course of their illness. Many patients become very thin, and this can increase the likelihood of pressure sores. Obesity occasionally occurs—particularly before the development of significant dysphagia. No published studies relating to weight change in MND have been found.

It is clearly important to maintain a reasonable calorie input and to ensure that dietary deficiencies do not occur. The advice of a dietician should be sought. The weight should be discreetly monitored, always remembering that rapid weight loss may cause the patient much anxiety and it is not helpful to over-emphasize the problem. Weight loss is an indication to consider endoscopic gastrostomy.

Constipation

Constipation can be a major problem in MND (Newrick and Langton Hewer 1984; Saunders et al. 1981). There are many precipitating factors, including weakness of the abdominal muscles, possible spasticity of the pelvic floor muscles, immobility, lack of fibre in the diet, dehydration, the use of anticholinergic drugs (for example, to control salivary drooling) and opiates.

Constipation results in abdominal discomfort and distension, and occasionally faecal impaction, with the development of spurious diarrhoea; and occasionally intestinal obstruction. The problem is, in our experience, not always well handled. From the point of view of the patient it can be one of the most distressing problems.

Management of Constipation

1. The amount of fibre in the diet should be increased
2. Fluid input should be maintained: at least 2 l per day (this may be a practical impossibility)
3. Anticholinergic drugs should be avoided if possible
4. Occasionally, patients are able to have a bowel action after relaxing in a warm bath
5. Medication. Bulk-forming agents, such as ispaghula and methyl cellulose, can be useful in patients who cannot tolerate bran. The latter is also a faecal softener. Lactulose, an osmotic laxative, can be useful. Irritant purgatives should usually be avoided
6. Suppositories and enemas. Glycerol or bisacodyl suppositories, or sodium citrate enemas, may be used
7. If faecal impaction occurs manual evacuation should be undertaken by a professional helper, and *not by the spouse.*

In practice, it is frequently impossible to institute many of the above measures. For instance, a patient with severe dysphagia may be quite unable to take in 2 l of fluid per day and similarly, anticholinergic drugs may be required for the management of severe drooling.

Constipation is one of the least well researched problems experienced by patients with chronic neurological disease. There is much scope for studies for this area.

Pain

Pain is common in motor neuron disease (Saunders et al. 1981; Caroscio et al. 1987; Newrick and Langton Hewer 1985). A study of 42 patients (Newrick and Langton

Hewer 1985) showed that 64% had pain as a major symptom. Saunders and colleagues (1981), studying 100 cases seen in a hospice, found that 45% complained of pain, and that this arose from stiff joints, muscle cramp, and skin pressure. Other factors included the inability to move without external help, severe spasticity, and constipation resulting in abdominal discomfort and colic.

In some instances it is not possible to identify precisely why the patient is experiencing pain. Many patients experience pain in more than one site.

Control of Pain

It is essential to ascertain why and where the patient is experiencing pain. However, this is not always possible and frequently the pain is ill-defined although very real. The following measures may be helpful:

1. Ensuring that the patient is sitting or lying in a comfortable position. The importance of properly fitting chairs cannot be over-emphasized
2. The use of a turning bed at night may do something to alleviate nocturnal pain and discomfort
3. Physiotherapists have a major role to play in advising on changing position, massaging the limbs, etc.
4. Control of spasticity. Spasticity may be at least partly controlled with medication, including baclofen, diazepam or dantrolene
5. During the last stages of the disease narcotic analgesics may be used (see p 401).

Decreasing Mobility

The patient's mobility needs may need to be assessed and reassessed at quite short intervals. The overall objectives of management are (1) to maintain mobility, however limited; and (2) to prevent fractures and other injuries.

The maintenance of maximum mobility is one of the main objectives of management at this stage of the disease. Full use should be made of locally available resources and of the district mobility team, if this exists. The keyworker is important in ensuring the appropriate supply of equipment, housing alterations, etc. (see Table 16.1).

Wheelchairs

The choice of a wheelchair depends upon an accurate assessment of the patient's medical conditions and needs, the circumstances in which the chair will be used (e.g. in transit, in the house, or outside) and the length of time that the patient will be expected to sit in the chair. If this amounts to several hours per day comfort becomes of major importance. It is also important to have information about the fitness of the carer, who may need to push the patient in the chair.

It should be noted that it may not be possible to achieve all objectives with only one wheelchair. Two or more may be needed. Wheelchair assessment and provision is an expert activity. Inappropriate wheelchairs can cause unnecessary discomfort

Table 16.1. Mobility problems in MND

Mild	Moderate	Severe
(Able to walk without assistance – with some difficulty)	(Patient able to walk in the house with assistance, but not usually able to walk more than a few steps outside)	(Patient unable to walk; may be able to stand for a short while. Major risk of falling)
Provision of a stick	Portable wheelchair for the car	Help and advice regarding maintenance of "transferring" ability
Advice on footwear	Foot drop support	Provision of wheelchairs
Advice regarding prevention of falls	Walking frame	Ensure access to toilet
Advice regarding car driving and the possibility of modifications to the car	Driving may need to be abandoned	Possibly move the bed downstairs
	Minor housing modification, e.g. ramp on steps, rails around toilet, etc.	Provision of a hoist (avoid if possible)
	Prevention of leg oedema	Housing alterations, e.g. door widening for wheelchair access
		Major importance of seating advice

and much frustration. Our experience is that wheelchair provision is frequently unsatisfactory.

Types of Wheelchair

1. Folding pushchair
2. Small rigid chair for indoor use; ideal if the arms are weak but the legs strong, as it can be propelled with the legs
3. Standard folding chair for propulsion by the patient or carer, e.g. Model 8
4. Special chair with modifications, which may include reclining seat, head support and foot support. This may be needed for the patient who cannot transfer independently and therefore needs to spend many hours each day in the chair.
5. Electrically propelled chair; needed for some patients when the arms are weak, and when the carer cannot push the patient.

Impairments which may be Important in Choosing a Wheelchair

The impairment	*What is needed*
Neck weakness	Reclining back and collar
Trunk instability	Side supports
Leg swelling	Leg elevators
Obesity	Robust chair
Risk of pressure sores	Special seating
Arm weakness	Ballbearing arm supports, possibly electrically propelled chair

Leg Swelling

Swelling of the legs occurs in virtually all patients if the limbs are maintained in a dependent position. The results include discomfort, breakdown in the skin, and impaired mobility.

1. Elevate the legs when the patient is sitting. This may need the addition of leg supports to the wheelchair
2. Elastic stockings: these should only be fitted when the legs are no longer swollen
3. Diuretics

Leg swelling is frequently unrecognised and can be a significant problem

Weakness of Muscles

All the skeletal muscles are liable to become affected eventually in motor neuron disease. Four groups are picked out for particular mention:

Weakness of the Neck Muscles

Weakness of the neck extensors allows the head to fall forward so that the chin rests on the chest. This results in discomfort and can eventually produce a contracture so that the head cannot be passively extended.

The results of weakness of the neck muscles include excessive dribbling, impaired communication (because the patient cannot achieve eye-to-eye contact), discomfort, and problems with eating and swallowing.

The overall objective of management is to maintain the head in a reasonable position of function. This may be achieved in two main ways. The patient may be seated in a chair with a reclining back. In some instances, a collar is needed.

Collar. We have previously found collars to be quite a major problem. The following criteria are suggested:

1. The collar must be socially acceptable; i.e. not too bulky or ugly.
2. It must be comfortable.
3. It must not impede chewing or swallowing.
4. It should not impede the patient's ability to look downwards and to read.

In some instances, a band around the head may be attached to an upright stay, which is attached either to the patient's trunk or to the wheelchair.

Trunk Muscles

Weakness of the trunk muscles may make it difficult for the patient to sit upright. There may, therefore, be a tendency for him to "flop" to one side. Side supports, or a moulded back cushion for the wheelchair may be required.

Foot Drop

In the earlier stages of MND foot drop may impair walking and increase the tendency to fall, with resultant risk of major injury. This problem may be dealt with either by a lightweight toe-raising spring, or by a moulded ankle/foot orthosis.

Weakness of Wrist and Finger Extensors

Lightweight splints may be worn during the day to improve the function of the hand. A night-time splint may be needed to prevent contractures of the wrist and fingers.

Sleeping Problems

Disturbed nights are a major cause of tension and frustration in the health of the carer. The author has experience of several instances in which the carer has physically assaulted the patient in the night, after having been disturbed several times. Such a strain on personal relationships should not occur with proper pre-emptive management.

Contributory factors to insomnia include the following:

1. Pain. Various types of pain may occur (see above). It is essential to ascertain why pain is occurring. Analgesics may be required.
2. Immobility. The inability to change position is a major cause of discomfort. The problem may be helped by the use of a firm mattress, and a lightweight duvet. Sometimes a "turning bed" or a powered bed elevator (see below) may be required.
3. Anxiety and depression. This problem may sometimes be helped by anti-depressant drugs. Clearly, the origins may be complex and related to the dismal outlook. However, an atmosphere of security and affection can do much to prevent the problem.

The bed. It is essential that the patient should have a suitable bed. This must be of the correct height (so that the patient can transfer with minimum difficulty). The mattress must be firm. Various types of turning bed are available. A powered elevator, which changes the height of the top half of the bed, can be useful.

Emotional Problems

Anxiety and depression frequently occur in motor neuron disease. To some extent, they can be alleviated by efficient management and ensuring that the carer is properly supported.

Emotional lability (inappropriate laughing and crying) sometimes occurs. The former can give an impression of spurious contentedness. This problem can be exacerbated by spasm of the lower facial muscles which make it appear that the patient is constantly smiling. Emotional lability frequently responds to imipramine in a dose of 25 mg two or three times per day. Its significance should be fully discussed with the patient and carer. Antidepressants and anxiolytic drugs may be needed in some cases. However, properly planned care and support can do much to alleviate these problems.

Equipment

Provision of the correct equipment, at the right time, can do much to preserve quality of life (Cochrane 1987). The following is required:

1. An accurate assessment of need
2. A knowledge of the cost and availability of funds
3. The availability of the equipment
4. Supply without delay
5. Training of the patient and carer in the use of equipment.

It is sometimes appropriate for aids and equipment to be borrowed rather than bought. In the UK the MNDA keeps a supply of some of the larger items such as beds and wheelchairs.

Experience indicates that some patients do not receive the right equipment at the right time (Newrick and Langton Hewer 1984). This is usually due to the lack of an appropriately trained person who is responsible for identifying the need (perhaps a keyworker) and then ensuring that the equipment is actually supplied.

It is not sensible to make plans too far ahead, as they may unsettle the patient. For example, the ordering of a wheelchair when the patient can still walk quite well may cause distress. A sensitive, intelligent professional can quietly ensure that a wheelchair, collar, and sucker are available when required.

Phase IV. The Terminal Stage

It is not known what proportion of patients with MND die in hospital. The Saunders study (1981) gives information about 100 patients seen in a hospice.

In the Saunders series (1981) 61% were described as having a "quick" death, i.e. within 24 hours. It is known that some patients die in their sleep. Others develop an upper respiratory infection and rapidly succumb within a matter of hours. We are not aware of any published study giving information about the manner of death in an unselected group of patients. In particular the incidence of fatal choking is unknown. In Saunders' study, of 94 patients who died only one did so in a choking attack.

The important objectives during this stage are to prevent unnecessary distress in the patient and relatives, and to maintain effective symptomatic relief.

Use of Narcotics

Narcotics may be given for distress of various kinds, including dyspnoea, misery and frustration, restlessness at night, pain, and occasionally feelings of hunger (Saunders et al. 1981). Some patients are given narcotics initially at night only. Morphine or diamorphine can be used. The initial dose may be only 2.5 mg. It should be gradually increased. Narcotics can be given orally, as a suppository, or by injection. Saunders' comments are worth quoting:

> Opiate drugs are unrivalled in the treatment of terminal motor neurone disease. Used skilfully, orally for preference, they ease distress as no other drug does, and the duration of treatment of some of these patients serves to refute the belief that this is a way of shortening life. Control of distress at times appears to lengthen life; certainly it makes it more tolerable.

As mentioned elsewhere it is essential that the nursing staff on a ward should understand the nature of motor neuron disease, and should not be baffled or frightened by the problems that occur. An efficient and well-trained nursing team can do much to allay anxiety and distress in these last stages.

Audit

The main problems identified previously (Newrick and Langton Hewer 1984) involve the poor control of bulbar symptoms, inaccessibility to specialist advice, lack of co-ordination, and poor symptom control. Any audit system must encompass these four areas.

The quality of care given to patients with motor neuron disease will depend on many factors, including the availability within the locality of *general* disability services (e.g. wheelchairs, orthotics and pressure sore prevention).

A Royal College of Physicians' Consensus Conference on the Management of Chronic Neurological Disorders (Consensus Conference 1990) has identified some of the problem areas which could be audited. These include the following:

1. Accessibility to specialist care (e.g. waiting times and distance to travel for a neurological consultation)
2. Accessibility to other aspects of care (e.g. availability of counselling, community physiotherapy, day hospitals, and waiting times for provision of aids, etc.)
3. Continuity of care (e.g. proportion of outpatient visits at which a patient is seen by the consultant himself)
4. Information (e.g. information given about patients' associations; patients' understanding of their diagnosis; and the family doctor's satisfaction with information received from the neurological team)

5. Adverse outcome (e.g. pressure sores, fractures, and evidence of drug-induced confusional state)
6. Failure to address significant symptoms (e.g. insomnia, pain, and dysphagia)
7. Satisfaction of patients (and if appropriate, their carers) with services
8. Use of keyworker, or other similar person
10. The availability of identifiable services dealing with bulbar problems, particularly dysphagia. This should include the availability of expert help with establishing a gastrostomy
11. Availability of identified provision for the management of respiratory difficulties, e.g. intermittent negative pressure ventilation.

It is remarkable that there have been so few attempts to audit the adequacy of services for MND patients. This matter should be remedied, and it is suggested that the MNDA should support research to develop validated methods of evaluating the quality of service provision, patient outcome, and the opinions of patients and their carers.

References

Campbell MJ, Enderby PM (1984) Management of motor neurone disease. J Neurol Sci 64:65–71

Carey JS (1986) Motor neurone disease – a challenge to medical ethics: discussion paper. J R Coll Med 79:216–220

Caroscio JT, Mulvihill MN, Stering R, Abrams B (1987) Amyotrophic lateral sclerosis: its natural history. Neurol Clin 5:1–8

Carus R (1980) Motor neurone disease: a demeaning illness. Personal paper. Br Med J 280:455–451

Cochrane GM (1987) The management of motor neurone disease. Churchill Livingstone, Edinburgh

Consensus Conference (1990) Standards of care for patients with neurological disease: a consensus. Report of a working group. J R Coll Phys 24(2):90–97

Crossner CG (1984) Salivary flow rate in children and adolescents. Swed Dent J 8:271–276

Enderby PM (1983) Frenchay dysarthria assessment. College-Hill Press, San Diego

Enderby P, Langton Hewer R (1987) Communication and swallowing: problems and aids. In: Cochrane GM (ed) The management of motor neurone disease. Churchill Livingstone, Edinburgh, pp 22–47

Enderby PM, Hathorne I, Servant S (1984) The use of intraoral appliances in the management of velopharyngeal disorders. Br Den J September: 157–60

Fallat RJ, Jewitt B, Bass M, Kamm B, Norris FH (1979) Spirometry in amyotrophic lateral sclerosis. Arch Neurol 36:74–80

Gauderer MWL, Ponsky JL, Izant RJ (1980) Gastrostomy without laparotomy: percutaneous endoscopic technique. J Pediatry Surg 15:872–875

Goldblatt D (1977) Treatment of amyotrophic lateral sclerosis. In: Griggs RC, Moxley RT (eds) Advances in neurology, vol 17. Raven Press, New York, pp 265–283

Howard RS, Wiles CM, Loh L (1989) Respiratory complications and their management in motor neuron disease. Brain 112:1155–1170

Langton Hewer R, Enderby PM (1990) Bulbar dysfunction. In: Clifford Rose F (ed) Amyotrophic lateral sclerosis, vol I. Demos, New York, pp 99–107

Lazarus BA, Murphy JB, Culpepper L (1990) Aspiration associated with long-term gastric versus jejunal feeding: a critical analysis of the literature. Arch Phys Med Rehabil 71:46–53

Lear CS, Flanagan Jr, Moorrees CFA (1965) The frequency of deglutition in man. Arch Oral Biol 10:83–99

Linden P, Siebens AA (1983) Dysphagia: predicting laryngeal penetration. Arch Phys Med Rehabil 64:26–28

Mulder DW (ed) (1979) The diagnosis and treatment of amyotrophic lateral sclerosis. Houghton Mifflin, Boston

Newrick PG, Langton Hewer R (1984) Motor neurone disease: can we do better? A study of 42 patients. BMJ 289:539–542

Newrick PG, Langton Hewer R (1985) Pain in motor neuron disease. J Neurol Neurosurg Psychiat 48:838–840

Norris FH, Smith RA, Denys EH (1985) Motor neurone disease: towards better care. BMJ 291:259–262

Russell TR, Brotman M, Norris FH (1984) Percutaneous gastrostomy: a new simplified and effective technique. Am J Surg 148:132–137

Saunders CM, Walsh TD, Smith M (1981) Hospice care in motor neuron disease. In: Saunders CM (ed) Hospice: the living idea. Edward Arnold, London, pp 126–155

Wade DT, Langton Hewer R (1987) Epidemiology of some neurological diseases. Int Rehabil Med 8:129–137

Wilson B (1982) Battling with motor neurone disease. BMJ 284:34–35

Zalin H, Cooney TC (1974) Chorda tympani neurectomy – a new approach to submandibular salivary obstruction. Br J Surg 61:391–394

Appendix I. MND Table of Responsibilities

Stage	Consultant	Keyworker	Comments
Stage I. State of Diagnosis	1. Make the diagnosis efficiently with minimum of investigations 2. Tell the diagnosis to the patient and spouse clearly 3. Arrange for second opinion if appropriate 4. Give information as appropriate 5. Introduce patient to keyworker 6. Establish, with keyworker, a plan for the next 3 years. Don't over-emphasize the likelihood of inexorable deterioration	1. Identify him/herself to patient and family 2. Provide information 3. Don't be too obtrusive at this stage	General practitioner should be in touch with the patient and family and offer support. He should be generally available *Note:* The consultant has the main role at this first stage
Stage II. Stage of deterioration without major disability	1. See the patient as often as seems required, but avoid unnecessary hospital attendances 2. Give information, if appropriate, about current progress in MND research 3. Make it easy for the patient to participate in drug trials, if appropriate 4. Help the patient to avoid experimenting with untried and expensive therapies which may cause harm	1. Keep in touch by phone (give patient keyworker's telephone number. 2. Avoid being too "pushy" 3. Suggest joining MNDA	As above

Appendix I. *(continued)*

Stage	Consultant	Keyworker	Comments
	5. Give advice regarding matters such as prognosis and employment, as appropriate 6. Overall, ensure that the patient knows that he (the consultant or his deputy) is available to give help if needed		
Stage III. Major disability with multiple problems	1. Avoid unnecessary attendances at hospital. If attendance is needed don't delegate to untrained junior doctors 2. Continue to be available as needed 3. Advice on drugs: (a) insomnia; (b) drooling; (c) spasticity; (d) pain 4. Arrange intermittent and crisis admission to hospital. Ensure that nursing (and other staff) are trained in the management of breathing and bulbar problems 5. Ensure that equipment is prescribed as necessary, e.g. collars, wheelchairs, arm supports, ventilator 6. Advise concerning NG tube, tracheostomy, gastrostomy	The keyworker must be familiar with all the common problems that occur in MND 1. Visit the patient and be in telephone contact as necessary 2. The exact role will depend on the availability of other help, e.g. domiciliary physiotherapy, occupational therapy and speech therapy 3. Liaise with the consultant, GP and speech therapist to give advice and help on: dysphagia; choking; dysarthria; drooling 4. Liaise with the physio- and occupational therapists on mobility, i.e. walking aid, wheelchair, avoidance of falls	The difficult stage –requiring time, imagination and sympathy The GP will have an important role at this time. He will need to liaise with the consultant and the keyworker. These three must work as a team. Watch particularly for: (a) breakdown in health of the spouse; (b) insomnia; (c) pain; (d) excessive agitation; (e) depression *Note*: (b)–(e) above may require drug therapy. The GP must be readily available at this stage

Appendix I. *(continued)*

Stage	Consultant	Keyworker	Comments
		5. Liaise with the occupational therapist on efficient delivery of equipment: wheelchairs, collars, splints, turning bed. Ensure efficient undertaking of house alterations, extra stair-rail, door widening, stairlift, provision of downstairs toilet 6. Liaise with social worker: (a) ensure that financial advice is given and appropriate benefits obtained; (b) involvement of Crossroads Scheme, if appropriate 7. Liaise with district nursing service —negotiate district nurse help with bathing, bowel problems, etc. Extended nursing, if necessary	
Stage IV. Terminal stage	1. Continue to offer immediate availability. Remain in contact with the GP regarding advice 2. Admit the patient to hospital or hospice, if appropriate 3. Continue with symptomatic management 4. Ensure that opiate relief is given as necessary	Keyworker continues in contact with the family, as above	As above. Remember: a stress-free death of a MND patient can be regarded as a medical triumph
Stage V. Post-terminal stage	1. Letter of condolence to spouse 2. Interview with spouse if appropriate	Keyworker keeps in contact with spouse for as long as appropriate	General practitioner also keeps in touch as necessary

17 Clinical Trial Methodology

T. L. Munsat, D. Hollander and L. Finison

Introduction

Since the cause of amyotrophic lateral sclerosis (ALS) or motor neuron disease is unknown and the pathogenesis still unclear, one might question the advisability of constructing elaborate clinical trials which are time-consuming and expensive. Similarly, the view has been expressed that costly trials are not appropriate since if and when a meaningful treatment becomes available, it will be self-evident to both patient and physician and will not require sophisticated trial methodology (Tyler 1982). This view may satisfy those who feel that anything short of a complete cure does not represent progress, but it does injustice to the equally valid perspective of the patient and their family who, understandably, believe that any form of effective and safe treatment and any measure of improvement is worth the effort. In our experience most patients express the view that any undertaking, successful or otherwise, is preferred to sitting passively as a helpless bystander to the inexorable progression of this terrible disease. In addition, this "all or none" view of judging the significance of therapeutic intervention in ALS fails to appreciate the history of cancer chemotherapy where only after many partial and at times equivocal successes with anti-tumour drugs was a substantive treatment outcome observed.

In recent years patients and their families have become more interested and increasingly vocal in taking an active role in the decision involved in design of clinical trials. This consumer activism first developed during clinical trials with AIDS and is now increasingly echoed by patients with other diseases, particularly those, like ALS, which are fatal and at present incurable. This new era of patient activism has raised interesting issues related to trial design such as the reluctance of patients to participate in placebo-controlled trials.

This chapter will not review the entire history of therapeutic trials in ALS as this has been covered in recent years by several other authors (Tandan and Bradley 1985; Festoff 1987; Mitsumoto et al. 1988; Stewart and Appel 1987). Rather we will focus on the elements of proper trial design in ALS that might be useful for those assessing the value of trials published in the literature and those contemplating their own trials.

Measurement Techniques

The first element in constructing an effective trial in ALS, or in any disease for that matter, is to select an appropriate measurement instrument. The technique used should be reliable (small difference between sequential intra- and inter-examiner measurements), valid (the technique measures what we wish to examine as directly as possible, i.e. the functional status of the voluntary motor system), time efficient (less than 1 hour per patient) and relatively inexpensive.

The first attempt to quantify neurological deficits in ALS was carried out in a study designed to evaluate the therapeutic efficacy of guanidine hydrochloride (Norris et al. 1974). Dr. Forbes Norris constructed an "ALS Score" consisting of a battery of functional ratings and reflex measurements, as well as an assessment of fasciculation, wasting, emotionality, fatigability and tone. Most of the 34 functions tested were rated on a 0–3 scale for a theoretical normal-subject total of 100 points. This Norris ALS Score was subsequently used by others to study the effects of transfer factor (Olarte et al. 1979) and levamisole (Olarte and Shafer 1985). The ALS Score was an important landmark as the first attempt to quantify deficit in ALS, but suffered from several weaknesses. Pulmonary function, the prime determinant of survival, was not measured. The protocol consisted of a mixture of signs, symptoms and subjective observations most of which were expressed in different measurement units. All the items were assessments of functional performance while strength, the primary variable in this disease, was not measured either by manual muscle testing (MMT) or by more direct quantitative means. The different elements of the ALS Score were not properly weighted so that certain items, often of unclear relationship to the disease, were given equal status to items of greater clinical and pathological importance. For example, the plantar responses, which are not self-evidently a manoeuvre of outcome in ALS, were given equal weighting with the ability to speak or swallow. Bladder and bowel function was rated although these are always spared in ALS. Reliability studies have never been published and the data generated is ordinal which restricts the range of statistical analysis that can be used.

Recently, major modifications in the concept of an ALS score have led to testing protocols that are both biologically and statistically more sound. Appel et al (1987) have significantly modified and refined the ALS Score and used this instrument to analyze both the natural history of ALS and the results of drug interventions. This Baylor ALS Score contains more individual items. Of particular importance is the inclusion of complete MMT and measures of pulmonary function. Items are grouped into subscores for bulbar, respiratory, upper and lower extremity function and these subscores can be assessed individually or grouped into a total ALS score. Inter-rater reliability was excellent with an average of 2.9% difference in mean total scores (Appel et al. 1987).

More recently Jablecki et al. (1989) have also modified the ALS Score adding functional rating, but not measurement of pulmonary function. They include manual muscle testing of selected muscle groups in the upper and lower extremities and have also eliminated some of the more subjective tests as well as those which have less biological significance to the disease outcome.

Over the past few years there has been a tendency to replace manual muscle testing and rating scales by instrumented quantitative strength measurement in a variety of neuromuscular disease studies as well as in ALS (Griggs et al. 1989; Aitkens et al. 1989; Festoff et al. 1988; Brooks 1989; Munsat et al. 1989; Munsat 1989). Certain protocols have utilized strain gauge measurement of maximum voluntary isometric contraction. Isometric testing has the advantage of eliminating velocity of contraction and muscle length as variables. However, the examiner must be carefully trained to assure proper limb position and adequate stabilization.

Handheld myometers have excellent test-retest reliability and they are easily portable by the patient, making them very useful in practice. However, they lack sensitivity when strength is well preserved since the maximum force measurable is limited by the strength of the examiner who must be able to break the patient's resistance. The device is therefore not suitable for detecting minor degrees of weakness in strong or heavily muscled adults. Also, it does not produce a hard copy of the test results.

Isokinetic measurements overcome certain deficiencies of maximal voluntary isotonic contraction (MVIC). The proximal muscles are more readily assessed and it is probably a better indicator of the degree of upper motor neuron (UMN) involvement. However, the equipment is very expensive and the number of muscle groups measurable is rather limited.

There have been several recent studies which attest to the lack of sensitivity of MMT in ALS (Andres et al. 1987), Duchenne muscular dystrophy (Edwards et al. 1984) and other neuromuscular disorders (Aitkens et al. 1989; John 1984; Miller et al. 1988; Cook and Glass 1987). In a recent unpublished study of measurement techniques in ALS we compared MMT, the ALS Score and MVIC in 20 patients with classical ALS and found MVIC to be much more sensitive and valid.

Because both UMN and lower motor neuron (LMN) deficits occur in ALS the measurement of motor deficit is more complex than a simple assessment of strength. Unfortunately there is still no practicable method available by which UMN deficit can be reliably and sequentially quantitated.

Natural History

An understanding of the natural history of the disease under study is an essential element of proper clinical trial design. The more that is known about the rate of progression, mortality, clinical manifestations and the character of the disease temporal profile the more appropriate the trial design. This is true particularly if this information can be expressed in accurate and reproducible statistical terms. Precise natural history information allows one to more accurately predict actual drop-out rates, define the variance of disease decline at entry (which significantly affects sample size), judge the length of time necessary to observe a predetermined therapeutic response, and design a proper testing instrument. This will result in a more precise and cost-effective trial.

Mortality statistics alone do not accurately define the natural history of ALS. Although they provide a clear and measurable end-point for the disease, they may not be an accurate representation of the rate of motor unit loss or the aggressiveness of the disease process because of the prominant regional character of ALS. In certain patients, intercostal muscle weakness and particularly diaphragmatic impairment leading to respiratory insufficiency may be an early or even presenting feature. These patients may die at a time when limb function is reasonably normal unless respiratory support is provided. In addition, the time of death is often significantly influenced by the presence of additional and unrelated factors, such as associated intrinsic pulmonary disease or smoking.

It is clear that deterioration rates vary considerably between patients; as much as 20-fold in those studies that have accurately determined this (Appel et al. 1987;

Munsat et al. 1988). Of even greater importance in trial design is the observation that once deterioration has begun, the rate of loss of strength is quite linear until very late in the disease when the process may "burn out" with only a very small number of motor units remaining (Munsat et al. 1988). This linear decline in muscle strength can be utilized very effectively in trial design. Thus, during the early and active phase of the disease a patient may serve as their own control.

Sample Size Determination

The determination of the *number of patients* ("n") required for an ALS study is one of the most important planning steps in trial design. Although several techniques are available for determining the proper sample size, the formula and its use as described by Colton (1974) is most widely accepted and serves extremely well for ALS studies (Table 17.1). In this construction "n" refers to each arm of a two-population-based trial, one receiving drug and the other acting as a control. Thus, for the usual blinded, placebo controlled study the true "n" will be twice the derived figure. Additionally, a correction factor must be made for anticipated drop-outs, a number which can best be estimated by a precise knowledge of the natural history of ALS. Drop-outs have approximated 50% in most studies – thus a not insignificant factor. This formula is also constructed for a two-tailed study. It will provide information not only about whether a drug works, but also if the drug has deleterious effects.

The "*alpha*" value in the sample size equation refers to the probability of committing a Type I error. This value, set by the investigator, defines the probability of being in error if the therapeutic intervention is declared to be effective. This value traditionally is set at 0.05 which states that there is one chance in 20 that if the drug is found to be beneficial this conclusion will be in error. In fact, the actual number used in the equation is the Z-converted value or 1.96. Using this traditional set-point for alpha we are expressing the view that we are willing to miss a certain number of drugs that have a beneficial effect to make quite certain that a drug declared effective will in fact be so. This view has scientific appeal and scientific validity but may be somewhat confining from the perspective of patients and their families.

The "*beta*" value in the sample size formula is the least well understood despite its importance. It designates the degree to which the investigator is willing to accept a Type II error, which sets the level at which one is willing to be wrong in declaring a drug ineffective. The "power" of a study is defined as 1-beta. In the typical investigation beta is set at 0.20 which means that the "power" of the study will be 0.80 or 80%. This indicates there is a one in five chance that a drug found to be ineffective does in fact work. "Power" is to a negative study what alpha is to a positive one. It informs the reader how certain authors ensure that a negative trial is indeed negative. Without this information one cannot be certain that the negative result was not caused by an insufficient number of patients studied.

"*Sigma*" indicates the standard deviation in decline rate of the population under study. It is a value which describes the variance of the patients being studied. It is a critical element in determining eventual sample size and even small changes in sigma result in significant changes in "n" (Fig. 17.1). Of all the factors in the sample

Table 17.1. Example of the effects of sigma and delta changes on sample size

$$n=2\left[\frac{(\Sigma\alpha-\Sigma\beta)\sigma}{\delta}\right]^2$$

$$n=2\left[\frac{(-1.96-0.84)\ 1}{0.5}\right]^2=63$$

$$n=2\left[\frac{(-1.96-0.84)\ 0.5}{0.5}\right]^2=16$$

$$n=2\left[\frac{-1.96-0.84)\ 0.5}{1}\right]^2=4$$

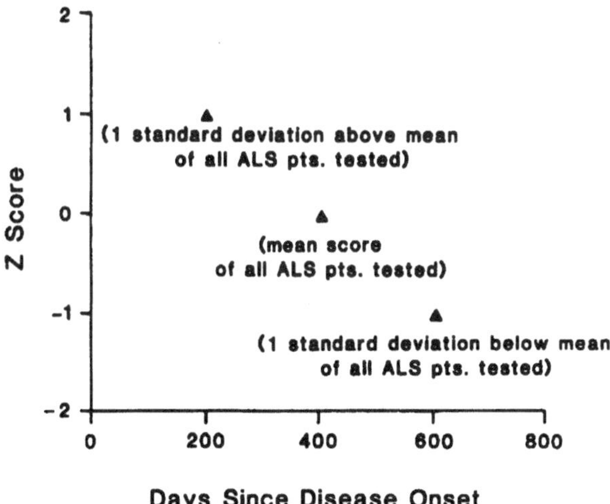

Fig. 17.1. Technique of plotting Z-scores against time.

size formula which control "n" sigma is by far the most important. Sigma can only be accurately determined by careful natural history studies performed in the same manner as the study itself.

The "*delta*" component of the sample size formula indicates the proposed degree of improvement expected from the drug under study. It is essential to set the outcome variables of the trial prior to, and not after, initiation of the trial. In most of our own ALS trials, we have set a predetermined outcome of a 50% reduction in the rate of deterioration. We judge this to be a meaningful and acceptable outcome expectation. Nonetheless, it should be noted that although a 50% slowing in the deterioration rate is seemingly a major change, it is unlikely that either patient or physician will be able to detect it because of the long time-course of the disease and because the change that occurs is a slowing of deterioration and not a functional improvement.

The Problem of Drop-outs

Almost all ALS drug trials have been plagued by a high drop-out rate. There are several reasons for this. ALS, when properly defined, is a uniformly fatal disease within about 3–5 years. Many drop-outs during a study are thus due to the death of the patient. The typical drug trial has a duration of 1–2 years. Many patients are referred to academic centres, where the majority of clinical trials are carried out, at a point at least midway in the course of their illness. There is an understandable tendency on the part of referring physicians, as well as patients themselves, to delay entry into an experimental trial until the diagnosis is certain and it is clear that the deficit is progressing despite everything being done. Although most trials do not enter patients who are in the advanced stages of their disease, few have established clear entry criteria based on rate of progression, either of neurologic deficit or disturbance of pulmonary function – the major factors which will determine the patient's longevity. It is necessary for referring physicians to understand the importance of transferring patients for trials as early as possible, even perhaps when the diagnosis is not yet entirely clear. Early entry is important from another perspective. If a therapy is effective this benefit would most easily be demonstrated in a patient with as large a responsive pool of motoneurons as possible. Motoneurons which have disappeared will, of course, be unresponsive to treatment. But in addition to sparing still unaffected cells an effective treatment might be of benefit to a population of neurons which, although damaged, may yet be amenable to repair and rescue. Entry criteria for ALS drug trials should include a slow enough course and adequate limb and pulmonary function. With good pulmonary capacity, proper limb strength and an appropriate rate of deterioration, the majority of drop-outs due to death in a clinical trial could be avoided.

In recent years the perspective of patients and their families about drug trials has changed considerably. They have become better informed about the nature of ALS and the various drug trials currently in progress. Patients have often made many inquiring calls, contacted physician friends, researched the literature and have become much better informed about the issues involved. This pattern of patient behaviour is appropriate and should be welcomed. It results in a better informed patient who can make considered decisions when these are required. It has also resulted in a more vocal patient population which now wants to take a more active role in many of the decisions that were previously left to the investigator alone. Patients are more critical and more careful about the trial they enter and will more readily drop out if the trial doesn't meet their expectation, or even if they have an opportunity to enter another more attractive trial elsewhere.

There has been an increasing reluctance, and at times frank refusal, of many patients to take part in placebo-controlled trials. This often leads to study drop-outs when the patient begins to realize that the treatment, active or placebo, is not influencing the course of their illness. More than ever before, we have encountered patients who will not co-operate with placebo arms of clinical trials. This is slowing if not seriously delaying at least two currently active ALS trails in the USA and will undoubtedly be an increasingly troublesome, even perhaps an insoluble future problem. Many patients have devised unique techniques of determining whether they are receiving placebo or active drug during a trial. Reluctance to enter a placebo-controlled trial is especially understandable when the drug under study can be obtained by prescription (deprenyl, for example) or is even available without prescription over the counter (dextromethorphan or branched chain amino acids, for example). Increasingly, we have found patients who surreptitiously continue to take other available drugs while in a placebo-controlled trial studying a different agent. There are indeed strategies to utilize controls other than placebo (see below) but the investigator is often hampered in using these techniques by restrictions posed by institutional committees and governmental regulatory agencies. These factors must all be considered in both the progress and the outcome of a clinical trial.

Drop-outs in ALS trials are not infrequently as a result of failures in communication between investigator and patient. The patient and his/her family may perceive the physician as more concerned with the research project than the welfare of the patient. This can occur particularly if the principal investigator assumes a supervisory role in the study and leaves most patient contact to Research Fellows or secretaries. It is important that the patient be treated as an ill person as well as a research participant and that proper, even greater attention be paid to their health care needs. The patient should view himself or herself as an active co-participant in the study, not a passive recipient of the physician's largesse.

Controls

The importance, indeed necessity, of utilizing a control population in any study evaluating the efficacy of a potential therapeutic intervention has been a cardinal principle of trial design for at least 40 years. The realization of the importance of using controls grew out of a substantial body of evidence which demonstrated that the mere process of entering a drug trial resulted in a number of measurable physiological and biochemical changes due to factors other than those attributable to the pharmacological effects of the drug under study. It became clear that in order to evaluate the true effectiveness or lack of effectiveness of a therapeutic intervention it would be necessary to compare this non-pharmacological effect with the true drug action. This understanding led to further exploration of the "placebo" effect as a variable which requires definition in any proper trial.

Although a number of control techniques are available and may be appropriate depending on the question at hand, the double-blind, placebo-controlled format has become established as the "gold standard" of trial design. In part, it has become institutionalized in the thinking of regulatory committees and granting agencies which are responsible for funding and approving drug trials. This view, however, may conflict with the growing reluctance and even resistance of patients to take part in placebo arms. In addition, placebo-controlled trials require a larger number of patients and are, therefore, much more time-consuming and expensive when compared to historical or self controls. Recently a multi-institutional trial of prednisone in Duchenne dystrophy demonstrated unequivocal therapeutic benefit (Brooke et al. 1987). However, because this was controlled by natural history data and not placebo, it was felt advisable to construct a repeat study, now placebo-controlled, which essentially demonstrated a similar result although with considerable additional time and expense.

The use of historical controls in clinical trials is strengthened if certain conditions are met (Bailar et al. 1984). First, the method of evaluation of the historical controls and the treatment group should be precisely defined, standardized and if possible performed by the same examiners. Historical controls obtained from studies performed by others and published in the literature are much less valuable. Second, the measurements should be precise with small variance on test-retest. Third, the disease should be predictable and, preferably, demonstrate linear progression.

Patients may also function as their own control, i.e. a self control (Louis et al. 1984); particularly if the natural history of the disease is well known and the rate of deterioration linear as is the case in ALS. In this regard, our own natural history studies have defined the mode of deterioration in ALS as being highly linear as determined by a variety of statistical measures. In addition, we have defined the variance of this linearity for different regions of the neuraxis.

Statistical Analysis

Many statistical procedures are currently available for the analysis of efficacy in drug trials. Our own experience in ALS trials and that in the literature provide certain guidelines. It is important to keep the number of outcome assessments limited to a small number of elements which are particularly pertinent to ALS. The more variables assessed, the greater the likelihood of a spurious positive finding. For every 20 outcome variables analyzed, there is a statistical probability of at least one spuriously positive outcome. Since most trials are designed to influence the rate of motoneuron drop-out and only secondarily to lead to improvement in muscle function, the primary outcome variable should be an assessment of any change in the rate of motor unit loss. In our judgement this can best be measured by MVIC testing. It is important to remember that ALS is a very regional disease and that analyses should include subgroup analysis of different areas of the neuraxis.

In any evaluation system it is important to transform raw scores in order to allow equal weighting of values measured in different units or capacities. In our studies we have elected to use a "Z" transformation which is a standard transformation technique (Colton 1974). This type of transformation will allow an equal contribution of items with relatively small dimension such as forced vital capacity, as well as items with large dimensions such as maximal voluntary ventilation. Similarly it will result in equal weighting for small muscles and large ones. This procedure relates an ALS population mean and standard deviation, for each test item, to the individual value generated. Z-transformed data is expressed in standard deviation units such that a score of -1 means that that the value obtained is one standard deviation below the mean of "all" ALS patients (Fig. 17.1). Z-transformed data can then be used in simple arithmetical combination.

In addition, it is appropriate to create composite scores representing various regions of the neuraxis as ALS has prominent regional characteristics. We have thus created "Megascores" which are simple algebraic sums of regional Z-transformed values. The Megascores are five in number representing *pulmonary function, bulbar function, arm function, hand function* and *leg function*. These Megascores can then be plotted against time to develop "Megaslopes" indicative of the rate and pattern of regional deterioration with time (Figs. 17.2, 17.3).

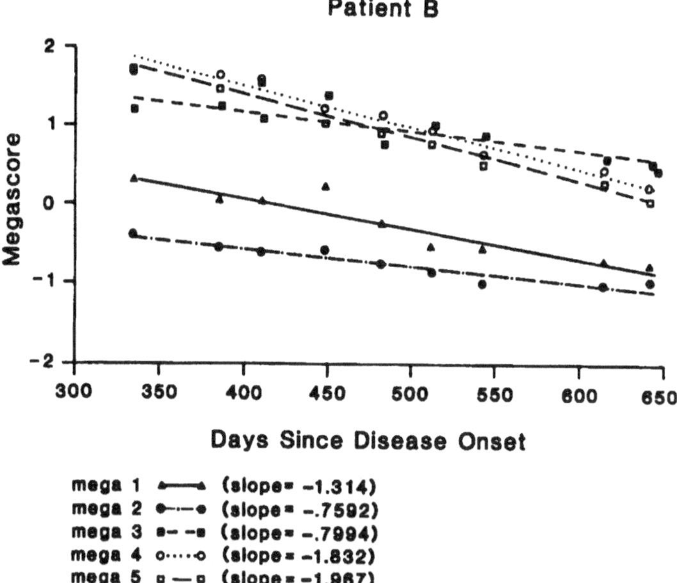

Fig. 17.2. Megascore change with time in a single patient with ALS.

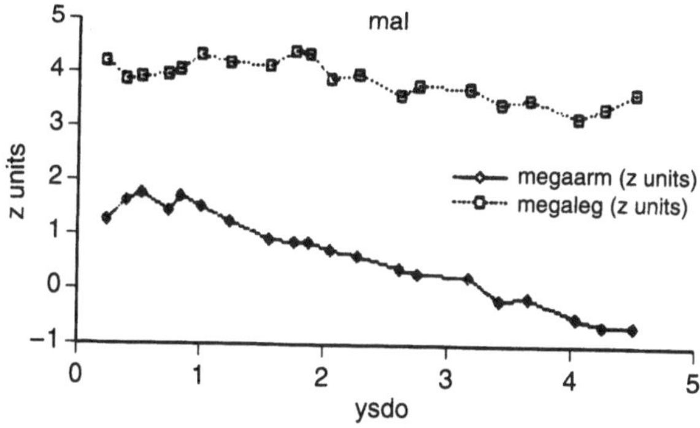

Fig. 17.3. Change in arm and leg Megascore in a single ALS patient; ysdo, years since disease onset.

Regulatory Processes

During the past 50 years there has been a proliferation and growth of agencies that carry out various regulatory functions to ensure that human research is carried out efficiently and in a manner consistent with the best interests of the participating patient. In the USA it has been suggested that the FDA is reluctant to approve trial designs other than the classical placebo-controlled, double-blind format. This also poses a problem, in the patient's view, as many patients are reluctant to take part in placebo-controlled trials. Also, it has been suggested that both corporate sponsored trials and Institutional Review Boards have also been reluctant to approve protocols that are not placebo-controlled. To a certain extent, serious life-threatening illnesses have been treated in a manner similar to more benign conditions. For example, regulatory agencies traditionally insist on a design "alpha" of 0.05 for sample size determinations regardless of the seriousness of the disease being studied. Others have suggested that it might be more appropriate to be less exacting with certain standards for diseases like ALS; but of course without additional hazard to the patient. Existing standards for drug acceptability may be scientifically satisfying but could exclude possibly effective therapies. As a result of increasing consumer activism the FDA has recently begun to relax some of the barriers which have made it so difficult for a new drug to be approved in more serious illnesses. This will undoubtedly greatly benefit the ALS population in the future.

ALS is a relatively uncommon disease which falls into the "orphan drug" catagery. This has lessened the enthusiasm of drug companies to foster and support research into its treatment. Despite recent legislation which provides certain tax benefits for companies developing drugs for orphan diseases as well as special FDA funding programmes, the problem still exists to a significant extent. Trials of drugs which have been approved for other purposes and may also be effective in ALS, such as cyclosporin and cyclophosphamide, are more readily supported by pharmaceutical companies in clinical trials.

In recent years the calibre of drug trials in ALS has improved considerably. When newer therapeutic agents with greater potential become available the technology and resources to evaluate these drugs quickly and efficiently will be ready.

References

Aitkens S, Lord J, Bernauer E, Fowler WM, Lieberman JS, Berck P (1989) Relationship of manual muscle testing to objective strength measurements. Muscle Nerve 12:173–177

Andres PL, Thibodeau IM, Finison L, Munsat TL (1987) Quantitative assessment of neuromuscular deficit in ALS. Neurol Clin 5:125–141

Appel V, Stewart SS, Smith G, Appel SH (1987) A rating scale for amyotrophic lateral sclerosis: description and preliminary experience. Ann Neurol 22:328–333

Bailar JC, Louis TA, Lavori PW, Polansky M (1984) Studies without internal controls. N Engl J Med 311:156–162

Brooke MH, Fenichel GM, Griggs RC et al. (1987) Clinical investigation of Duchenne muscular dystrophy. Interesting results in a trial of prednisone. Arch Neurol 44:812–817

Brooks BR (1989) Multicenter controlled trial: no effect of alternate-day 5 mg/kg subcutaneous thyrotropin-releasing hormone (TRH) on isometric strength decrease in amyotrophic lateral sclerosis. Neurology 39:322

Colton T (1974) Statistics in medicine. Little, Brown and Co, Boston

Cook JD, Glass DS (1987) Strength evaluation in neuromuscular disease. Neurol Clin 1987 5:101–123

Edwards RHT, Wiles CM, Mills KR (1984) Quantitation of muscle contraction and strength. In: Dyck PJ et al. (eds) Peripheral neuropathy. Saunders, Philadelphia, p 1093

Festoff BW (1987) Amyotrophic lateral sclerosis. In: Davidoff RA (ed) Handbook of the spinal cord. Marcel Dekker, New York, pp 607–664

Festoff BW, Melmed S, Smith R (1988) Therapeutic trial of recombinant human growth hormone in amyotrophic lateral sclerosis. In: Seratrice G et al. (eds) Advances in neuromuscular diseases. Expansion Scientifique Française, Paris, pp 117–126

Griggs RC, Pandya S, Florence JM et al. (1989) Randomized controlled trial of testosterone in myotonic dystrophy. Neurology 39:219–222

Jablecki CK, Berry C, Leach J (1989) Survival prediction in amyotrophic lateral sclerosis. Muscle Nerve 12:833–841

John J (1984) Grading of muscle power: comparison of MRC and analogue scales by physiotherapists. Int J Rehab Res 7:173–181

Louis TA, Lavori PW, Bailar JC, Polansky M (1984) Crossover and self-controlled designs in clinical research. N Engl J Med 310:24–31

Miller LC, Michael AF, Baxter TL, Young K (1988) Quantitative muscle testing in childhood dermatomyositis. Arch Phys Med Rehabil 69:610–613

Mitsumoto H, Hanson M, Chad D (1988) Amyotrophic lateral sclerosis. Recent advances in pathology and therapeutic trials. Arch Neurol 45:189–202

Munsat TL (1989) The use of quantitative techniques to define amyotrophic lateral sclerosis. In: Munsat TL (ed) Quantification of neurologic deficit. Butterworth, London

Munsat TL, Andres PL, Finison L, Conlon T, Thibodeau L (1988) The natural history of motoneuron loss in ALS. Neurology 38:452–458

Munsat TL, Andres PL, Skerry IM (1989) Quantitative assessment of deficit in amyotrophic lateral sclerosis. In: Fowler W (ed) Physical medicine and rehabilitation: state of the art reviews. Hanley & Belfus, Philadelphia

Norris FH, Calanchini PR, Fallat RJ, Panchari S, Jewett B (1974) The administration of guanidine in amyotrophic lateral sclerosis. Neurology 24:721–728

Olarte MR, Gersten JC, Zabriskie J, Rowland LP (1979) Transfer factor is ineffective in amyotrophic lateral sclerosis. Ann Neurol 5:385–388

Olarte MR, Shafer SQ (1985) Levamisole is ineffective in the treatment of amyotrophic lateral sclerosis. Neurology 35:1063–1066

Stewart SS, Appel SH (1987) The treatment of amyotrophic lateral sclerosis. In: Appel SH (ed) Current neurology. Yearbook Medical, Chicago, pp 51–90

Tandan R, Bradley WG (1985) Amyotrophic lateral sclerosis. III. Etiopathogenesis. Ann Neurol 18:419–431

Tyler HR (1982) Advances in neurology, vol 36. Human motor neuron diseases. Raven Press, New York

18 New Therapeutic Approaches: Rationale and Results

H. Mitsumoto

Introduction

Although amyotrophic lateral sclerosis (ALS) remains an enigmatic disease, advances in understanding its pathogenesis have accelerated in recent years. Attempts to treat patients with ALS have become equally vigorous in the past decade. Although it is possible that specific treatment of ALS will become available only after the pathogenesis is discovered, therapeutic trials could also provide clues to the aetiology of this disorder (Mitsumoto et al. 1988). Progress in therapeutic trials also has been accelerated by newly developed quantitative techniques to assess objectively the possible therapeutic effects of new treatments (Andres et al. 1987, 1988; Brooks et al. 1991; Sufit et al. 1987). At the same time the need for critical understanding of the natural course of ALS has been emphasized (Munsat et al. 1988; Ringel et al. 1993). Public awareness of the disease has increased and patients' wishes to try new therapeutic modalities have also facilitated the use of therapeutic trials. We expect to have a greater number of therapeutic trials in the coming years. In this chapter we shall review the rationale and results of recent therapeutic trials.

Rationale and Results

The Use of Thyrotropin-Releasing Hormone (TRH)

TRH treatment of ALS has made a definite impact on ALS research because TRH trials have triggered (or coincided with) an enormous number of research publications on TRH in general (more than 5000 publications in the past 10 years). As a therapeutic agent, no other compound in the history of ALS therapy (Festoff and Crigger 1979) has ever shown clinical improvement of the magnitude demonstrated with TRH (Engel et al. 1983).

The initial report of clinical improvement with TRH was followed by a large series of independent open or controlled clinical trials (Table 18.1). However, the results of these studies were contradictory and thus caused controversy (Committee on Health Care Issues 1987; Munsat 1988; Munsat and Brooks 1987; Rowland 1988a). Although TRH can provide subjective as well as objective, mostly transient, clinical benefits to some patients, it is agreed that the course of the disease has not been altered by TRH treatment.

Table 18.1. TRH trials in ALS

Investigator	Route	Dose/d	Dosing	Conclusion[1]
Brooks et al. 1987	IV[2]	10 mg/kg	Controlled	Beneficial
Caroscio et al. 1986	IV	500 mg	Controlled	No benefits
Mitsumoto et al. 1986	IV	500 mg	Controlled	No benefits
Caroscio et al. 1986	SC[3]	150 mg	Controlled	No benefits[4]
Mitsumoto et al. 1986	SC	25 mg	Controlled	No benefits
The Nat. TRH 1989	SC	5 mg/kg[5]	Controlled	No benefits
Brooke et al. 1986	IM[6]	150 mg	Controlled	No benefits[7]
Imoto et al. 1984	IM	8 mg	Controlled	No benefits
Engel et al. 1983	IV	500 mg	Open	Beneficial
Serratrice et al. 1986	IV	240 mg	Open	Beneficial
Engel and Spiel 1985	SC	100–150 mg	Open	Beneficial
Jerusalem and Fresman 1985	SC	80–150 mg	Open	Some benefits
Brooks et al. 1987	SC	1–6 mg	Open	Beneficial
Mitsumoto 1988	SC	125 mg[5]	Open	No benefits
Yamane et al. 1986	IM	0.5–2 mg	Open	Beneficial
Saida et al. 1988	IM	8 mg	Open	No benefits
Stober et al. 1985	IT[8]	0.8–4 mg	Open	No benefits
Saida et al. 1988	IT	0.5–1 mg	Open	No benefits[9]
Serratrice et al. 1985	IT	0.2–0.8 mg	Open	Some benefits

[1]The investigators' own overall conclusion. [2]Intravenous. [3]Subcutaneous. [4]Dynamometer showed only significant improvement 1 hr after SC injection. [5]Every other day schedule. [6]Intramuscular. [7]Transient proximal muscle improvement. [8]Intrathecal. [9]A few patients improved spasticity but not strength.

The rationale for the use of TRH in ALS has been fully reviewed by Engel (1989) and Munsat et al. (1989): it is based on the fact that extra-hypothalamic TRH is found in the anterior grey matter of the spinal cord (Hökfelt et al. 1975; Kardon et al. 1977; Manaker et al. 1985), although TRH receptor is also highly concentrated in the substantia gelatinosa of the spinal cord (Manaker et al. 1988). In the grey matter TRH is concentrated in synaptic terminals adjacent to alpha motor neurons (Manaker et al. 1985), suggesting that TRH may act as a neuromodulator or as a neurotrophic factor to motor neurons. Furthermore, Faden et al. (1981, 1984) demonstrated that cats with acute spinal cord injuries functioned better when given TRH after the injury.

Coincident with clinical trials, however, basic knowledge of the role of TRH in ALS has been accumulated: cerebrospinal fluid (CSF) concentrations of TRH and the tissue content of TRH in the spinal cord are normal in ALS (Jackson et al. 1986). An experimental depletion of TRH failed to induce motor neuron degeneration in animal models (Van Den Bergh et al, 1987). Based on these observations Munsat et al. (1989) have concluded that TRH alone is not essential to the viability of the motor system but that it may play a role as one of several trophic factors. In cultured spinal ventral horn cells, TRH has some trophic effects (Schmidt-Achert et al. 1984). Electrophysiologically, TRH increases F-wave responses (Clarke et al. 1984), resulting from the primarily stimulatory effects of TRH on alpha motor neurons (Lechan et al. 1984).

Since the initial report by Engel et al. (1983), who used a dose of 500 mg intravenous TRH, a variety of dosages and routes of administration have been investigated (Table 18.1, and also see reviews by Bradley 1990 and Brooks 1989). For example, we (Mitsumoto et al. 1986) studied the effect of 500 mg intravenous

TRH (one dose per patient) according to the method described by Engel et al. (1983) and then the effect of daily 25 mg subcutaneous TRH injections for a period of 3 months, both by a controlled cross-over design. We demonstrated that TRH crosses the blood–brain barrier even with 25 mg of subcutaneous TRH injection (up to 59-fold elevation over control at 40 minutes). The pulse rate is significantly increased with subcutaneous TRH, suggesting that even a small amount of TRH exerts a significant cardiovascular effect. The majority of patients experienced typical TRH side effects, which made a double-blind, controlled study difficult, as has been reported by others (Brooke 1989).

In our studies only isometric quadriceps strength was found to be increased and then only marginally. Ten patients experienced subjective clinical benefit and 10 patients complained of worsening of symptoms after TRH was discontinued. However, our study failed to demonstrate overall beneficial effects in patients with ALS. Our subsequent open study with alternate-day 125 mg subcutaneous TRH showed no clinical improvement in 10 patients for a period of up to 21 months' observation (Mitsumoto 1988). All controlled studies except the study by Brooks et al. (1987) found no clinical benefit (Table 18.1), although a transient, statistically significant improvement in at least some muscles has been noted in some of the studies (Brooke et al. 1986; Brooks et al. 1987; Caroscio et al. 1986). The most recent multicentre study, however, revealed no benefits with alternate-day 5 mg/kg subcutaneous TRH on isometric muscle strength in 108 randomized patients with ALS (The National TRH Study Group 1989).

The clinical effects of several TRH analogues, which are found to be pharmacologically more potent and stable than TRH, have also been tested. Open studies with intramuscular DN-1417 injection showed no clinical improvement (Hawley et al. 1987; Saida et al. 1988), while those with intravenous CG3509 showed transient improvement in some muscles (Jerusalem and Fresmann 1985). A controlled study with intravenous MK-771 (Neville et al. 1986) failed to show any changes, whereas Guiloff et al. (1987a) showed significant improvement with RX77368 in some muscles in their controlled studies. This was followed by another open study showing some benefits (Modarres-Sadeghi et al. 1988). Electrophysiologically acute intravenous RX77368 injection increases fibre density in single fibre EMG and mean amplitude in macro EMG; these changes have been considered to be "central" effects on motor neurons (Guiloff et al. 1987b, 1988).

In recent critical reviews on TRH clinical trials, Bradley (1990) warned that we have to be very cautious in interpreting our own results, and Brooke (1989) stated that there had been no "perfect" study. Brooke (1989) felt it reasonable to conclude that the effect of TRH in ALS is a definite, acute, and transient response, of unclear mechanism. Furthermore, the reason why only a few muscles improve remains unexplained. Brooks (1989) pointed out that appropriate subject selection is crucial, and in fact Brooks et al. (1987) carefully preselected subjects who responded to TRH for their controlled studies. The lack of overall benefits from TRH found in the previous studies is now claimed to be due to the fact that male and female patients were pooled for statistical analyses (Miller and Warnick 1989): all negative results but one (Brooke et al. 1986) contained relatively more female patients in their pooled population. However, an open study containing more female patients (4 males and 5 females) showed positive results (Serratrice et al. 1985, 1986). Brooks (1989) and Miller and Warnick (1989) believe that the effect of TRH may be specific to gender, more to male than female, and to those having bulbar involvement more than other types of weakness.

Approaches to Counteract Excitatory Amino Acids

Glutamate, which is the most abundant free amino acid in the central nervous system, has been considered as one of the major excitatory neurotransmitters (Greenamyre 1986). The excitatory electrophysiological effects are mediated by glutamate receptors, which are categorized into two major groups: (1) ionotropic and (2) metabotropic receptors. The ionotropic receptors contain cation-specific ion channels, which are subclassified into the N-methyl-D-aspartate (NMDA) receptor and AMPA-kainate receptor. There are several subtypes in each receptor group. The metabotropic receptors modulate the production of intracellular messengers and include at least six subtypes depending on response to specific agonist, such as AP4 and quisqualate (Nakanishi 1992).

Among neurotoxins those particularly mediated by glutamate receptors are called excitotoxins. These have gained increasing attention since these excitotoxins may be related to various acute neuronal cell injuries and even chronic neuronal cell loss in neurodegenerative diseases including Huntington's chorea and ALS. Choi et al. (1988) have shown that glutamate neurotoxicity is predominantly mediated by the activation of the NMDA receptor, leading to excessive free calcium entry to the cell. The toxic effect of intrinsic glutamate in the CNS, however, is normally prevented by the high-affinity uptake system at the pre-synaptic axon terminal. However, if there is inhibition or removal of this uptake system, the neurotoxic effect of glutamate may appear (Albin and Greenamyre 1992; Albers et al. 1992; Buchan 1992; Greenamyre 1986).

Recently attention has been given to exogenous excitotoxins causing clinically significant neurotoxicity, supporting further the idea that some neurodegenerative disorders may be mediated by glutamate receptors. A hypothesis that a plant neurotoxin in cycad seed may cause the endemic ALS-Parkinson's disease-dementia (ALS-PD) complex of Guam was suggested by a previous epidemiological study (Kurland 1972). Spencer et al. (1987) showed that an unusual amino acid in cycad seed, known as beta-N-methylamino-L-alanine (BMAA), can experimentally produce neuronal degeneration in primates, suggesting that ALS-PD may be caused by BMAA, which might act as a long-latency neurotoxin (Spencer 1987). Around the same time other examples of neuro-excitotoxins were demonstrated. Beta-N-oxalylamino-L-alanine (BOAA), a potent neurotoxin found in chickling pea (*Lathyrus sativus*), was shown to produce experimental lathyrism in primates, suggesting that BOAA is in fact the cause of lathyrism (Spencer et al. 1986). Ross et al. (1987) have shown that the effect of BMAA is mediated by the NMDA receptor, whereas BOAA is found to stimulate non-NMDA receptors. Another example of neurotoxicity due to an extrinsic excitotoxin is domoic acid which is found in mussel and is a potent kainate receptor agonist, causing acute intoxication followed by chronic dementia and motor neuronopathy in some patients; an epidemic occurring in Canada was described by Teitelbaum et al. (1990).

The possibility that cycad seed used as food might have caused ALS-PD, appeared to be supported by the observation that the incidence of Guamanian ALS-PD has markedly declined in the recent years (Garruto et al. 1985), coinciding with improvement in the food supply in Guam. However, Steele et al. (at ALS/MND Research Foundation Workshop, Santa Cruz, 1989) reported the result of retrospective dietary studies among elderly Chamorro residents of three Guam villages with different rates of ALS-PD. The studies showed no differences in pre-war or wartime foods or diet among those villages. Furthermore, Duncan et al. (1988) have shown that BMAA is in fact present only in low concentration in cycad seed and is even lower in flour prepared from cycad seed. Furthermore, it is less potent in neuron culture studies, unable to cross the blood–brain barrier effectively, and has a short half-life in vivo, casting doubt upon the BMAA hypothesis as the cause of ALS-PD complex. Spencer et al. (1991) also recently concluded that BMAA did not seem to be the cause of ALS-PD complex.

While a hypothesis of exogenous excitotoxin causing ALS remains to be established, knowledge on endogenous excitotoxins has been progressively accumulated. Increased levels of glutamate, aspartate, or both have been noted in the serum and CSF of ALS patients (Iwasaki et al. 1991; Patten et al. 1978; Plaitakis and Caroscio 1987; Rothstein et al. 1989). In contrast to these changes, however, Perry et al. (1990) found no such increase. Technical differences may have been the reason for such discrepancies, and thus a need for further studies is suggested (Young 1990). In the autopsy studies, the CNS tissue glutamate and aspartate levels and the glutamate to glutamine ratio are significantly decreased in ALS, suggesting that there is a generalized defect in glutamate metabolism (Malessa et al. 1991; Plaitakis 1990; Plaitakis et al. 1988a; Rothstein et al. 1990). Rothstein et al. (1992) recently investigated the high-affinity glutamate uptake in synaptosomes in various brain cortex and spinal cords from controls and patients who died of ALS. They found a marked decrease in the uptake in the motor cortex and spinal cord from ALS patients, suggesting that a faulty transport mechanism of extracellular glutamate may lead to neurotoxic levels of extracellular glutamate in ALS. With quantitative autoradiography, Allaoua et al. (1992) measured various glutamate receptors binding in the spinal cords of patients who died of ALS. Only NMDA binding decreased in the ventral and dorsal horn, suggesting that the NMDA may indeed be the target if excitotoxins are involved in the pathogenesis of ALS.

A glutamate loading test results in a greater increase in plasma glutamate and aspartate in ALS patients than in healthy controls, suggesting an abnormal systemic metabolism of excitatory amino acids in these patients (Plaitakis and Caroscio 1987). Further, glutamate dehydrogenase, which reversibly transforms glutamate to 2-oxoglutarate is found partially, but significantly, diminished in multi-system atrophy and atypical ALS (Plaitakis et al. 1982, 1984), and recently even in typical ALS (Hugon et al 1989). In contrast, Malessa et al. (1991) found normal glutamate dehydrogenase activity in the spinal cords of ALS patients but an increased activity in other CNS areas, supporting again the idea that glutamate metabolism is abnormal in ALS.

Despite the increasing amount of data supporting the excitatory amino acid hypothesis in ALS, many questions remain to be answered (Young 1990). For example, how does the glutamate neurotoxicity lead to selective neuronal involvement? How does glutamate over-activity produce the late onset and the slowly progressive course commonly seen in neurodegenerative disorders? Albin and Greenamyre (1992) recently proposed hypotheses explaining how excitatory amino acids could cause the cell degeneration in neurodegenerative disorders: The *altered receptor hypothesis* explains the regional pattern of neuronal loss that would be a function of the regional distribution of altered receptor subtypes. The *weak excitotoxic hypothesis* explains the mechanisms of chronic neuronal degeneration, in which a primary metabolic disease of neurons would enhance or trigger glutamate receptor activity, causing final cell death.

Since branched chain amino acids (BCAA) are known to activate glutamate dehydrogenase activity (Yielding and Tomkins 1961), a combination of L-leucine,

L-isoleucine, and L-valine was given orally for a period of 1 year in double-blind fashion to 22 patients with ALS (Plaitakis et al. 1988b). This study showed a significant benefit in some muscles and for gait. The subsequent study confirmed the earlier results that BCAA slowed the disease in treated patients, although the number of patients studied was still small (Plaitakis et al. 1992). Testa et al. (1989) investigated the clinical benefits in 30 ALS patients who were treated with either BCAA or no treatment in a randomized but open study for 12 months. They found no difference in Norris scores. An Italian study group decided to discontinue the treatment trial with BCAA because they found that more patients on BCAA died than those on placebo (Beghi et al. 1993). A similar, large-scale, long-term double-blind study is currently under way in the United States and Europe. Patient recruitment has proved difficult since BCAA products (of unknown purity) are readily available at any health food store.

L-threonine was another amino acid tried for ALS patients to counteract excitatory amino acid because threonine is known to increase glycine, one of the inhibitory amino acids (Patten et al. 1978; Patten and Klein 1988). In fact, glycine receptors are found to be reduced in ALS spinal cord (Hayashi et al. 1980). However, glycine may even potentiate the effects of glutamate receptors (Plaitakis 1990). Subsequent studies have not confirmed the initial observation either by an open clinical trial (Testa et al. 1992) or by placebo-controlled, double-blind studies (Blin et al. 1991).

In contrast to the use of amino acids to reverse a presumed "imbalance" of intrinsic excitotoxins, the use of an NMDA antagonist has been proposed. This notion is based on the finding that dextromethorphan attenuates glutamate neurotoxicity (Choi 1987). The few known NMDA antagonists are generally not tolerated in vivo, but dextromethorphan is widely used as a cough medicine. Appelbaum et al. (1991) used low-dose dextromethorphan in 18 ALS patients in a cross-over design. They found no statistically significant differences between the two phases. Hollander et al. (1992) showed the feasibility of long-term, high-dose (4.75 to 10 mg/kg/day) dextromethorphan treatment. Currently studies with higher dosages are under way.

A French study group recently investigated the effects of riluzole, a presynaptic glutamate-release blocking agent, and found in their initial studies that fewer patients with bulbar ALS died while on rilzuzole than those taking placebo (Bensimon et al. 1994). A large multinational study to confirm this observation is under way.

Treatment of Immunological Abnormalities

Extensive immunological investigations to date have shown no consistent immunological abnormalities in ALS patients (see reviews by Appel et al. 1986; Drachman and Kuncl 1989; Mitsumoto et al. 1988; Rowland 1988b). However, paraproteinaemia has been found in some patients with lower motor neuron syndrome or with both upper and lower motor neuron syndrome, which clinically resembles ALS (Krieger and Melmed 1982; Peters and Clatanoff 1968; Rowland et al. 1982; Rudnicki et al. 1987; Shy et al 1986). Recently patients with motor neuron disease associated with paraproteinaemia were reported to have increased CSF protein, suggesting that paraproteinaemia may be associated with a disease process in the central nervous system (Younger et al. 1990). Although the significance of this paraproteinaemia in motor neuron disease is still unclear, it is important to identify such paraproteins in patients who apparently suffer from an otherwise untreatable disease, since some of these patients may respond to aggressive immunosuppressive therapy (Parry et al. 1986; Patten 1984). By using sensitive immunofixation electrophoresis, 4.8% of a series of 206 patients with the motor neuron syndrome were found to have paraprotein, a higher frequency compared to that seen in a general, healthy population (Kyle et al. 1972). By using Western blot techniques, Duarte et al. (1991) found two or three monoclonal immunoglobulin components in 60% of 30 ALS patients compared to only 13% of 30 matched controls, again supporting the idea that monoclonal immunoglobulins may have some significance in ALS.

Schluep and Steck (1988) found that IgM paraprotein from a patient with motor neuron disease reacts at the pre-synaptic terminal of the neuromuscular junction, suggesting that this site may be the immunogenic target in this motor neuron disease. Furthermore, IgM paraprotein found in two patients with lower motor neuron syndrome was found to react against gangliosides GM_1 and GD_{1b} (Freddo et al. 1986; Latov et al. 1988), and a patient who had a disease resembling motor neuron disease was electrodiagnostically found to have motor neuropathy secondary to multifocal conduction block (Parry and Clarke 1988). Pestronk et al. (1988b) reported that the two patients with motor neuropathy and multifocal conduction block were found to have the elevated anti-GM_1 antibodies and responded with cyclophosphamide treatment. Feldman et al. (1991) reported their experience with 13 patients with multifocal motor neuropathy and elevated titres of antiganglioside antibodies. All patients failed to respond to oral prednisone, and four patients did not improve with plasmapheresis; however, all 9 patients who received cyclophosphamide improved clinically. Kaji et al. (1992) reported that 2 patients with multifocal motor neuropathy, one with elevated anti-GM_1 antibodies and the other without antibody, improved with intravenous immunoglobulin treatment.

Recent studies further provide a scientific basis for the idea that the anti-GM_1 antibodies may play a major role in immunologically injuring motor neurons or peripheral nerves: anti-GM_1 antibodies react with animal spinal cord motor neurons (Corbo et al. 1992); IgM deposit is found at the node of Ranvier in a patient with motor neuron disease with high anti-GM_1 antibody titres (Santoro et al. 1990); and passive transfer of sera from a patient with high anti-GM_1 antibodies and multifocal conduction block reproduces conduction block and immunoglobulin deposit in animal nerves (Santoro et al. 1992).

The significance of anti-GM_1 antibody in ALS patients has also been studied. More than 50% of patients and, if anti-GM_1 and -GD_{1b} antibodies are combined, more than 75% of patients with ALS are found to have anti-ganglioside antibodies (Pestronk et al. 1988a; Pestronk et al. 1989). However, the titres of anti-ganglioside antibodies in patients with ALS are much lower than those in patients with multifocal motor neuropathy (multifocal conduction block). Motor neuron disease with very high titres of serum anti-ganglioside antibodies, which manifests otherwise as typical ALS syndrome, may represent a subset or even a different disease (Pestronk 1991).

While anti-ganglioside antibodies may be an important immunological marker of various motor neuron diseases, the specificity of these antibodies remains unsettled. Disease controls such as multiple sclerosis, idiopathic demyelinating neuropathy, and lupus erythematosus may carry the same antibodies (Endo et al. 1984; Pestronk et al. 1989). Sadiq et al. (1990) reviewed anti-ganglioside antibodies in various neuromuscular diseases and found increased titers in those with pure motor neuropathy, sensorimotor neuropathy, or lower motor neuron syndrome. Moreover, the presence of a low titer of anti-ganglioside antibody in healthy people suggests that this antibody may be a normal constituent of the human antibody repertoire (Nobile-Orazio et al. 1990; Shy et al. 1989). The significance of multifocal conduction block appears equally confusing. A recent study with electrodiagnostic analyses in 169 patients with motor neuron disease showed too many exceptions in electrodiagnostic changes to identify a unique syndrome, although motor neuron disease with multifocal conduction block appears to be a uniform disorder (Lange et al. 1992).

An immune-mediated motor neuron degeneration has been experimentally produced in guinea pigs by injecting isolated motor neurons from the swine spinal cord (Engelhardt and Joo 1986). Engelhardt et al. (1989) further developed experimental motor neuron disease or experimental grey matter disease in guinea pigs by injecting either isolated bovine motor neurons or bovine grey matter. In these models, all the key features of autoimmune disease were present, including neuronophagia, perivascular mononuclear inflammation, positive IgG deposits in the motor neurons and neuromuscular junctions, and elevated antimotor neuron antibody. The lack of any significant inflammatory cell infiltration in spinal cord has been one of the hallmarks in the neuropathology of ALS. However, Troost et al. (1989) recently found mononuclear infiltrate in 38 of 48 spinal cords from ALS patients. Engelhardt and Appel (1990) identified reactive microglia, macrophages, or both, in degenerating pyramidal tracts and the ventral horn. The surface of most of these cells stained positively for IgG, and 50% stained positively for HLA-DR. Furthermore, a population of both upper and lower motor neuron cells had granular IgG deposits corresponding to the granular endoplasmic reticulum. Although the significance of this intraneuronal IgG deposit needs further clarification, the authors speculate that IgG is internalized at the axon terminal, where an autoimmune process may take place in ALS, and that internalized IgG is transported back to the cell body (Engelhardt and Appel 1990).

Other evidence supporting the autoimmune theory of ALS is that immunoglobulins obtained from ALS patients and experimental animals with autoimmune motor neuron disease cause various abnormalities at the calcium channel: the spontaneous neurotransmitter release at the neuromuscular junction as manifested by increased miniature end plate potential (MEPP) frequency (Appel et al. 1991; Uchitel et al. 1988; Uchitel et al. 1992) and inhibition of calcium entry on the dihydropyridine (DPH)-sensitive skeletal muscle calcium channel tested (Delbono et al. 1991). They speculated that an increased Ca^{2+} permeation at the axon terminal resulting from an abnormal immune reaction may lead to increased MEPP frequency and subsequent morphological changes. Appel's team reported new evidence that serum samples from 36 of 48 ALS patients contained IgG that reacted with the rabbit muscle L-type calcium-channel protein (Smith et al. 1992). These antibodies were found in only one of 25 controls, in 6 of 9 patients with Lambert-Eaton syndrome, and in 3 of 15 patients with myasthenia gravis. This report again supports the idea that the IgG from ALS patients may contain anti-calcium channel autoantibodies. We need additional studies to understand the significance of these autoantibodies in ALS (Rowland 1992).

The idea that ALS serum has cytotoxic effects has been tested on many occasions since Wolfgram and Myers (1972) first reported such toxic effects on tissue culture (see also a review by Mitsumoto et al. 1988). Conradi and Ronnevi (1985) showed that ALS plasma has in vitro erythrocyte fragility effects. The degree of erythrocyte fragility of ALS plasma was reduced when the patients were treated with prednisolone and azathioprine (Conradi and Ronnevi 1987). Recently the same observation was made by Danish investigators (Overgaard et al. 1991).

In ALS patients, various immunosuppressive treatments have been attempted. A series of trials with plasma exchange failed to show any clinical improvement (Conomy et al. 1980; Dau 1984; Kelemen et al. 1983; Lisak 1984; Olarte et al. 1980). Brown et al. (1986) tried intensive immunosuppression with cyclophosphamide in 6 patients with ALS but failed to find any improvement after 18 months' follow-up. Appel et al. (1988) performed a controlled study with cyclosporine in 36 ALS patients. Men who entered this study within 18 months after the onset of the symptoms appeared statistically to progress more slowly than other groups of patients. Based on this result a multicentre controlled study investigating only male patients has been performed. However, the preliminary results did not confirm the initial observation of Appel et al. (1988) (S. Appel, personal communication 1989).

In a different direction, Drachman et al. (1994) reported the results of a controlled study with total lymphoid irradiation in order to thoroughly suppress the immune system in ALS patients. Even with this degree of "complete" immunosuppression, the radiation treatment failed to alter the course of ALS. Werdelin et al. (1990) tried long-term (1 year), high-dose corticosteroid and azathioprine treatment in patients with either ALS or motor neuron disease but again found no clinical improvement. Intravenous immunoglobulin (IV-Ig) has been effective in treating some immune-mediated neurological diseases, such as Guillain-Barré syndrome and chronic idiopathic demyelinating neuropathy. We studied the clinical effects of monthly 400 mg/kg IV-Ig treatments in 19 ALS patients (Mitsumoto et al. 1992). The rate of monthly deterioration of isometric muscle strength in the upper extremities slowed significantly with IV-Ig treatment. Further studies are necessary to clarify these curious findings.

Drachman and Kuncl (1989) feel that ALS is an "unconventional" autoimmune disease and that the negative results from currently available treatment modalities do not preclude the possibility of such autoimmune mechanisms in ALS. They suggested that more clinical trials will be needed because novel immunosuppressive treatments may be beneficial.

Approaches to Trace Metal Intoxication

Since Currier and Haerer (1968) first reported heavy metal exposure in ALS, this association has been suspected by others (Yase 1984; Roelofs-Iverson et al. 1984). In particular, lead levels have repeatedly been found to be raised in blood and CSF in ALS, by various techniques (Conradi et al. 1982b). However, clinical trials of chelation with d-penicillamine in ALS patients has been disappointing, although a sustained urinary lead excretion has been noted (House et al. 1978; Conradi et al. 1982a). Another metal which recently has received more attention is aluminium, since it is a common element in the earth's crust, including the soil of Guam, and deposited in motor neurons together with silicon and other metals (Yase 1984; Yanagihara 1982; Pearl et al. 1982; Garruto et al. 1986). Strong and Garruto (1991) have shown that aluminium induces neurofibrillary changes in cultured motor neurons. Furthermore, there is an anecdotal report of aluminium intoxication manifested by motor neuron syndrome, which improved with desferrioxamine (Patten 1988). We have managed a similar patient with the clinical syndrome of ALS, who had a history of a long-term aluminium exposure, but who did not improve with desferrioxamine treatment despite increased aluminium excretion in the urine (R. Shields, M.D., Cleveland Clinic, personal communication, 1989).

Antiviral Agents

Viral infection has always been among the most attractive hypotheses of the cause of ALS. The polio virus has been a prime suspect among various other viruses since lower motor neurons are affected by both ALS and in polio infection (Cwik and Mitsumoto 1992). Progressive weakness is the main feature of ALS, and occurs also in post-polio syndrome. However, many attempts to identify evidence of viral infection specific to ALS have been unsuccessful (Fallis and Weiner 1982).

Despite the unsuccessful attempts to find evidence of viral infection in ALS, antiviral treatment has been given to patients with the disease. Controlled studies with guanidine or amantadine have been negative (Norris et al. 1974; Munsat et al. 1981). Levamisole, a drug with "immunostimulatory" properties, has been studied in controlled trials in ALS patients, but found to be ineffective (Olarte and Shafer 1985). Another form of antiviral treatment studied in ALS is human leukocyte alpha-interferon, which is a naturally occurring glycoprotein believed to trigger a protective and therapeutic antiviral state in cells. The safety and efficacy of alpha-interferon have been examined in open fashion by subcutaneous, intravenous, intraventricular, and intrathecal routes (Dalakas et al. 1986; Färkkilä et al. 1984; Mora et al. 1986; Smith and Norris 1988). Only Färkkilä et al. (1984) claimed that reversible clinical improvement occurred with intravenous alpha-interferon in 7 patients with ALS. All other studies found no clinical improvement.

Enhancing Axonal Sprouting and Regeneration

Gangliosides are present in the neuronal cell membrane, particularly concentrated at the synaptic membrane. Exogenous gangliosides are suspected of being neurotrophic (Lipton 1989) and have been found to stimulate axonal sprouting (Gorio et al. 1982). Therefore, they are expected to improve axonal regeneration in remaining neurons in ALS so that the muscle reinnervation process can be facilitated. Purified bovine brain gangliosides have been tested in ALS patients under controlled conditions but have failed to show any clinical benefits (Bradley 1990; Bradley et al. 1984; Harrington et al. 1984; Lacomblez et al. 1989).

Neurotrophic Factors

The concept of neurotrophism in nerve regeneration (Ramon y Cajal 1928) and the concept of abiotrophy in neurodegenerative disorders (Gowers 1902) are old. Appel (1981) proposed the hypothesis that the lack of specific trophic factors can cause ALS and other neurodegenerative diseases. However, until recently neurotrophic factor for motor neurons has not been identified, but of novel neurotrophic factors, ciliary neurotrophic factor (CNTF) has recently emerged as an example. CNTF is specifically required for the survival of embryonic chick ciliary ganglion neuron (Manthorpe et al. 1980). It is now recognized as a key motor neuron trophic factor or motor neuron surviving factor (Snider and Johnson 1989; Thoenen 1991). More detailed discussion on neurotrophic factors in motor neuron diseases is available elsewhere (Mitsumoto 1993).

In in vitro culture studies, CNTF prolongs survival of motor neurons (Wong et al. 1990, 1991; Arakawa et al. 1990; Sendtner et al. 1991). In in vivo studies, CNTF significantly reduces the naturally occurring neuronal cell death, specifically in spinal motor neurons (Oppenheim et al. 1991). When it is applied to the distal stump of transected rat neonatal facial nerves, it can rescue 80% of facial motor neurons that normally degenerate following axotomy (Sendtner et al. 1990). Recently, Hagg et al. (1992) showed that CNTF can prevent neuronal degeneration in adult rats, which suggests that CNTF has biological effects not only on embryonic or neonatal animals, but also on mature animals. CNTF also induces motor neuron sprouting in mouse gluteus muscle when injected subcutaneously over the gluteus muscle daily for 7 days (Gurney et al. 1992).

CNTF does not easily cross the blood–brain barrier (Thoenen 1991). However, it is transported to neurons by retrograde axonal transport. Furthermore, it is suspected that the CNTF receptor exerts specific signal transduction on the neurons, perhaps at the axon terminals (Ip et al. 1992; Ip and Yancopoulos 1994). The CNTF receptor is not a common transmembrane receptor complex, but it has an unusual glycophosphatidylinositol linkage that extends outside the cell membrane. The CNTF receptor protein appears to interact with other transmembrane signalling protein(s) to transduce the CNTF signal. This feature is in contrast to known neurotrophic factors, which belong to the neurotrophin family. Furthermore, the CNTF binding protein appears to participate in a multimolecular receptor complex involving leukaemia inhibitory factor (LIF)/cholinergic differentiation factor, and interleukin-6 (IL-6). These findings suggest that the action of CNTF might be similar to that of a haemopoietic cytokine (Patterson 1992; Kishimoto et al. 1992). Thus, the CNTF receptor belongs to the family of cytokine receptors. However, the restricted distribution of the CNTF receptor to the neurons and skeletal muscles confers a high degree of specificity on the neurobiological actions of CNTF in vivo (Davis et al. 1991).

The effects of CNTF have been studied in three different natural motor neuron disease models. In mice with autosomal-dominant, late-onset motor neuron degeneration (*Mnd* mice) (Messer et al. 1987), CNTF is capable of reducing hind limb dysfunction (Helgren et al. 1992). The second model is a recently discovered mouse mutant, progressive motor neuronopathy (*pmn*), in which severe motor neuron depletion develops very early, resulting in death at 6 to 7 weeks of age (Schmalbruch et al. 1991). CNTF treatment significantly improved the motor function and rescued degenerating motor neurons (Sendtner et al. 1992). The third model is the wobbler mouse which has been most extensively investigated as a model of motor neuron disease (Duchen et al. 1968; Mitsumoto and Bradley 1982). CNTF treatment slows the disease progress and increases the number of motor neurons immuno-reactive to calcitonin-gene-related peptide (Holmlund et al. 1992). Subsequent studies showed that in vivo and in vitro muscle twitch tensions are significantly greater and denervation muscle atrophy is significantly reduced in CNTF-treated wobbler mice than untreated mice (Mitsumoto et al. 1994; Ikeda et al. 1994).

Studies investigating whether the degeneration of motor neurons in ALS results from the lack of neurotrophic factors have been limited and so far have found no such evidence (Braunstein and Reviczky 1987; Ebendal et al. 1989; Kerkhoff et al. 1991; Aquilonius et al. 1992). However, neurotrophic factors having surviving effects on motor neurons have only recently been discovered. Now, the genes for human CNTF are cloned (Lin et al. 1989; Masiakowski et al. 1991), and recombinant human CNTF (rHCNTF) is manufactured for clinical trials. The phase I clinical trials to determine the safety and toxicity of rHCNTF in ALS patients have been completed (ALS CNTF Treatment Study Group, 1992). The phase II and III, placebo-controlled, double-blind, multicentre studies to determine long-term safety and clinical efficacy of rHCNTF in ALS patients are in progress. Furthermore, it is very likely that clinical effects of other neurotrophic factors such as brain-derived neurotrophic factors (BDNF) will be tested for the treatment of ALS.

Other Approaches

Growth Hormone

Patients with ALS often undergo uncontrolled massive weight loss despite normal oral intake. This situation is called ALS cachexia, the mechanism of which is poorly understood but obviously related, in part, to loss of muscle bulk due to denervation and disuse (Norris et al. 1979). Muscle atrophy in ALS is due to increased skeletal muscle catabolism (Corbett et al. 1982). Further, serum alpha-2-macroglobulin is increased, suggesting that systemic catabolism is increased in ALS (Festoff 1983). Smith et al. (1993) investigated the effect of growth hormone to counteract this catabolic process in more than 70 ALS patients in a controlled trial. They found no significant clinical benefits in the treated patients.

Insulin-Like Growth Factor-I (IGF-I)

The IGFs are among the many neurotrophic growth factors implicated in brain development. Particularly, the IGF-I (somatomedin C), which mediates many effects of growth hormone, is ubiquitous and important for normal human growth and development (LeRoith et al. 1992), including the growth and differentiation of muscle cells (Florini 1987) and differentiation of oligodendrocytes (McMorris et al. 1986). In ALS patients, the serum IGF-I levels did not differ from those in control patients (Braustein and Reviczky 1987). A new clinical trail with IGF-I is now under way (Dr. S Appel, Baylor University).

N-acetylcysteine and N-acetylmethionine

These agents are chelators for bivalent lead, cadmium, and gold and supply cysteine for free radical scavengers such as glutathione (de Jong et al. 1988). An open study with a large daily subcutaneous injection (up to 100 ml) of either 5% N-acetylcysteine or 5.85% N-acetylmethionine solution to ALS patients for a period of 6 months has shown modest clinical improvement (de Jong et al. 1988). However, another study failed to find any benefit (Kuther and Struppler 1986).

Anti-oxidant and Anti-free Radicals

Iron-induced oxidant stress has been implicated in the pathogenesis of Parkinson's disease (Olanow 1992). The observation that ALS and Parkinson's disease occur in the same patients in Guam suggests that the pathogenetic mechanisms of these two diseases might be associated. Selegiline (deprenyl) is a selective, non-competitive inhibitor of the monoamine oxidase enzyme that is primarily responsible for the degradation of dopamine in the striatum, but it also has been claimed by some to protect against neurodegeneration by reducing the formation of free radicals and oxidative stress. Dr. Forbes Norris and colleagues (ALS Research Foundation, San Francisco, personal communication) investigated the effects of selegiline in 25 ALS patients in a double-blind, cross-over trial. Their findings suggest possible benefits that justify further study.

The recent discovery of the abnormal gene responsible for superoxide dismutase in some familial ALS patients opened an entirely new area of research into the pathogenesis of ALS and potential treatment modalities (Deng et al. 1993).

Octacosanol

Octacosanol is one of several long chain alcohols that can be extracted from wheat germ oil. In healthy volunteers it has been shown to improve athletic performance (Cureton 1972). A small controlled study with patients with Parkinson's disease has revealed some beneficial effects (Snider 1984), and octacosanol had been claimed anecdotally to provide some benefits in ALS patients. Norris et al. (1986) thus performed a controlled trial with octacosanol but found no clinical benefit.

Lecithin

Lecithin is a precursor of acetylcholine, the key neuromuscular transmitter of lower motor neurons and also essential for synthesis of lipoprotein neuronal cell membrane (Growdon et al. 1977). There is evidence that cholinergic activity is reduced in ALS, since cholinergic receptors and choline acetyltransferase, the enzyme responsible for acetylcholine synthesis, are both diminished in activity in the spinal cord in ALS (Whitehouse et al. 1983; Gillberg et al. 1982). These data provide a rationale for lecithin treatment for ALS patients, although the decreased cholinergic activity may be simply secondary to the neuronal cell loss in ALS. An open study performed by Kelemen et al. (1982) did not find any clinical benefits.

Testosterone

Androgen receptors are absent in the motor neurons controlling external ocular muscles, which are generally spared in ALS. In contrast, other motor neurons in the brain stem and spinal cord, which are always affected in ALS, are found to have androgen receptors. Moreover, the observation that ALS is more common in males than females but equal in incidence between males and post-menopausal females suggests that an impaired androgen receptor may play a role in ALS (Weiner 1980). Testosterone has been given in open fashion to four male patients with ALS but there was no change in the clinical course. Their hypothalamic-pituitary function was in fact found to be normal (Jones et al. 1982).

L-cysteine

Abnormal cyanide metabolism has been shown in both smoking and nonsmoking ALS patients who have increased cyanide levels in blood, urine, and saliva as compared to controls (Kameyama 1980; Kato et al. 1985). L-cysteine has been known to be an effective treatment in exogenous cyanide intoxication. Oral L-cysteine has been given to 90 ALS patients in an open study, and in those with initial high cyanide levels the cyanide levels normalized. However, this treatment did not change the course of the disease (Akiguchi et al. 1984).

Bestatin

Bestatin is a microbial product which has been shown to augment cell-mediated immune responses, to stimulate the DNA-synthesizing system, and to inhibit aminopeptidase involved with initial interaction between serum growth factor and quiescent cells (Jerusalem et al. 1988). Since bestatin is reported to improve murine muscular dystrophy, its clinical effect has been investigated in ALS patients in controlled fashion by Jerusalem et al. (1988), who found no significant clinical improvement in these patients. This study revealed a remarkable frequency of "placebo effects."

Conclusion

The results of a number of diverse clinical trials performed in recent years can be summarized as follows: (1) none of the clinical trials convincingly improved ALS patients; (2) the clinical trials with TRH and its analogues have caused continued controversy, because acute, transient improvement has been shown in some muscles. However, it is not known what the mechanism for such improvement is and why only a few muscles, different muscles depending on the study, improve with TRH; and (3) clinical trials with TRH analogues, amino acids counteracting excitotoxins, glutamate receptor antagonists, immunosuppressive agents, and neurotrophic factors are currently under way, being carried out mostly in controlled fashion.

In the future, it is clear that more clinical trials will be performed. When conducting any clinical trials, except for open phase I feasibility tests, investigators should carry out their trials under the best placebo-controlled, double-blinded conditions. It is important to be cautious in every step of a clinical trial, particularly when we interpret our own results. Furthermore, it is our responsibility to take care of our patients first in the best way possible and not to lose them in our "scientific path".

Although it is essential to study any new, potentially beneficial treatment modalities in patients afflicted with this devastating disease, the importance of basic research to elucidate the underlying aetiology and pathogenesis of ALS cannot be over-emphasized, since knowledge from such investigations should lead us to formulate more appropriate treatments.

References

Akiguchi I, Kato T, Sugiyama H (1984) Trials of L-cysteine, TRH and 5-HTP in amyotrophic lateral sclerosis. Clin Neurol (Tokyo) 24:1270–1273

Albers GW, Goldberg MP, Choi DW (1992) Do NMDA antagonists prevent neuronal injury? Yes. Arch Neurol 49:418–420

Albin RL, Greenamyre JT (1992) Alternative excitotoxic hypotheses. Neurology 42:733–738

Allaoua H, Chaudieu I, Krieger C et al. (1992) Alterations in spinal cord excitatory amino acid receptors in amyotrophic lateral sclerosis patients. Brain Res 579:169–172

ALS CNTF Treatment Study (ACTS) Group (Brooks BR, Cedarbaum JM, Mitsumoto H, Neville H, Pestronk A 1992) A phase I ascending-dose tolerability and pharmacokinetic study of recombinant human ciliary neurotrophic factor (rHCNTF) in patients with amyotrophic lateral sclerosis (abstract). Neurology

Andres PL, Thibodeau LM, Finison LJ, Munsat L (1987) Quantitative assessment of neuromuscular deficit in ALS. Neurol Clin 5:125–141

Andres PL, Finison LJ, Conlon T, Thibodeau LM, Munsat TL (1988) Use of composite scores (megascores) to measure deficit in amyotrophic lateral sclerosis. Neurology 38:405–408

Appel SH (1981) A unifying hypothesis for the cause of amyotrophic lateral sclerosis, parkinsonism, and Alzheimer disease. Ann Neurol 10:499–505

Appel SH, Stockton-Appel V, Stewart SS, Kerman RH (1986) Amyotrophic lateral sclerosis: associated clinical disorders and immunological evaluations. Arch Neurol 43:234–238

Appel SH, Stewart SS, Appel V et al. (1988) A double-blind study of the effectiveness of cyclosporine in amyotrophic lateral sclerosis. Arch Neurol 45:381–386

Appel SH, Engelhardt JI, Garcia J, Stefani E (1991) Immunoglobulins from animal models of motor neuron disease and from human amyotrophic lateral sclerosis patients passively transfer physiological abnormalities to the neuromuscular junction. Proc Natl Acad Sci USA 88:647–651

Appelbaum JS, Salazar-Grueso EF, Richard JG et al. (1991) Dextromethorphan in the treatment of ALS: a pilot study. Neurology 41(Suppl 1):393

Aquilonius S-M, Askmark H, Ebendal T, Gilberg P-G (1992) No re-expression of high-affinity nerve growth factor binding sites in spinal motor neurons in amyotrophic lateral sclerosis. Eur Neurol 32:216–218

Arakawa Y, Sendtner M, Thoenen H (1990) Survival effect of ciliary neurotrophic factor (CNTF) on chick embryonic motoneurons in culture: comparison with other neurotrophic factors and cytokines. J Neurosci 10:3507–3515

Beghi E, Fiordelli E, Gerardo OS et al. (The Italian ALS Study Group) (1993) Branched-chain amino acids and amyotrophic lateral sclerosis: a treatment failure? Neurology 43:2466–2470

Bensimon G, Lacomblez, L, Meininger V and the ALS/Riluzole Study Group (1994). A controlled trial of riluzole in amyotrophic lateral sclerosis. N Engl J Med 330:585–591

Blin O, Pouget J, Aubrespy G et al. (1991) A double-blind placebo-controlled trial of L-threonine in amyotrophic lateral sclerosis. J Neurol 239:79–81

Bradley WG (1990) Critical review of gangliosides and thyrotropin-releasing hormone in peripheral neuromuscular diseases. Muscle Nerve 13:833–842

Bradley WG, Hedlund W, Cooper C (1984) A double-blind controlled trial of bovine brain gangliosides in amyotrophic lateral sclerosis. Neurology 34:1079–1082

Braunstein GD, Reviczky AL (1987) Serum insulin-like growth factor-I levels in amyotrophic lateral sclerosis. J Neurol Neurosurg Psychiatry 50: 792–794

Brooke MH (1989) Thyrotropin-releasing hormone in ALS: are the results of clinical studies inconsistent? Ann NY Acad Sci 553:422–430

Brooke MH, Florence JM, Heller SL et al. (1986) Controlled trial of thyrotropin-releasing hormone in amyotrophic lateral sclerosis. Neurology 36:146–151

Brooks BR (1989) A summary of the current position of TRH in ALS therapy. Ann NY Acad Sci 553:431–460

Brooks BR, Sufit RL, Montgomery GK, Beaulieu DA, Erickson LM (1987) Intravenous thyrotropin-releasing hormone in patients with amyotrophic lateral sclerosis: Dose-response and randomized concurrent placebo-controlled pilot studies. Neurol Clin 5:143–158

Brooks BR, Sufit RL, DePaul R et al. (1991) Design of clinical therapeutic trials in amyotrophic lateral sclerosis. Adv Neurol 56:422–430

Brown RH, Hauser SL, Harrington H, Weiner HL (1986) Failure of immunosuppression with a 10 to 14-day course of high-dose intravenous cyclophosphamide to alter the progression of amyotrophic lateral sclerosis. Arch Neurol 43:383

Buchan AM (1992) Do NMDA antagonists prevent neuronal injury? No. Arch Neurol 49:420–421

Caroscio JT, Cohen JA, Zawodniak J et al. (1986) A double-blind, placebo-controlled trial of TRH in amyotrophic lateral sclerosis. Neurology 36:141–145

Choi DW (1987) Dextrorphan and dextromethorphan attenuate glutamate neurotoxicity. Brain Res 403:333–336

Choi DW, Koh J, Peters S (1988) Pharmacology of glutamate neurotoxicity in cortical cell culture: attenuation by NMDA antagonists. Neuroscience 8:185–196

Clarke KA, Parker AJ, Stirk GC (1984) Motoneuron excitability during antidromically evoked inhibition after administration of a thyrotrophin-releasing hormone (TRH) analogue. Neuropeptides 4:403–411

Committee on Health Care Issues (1987) Current status of thyrotropin-releasing hormone therapy in amyotrophic lateral sclerosis. Ann Neurol 22:541–543

Conomy JP, Gerhard G, Goren H, Braun W, Barna B (1980) Plasmapheresis in the treatment of amyotrophic lateral sclerosis. Neurology 30:356

Conradi S, Ronnevi LO (1985) Cytotoxic activity in the plasma of amyotrophic lateral sclerosis (ALS) patients against normal erythrocytes: quantitative determinations. J Neurol Sci 68:135–145

Conradi S, Ronnevi LO (1987) Effect of immunosuppression on cytotoxic activity in amyotrophic lateral sclerosis. Clin Neuropharmacol 10:280–286

Conradi S, Ronnevi LO, Nise G, et al. (1982a) Long-term penicillamine-treatment in amyotrophic lateral sclerosis with parallel determination of lead in blood, plasma and urine. Acta Neurol Scand 65:203–211

Conradi S, Ronnevi LO, Norris IH (1982b) Motor neuron disease and toxic metals. Adv Neurol 36:201–231

Corbett AJ, Griggs RC, Moxley RT III (1982) Skeletal muscle catabolism in amyotrophic lateral sclerosis and chronic spinal muscular atrophy. Neurology 32:550–552

Corbo M, Quanttrini A, Lugarese A et al. (1992) Patterns of reactivity of human anti-GM_{1a} antibodies with spinal cord and motor neurons. Ann Neurol 32:487–493

Cureton TK (1972) The physiological effects of wheat germ oil on humans in exercise. Forty-two physical training programs using 894 humans. CC Thomas, Springfield IL

Currier RD, Haerer AF (1968) Amyotrophic lateral sclerosis and metallic toxins. Arch Environ Health 17:712–719

Cwik VA, Mitsumoto H (1992) Postpoliomyelitis syndrome. In: Smith RA (ed) Handbook of amyotrophic lateral sclerosis. Marcel Dekker, New York, pp 77–92

Dalakas MC, Aksamit AJ, Madden DL, Sever JL (1986) Administration of recombinant human leukocyte $alpha_2$-interferon in patients with amyotrophic lateral sclerosis. Arch Neurol 43:933–935

Dau PC (1984) Plasmapheresis, therapeutic or experimental procedure? Arch Neurol 41:647–653

Davis S, Aldrich TH, Valenzuela DM et al. (1991) The receptor for ciliary neurotrophic factors. Science 253:59–63

de Jong JMBV, Den Hartog Jager WA, Posthumus Meyjes FE et al. (1988) N-acetylcysteine and N-acetylmethionine treatment of amyotrophic lateral sclerosis (ALS) and of rapidly progressive motor neuron disease (MND). In: Tsubaki T, Yase Y (eds) Amyotrophic lateral sclerosis. Elsevier, Amsterdam, pp 313–318

Delbono O, Garcia J, Appel SH (1991) IgG from amyotrophic lateral sclerosis affects tubular calcium channels of skeletal muscle. Am J Physiol 260:C1347–C1351

Deng, HX, Hentati A, Tainer JA et al. (1993) Amyotrophic lateral sclerosis and structural defects in Cu,Zn superoxide dismutase. Science 261:1047–1051

Drachman DB, Chaundhry V, Cornblath D et al. (1994) Trial of immunosuppression in amyotrophic lateral sclerosis using total lymphoid irradiation. Ann Neurol 35:142–150

Drachman DB, Kuncl RW (1989) Amyotrophic lateral sclerosis: an unconventional autoimmune disease? Ann Neurol 26:269–274

Duarte F, Binet S, Lacomblez L et al. (1991) Quantitative analysis of monoclonal immunoglobulins in serum of patients with amyotrophic lateral sclerosis. J Neurol Sci 104:88–91

Duchen LW, Strich SJ, Falconer DS (1968) An hereditary motor neurone disease with progressive denervation of muscle in the mouse: The mutant 'wobbler'. J Neurol Neurosurg Psychiatry 31:535–542

Duncan MW, Kopin IJ, Garruto RM, Lavine L, Markey SP (1988) 2-amino-3(methylamino)-propionic acid in cycad-derived foods is an unlikely cause of amyotrophic lateral sclerosis/parkinsonism. Lancet ii:631–632

Ebendal T, Asmark H, Aquilonius S-M (1989) Screening for neurotrophic disturbances in amyotrophic lateral sclerosis. Acta Neurol Scand 79:188–193

Endo T, Scott DD, Stewart SS et al. (1984) Antibodies to glycosphingolipids in patients with multiple sclerosis and SLE. J Immunol 132:1793–1797

Engel WK (1989) High-dose TRH treatment of neuromuscular disease: summary of mechanisms and critique of clinical studies. Ann NY Acad Sci 553:462–472

Engel WK, Siddique T, Nicholoff JT (1983) Effect on weakness and spasticity in amyotrophic lateral sclerosis of thyrotropin-releasing hormone. Lancet ii:73–75

Engel WK, Spiel RH (1985) Prolonged at-home treatment of motor neuron disorders with self-administered subcutaneous (sc) high-dose TRH. Neurology 35 (Suppl 1):106–107

Engelhardt JI, Appel SH (1990) IgG reactivity in the spinal cord and motor cortex in amyotrophic lateral sclerosis. Arch Neurol 47:1210–1216

Engelhardt JI, Appel SH, Killian JM (1989) Experimental autoimmune motorneuron disease. Ann Neurol 26:368–376

Engelhardt J, Joo F (1986) Immune-mediated guinea pig model for lower motor neuron disease. J Neuroimmunol 12:279–290

Faden A, Holaday JW, Jacobs TP (1981) Thyrotropin-releasing hormone improves neurological recovery after spinal trauma in cats. N Engl J Med 305:1063–1067

Faden AI, Jacobs TP, Smith MT (1984) Thyrotropin-releasing hormone in experimental spinal injury: Dose response and late treatment. Neurology 34:1280–1284

Fallis RJ, Weiner LP (1982) Further studies in search of a virus in amyotrophic lateral sclerosis. Adv Neurol 36:355–361

Färkkilä M, Iivanainen M, Roine R (1984) Neurotoxic and other side effects of high-dose interferon in amyotrophic lateral sclerosis. Acta Neurol Scand 69:42–46

Feldman EL, Bromberg MB, Albers JW et al. (1991) Immunosuppressive treatment of multifocal motor neuropathy. Ann Neurol 30:397–401

Festoff BW (1983) Occurrence of reduced alpha$_2$-macroglobulin and lowered protease inhibiting capacity in plasma of amyotrophic lateral sclerosis patients. Ann NY Acad Sci 421:369–376

Festoff BW, Crigger NJ (1979) Therapeutic trials in amyotrophic lateral sclerosis: a review. In: Mulder DW (ed) The diagnosis and treatment of amyotrophic lateral sclerosis. Houghton Mifflin, Boston, pp 337–366

Florini JR (1987) Hormonal control of muscle growth. Muscle Nerve 10:577–598

Freddo L, Yu RK, Latov N et al. (1986) Gangliosides GM$_1$ and GD$_{1b}$ are antigens for IgM M-protein in a patient with motor neuron disease. Neurology 36:454–458

Garruto RM, Yanagihara R, Gajdusek DC (1985) Disappearance of high-incidence amyotrophic lateral sclerosis and parkinsonism-dementia on Guam. Neurology 35:193–198

Garruto RM, Swyt C, Yanagihara R et al. (1986) Intraneuronal co-localization of silicon with calcium and aluminum in amyotrophic lateral sclerosis and parkinsonism with dementia of Guam. N Engl J Med 317–711

Gillberg PG, Aquilonius SM, Eckernas SA et al. (1982) Choline acetyltransferase and substance P-like immunoreactivity in the human spinal cord: changes in amyotrophic lateral sclerosis. Brain Res 250:394–397

Gorio A, Carmignoto G, Ferrari G (1982) Axon sprouting stimulated by gangliosides: a new model for elongation and sprouting. In: Tapport MM, Gorio A (eds) Gangliosides in neurological and neuromuscular function, development and repair. Raven Press, New York, pp 177–195

Gowers WR (1902) A lecture on abiotrophy. Lancet i:1003–1007

Greenamyre JT (1986) The role of glutamate in neurotransmission and in neurologic disease. Arch Neurol 43:1058–1063

Growdon JH, Cohen EL, Wurtman RJ (1977) Treatment of brain disease with dietary precursors of neurotransmitters. Ann Intern Med 86:337–339

Guiloff RJ, Eckland DJA, Demaine C et al. (1987a) Controlled acute trial of a thyrotropin-releasing hormone analogue (RX77368) in motor neuron disease. J Neurol Neurosurg Psychiatry 50:1359–1370

Guiloff RJ, Stålberg E, Eckland DJA, Lightman SL (1987b) Electrophysiological observations in patients with motor neuron disease receiving a thyrotropin-releasing hormone analogue (TX77368). J Neurol Neurosurg Psychiatry 50:1633–1640

Guiloff RJ, Modarres-Sadeghi H, Stålberg, Rogers H (1988) Short-term stability of single motor unit recordings in motor neuron disease: a macro EMG study. J Neurol Neurosurg Psychiatry 51:671–676

Gurney ME, Yamamoto H, Kwon Y (1992) Induction of motor neuron sprouting in vivo by ciliary neurotrophic factor and basic fibroblast growth factor. J Neurosci 12:3241–3247

Hagg T, Quon D, Higaki J et al. (1992) Ciliary neurotrophic factor prevents neuronal degeneration and promotes low affinity NGF receptor expression in the adult rat CNS. Neuron 8:145–158

Harrington H, Hallett M, Tyler HR (1984) Ganglioside therapy for amyotrophic lateral sclerosis: a double-blind controlled trial. Neurology 34:1083–1085

Hawley RJ, Kratz R, Goodman RR et al. (1987) Treatment of amyotrophic lateral sclerosis with TRH analog DN1417. Neurology 37:715–717

Hayashi H, Suga M, Satake M, Tsubaki T (1980) Reduced glycine receptor in the spinal cord in amyotrophic lateral sclerosis. Ann Neurol 9:292–294

Helgren ME, Friedman B, Kennedy M et al. (1992) Ciliary neurotrophic factor (CNTF) delays motor impairments in the Mnd mouse, a genetic model for motor neuron disease. Soc Neurosci Abstr 18:618

Hökfelt T, Fuxe K, Johannsson O, Jeffcoate S, White N (1975) Thyrotropin-releasing hormone (TRH)-containing nerve terminals in certain brain stem nuclei and in the spinal cord. Neurosci Lett 1:133–139

Hollander D, Pradas J, Kaplan R et al. (1992) Long-term high-dose dextromethorphan in amyotrophic lateral sclerosis. Ann Neurol 32:201

Holmlund TH, Mitsumoto H, Greene T et al. (1992) The effect of ciliary neurotrophic factor (CNTF) on spontaneously degenerating motor neurons in wobbler mice. Neurology 42(Suppl 3):369

House AO, Abbott RJ, Davidson DLW et al. (1978) Response to penicillamine of lead concentration in CSF and blood in patients with motor neurone disease. BMJ iii:1684

Hugon J, Tabaraud F, Rigaud M, Vallat JM, Dumas M (1989) Glutamate dehydrogenase and aspartate aminotransferase in leukocytes of patients with motor neuron disease. Neurology 39:956–958

Ikeda K, Mitsumoto H, Wong V, Lindsay RM, Cedarbaum JM (1994) Morphological changes correlate with functional improvement following ciliary neurotrophic factor (CNTF) treatment of motor neuron disease in the wobbler mouse (abstract). Neurology (in press)

Imoto K, Saida K, Iwamura K, Saida T, Nighitani (1984) Amyotrophic lateral sclerosis: a double-blind crossover trial of thyrotropin-releasing hormone. J Neurol Neurosurg Psychiatry 47:1332–1334

Ip NY, Nye SH, Boulton TG et al. (1992) CNTF and LIF act on neuronal cells via shared signaling pathways that involve the IL-6 signal transducing receptor component gp 130. Cell 69:1121–1123

Ip NY, Yancopoulos GD (1994) Ciliary neurotrophic factor and its receptor complex. Prog Growth Factor Res 4 (in press)

Iwasaki Y, Ikeda K, Kinoshita M (1991) Plasma amino acid levels in patients with amyotrophic lateral sclerosis. J Neurol Sci 105:1–4

Jackson IM, Adelman LS, Munsat TL, Forte S, Lechan R (1986) Amyotrophic lateral sclerosis: thyrotropin-releasing hormone and histidyl proline diketopiperazine in the spinal cord and cerebrospinal fluid. Neurology 36:1218–12223

Jerusalem F, Fresmann J (1985) Comparison of treatment trials in ALS between thyrotrophin-releasing hormone (TRH) and TRH-derivative orotyl-L-histidyl-L-prolineamide. J Neurol 232(Suppl):35

Jerusalem F, Fresmann J, Sayn H, Schulz G (1988) A double-blind, placebo-controlled trial of bestatin in amyotrophic lateral sclerosis (ALS). In: Tsubaki T, Yase Y (eds) Amyotrophic lateral sclerosis. Elsevier, Amsterdam, pp 333–338

Jones TM, Yu R, Antel JP (1982) Response of patients with amyotrophic lateral sclerosis to testosterone therapy: endocrine evaluation. Arch Neurol 39:721–722

Kaji R, Shibasaki H, Kumura J (1992) Multifocal demyelinating motor neuropathy: cranial nerve involvement and immunoglobulin therapy. Neurology 42:506–509

Kameyama M (1980) Cyanide metabolism in the nervous disorder, with special reference to motor neuron disease. Clin Neurol (Tokyo) 20:99–1007

Kardon FC, Winokur A, Utiger RD (1977) Thyrotrophin-releasing hormone (TRH) in rat spinal cord. Brain Res 122:578–581

Kato T, Kameyama M, Nakamura S, Inada M, Sugiyama H (1985) Cyanide metabolism in motor neuron disease. Acta Neurol Scand 72:151–156

Kelemen J, Hedlund W, Murray-Douglas P, Munsat TL (1982) Lecithin is not effective in amyotrophic lateral sclerosis. Neurology 32:315–316

Kelemen J, Hedlund W, Orlin JB, Berkman EM, Munsat TL (1983) Plasmapheresis with immunosuppression in amyotrophic lateral sclerosis. Arch Neurol 40:752–753

Kerkhoff H, Jennekens FGI, Troost D, Veldman H (1991) Nerve growth factor receptor immunostaining in the spinal cord and peripheral nerves in amyotrophic lateral sclerosis. Acta Neuropathol 81:649–656

Kishimoto T, Akira S, Taga T (1992) Interleukin-6 and its receptor: A paradigm for cytokines. Science 258:593–596

Krieger C, Melmed C (1982) Amyotrophic lateral sclerosis and paraproteinemia. Neurology 32:896–898

Kurland LT (1972) An appraisal of the neurotoxicity of cycad and the etiology of amyotrophic lateral sclerosis on Guam. Fed Proc 31:1540–1542

Kuther G, Struppler A (1986) Therapieversuch der amyotrophischen lateralsklerose mit N-acetylcystein. Fortschr Myol 8:51–57

Kyle RA, Finkelstein S, Elvback LR, Kurland LT (1972) Incidence of monoclonal proteins in a Minnesota community with a cluster of multiple myeloma. Blood 40:719–724

Lacomblez L, Bouche P, Bensimon G, Meininger V (1989) A double-blind, placebo-controlled trial of high doses of gangliosides in amyotrophic lateral sclerosis. Neurology 39:1635–1637

Lange DJ, Trojaborg W, Latov N et al. (1992) Multifocal motor neuropathy with conduction block: is it a distinct clinical entity? Neurology 42:497–505

Latov H, Hays AP, Donofrio PD et al. (1988) Monoclonal IgM with unique specificity to gangliosides GM1 and GD1b and to lacro-N-tetraose associated with human motor neuron disease. Neurology 38:763–768

Lechan RM, Snapper SB, Jacobson S et al. (1984) The distribution of thyrotropin-releasing hormone (TRH) in the Rhesus monkey spinal cord. Peptides 5:185–194

LeRoith D, Clemmons D, Nissley P, Rechler MM (1992) Insulin-like growth factors in health and disease. HIN conference. Ann Intern Med 116:854–862

Lin LFH, Mismer D, Lile JD et al. (1989) Purification, cloning and expression of ciliary neurotrophic factors (CNTF). Science 246:47–56

Lipton SA (1989) Growth factors for neuronal survival and process regeneration: implications in the mammalian central nervous system. Arch Neurol 46:1241–1248

Lisak RP (1984) Plasma exchange in neurologic diseases. Arch Neurol 41:654–657

Malessa S, Leigh PN, Bertel O et al. (1991) Amyotrophic lateral sclerosis: glutamate dehydrogenase and transmitter amino acids in the spinal cord. J Neurol Neurosurg Psychiatry 54:984–988

Manaker S, Winokur A, Rhodes CH, Rainbow TC (1985) Autoradiographic localization of thyrotrophin-releasing hormone receptors in the rat central nervous system. J Neurosci 5:167–174

Manaker S, Caine SB, Winokur A (1988) Alterations in receptors for thyrotropin-releasing hormone, serotonin, and acetylcholine in amyotrophic lateral sclerosis. Neurology 38:1464–1474

Manthorpe M, Skaper S, Adler R et al. (1980) Cholinergic neurotrophic factors: Fractionation properties of an extract from selected chick embryonic eye tissues. J Neurochem 34:69–75

Masiakowski P, Liu H, Radziejewski C et al. (1991) Recombinant human and rat ciliary neurotrophic factors. J Neurochem 57:1003–1012

McMorris FA, Smith TM, DeSalvo S et al. (1986) Insulin-like growth factor-I/somatomedin C: a potent inducer of oligodendrocyte development. Proc Natl Acad Sci USA 83:822–826

Messer A, Strominger NL, Mazurkiewicz JE (1987) Histopathology of the late-onset motor neuron degeneration (Mnd) mutant in the mouse. J Neurogen 4:201–213

Miller SC, Warnick Je (1989) Protirelin (thyrotropin-releasing hormone) in amyotrophic lateral sclerosis. Arch Neurol 46:330–335

Mitsumoto H (1988) Clinical trials of thyrotropin-releasing hormone (TRH) in patients with amyotrophic lateral sclerosis (ALS) and in motor neuron disease of wobbler mice. In: Tsubaki T, Yase Y (eds) Amyotrophic lateral sclerosis. Elsevier, Amsterdam, pp 295–300

Mitsumoto H (1993) Motor neuron disease: the path to clinical trials. Course syllabus. In: Brooke MH (course director) Update on neuromuscular disease. American Academy of Neurology, New York

Mitsumoto H, Bradley WG (1982) Murine motor neuron disease (the wobbler mouse). Degeneration and regeneration of the lower motor neuron. Brain 105:811–834

Mitsumoto H, Salgado ED, Negrosky D et al. (1986) Amyotrophic lateral sclerosis: effects of acute intravenous and chronic subcutaneous administration of thyrotropin-releasing hormone in controlled trials. Neurology 36:152–159

Mitsumoto H, Hanson MR, Chad DA (1988) Amyotrophic lateral sclerosis: recent advances in pathogenesis and therapeutic trials. Arch Neurol 45:189–202

Mitsumoto H, Kumar S, Levin KH et al. (1992) Intravenous immunoglobulin treatment in amyotrophic lateral sclerosis. Ann Neurol 32:252

Mitsumoto H, Ikeda K, Greene T, Wong V, Lindsay RM, Cedarbaum JM (1994) The effects of ciliary neurotrophic factor (CNTF) in wobbler mouse motor neuron disease. Ann Neurol (in press).

Modarres-Sadeghi H, Rogers H, Emami J, Guiloff RJ (1988) Subacute administration of a TRH analogue (RX77368) in motoneurone disease. An open study. J Neurol Neurosurg Psychiatry 51:1146–1157

Mora JS, Munsat TL, Kao Ko-Pei et al. (1986) Intrathecal administration of natural human interferon alpha in amyotrophic lateral sclerosis. Neurology 36:1137–1140

Munsat TL (1988) Negative drug trial bias. Letter to editor. Neurology 38:165

Munsat TL, Schmidt-Easterday C, Levy S, Wolff SM, Hiatt R (1981) Amantadine and guanidine are ineffective in ALS. Neurology 31:1054–1055

Munsat TL, Brooks BR (1987) Don't throw out the baby with the bathwater. Letter to the editor. Neurology 37:544

Munsat TL, Andres PL, Finison L, Conlon T, Thibodeau L (1988) The natural history of motoneuron loss in amyotrophic lateral sclerosis. Neurology 38:409–413

Munsat TL, Lechan R, Taft JM, Jackson IMD, Reichlin S (1989) TRH and diseases of the motor system. Ann NY Acad Sci 553:388–398

Nakanishi S (1992) Molecular diversity of glutamate receptors and implications for brain function. Science 258:597–603

The National TRH Study Group (1989) Multicenter controlled trial: no effect of alternate-day 5 mg/kg subcutaneous thyrotropin-releasing hormone (TRH) on isometric-strength decrease in amyotrophic lateral sclerosis. Neurology 39(Suppl 1):478

Neville HE, Roelofs JI, Smith SA et al. (1986) Treatment of ALS patients with intravenous MK-711 (A TRH analog) does not improve muscle strength. Muscle Nerve 9:104

Nobile-Orazio E, Carpo M, Legname G et al. (1990) Anti-GM1 IgM antibodies in motor neuron disease and neuropathy. Neurology 40:1747–1750

Norris FH, Calanchini PR, Fallat RJ, Panchari S, Jewett B (1974) The administration of guanidine in amyotrophic lateral sclerosis. Neurology 24:721–728

Norris FH, Denys EH, Sang UK (1979) Old and new clinical problems in amyotrophic lateral sclerosis. In: Tsubaki T, Toyokura Y (eds) Amyotrophic lateral sclerosis. University Park Press, Baltimore, pp 3–26

Norris FH, Denys EH, Fallat RJ (1986) Trial of octacosanol in amyotrophic lateral sclerosis. Neurology 36:1263–1264

Olanow CW (1992) An introduction to the free radical hypothesis in Parkinson's disease. Ann Neurol 32:S2–S9

Olarte MR, Schoenfeldt RS, McKiernan G, Rowland LP (1980) Plasmapheresis in amyotrophic lateral sclerosis. Ann Neurol 8:644

Olarte MR, Shafer SQ (1985) Levamisole is ineffective in the treatment of amyotrophic lateral sclerosis. Neurology 35:1063–1066

Oppenheim RW, Prevette D, Qin-Wei Y et al. (1991) Control of embryonic motor neuron survival in vivo by ciliary neurotrophic factor (CNTF). Science 251:1616–1618

Overgaard K, Werdelin L, Sorensen H et al. (1991) Cytotoxic activity in plasma from patients with amyotrophic lateral sclerosis. Neurology 41:925–927

Parry GJ, Clarke S (1988) Multifocal acquired demyelinating neuropathy masquerading as motor neuron disease. Muscle Nerve 11:103–107

Parry GJ, Holtz SJ, Ben-Zeev D, Drori JB (1986) Gammopathy with proximal motor axonopathy simulating motor neuron disease. Neurology 36:273–276

Patten BM (1984) Neuropathy and motor neuron syndromes associated with plasma cell disease. Acta Neurol Scand 69:47–61

Patten BM (1988) ALS associated with aluminum intoxication. In: Tsubaki T, Yase Y (eds) Amyotrophic lateral sclerosis. Elsevier, Amsterdam, pp 51–58

Patten BM, Harati Y, Acosta L, Lung S, Felmus MT (1978) Free amino acid levels in amyotrophic lateral sclerosis. Ann Neurol 3:305–309

Patten BM, Klein LM (1988) 1-Threonine and modification of ALS. Neurology 38(Suppl 1):355

Patterson PH (1992) The emerging neuropoietic cytokine family: first CDF/LIF, CNTF and IL-6; next ONC, MGF, GCSF? Curr Opin Neurobiol 2:94–97

Perl DP, Gajdusek DC, Garruto RM et al. (1982) Intraneuronal aluminum accumulation in amyotrophic lateral sclerosis and parkinsonism-dementia of Guam. Science 217:1053–1055

Perry TL, Krieger C, Hansen S et al. (1990) Amyotrophic lateral sclerosis amino acid levels in plasma and cerebrospinal fluid. Ann Neurol 28:12–17

Pestronk A (1991) Invited review. Motor neuropathies, motor neuron disorders, and antiglycolipid antibodies. Muscle Nerve 14:927–936

Pestronk A, Adams RN, Clawson L (1988a) Serum antibodies to GM_1 ganglioside in amyotrophic lateral sclerosis. Neurology 38:1457–1461

Pestronk A, Cornblath DR, Ilyas AA et al. (1988b) A treatable multifocal motor neuropathy with antibodies to GM1 ganglioside. Ann Neurol 24:73–78

Pestronk A, Adams RN, Cornblath D et al. (1989) Patterns of serum IgM antibodies to GM1 and GD1a gangliosides in amyotrophic lateral sclerosis. Ann Neurol 25:98–102

Peters HA, Clatanoff DV (1968) Spinal muscular atrophy secondary to macroglobulinemia: reversal of symptoms with chlorambucil therapy. Neurology 18:101–108

Plaitakis A (1990) Glutamate dysfunction and selective motor neuron degeneration in amyotrophic lateral sclerosis: a hypothesis. Ann Neurol 28:3–8

Plaitakis A, Berl S, Yahr MD (1982) Abnormal glutamate metabolism in an adult-onset neurological disorder. Science 215:193–96

Plaitakis A, Berl S, Yahr MD (1984) Neurological disorders associated with deficiency of glutamate dehydrogenase. Ann Neurol 15:144–153

Plaitakis A, Caroscio JT (1987) Abnormal glutamate metabolism in amyotrophic lateral sclerosis. Ann Neurol 22:575–579

Plaitakis A, Constantakakis E, Smith J (1988a) The neuroexcitotoxic amino acids glutamate and aspartate are altered in the spinal cord and brain in amyotrophic lateral sclerosis. Ann Neurol 24:446–449

Plaitakis A, Smith J, Mandeli J, Yahr MD (1988b) Pilot trial of branched-chain amino acids in amyotrophic lateral sclerosis. Lancet i:1015–1018

Plaitakis A, Sivak M, Fesdjian CO Mandeli J (1992) Treatment of amyotrophic lateral sclerosis with branched chain amino acids (BCAA): results of a second study. Neurology 42(Suppl 3):454

Ramon y Cajal S (1928) Degeneration and regeneration of the nervous system, vol 1. Hafner, New York

Ringel SP, Murphy JR, Alderson MK et al. (1993) The natural history of ALS. Neurology 43:1316–1322

Roelofs-Iverson RA, Mulder DW, Elveback LR, Kurland LT, Molgaard CA (1984) ALS and heavy metals: a pilot case–control study. Neurology 34:393–395

Ross SM, Seelig M, Spencer PS (1987) Specific antagonism of excitotoxic action of "uncommon" amino acids assayed in organotypic mouse cortical cultures. Brain Res 425:120–127

Rothstein JD, Kuncl RW, Claawson L et al. (1989) Cerebrospinal fluid excitatory amino acids in amyotrophic lateral sclerosis. Ann Neurol 26:144

Rothstein JD, Tsai G, Kuncl RW et al. (1990) Abnormal excitatory amino acid metabolism in amyotrophic lateral sclerosis. Ann Neurol 28:18–25

Rothstein JD, Martin LJ, Kuncl RW (1992) Decreased glutamate transport by the brain and spinal cord in amyotrophic lateral sclerosis. N Engl J Med 326:1464–1468

Rowland LP (1988a) To the Editor (in reply to letter to editor by Munsat TL: Negative drug trial bias). Neurology 38:165–166

Rowland LP (1988b) Research progress in motor neuron diseases. Rev Neurol 144:623–629

Rowland LP (1992) Amyotrophic lateral sclerosis and autoimmunity. N Engl J Med 327:1752–1753

Rowland LP, Defendini R, Sherman W et al. (1982) Macroglobulinemia with peripheral neuropathy simulating motor neuron disease. Ann Neurol 11:532–536

Rudnicki S, Chad DA, Drachman DA et al. (1987) Motor neuron disease and paraproteinemia. Neurology 37:335–337

Sadiq SA, Thomas EP, Kilidireas K et al. (1990) The spectrum of neurological disease associated with anti-GM antibodies. Neurology 40:1067–1072

Saida T, Saida K, Imoto K, Nishitani H, Sobue I (1988) Trials of thyrotropin-releasing hormone treatment of amyotrophic lateral sclerosis in Japan. In: Tsubaki T, Yase Y (eds) Amyotrophic lateral sclerosis. Elsevier, Amsterdam, pp 301–306

Santoro M, Thomas FP, Fink ME et al. (1990) IgM deposits at nodes of Ranvier in a patient with amyotrophic lateral sclerosis, anti-GM1 antibodies, and multifocal motor conduction block. Ann Neurol 28:373–377

Santoro M, Uncini A, Corbo M et al. (1992) Experimental conduction block induced by serum from a patient with anti-GM1 antibodies. Ann Neurol 32:487–493

Schluep M, Steck AJ (1988) Immunostaining of motor nerve terminals by IgM M protein with activity against gangliosides GM1 and GD1b from a patient with motor neuron disease. Neurology 38:1890–1892

Schmalbruch H, Jensen H-JS, Bjaerg M et al. (1991) A new mouse mutant with progressive motor neuronopathy. J Neuropathol Exp Neurol 50:192–204

Schmidt-Achert KM, Askanas V, Engel WK (1984) Thyrotropin-releasing hormone enhances choline acetyltransferase and creatine kinase in cultured spinal ventral horn neurons. J Neurochem 43:586–589

Sendtner M, Kreutzberg GW, Thoenen H (1990) Ciliary neurotrophic factor prevents the degeneration of motor neurons after axotomy. Nature 345:440–441

Sendtner M, Arakawa Y, Stockli KA et al. (1991) Effect of ciliary neurotrophic factor (CNTF) on motoneuron survival. J Cell Sci 15:103–109

Sendtner M, Schmalbruch H, Stockli KA et al. (1992) Ciliary neurotrophic factor prevents degeneration of motor neurons in mouse mutant progressive motor neuronopathy. Nature 358:502–504

Serratrice G, Desnuelle C, Guelton C et al. (1985) Clinical evaluation of thyrotropin releasing hormone in amyotrophic lateral sclerosis. Press Med 14:487–488

Serratrice G, Desnuelle C, Crevat A, Guelton C, Meyer-Dutour A (1986) Treatment of amyotrophic lateral sclerosis with thyrotropin releasing hormone (TRH). Rev Neurol 142:133–139

Shy ME, Rowland LP, Smith T et al. (1986) Motor neuron disease and plasma cell dyscrasia. Neurology 36:1429–1436

Shy ME, Evans VA, Lublin FD et al. (1989) Antibodies to GM1 and GD1b in patients with motor neuron disease without plasma cell dyscrasia. Ann Neurol 25:511–513

Smith RA, Norris FH (1988) Treatment of amyotrophic lateral sclerosis with interferon. In: Smith RA (ed): Interferon treatment of neurologic disorders. Marcel Dekker, New York, pp 265–275

Smith RA, Melmed S, Sherman B, Frane J, Munsat TL, Festoff BW (1993) Recombinant growth hormone treatment of amyotrophic lateral sclerosis. Muscle Nerve 16:624;633

Smith RG, Hamilton S, Hofmann F et al. (1992) Serum antibodies to l-type calcium channels in patients with amyotrophic lateral sclerosis. N Engl J Med 327:1271–1728

Snider SR (1984) Octacosanol in parkinsonism. Ann Neurol 16:723

Snider WD, Johnson EM Jr (1989) Neurotrophic molecules. Ann Neurol 26:489–506

Spencer PS (1987) Guam ALS/parkinsonism-dementia: a long-latency neurotoxic disorder caused by "slow toxin(s)" in food? Can J Neurol Sci 14:347–357

Spencer PS, Ludolph A, Dwivedi MP et al. (1986) Lathyrism: evidence for role of the neuroexcitatory amino acid BOAA. Lancet ii:1066–1067

Spencer PS, Nunn PB, Hugon J et al. (1987) Guam amyotrophic lateral sclerosis-parkinsonism-dementia linked to a plant excitant neurotoxin. Science 237:517–522

Spencer PS, Allen CN, Kisby GE et al. (1991) Lathyrism and Western Pacific amyotrophic lateral sclerosis: etiology of short and long latency motor system disorders. Adv Neurol 56:287–299

Stober T, Schmrigk K, Dietzsch S, Thielen T (1985) Intrathecal thyrotropin-releasing hormone therapy of amyotrophic lateral sclerosis. J Neurol 32:13–14

Strong MJ, Garruto RM (1991) Chronic aluminum-induced motor neuron degeneration: clinical, neuropathological and molecular biological aspects. Can J Neurol Sci 18:428–431

Sufit SR, Clough JA, Schram M et al. (1987) Isokinetic assessment in ALS. Neurol Clin 5:197–212

Teitelbaum JS, Zatorre RJ, Carpenter S (1990) Neurological sequelae of domoic acid intoxication due to ingestion of mussels from Prince Edward Island. N Engl J Med 322:1781–1787

Testa D, Caraceni T, Fetoni V et al. (1989) Branched-chain amino acids in the treatment of amyotrophic lateral sclerosis. J Neurol 236:445–447

Testa D, Caraceni T, Fetoni V et al. (1992) Chronic treatment with L-threonine in amyotrophic lateral sclerosis: a pilot study. Clin Neurol Neurosurg 94:7–10

Thoenen H (1991) The changing scene of neurotrophic factors. Trends in Neurosci 14:165–170

Troost D, van den Oord JJ, deLong JM, Swaab DF (1989) Lymphocytic infiltration in the spinal cord of patients with amyotrophic lateral sclerosis. Clin Neuropathol 8:289–294

Uchitel OD, Appel SH, Crawford F, Sczupak L (1988) Immunoglobulins from amyotrophic lateral sclerosis patients enhance spontaneous transmitter release from motor-nerve terminals. Proc Natl Acad Sci USA 85:7371–7374

Uchitel OD, Scornik F, Protti DA et al. (1992) Long-term neuromuscular dysfunction produced by passive transfer of amyotrophic lateral sclerosis immunoglobulins. Neurology 42:2175–2180

Van Den Bergh P, Kelly JJ, Adelman L, Munsat TL, Jackson IMD (1987) Effect of spinal cord TRH deficiency on lower motor neuron function in the rat. Muscle Nerve 10:397–405

Weiner LP (1980) Possible role of androgen receptors in amyotrophic lateral sclerosis: a hypothesis. Arch Neurol 37:129–131

Werdelin L, Boysen G, Jensen TS et al. (1990) Immunosuppressive treatment of patients with amyotrophic lateral sclerosis. Acta Neurol Scand 82:132–134

Whitehouse PJ, Wamsley JK, Zarbin MA et al. (1983) Amyotrophic lateral sclerosis: alterations in neurotransmitter receptors. Ann Neurol 14:8–16

Wolfgram F, Myers L (1972) Toxicity of serum from patients with amyotrophic lateral sclerosis for anterior horn cells in vitro. Trans Am Neurol Assoc 97:19–23

Wong V, Arraiga R, Lindsay RM (1990) Effects of ciliary neurotrophic factor (CNTF) on ventral spinal cord neurons in culture. Soc Neurosci Abstr 16:484

Wong V, Arraiga R, Lindsay RM et al. (1991) Ciliary neurotrophic factor (CNTF) supports survival of motor neurons in culture. Neurology 41 (Suppl 1):312

Yamane K, Osawa M, Kobayashi I, Maruyama S (1986) Treatment of amyotrophic lateral sclerosis with thyrotropin releasing hormone. Jpn J Psychiatry Neurol 40:179–187

Yanagihara RT (1982) Heavy metals and essential minerals in motor neuron disease. Adv Neurol 36:235–247

Yase Y (1984) Metal metabolism in motor neuron disease. In: Chem KM, Yase Y (eds) Amyotrophic lateral sclerosis in Asia and Oceania. National Taiwan University Press, Taipei, pp 337–356

Yielding KL, Tomkins GM (1961) An effect of L-leucine and other essential amino acids on the structure and activity of glutamate dehydrogenase. Proc Natl Acad Sci USA 47:983

Young AB (1990) What's the excitement about excitatory amino acids in amyotrophic lateral sclerosis? Ann Neurol 28:9–11

Younger DS, Rowland LP, Latov N et al. (1990) Motor neuron disease and amyotrophic lateral sclerosis: relation of high CSF protein content to paraproteinemia and clinical syndromes. Neurology 40:595–599

19 Living with Motor Neuron Disease

Editor's note. This chapter is a personal account of motor neuron disease, describing the advent of the disease, its relentless course, and the brave and continual efforts made by the patient, and his wife and family, to cope with its debilitating effects. This patient's illness occurred during the early 1980s, but despite the decade that has elapsed, treatment remains inefficacious, and the tragedy is thus repeated.

Sidney and I first met in November 1953. We were both attending a charity ball and on arrival the first person to whom my escort introduced me was Sidney. For the next two years although our paths crossed from time to time at various other functions, we never conversed except for a polite "Hello, how are you?" Then in January 1955 we were both at a Harold House (the Liverpool Jewish Youth Club) interfunction and since there were relatively few people there we drifted together during the evening and made a date to go to the theatre on the following Wednesday. Little did I know how much Sidney disliked staged drama, and after we were married we never went again! However, the evening went well and we dated again and got engaged at the end of 10 weeks. We did not get married until February 1965 because there was a home to be put together and also Sidney's business was very new and as yet unestablished and needed a great deal of his attention.

Sidney was born in 1924 in Liverpool, the elder of two sons by eight years. He attended Boaler Street Council School, followed by St. Margarets, Anfield, and was not very academic, although he was exceptionally good at mathematics, and showed a great deal of musical ability. However, he was not greatly encouraged by his parents in either of these fields and on leaving school attended Skerry's secretarial college where he took courses in shorthand, typing and book-keeping, passing all his examinations with distinction.

Shortly after World War II broke out the family rented a house in Prestatyn, North Wales, where his mother and brother lived, with Sidney and his father travelling to and fro at weekends. The family business was retailing wallpaper and paint, but during the war his father worked in censorship. On leaving Skerry's Sidney's first job was with American Express but he was then drafted briefly into munitions until he was called up for Army service, in which he served in the Royal Tank Regiment for 4 years. He landed on the beaches of Normandy on D-Day plus 12 and had many tales to tell of his experiences! When the war ended, being part of the occupation forces in Germany, but with no fighting, Sidney, because of his qualifications was made an instructor of book-keeping, a position he held until he was demobbed in 1947.

Not having any specific plan for his future, Sidney joined his father in the wallpaper business which was situated in Birkenhead, and travelled each day with his father with whom he got on very well, and as they were of similar nature they had a lot in common. They were not only "gentlemen", but also gentle men and rather shy. However, after Sidney's brother had done his National Service he too joined the business and shortly after that, on the suggestion of a close friend, Sidney opened his own establishment serving wallpaper and paint to the credit trade, something which had never been done before, making Sidney's business the first of its kind on Merseyside. He had been open for only a few weeks when we met at Harold House.

Our first child, Vikki, was born in December 1956 and because she was such a model baby we decided not to wait but to try for another so there would not be too great a gap between her and the next. In fact, there were only 20 months, fine for one baby but in fact the next turned out to be twin boys, Neil and Lawrence. Sidney was an incredible help to me, incredible because he had never had anything to do with babies previously. Because he was a good early riser he did the 6 a.m. feed and I used to hear father and daughter "talking" to each other in the next room! By the time the boys came along he was an old hand, and helped me every inch of the way, making up bottles (12 at a time!) feeding and hanging out nappies. There wasn't one aspect of their care in which he didn't participate.

The children grew up and on Sundays we often went to Prestatyn for the day, pitching our tent on the Ffrith Beach, armed with picnics and "potties" and hearing tales of the war-time evacuation. The business was flourishing and I helped when needed to begin with, and when the children went to school from 10.30 a.m. until 4.00 p.m. They were heady days. However, after 17 years the business world changed, due to various Government edicts and the advent of DIY and Sidney, not being a speculating type of business man, closed his business and joined in a venture which his brother had started a short while previously, and which needed an administrator. This he did for seven years until, after a heart attack due to stress and constant differences with his brother, he had to give up. He was lucky that a friend of very long standing offered him a job as a sales representative for glassware, which one would expect to be even more stressful, but Sidney did not find it so because at least there were no personality conflicts and he just got on with the job, at which he was very good. He carried on in this position until he was struck down with Motor Neuron Disease, and referred to his time "on the road" as "two and a half of the happiest years of my life".

Although a modest person Sidney also involved himself in communal life. In the 1950s he was a club officer of Harold House and in the late 60s joined the Parents Committee of the Jewish Lads' & Girls' Brigade after which he became the Commanding Officer of the Merseyside Battalion of the JLGB from 1970–1973. The children were members of the Brigade during that time and Monday night parades, summer camps and winter camps became a way of life of three years. At the age of 15 the boys left, but at the time of writing Vikki is London Regiment Staff Secretary. The children did well at school, and Vikki shone as a pianist, winning many cups at music festivals and obviously inheriting Sidney's musical ability. She works in a bank, while Neil and Lawrence went to Liverpool and Manchester Universities respectively and are now solicitors.

The writing of this account has been prompted by two things. First what appear to be some anomalies in the disease or more to the point in my husband's handling of it; second is the ever-recurring thought that our doctors, whether the consultant or GP, having admitted that comparatively little is known about the ailment, and only seeing their patients at intervals, may benefit from having an insight, albeit secondhand, into the daily life of sufferers. A pattern may emerge which might help in further research, or at the very least, might be of use to fellow sufferers in dealing with "the first year".

Early in May 1981 Sidney came home one day and said that his left leg wouldn't "behave itself". He described it by saying "If anyone had been walking behind me, they would have thought I was drunk!". This happened several times, but as he was waiting to have a hernia operation, we thought it might be just something to do with that. In June he had the operation, in any case not walking to excess, so although he said his leg felt "as though it was attached by a piece of string" I personally never saw him walk strangely. He did not mention it to his doctor, however, in case he would require hospital tests to ascertain the cause of his "funny legs" which might stop him from attending our son's wedding due to take place at the beginning of September. At the end of August, he fell down a flight of stairs injuring his right shoulder. I went myself to the doctor to see whether it would be possible for him to have a pain-killing injection to get us over the weekend of the wedding and it was then that this particular doctor heard about the "funny legs", the right leg now being affected also. We were at that time going through some family problems, and the doctor suggested that Sidney's problems might be psychosomatic – Sidney should come and see him when the wedding was over.

In October, Sidney did see his doctor, who referred him to a specialist who admitted him to hospital for tests and, after a week of negative results, transferred him to the neurological department of another hospital. They drew a blank there too, and he was discharged with an appointment for a clinic in February to see whether anything further would develop. By now his left arm was affected, and from then on the condition progressed rapidly until he was unable to completely dress himself (socks and buttons gave him trouble), could no longer get in and out of the bath, and was falling fairly frequently.

From here on it's not really the first year that I'm writing about, but the second year. From May 1981, when the first symptom appeared until April 1982 when it was diagnosed, we were leading a perfectly normal life, and my husband was working. However in January 1982 I started to work alongside him and therefore even I learned something from closer observation. One of the first things which became apparent, was that temperature, both indoors and outdoors, made, and still makes, a great deal of difference to his well-being and performance. For example, in January 1982 when the weather was bitterly cold getting in and out of the car was a trial. This was eased by wearing leg warmers up to the thighs (thermal underwear would have done, but the restrictions of the pants between the thigh and bottom irritated and hindered his walking), but even in the house if a door was left ajar he apparently felt the slightest draught and yet it may well have been up to 15 degrees warmer indoors than out.

After the diagnosis, when his condition became "official", he took to sitting with a blanket over his legs, again with the emphasis on the thighs. This went on for quite a time even into the hot weather of the summer (and there was a great deal of hot weather in the summer of 1982). Again both indoors and out, when sitting in the sunshine, the perspiration would be rolling off him! He put it down to the wasting of his muscles and in one week in early May, in fact he lost 7 pounds in weight and his thighs became very thin. With the intensive physiotherapy he was doing, the muscles built up again within a matter of weeks and so did the thighs and by the end of June the blanket had disappeared. Was this perhaps because I showed my own distaste for it in no uncertain terms, mainly because it made him look like an invalid, a term which we both refuse to use in his case. He had a condition which prevented him using his legs, arms and hands to the full but we didn't consider him as a "sick man", a term which we would use to describe someone who frequently has to have days in bed, injections, constant nursing, periods of hospitalisation.

When Sidney first stopped work we had to adapt to a completely new way of life, first due to his disablement, second, because he was now a "retired" man, and retirement is supposed to be traumatic for a man, especially if he's unprepared for it, as in Sidney's case. To begin with, although his hands were rapidly weakening, he could still write, and as his speech was not noticeably affected he could speak on the telephone, and he busied himself sorting out matters to do with the Social Services vis-à-vis benefits and equipment available to help us live more comfortably. At that time I described our changed status by saying that "life is full of rituals, and we are simply changing some of ours". Physically, however, he had immense difficulty in climbing into bed, coupled with a rapidly diminishing ability to dress and undress himself. He was by then also unable to sit up in bed unaided and a back support became necessary. The going to bed process at that time took approximately an hour, after which we were both exhausted. I never helped him to do something until it became *absolutely* necessary, and some things took him a long time. At this stage, he was dragging himself upstairs still, but by the end of May the punishment was so great that we moved the bed downstairs and we doubled our efforts with the Social Services to get alterations to the house including the installation of a lift so that we could get back upstairs and return to living "normally". To my mind, sleeping downstairs smacks of invalidism and Sidney is not an invalid!

From the time Sidney found writing getting more difficult he also found it difficult to concentrate on reading books, or even newspapers. He had no interest in what he wore and, warm weather or otherwise, would not wear a tie. When I myself became aware of this I encouraged him to choose a shirt himself, in order to get him back to making decisions, and eventually he himself realised that wearing a tie, which he had always done all his working life, was much neater and smarter. I consulted him on all matters, great and small, where decisions had to be made, for two reasons, firstly because I had enough physical things to do anyway, and also because the man of the house must remain just that, and not heave his responsibilities on to someone else. He was also somewhat sleepy and with the weather being fine he sat outside every afternoon and snoozed a great deal. I put it down to the fact that his lifestyle had changed completely and also to the prednisolone which he had been prescribed in large doses. He also became tired by about 7 p.m. and was disturbed enough about it to mention it to his consultant. Nevertheless, having mentioned it (I think this was round about May or June) he started to pick up and even though he nodded off in the afternoon, as we had a lot of visitors calling in the evenings by this time, within a comparatively short time he was able to sustain himself through the evening until about 9.30 p.m. and gradually later still until eventually, if unthinking visitors stayed particularly late, it didn't cause distress even up to midnight. At the time of writing, our average time for going to bed if there are no visitors is about 10.30 p.m. to 10.45 p.m. Between the evening meal and bed he usually nods off for maybe half an hour in front of the TV but then many men in perfect health do that anyway!

At the same time, having been on steroids now for some weeks, he was obliged to also take a diuretic and this gave rise to bladder problems, in that if Sidney needed to pass water fast he couldn't exactly *run* to the toilet (which we don't have downstairs anyway) and this coincided with the fact that, in any case, he could no longer use a conventional toilet. He had difficulty getting his penis out, firstly because it started to "disappear", and secondly because his hands could not manipulate it properly. Having found it, he could no longer aim straight! There were numerous panics, accidents, disasters about which we were able to laugh, both then and now (the ability to laugh at difficulties is imperative) so we started timing the passing of water with the aid of a bottle which is somewhat restricting if we wanted to go out for more than one hour but at least Sidney regained his confidence and realised that he was not incontinent which he had thought, with horror, was the case. We also got him fitted with a penile sheath device so that we could go on longer car journeys and his confidence grew to the extent that his penis never again disappeared.

Bowels, too, were a problem, in that he was unable to sit on a conventional toilet since his poor balance would not allow him to lower himself on the seat satisfactorily, nor could he lift himself off without the aid of a frame with arms because he had lost a lot of leg muscle power. Coupled with this was the fact that he could no longer "push" and so in order to make a motion easier we started to use suppositories, at the same time fixing bowel movement for every third day in order that the bowel could be trained so that he could not be taken unawares. This situation remains unchanged and quite satisfactory although now the process takes much longer.

Around July he suddenly developed an intense sensitivity to sound and his teeth became sensitive to cold. My husband who for 26 years rarely heard the phone ring, or the doorbell, suddenly began to hear them even before I did, and I have exceptionally sensitive hearing. With regard to this, I wondered if he was compensating for not being able to move to answer either of these sounds by at least making himself aware of them. Also, I have unfortunately a somewhat strident voice and my "shouting" all the time almost became a bone of contention. The sound of cutlery being put on our formica-topped table almost deafened him. This sensitivity, coupled with the teeth, lasted for perhaps 3 months or so and then seemed to wear off, at least they were never mentioned again.

When Sidney stopped working in April he was then able to walk, at most about 50 yards, with a walking stick but also aided by me on the other side. Since his balance was very bad, within weeks he had transferred to a Zimmer which gave him a great deal more independence. Before he was diagnosed he had fallen several times without knowing why, but until now, with help, was able to raise himself as he still had quite a lot of muscle power in his thighs. He still fell from time to time but since he was no longer able to help himself at all, and he is too heavy for me to lift off the floor, we were advised by our OT that we should phone 999 Police emergency, tell them that my husband was unhurt, but on the floor, and ask if they could send someone to pick him up. This service is magnificent, and it has never taken more than 15 minutes for a policeman (or even two) to arrive. However, apart from "necessary" walking, i.e. from one room to another, or through the back door to sit outside, he did no walking at all. Then one day, late July or early August I think, he decided to try a walk around the garden. Bearing in mind that grass is one of the most difficult surfaces on which to walk and also that the ground under the grass is inevitably uneven, this really took a great deal of courage. He started by doing one circuit each afternoon (we were lucky that the summer of 1982 was so wonderful), working his way up to three or four circuits, and it proved to be extremely therapeutic, not only because he *had* to lift his feet, but because it also represented a challenge. He only fell once, but continued walking on grass until September when the ground became too damp and soggy.

We considered what alternative there was to the garden and Sidney decided that the time had come to try the pavement outside the house. Again, hazardous because of uneven flagstones, but nothing ventured nothing gained, and off we went, Sidney going *backwards* down the sloping ramp, then into the street, with me following behind with the wheelchair in case he got tired. With great courage and tenacity, within a month he increased his distance to 300 yards!

Within days of this wonderful event, however, the weather became wetter and windy. It was actually the wind that caused most distress because on account of the balance problem even one small gust of a light wind could blow him over. On many occasions we had walked maybe down the path and no more than perhaps 10 yards when a breeze would quite suddenly blow up and quickly become quite gusty and we had to go back, Sidney's confidence having become more than a little shaken! This, then, was the pattern, either too cold making his legs stiff, too wet, making it impossible to even go out at all, or too windy. And so, by the middle of November, walking in the street became a thing of the past. We still went out most days, however, Sidney accompanying me in the car when I went shopping or visiting people for afternoon tea, but for actual walking exercise, which after all, the daily physiotherapy was being done to achieve, Sidney had to resort to walking round the hall in the house which he did with his usual dedication, again clocking up 100 yards a day, split up in several parts, because for someone who already has a balance problem, walking in circles or rather "ovals" is far from ideal and yet another confidence breaker. By mid-December, we had to make a decision on how best to deal with this situation, i.e. winter weather, lack of exercise, and we decided to try to get away from the English winter to a mild climate, and hence the decision to go to Israel. Suffice to say that the last time Sidney walked out in the street, a distance of approximately 40 yards, was on Christmas Day. After that, in anticipation of Israel totally reversing the whole situation, even circling the hall diminished until the maximum he ever did was about five rounds, equalling approximately 45 yards.

From the very beginning Sidney's days have been full of activity of one sort or another, passing with incredible speed for both of us. In fact, for me, life is one continuous exercise in time and motion! We honestly don't know where the day goes, and sometimes they're not long enough to cram in all we want to do. Right up until the cold weather in November on most days Sidney accompanied me in the car (although getting in and out was difficult) to do shopping and as my parents live in Southport, and at that time my father was desperately ill, we also used, at least once a week, to visit him in hospital where he was confined for four months. We would start getting up, having breakfast, etc., at about 7.30 a.m., go on to physiotherapy, washing, dressing. At this time, in fact, up until September 1982 Sidney could still wash and shave himself and walk around the house on the Zimmer without any sort of aid or supervision, opening and closing doors, switching on TV. Now, being on the "retired" list, we were often invited out for tea, just as often as people came to us, and there was a certain novelty in being able to do *whatever* we wanted, *whenever* we wanted, once the "getting up" routine was over. It no longer mattered when we had lunch, or dinner, and I took the attitude that, as long as our friends were coming to see us, they were, and still are, the most important people in the world, and if I don't dust today, I'll do it tomorrow, or if the washing or ironing gets left, so be it – we have the rest of our lives to live at a pace which suits *us*. A routine is vital, but it must also be flexible. We *never* refuse an invitation, nor are we ever "not at home" to guests. Unfortunately, at the time of writing – in fact since November – the cold weather has precluded any night-time outings, and equally regrettably, the weather has still not improved sufficiently to resume these.

Physiotherapy plays a very great part in Sidney's life, especially since it appears to be the only treatment available in order to keep the sufferer mobile. In fact, Sidney started doing physiotherapy five months before he was diagnosed, as the result of a chance conversation with one of the housemen in the first hospital to which he was admitted for tests to see what ailed him (October 1981). This doctor had said that all that the tests had shown so far was that there was a definite lack of muscle power in the thighs and that some of the muscles may have wasted. When I asked him what could be done about it he told me that re-training other muscles to compensate through physiotherapy was the answer. As a result, when Sidney was discharged from Walton Hospital to which he was subsequently transferred for further tests, again, in the absence of a diagnosis he decided to at least try and get his legs right as quickly as possible, and on request was given a regime of exercises to follow, including some arm exercises to help his (by then) arthritic shoulders. He did these exercises morning and evening (bearing in mind that he was still a working man) and there is no doubt in my mind (or his) whatsoever that these kept him going through the winter.

After he had been diagnosed in April 1982 he asked his consultant if he could have supervised physiotherapy on a regular basis, because there were some exercises that he was unable to do any longer, because of his lack of balance, and others which had become difficult. At this stage, because he had great difficulty getting into bed, turning over in bed and also sitting up in bed, even with support he thought he might get some tips on how to achieve these more satisfactorily. As a result, he was referred to Sefton General physiotherapy department and in June started by attending at first three times a week, then after two weeks, once a week, then after a further four weeks, once a month until January 1983 when it became once in six weeks. However, after the initial visit when he was shown how to do some exercises which he would find easier but just as effective, in other words, adapted to his current capabilities, he was greatly motivated and continued to do them at home with enthusiasm, and they created in him an intense feeling of achievement and well-being. However, possibly due to the cold weather, or whether the disease has progressed, at the time of writing although the physiotherapy goes well (he does it every single day) the walking does not. And yet I have been very much aware that, if he had difficulty lifting "heavy feet", once he has made up his mind to "move the bloody thing" it *has* moved. When I am watching, and "willing" him to lift up his feet, they lift! However, it is a gruelling exercise for both of us, and to become exhausted, physically and mentally, does nobody any good, especially if the extent of his walking does not give him any independence, nor me any freedom because I have to accompany him on account of the balance problem.

His hands have become almost useless. The wrists have gone limp and the fingers have curled in, and typing, even with pegs on his fingers, is now very difficult, and raising his arm, on the end of which is a limp hand is rather difficult. For the same reason, holding a fork and spoon in the right hand (he has been unable to use a knife since June 1982) for eating is a problem. If the food on the fork is not too heavy to lift, then the angle at which it reaches the mouth is wrong. Similarly, with a spoon (only thick soup), and none of the recommended cutlery for disabled people seems to fit the bill. Lifting a cup to his lips is also difficult, using two hands. The cup must be only half full (weight) and the liquid must not be too hot, first of all because it will burn his hands which are sensitive, and also because he's sensitive to drinking hot liquid. Nevertheless, he is able to do things which I feel may be against the "rules" of the disease, i.e. he can do a perfectly controlled "sit down" and has no problem co-ordinating. The problems seem to stem from weakness of the extremities though not of the body) rather than lack of co-ordination. He has at times had a problem leaning forward in a chair to read a newspaper (he turns the pages with a wet finger) from a lectern but this is also intermittent.

Even before Sidney stopped working, he felt that his speech was showing signs of slurring, but I truly never heard it. However, I was advised that "the patient always knows. . ." and in fact, in June 1982 there was noticeable, if slight, slurring in speech. The voice was at times "thin" or as Sidney described it, nasal, but the slurring was intermittent. However, I noticed that when it did happen the shape of his mouth changed and it became a little lop-sided and I could always tell when it was *going* to happen. His tongue, he said, felt "strange", and in July, he mentioned it to the physiotherapist who arranged for him to see a speech therapist. As is often the case, on the first occasion on which he saw the therapist, his speech was almost perfectly sustained throughout the session during which he spoke non-stop for over half an hour. However, as she listened, she became aware of the sounds with which he had difficulty, and gave him a set of exercises to do, which he still does daily. Unfortunately, he has always suffered from chronic catarrh, which doesn't help, and for many weeks the speech remained the same although it became slower, and hesitant, rather than slurred. He has always, understandably, been very self-conscious about his speech, and in company, he sometimes starts off rather badly, but as he realises that people are understanding him, his confidence grows and the speech improves considerably. The voice no longer goes "thin" or nasal.

Sidney sleeps incredibly well. Since the day we married, I have known him to be a poor sleeper, who *had* to read before going to sleep, no matter what time we retired. He was also an early riser, particularly in the summer. Many times, being unable to sleep beyond 5 a.m., he would be wide awake until 6. a.m. and walk the dog (poor thing!) around to the newsagent to collect the newspaper. He snored loudly and frequently. Now he goes to sleep without reading. For a brief period he continued to go through the motions, sitting up in bed, leaning against a back rest, right up to our return from Israel. He had a problem turning the pages of his book, but the strangeness is that he no longer needs to read in order to be able to sleep! Also, even lying on his back, still suffering from catarrh, he *never* snores, and his breathing is so quiet that I have to watch his chest rise and fall in order to ascertain whether or not he's breathing at all.

On the day on which Sidney was informed of the nature and name of his complaint he was prescribed steroids and we were advised that there were various possible side-effects, at the same time knowing that the treatment was only a trial and might not do any good anyway, but Sidney felt he had to try anything and everything. In fact, the only side-effects were increase of appetite and retention of fluid. Accordingly, he put on a lot of weight (about half a stone). For a while, he gave in to his appetite then, in the knowledge that (a) he didn't look good and (b) it wouldn't help his walking or his balance to be overweight, he put himself on a very strict diet of approximately 800 calories a day, certainly a crash diet (supplemented by vitamins) but bearing in mind also that he was getting very little exercise. It was high protein, low carbohydrates, and a lot of vegetables. He cut out bread, potatoes, cake and chocolates, nothing very vital. He lost 21 pounds in six months which, when you consider that he was battling against the cortisone too, was probably equivalent to double. On finishing with the steroids he resumed eating normally but in moderation, and has not put back on any of the weight he lost.

In April 1983 Sidney was admitted to hospital for physiotherapy to try to improve his walking. There is no doubt that the physiotherapy he had was beneficial, and he has continued to do his exercises at home, as well as at the hospital. However, in view of the fact that overall he was less active in hospital than he is at home, and that he had eating problems in hospital because his chair and table didn't marry up to give him the correct height he needed in order to feed himself (at home we had a chair which was made to suit the height of our table) I'm not sure that in the end there were not more minuses than pluses. His walking, which was what he actually went in for, did not in fact improve, and within a week of discharge Sidney made the decision that he'd punished himself enough with the effort of trying, to no avail, and for all practical purposes, his walking days were over. In any case, his wrists had become limp and he was unable to throw forward the Zimmer successfully to make walking a practical proposition. It's sad that there are no outpatient facilities in the unit, *nor is there any provision for physiotherapists to make domiciliary visits.* So it is left for Sidney and I to get on with it ourselves. It is interesting to note that with reference to Sidney's reactions to temperature at home, he did not worry about the large airy ward, nor the fact that the ward door was never shut.

We have nothing but praise for the nursing staff who eventually became familiar with Sidney's needs and for the Ward Sister who had the perception and compassion to allow me to come in every day as I did and make me so welcome. We came away having made wonderful friends and that alone was a heartwarming experience.

May 1983

Several things have changed since April and our timetable needed to be updated. Sidney is no longer walking, except from one chair to another, nor is he able to eat unassisted. I have also been fortunate in getting a voluntary helper on Wednesday and Thursday evenings. As a result of the first, I no longer need to lift heavy weights into the lift, which has been rebalanced, nor carry the wheelchair up and downstairs. Also, a great deal of time which was hitherto devoted to accompanying Sidney when he was walking can now be devoted to other things, or at least I can take a little more time over them.

An obvious observation is why don't we use the services of the district nurse. Well, we know that they cannot guarantee at what time they're going to come each day, and we understand that perfectly. Also, we realise that they are not going to hang around while Sidney does his leg physiotherapy (40 minutes) before getting washed, etc. I, meanwhile, would be a bag of nerves wondering at what time the nurse was going to come, should I go out shopping, or wait, or what; and if they come late, it means we can no longer be ready to face the public by a reasonable time in the morning, no time for coffee, and some days I might be supervising physiotherapy after lunch, and in the end nothing useful would have been achieved by me. Add to that the fact that Sidney can do nothing in bed until they come, except just be there and wait, not an ideal situation. Similarly at the end of the day he certainly doesn't want to go to bed at 4 p.m.!

November 1983

Once more, several things have changed. We now have a hoist to help Sidney into bed. Also, if the weather is in any way fit, I must drop everything to take advantage of it and take him out. To wait for the afternoon is risking a complete change of weather conditions. Also, in July Sidney was still able to eat crispbread and hold a cup and drink without assistance. Although since the diagnosis I have always helped to turn him over during the night, I never mentioned it before because he was able to move a little better and I almost did it in my sleep. Now it's more complicated in that I also have to "re-arrange" various limbs, so I need to be more alert.

Tribute to the National Health Service and Social Services

The people to whom we owe so much:

General Practitioner
Consultant Neurologist
Physiotherapists
Speech Therapist
Occupational Therapists
District Nurse
Ward Sister
Lift Maintenance
Liverpool Personal Service
POSSUM

It goes without saying that Sidney's doctors are exceptional people. However, we both want it to go on record, in print, that this is precisely what they are – sympathetic, compassionate and supportive, and we know that if some drug or treatment is discovered, we will be the first to hear about it, therefore relieving us of the added burden of having to keep searching. Hospital staff at both the Royal Liverpool and Walton Hospitals are to be commended, and our district nurse is a treasure. The physiotherapists have been wonderful, but we feel that as physiotherapy is the *only* treatment for MND, there should be some domiciliary care available. After all, if there were any effective drugs, *they* would be prescribed without consideration of cost. On the other hand, Sidney's speech therapist does home visits, for which we are grateful, and they are a source of inspiration and comfort.

Our first contact with the Social Services was in December 1981, before Sidney was diagnosed, but was already having difficulty getting in and out of the bath, and it was suggested by my parents that we get a handrail from the Social Services, to go over the bath taps. This was our first introduction to our social worker, Mrs Dunne, who duly arrived with the rail, and suggested a folding stool to sit in the bath to make getting out even easier. Then came a walking stick, and the thing we remember best was the speed with which these items arrived, and always with a cheery word and a smile.

After the diagnosis, when Sidney had to give up working, it became apparent that we were going to need a great deal more help, and after Sidney had sorted out our financial situation vis-à-vis sick pay, mobility and attendance allowances (advice from booklets supplied by Mrs. Dunne) it was suggested that we visit the Merseyside Disabled Aid Centre to see what was available to make life easier for Sidney. We saw several items for which we thought we would have to pay, we were astounded to discover that they would be provided free of charge by the Social Services! We were truly amazed. We then set about having alterations made to our home and discovered, again to our amazement, that these could be made through the Home Improvements Scheme, Government sponsored, and because of Sidney's disablement we would receive a 90% grant! We were advised to contact the Senior OT for Liverpool whose office is in Hatton Garden, who within days came to assess our needs and make suggestions about some things we'd never even thought of, like widening doorways for possible future use of a wheelchair, and because we were by then sleeping downstairs, a stair-lift to take Sidney upstairs again. Unfortunately, the stair-lift was turned down by Home Improvements because of cost, and we were offered the alternative of a small bathroom and toilet combined to be built on the back of the house if we were prepared to remain sleeping downstairs in what is our dining room. Strangely enough, the only real trouble we had at any time was with the Home Improvements inspector who couldn't understand why we should possibly want to go back upstairs when he was offering the wonder of a downstairs bathroom, and it took a lot of tears and talking to convince him that living downstairs was not "normal" and it didn't make sense to cram everything downstairs when there are four perfectly good rooms upstairs which would remain unused, not to mention fitted wardrobes which couldn't be moved. Was I going to spend the future running up and down stairs for clothing, etc. on top of everything else I had to deal with? The man just couldn't comprehend it. However, we were introduced to a Terry lift, and after some frantic measuring, discovered that it could operate from the cloakroom to the small front bedroom (next to our own) and this could also accommodate a washbasin and a walk-in shower. After submitting estimates for these things, and also a ramp from the street to the front door, with a little "encouragement" from us, everything was agreed and work began at the end of September and was finished at the end of November.

During this period we were also supplied with a self-propelling lightweight folding wheelchair, an upholstered armchair with a high seat and back, a chemical toilet with stand (to raise the height of the seat) and several other items which we tried but found to be unsuitable, and therefore returned.

Most of the information on what is available and where to get it, apart from Invalidity Allowance, etc. came to us by way of *chance* remarks, mainly by occupational therapists. For example, the existence of a firm that not only does DHSS repairs to wheelchairs and invalid cars, but can adapt one's own car to suit one's own individual need on a private basis; also the existence of REMAP, a voluntary organisation composed of professional people will make "one off" items to suit the specific needs to one disabled individual. There is of course the Motor Neurone Disease Association, a self-help organisation, and the Liverpool Personal Service, which has been rather a life saver in sending voluntary helpers at times when I needed them most.

Conclusions (December 1983)

First let me say that my husband is an exceptional man and that it is a privilege to help him and take care of him. He is mostly very receptive to ideas and changes in routine which will help our joint situation and allow us to live as normal a life as possible under the circumstances. Also our natures are quite different and complementary to one another, Sidney being more passive and therefore not impatient, and I being more active. Sidney has the spirit, and I have the will, to make things better. We are a team. There is not much, if anything, that we won't try in order to enrich our lives. Sidney is also undemanding, gracious and acquiescent, the last being very important in that I am obliged to see to some very personal bodily functions and I am frequently surprised that he can accept what I have to do and let me get on with it. The fact that he doesn't fight me all the time makes it easier for me.

One learns to recognise the little nuances which will make a good day or not such a good day; there are no "bad" days. If the physiotherapy goes well in the morning it is generally the prelude to a good day. If I am rushed or show the slightest impatience he performs badly. He also performs better when he concentrates hard. Again, with reference to the physiotherapy, we discovered quite early that even though it's a blow to the morale, if the exercises don't go particularly well, not only on one day but over a period of several weeks, one must never give up trying. Suddenly, if the will is there, the ability to do the failed exercise will return. Inevitably, the disease progresses, but one must never give in too quickly, neither the sufferer nor the helper, who must always be quick with encouragement since success breeds success.

I have become very much aware that the psyche has a great deal to do with performance. We all have fixations and mental blocks about certain things, even in perfect health, and I feel that with MND or similar conditions, these should not be confused with the *ability* to do certain things, e.g. Sidney could (when he was able), hold a *full* cup of water, but only a *half* cup of tea or vitamin drink. I feel that his sensitivity to heat and cold also comes into this category. It's also very easy to fall into the trap of relating everything to the condition. For example, shortly after the diagnosis, Sidney had middle-ear trouble which we immediately thought was another manifestation of the disease and we discussed how to deal with future indispositions in order not to fall into the same trap again causing unnecessary despondency. I know comparatively little about the physiology of the nervous system and this particular disorder, but for some unknown reason I instinctively feel having lived with it, that possibly some form of hypnotherapy could be useful, although Sidney says he would not submit to it himself. On the other hand, I suspect that had it been suggested earlier, by a *medical* authority as having been found to be useful, he may not have had a closed mind.

A few weeks ago, in October, we took a trip to London where our three children all live, and stayed in the London Tara Hotel. This first class hotel with over 800 rooms has recently been refurbished with ten rooms on the mezzanine floor specially for disabled people and the special facilities were wonderful. The staff in general, and the doorman in particular were fantastic and very much geared to our needs, and the fact that we were never ready to leave our room before 1 p.m. We stayed for five days and my husband's performance actually *improved* as a result of this very pleasant and relaxed break. Therapeutically, it was far more beneficial than the month in Israel!

One of Sidney's main occupations is typing, and he does a great deal of work for charitable organisations, as well as seeing to our own correspondence. He used pegs on the end of two fingers, but because the fingers are curling in, he needed to find an alternative to the pegs. We now have a POSSUM typewriter and also a POSSUM environmental control (PSU 3). He operates the typewriter with his head, although there are various other choices of operation. The environmental control is operated by "sucking" and Sidney can now switch on and off TV, radio, tape recorder, light, electric fires in two different rooms, call me from upstairs or downstairs and open the front door, where there is an intercommunication speaker in order that callers can announce themselves.

I leave Sidney alone at some period most days, to go out shopping, sometimes for perhaps an hour or even longer now that he has POSSUM. Otherwise we would be together, mostly in the same room even, for 24 hours a day! It's a good thing we quite like each other's company! Apart from the fact that it's necessary for me to go out, it's just as vital for Sidney to retain his feeling of independence, albeit limited, by being alone in the house. He also needs a certain amount of solitude, knowing that for a time he has the total freedom of being unobserved and possible to express his emotions should he desire to do so. It is important to retain the freshness of spending time together by also being apart.

We are fortunate in that we have a great many good and caring friends and relatives who visit us frequently. In fact, hardly a day goes by without somebody calling in, if only for 10 minutes. If we're invited to somebody else's house we make every possible effort to go, and rarely turn down an invitation, provided the steps and width of doors allow us into the house. We feel deeply the support and kind thoughts emanating from these incredibly understanding people, even though Sidney has a speech problem, the one symptom which dominates his life more than any other. Although we have lost a great deal, we have also made gains in different directions. Part of our new philosophy is that for every minus situation, there is also a plus, and therefore it makes sense to concentrate on the things we *can* do rather than bemoan the things we can't. Since July there have been no changes in Sidney's condition, and for the first time in nearly two years I have a better idea of what each day will bring. If our routine has changed at all, it's mainly to do with weather conditions, e.g. no matter how warm the room, it is still too cold for Sidney to use the shower in winter; and even though I'm often very tired, I'm never dispirited!

Finally, it is a fact that everyone sees just a little bit of Sidney, except I, who see the whole man. His consultant sees him only at clinics, his GP when he calls may find him still getting washed, or doing physiotherapy and is therefore unable to converse, the physiotherapist, for maybe 45 minutes and does not really know the nature of the man at all, our district nurse, who comes only to see if we need anything, and the occupational therapists who only see him briefly. The only knowledge all these good people have in common with reference to Sidney is the fact that they have never been greeted with anything but a smile . . .

The Last Chapter

January was much the same as December with no discernible changes, except perhaps one or two foods which became difficult to eat, not to swallow, but they tended to stick to the roof of his mouth. However, I did get a physiotherapist to come in privately on one day a week not so much for Sidney's benefit, but to give me a break. Also, he had for some time complained that his heels "burned" through the night, and although we had tried sheepskin underneath and a cage to take the weight of the blankets neither helped, so I massaged his heels nightly before settling, which gave a slight relief.

At the end of January, Sidney's mother died, and this coincided with his neck muscles weakening, and his breathing becoming more shallow, the latter more apparent through his speech which was becoming more difficult. He continued with limited speech therapy. February brought a noticeable dropping of the head, especially when I wheeled him out and the chair went over the slightest bump. It occurred to me that "walks" might soon become very difficult, and he started wearing a collar when the occasion demanded. The left arm became weaker vis-à-vis physio with the "floating" arms. On 24th February, a Friday, during the night he tried to speak but couldn't which frightened him. On Saturday morning his voice returned, but he said his throat felt "funny". We talked at length, discussing things like portable ventilators, and in the afternoon called the doctor who could discern nothing unusual but gave him an antibiotic "in case" he'd picked up an infection. Sunday was a good day, although he sneezed a lot and I realised the "funny" throat was the beginning of a cold. He started to cough, especially through Sunday night. Monday, he seemed not too bad, certainly no cause for alarm, but after lunch suggested lying on the bed because that's when he seemed to cough most, and we could apply dorsal drainage to help him cough up phlegm which seemed to be bubbling around but not moving. In the event he hardly coughed at all and got up at 4.30 p.m. to do his usual standing and stretching exercise on the Zimmer.

Sidney slept on Monday night until the small hours when the coughing started again, and by 5 a.m. I was kneeling beside him banging his back to help each spasm. At 7 a.m. he said he thought the doctor should know about it, although he didn't actually feel "ill". The doctor came and suggested hospital for some physio and suction to remove the phlegm and he'd probably be home by the weekend. He was admitted at 11 a.m. accompanied by me, had two sessions of suction and was in excellent spirits. At 3 p.m. his consultant sent for me and said it was "all over"; I did not immediately comprehend what he was saying. The lungs were filling up and whatever followed, even if he made a spontaneous recovery, he would never be able to come home again . . .

After some further discussion, I had to return to Sidney with this dreadful knowledge, alone, be myself, just me and my "happy face", the mask I had worn for three years. And my sweetheart didn't know that we had reached the parting of the ways, with him going ahead of me and I being left behind to live without him, unable to reach out and touch him ever again. I asked him if he'd like me to stay the night in the nurses home and he replied that he would so I said I'd go and get my night things and I'd be back to help him with the evening meal (which he didn't eat).

I don't know how I managed to drive home and back. My main concern now was that it would happen quickly. I knew from past conversations with Sidney that dying was not in his scheme of things and because he was nervous of it he refused to acknowledge it. It was something that would happen (probably from a heart attack) but not now. I did not want him to have time to be afraid. I wanted him to be spared the shock of realisation that he had deteriorated irrevocably and that he would never see his home again. And so I had to behave as though everything was "normal" not to arouse his suspicion. I could not even stay with him through the night because once it was dark and the night staff came on he would wonder why I was still there.

With heavy heart I left him. I felt guilty about knowing something that he didn't know. Something so very personal and vital and he denied the knowledge. Why? Because he was not ready to die yet. At least, he had given no indication that he was ready to go in spite of what he must have known the future held. And I didn't want him to be afraid or distressed. We had never found it easy to say good-bye to each other, even for brief separations. I couldn't do anything more for him and I couldn't "do" for him anymore. We'd reached the end of the line. I sent for the children but couldn't let him see them for all the same reasons – he'd have been suspicious. They saw him through the glass walls of the ward.

My precious Sidney died at 1.15 a.m. Thursday 1 March 1984, peacefully. Inside I was full of unshed tears even greater than the tears I had already shed to see my treasure melting away over three years.

Going through a bereavement is like having your first baby. No-one ever tells you what it's like or what to expect, and, in any case, everyone reacts differently. One wonders what the "very last day" will be like; I was unbelievably calm considering that the light was about to go out of my life forever. I was overtaken by the shock of the event and was protected by that shock. I was numb without realising it. It took six and a half months for me to even acknowledge that February showed any deterioration whatever. I was able to handle Sidney's clothes, look at his photograph, speak about him, and feel nothing. But seven weeks later he was everywhere and when the weather became warm and he could have sat outside and the real grieving began. I went through the equivalent of post-natal depression. Our children, however, have been a tower of strength and our friends have continued to support me by including me in many of their activities.

Since Sidney died our second son has married and perhaps Sidney had the pleasure of being at Neil's wedding in spirit instead of in a wheelchair. Our other married son will shortly become a father, and so there is much to occupy my mind. I could have turned inward, but I am still looking for the "pluses" rather than the "minuses", and trying to be better rather than bitter. It is not always easy but since we met 30 years ago Sydney has always been my inspiration and even though he no longer shares my life he inspires me still with the memory of his goodness. If dying for him would have helped, I would have done that. Instead, I tried to give him a high quality of life, and make the last years worth something of value. Dear Sidney, whatever becomes of me in the future no-one will ever love and cherish me as you did.

Subject Index